OXIDATIVE STRESS AND DISEASE

Series Editors

LESTER PACKER, PH.D.
ENRIQUE CADENAS, M.D., PH.D.

University of Southern California School of Pharmacy
Los Angeles, California

Oxidative Stress, Inflammation, and Health

Oxidative Stress, Inflammation, and Health

edited by

Young-Joon Surh
Lester Packer

Taylor & Francis
Taylor & Francis Group

Boca Raton   London   New York   Singapore

A CRC title, part of the Taylor & Francis imprint, a member of the
Taylor & Francis Group, the academic division of T&F Informa plc.

Published in 2005 by
CRC Press
Taylor & Francis Group
6000 Broken Sound Parkway NW, Suite 300
Boca Raton, FL 33487-2742

© 2005 by Taylor & Francis Group, LLC
CRC Press is an imprint of Taylor & Francis Group

No claim to original U.S. Government works
Printed in the United States of America on acid-free paper
10 9 8 7 6 5 4 3 2 1

International Standard Book Number-10: 0-8247-2733-9 (Hardcover)
International Standard Book Number-13: 978-0-8247-2733-8 (Hardcover)

Library of Congress Cataloging-in-Publication Data

Catalog record is available from the Library of Congress

Taylor & Francis Group
is the Academic Division of T&F Informa plc.

Visit the Taylor & Francis Web site at
http://www.taylorandfrancis.com

and the CRC Press Web site at
http://www.crcpress.com

Series Introduction

Oxygen is a dangerous friend. Through evolution, oxygen—
itself a free radical—was chosen as the terminal electron
acceptor for respiration. The two unpaired electrons of oxygen
spin in the same direction; thus, oxygen is a biradical. Other
oxygen-derived free radicals, such as superoxide anion or
hydroxyl radicals, formed during metabolism or by ionizing
radiation are stronger oxidants, i.e., endowed with a higher
chemical reactivity. Oxygen-derived free radicals are generated during oxidative metabolism and energy production in
the body and are involved in regulation of signal transduction
and gene expression, activation of receptors and nuclear transcription factors, oxidative damage to cell components, the
antimicrobial and cytotoxic action of immune system cells,
neutrophils and macrophages, as well as in aging and age-
related degenerative diseases. Overwhelming evidence indicates that oxidative stress can lead to cell and tissue injury.
However, the same free radicals that are generated during
oxidative stress are produced during normal metabolism
and, as a corollary, are involved in both human health and
disease.

In addition to reactive oxygen species, research on reactive nitrogen species has been gathering momentum to develop an area of enormous importance in biology and medicine. Nitric oxide or nitrogen monoxide (NO) is a free radical generated by nitric oxide synthase (NOS). This enzyme modulates physiological responses in the circulation such as vasodilation (eNOS) or signaling in the brain (nNOS). However, during inflammation, a third isoenzyme is induced, iNOS, resulting in the overproduction of NO and causing damage to targeted infectious organisms and to healthy tissues in the vicinity. More worrisome, however, is the fact that NO can react with superoxide anion to yield a strong oxidant, peroxynitrite. Oxidation of lipids, proteins, and DNA by peroxynitrite increases the likelihood of tissue injury.

Both reactive oxygen and nitrogen species are involved in the redox regulation of cell functions. Oxidative stress is increasingly viewed as a major upstream component in the signaling cascade involved in inflammatory responses and stimulation of adhesion molecule and chemoattractant production. Hydrogen peroxide decomposes in the presence of transition metals to the highly reactive hydroxyl radical, which by two major reactions –hydrogen abstraction and addition– accounts for most of the oxidative damage to proteins, lipids, sugars, and nucleic acids. Hydrogen peroxide is also an important signaling molecule that, among others, can activate NF-κB, an important transcription factor involved in inflammatory responses. At low concentrations, hydrogen peroxide regulates cell signaling and stimulates cell proliferation; at higher concentrations it triggers apoptosis and, at even higher levels, necrosis.

Virtually all diseases thus far examined involve free radicals. In most cases, free radicals are secondary to the disease process, but in some instances free radicals are causal. Thus, there is a delicate balance between oxidants and antioxidants in health and disease. Their proper balance is essential for ensuring healthy aging.

The term oxidative stress indicates that the antioxidant status of cells and tissues is altered by exposure to oxidants.

The redox status is thus dependent on the degree to which cell's components are in the oxidized state. In general, the reducing environment inside cells helps to prevent oxidative damage. In this reducing environment, disulfide bonds (S—S) do not spontaneously form because sulfhydryl groups are maintained in the reduced state (SH), thus preventing protein misfolding or aggregation. This reducing environment is maintained by oxidative metabolism and by the action of antioxidant enzymes and substances, such as glutathione, thioredoxin, vitamins E and C, and enzymes such as superoxide dismutases, catalase, and the selenium-dependent glutathione reductase and glutathione and thioredoxin hydroperoxidases, which serve to remove reactive oxygen species (hydroperoxides).

Changes in the redox status and depletion of antioxidants occur during oxidative stress. The thiol redox status is a useful index of oxidative stress mainly because metabolism and NADPH-dependent enzymes maintain cell glutathione (GSH) almost completely in its reduced state. Oxidized glutathione (glutathione disulfide, GSSG) accumulates under conditions of oxidant exposure and this changes the ratio GSSG/GSH; an increased ratio is usually taken as indicating oxidative stress. Other oxidative stress indicators are ratios of redox couples such as NADPH/NADP, NADH/NAD, thioredoxin$_{reduced}$/thioredoxin$_{oxidized}$, dihydrolipoic acid/α-lipoic acid, and lactate/pyruvate. Changes in these ratios affects the energy status of the cell, largely determined by the ratio ATP/ADP + AMP. Many tissues contain large amounts of glutathione, 2–4 mM in erythrocytes or neural tissues and up to 8 mM in hepatic tissues. Reactive oxygen and nitrogen species can oxidize glutathione, thus lowering the levels of the most abundant nonprotein thiol, sometimes designated as the cell's primary preventative antioxidant.

Current hypotheses favor the idea that lowering oxidative stress can have a health benefit. Free radicals can be overproduced or the natural antioxidant system defenses weakened, first resulting in oxidative stress, and then leading to oxidative injury and disease. Examples of this process include heart disease, cancer, and neurodegenerative disorders. Oxidation of human low-density lipoproteins is considered an early step in the progression and eventual development of atherosclerosis, thus leading to cardiovascular disease. Oxidative DNA damage may initiate carcinogenesis. Environmental sources of reactive oxygen species are also important in relation to oxidative stress and disease. A few examples: UV radiation, ozone, cigarette smoke, and others are significant sources of oxidative stress.

Compelling support for the involvement of free radicals in disease development originates from epidemiological studies showing that an enhanced antioxidant status is associated with reduced risk of several diseases. Vitamins C and E and prevention of cardiovascular disease are a notable example. Elevated antioxidant status is also associated with decreased incidence of cataracts, cancer, and neurodegenerataive disorders. Some recent reports have suggested an inverse correlation between antioxidant status and the occurrence of rheumatoid arthritis and diabetes mellitus. Indeed, the number of indications in which antioxidants may be useful in the prevention and/or the treatment of disease is increasing.

Oxidative stress, rather than being the primary cause of disease, is more often a secondary complication in many disorders. Oxidative stress diseases include inflammatory bowel diseases, retinal ischemia, cardiovascular disease and restenosis, AIDS, adult respiratory distress syndrome, and neurodegenerative diseases such as stroke, Parkinson's disease, and Alzheimer's disease. Such indications may prove amenable to antioxidant treatment (in combination with conventional therapies) because there is a clear involvement of oxidative injury in these disorders.

In this series of books, the importance of oxidative stress and disease associated with organ systems of the body is highlighted by exploring the scientific evidence and the medical applications of this knowledge. The series also highlights the major natural antioxidant enzymes and antioxidant substances such as vitamins E, A, and C, flavonoids, polyphenols, carotenoids, lipoic acid, coenzyme Q_{10}, carnitine, and other micronutrients present in food and beverages. Oxidative stress is an underlying factor in health and disease. More and more evidence indicates that a proper balance between oxidants and antioxidants is involved in maintaining health and longevity and that altering this balance in favor of oxidants may result in pathophysiological responses causing functional disorders and disease. This series is intended for researchers in the basic biomedical sciences and clinicians. The potential of such knowledge for healthy aging and disease prevention warrants further knowledge about how oxidants and antioxidants modulate cell and tissue function.

Besides oxidative stress, proinflammatory damage is a underlying factor in the cause of many human disorders. Inflammation and inflammatory processes have been revisited because of their implications in the pathophysiology of the wide array of human diseases including cancer, neurodegenerative disorders, diabetes, as well as aging. A number of environmental agents that induce inflammatory as well as oxidative cell death and tissue damage have been identified. Therefore, dietary and pharmaceutical management of cellular anti-inflammatory and antioxidative capacity is important in maintaining the body's defense the against proinflammatory and prooxidative insults.

Specifically focusing on the redox regulation of cell signaling responsible for oxidative stress and inflammatory cell death and tissue damage, *Oxidative Stress, Inflammation, and Health* provides a comprehensive overview of cutting-edge research on the intracellular events mediating or preventing oxidative stress and proinflammatory processes induced by endogenous and xenobiotic factors.

Lester Packer
Enrique Cadenas

Preface

Pro-oxidative and pro-inflammatory process are pathophysiological factors in the cause of a vast variety of human disorders. Thus, oxidative stress and inflammatory damage are
underlying factors implicated in cancer, neurodegenerative
disorders, diabetes, etc. as well as aging. A number of xenobiotic and endogeneous substance have been identified to
induce oxidative and inflammatory cell and tissue damage.

The eukaryotic cell contains a multitude of signal transduction pathways coupling noxious stimuli to specific regulation of gene expression for adaptive responses. After
exposture to pro-oxidant or pro-inflammatory insults, cell
respond to potentiate the antioxidant/anti-inflammatory
mechanisms to minimize cellular damage. Transient changes
in redox status, communicated via a series of cellular signaling systems, initiate de novo synthesis of a distinct set of cytoprotective or stress-responsive proteins and enzymes that are
responsible for adaptive cellular responses to noxious stimuli.

Several components of the signaling pathways that sense
oxidative and inflammatory injury have recently been identified in higher eukaryotes. Certain stress-activated protein

kinases or other upstream signaling enzymes can activate the redox-sensitive transcription factors, thereby upregulating the expression of early-response genes to adapt and survive the subsequent injury. In this context, pharmacological and dietary manipulation of the cellular antioxidant and anti-inflammatory capacity represents and important strategy to cope with oxidative stress and/or inflammation-mediated pathophysiological disorders.

This volume is specifically focused on the redox regulation of intracellular signaling responsible for oxidative stress and inflammatory tissue damage. The chapters provide a comprehensive overview of cutting-edge research on the intracellular events mediating or preventing pro-oxidative and pro-inflammatory processes. The book is based on a recent symposium, "Inflammatory and Oxidative Cell Death and Tissue Damage: Molecular Mechanisms, Biomarkers and Chemoprevention" held at the Seoul National University, South Korea.

The chapters have been collated into three parts: (1) addressing the redox-regulation of gene expression (Chapters 1 to 8); (2) pathophysiological implications of oxidative and inflammatory injury (Chapters 9 to 17); and (3) cytoprotective effects of natural antioxidants and their therapeutic potential (Chapters 18 to 23). The editors thank the leading experts who have provided vital information on the state-of-the-art research of their own specialty, making this treatise a valuable reference source for the research community. We hope that this volume will contribute to the development and progress of this highly interesting and important area of biomedical research.

Young-Joon Surh
Lester Packer

Contributors

Amit Agarwal Natural Remedies Research Center, Bangalore, India

Bharat B. Aggarwal Cytokine Research Section, Department of Experimental Therapeutics, The University of Texas M. D. Anderson Cancer Center, Houston, Texas, U.S.A.

Yukihiro Akao Gifu International Institute of Biotechnology, Gifu, Japan

Okezie I. Aruoma Department of Applied Science, London South Bank University, London, U.K.

Debasis Bagchi Department of Pharmacy Sciences, School of Pharmacy and Health Professions, Creighton University Medical Center, Omaha, Nebraska, U.S.A.

Manashi Bagchi Interhealth Research Center, Benicia, California, U.S.A.

Theeshan Bahorun Department of Biological Sciences, Faculty of Sciences, University of Mauritius, Redúit, Mauritius

Aalt Bast Department of Pharmacology and Toxicology, University of Maastricht, Maastricht, The Netherlands

Young-Nam Cha Department of Pharmacology and Toxicology, College of Medicine, Inha University, Inchon, South Korea

Soo-Cheon Chae Department of Microbiology and Immunology, School of Medicine, Wonkwang University, Iksan, South Korea

Min Kyung Cho National Research Laboratory, College of Pharmacy and Research Institute of Pharmaceutical Sciences, Seoul National University, Seoul, South Korea

Sung Won Cho Genomic Research Center for Gastroenterology, Ajou University School of Medicine, Suwon, South Korea

Byung-Min Choi Department of Microbiology and Immunology, School of Medicine, Wonkwang University, Iksan, South Korea

So Yeon Choi National Research Laboratory, College of Pharmacy and Research Institute of Pharmaceutical Sciences, Seoul National University, Seoul, South Korea

An-Sik Chung Department of Biological Sciences, Korea Advanced Institute of Science and Technology, Daejeon, South Korea

Hae Young Chung College of Pharmacy, Aging Tissue Bank, Longevity Life Science and Technology Institutes, Pusan National University, Pusan, South Korea

Hun-Taeg Chung Department of Microbiology and Immunology, School of Medicine, Wonkwang University, Iksan, South Korea

Myung Hee Chung Department of Pharmacology, Seoul National University College of Medicine, Seoul, South Korea

Dipak K. Das Cardiovascular Research Center, University of Connecticut, School of Medicine, Farmington, Connecticut, U.S.A.

An De Naeyer Laboratory of Eukaryotic Gene Expression and Signal Transduction (LEGEST), Department of Molecular Biology, University of Gent, Gent, Belgium

Nathalie Dijsselbloem Laboratory of Eukaryotic Gene Expression and Signal Transduction (LEGEST), Department of Molecular Biology, University of Gent, Gent, Belgium

D. Neil Granger Department of Molecular and Cellular Physiology, Louisiana State University Health Sciences Center, Shreveport, Louisiana, U.S.A.

Guy Haegeman Laboratory of Eukaryotic Gene Expression and Signal Transduction (LEGEST), Department of Molecular Biology, University of Gent, Gent, Belgium

Guido R. M. M. Haenen Department of Pharmacology and Toxicology, University of Maastricht, Maastricht, The Netherlands

Ki-Baik Hahm Genomic Research Center for Gastroenterology, Ajou University School of Medicine, Suwon, South Korea

Sang Uk Han Genomic Research Center for Gastroenterology, Ajou University School of Medicine, Suwon, South Korea

Jin Won Hyun Department of Biochemistry, Cheju National University College of Medicine, Jeju, Jeju-do, South Korea

Jung-Hee Jang Laboratory of Biochemistry and Molecular Toxicology, College of Pharmacy, Seoul National University, Seoul, South Korea

Kyung Jin Jung College of Pharmacy, Aging Tissue Bank, Longevity Life Science and Technology Institutes, Pusan National University, Pusan, South Korea

Keon Wook Kang National Research Laboratory, College of Pharmacy and Research Institute of Pharmaceutical Sciences, Seoul National University, Seoul, South Korea

Yoji Kato School of Humanity for Environmental Policy and Technology, Himeji Institute Technology, Hyogo, Japan

Dae Young Kim Genomic Research Center for Gastroenterology, Ajou University School of Medicine, Suwon, South Korea

Sang Geon Kim National Research Laboratory, College of Pharmacy and Research Institute of Pharmaceutical Sciences, Seoul National University, Seoul, South Korea

Chang Ho Lee Department of Pharmacology and Institute of Biomedical Science, College of Medicine, Hanyang University, Seoul, South Korea

Wakako Maruyama Laboratory of Biochemistry and Metabolism, Department of Basic Gerontology, National Institute for Longevity Sciences, Aichi, Japan

Nilanjana Maulik Molecular Cardiology Laboratory, Cardiovascular Research Center, University of Connecticut, School of Medicine, Farmington, Connecticut, U.S.A.

Thomas Müller Philip Morris Research Laboratories GmbH, Köln, Germany

Hajime Nakamura Translational Research Center, Kyoto University Hospital, Kyoto, Japan

Takayuki Nakamura Translational Research Center, Kyoto University Hospital, Kyoto, Japan

Ki Taik Nam Genomic Research Center for Gastroenterology, Ajou University School of Medicine, Suwon, South Korea

Makoto Naoi Department of Neurosciences, Gifu International Institute of Biotechnology, Kakamigahara, Gifu, Japan

Atsumi Nitta Department of Pharmacy, Nagoya University School of Medicine, Nagoya, Japan

Takashi Okamoto Department of Molecular and Cellular Biology, Nagoya City University, Nagoya, Japan

Lester Packer Department of Molecular Pharmacology and Toxicology, School of Pharmacy, University of Southern California, Los Angeles, California, U.S.A.

Hyun-Ock Pae   Department of Microbiology and Immunology, School of Medicine, Wonkwang University, Iksan, South Korea

Jong-Min Park   Department of Biological Sciences, Korea Advanced Institute of Science and Technology, Daejeon, South Korea

Irfan Rahman   Department of Environmental Medicine, Division of Lung Biology and Disease, University of Rochester Medical Center, Rochester, New York, U.S.A.

Vinod S. Saxena   Natural Remedies Research Center, Bangalore, India

Masayo Shamoto-Nagai   Laboratory of Biochemsitry and Metabolism, Department of Basic Gerontology, National Institute for Longevity Sciences, Aichi, Japan

Muhammad Soobrattee   Department of Biological Sciences, Faculty of Sciences, University of Mauritius, Redúit, Mauritius

Klaokwan Srisook   Department of Pharmacology and Toxicology, College of Medicine, Inha University, Inchon, South Korea

Karen Y. Stokes   Department of Molecular and Cellular Physiology, Louisiana State University Health Sciences Center, Shreveport, Louisiana, U.S.A.

Young-Joon Surh   National Research Laboratory, College of Pharmacy, Seoul National University, Seoul, South Korea

Anitaben Tailor   Department of Molecular and Cellular Physiology, Louisiana State University Health Sciences Center, Shreveport, Louisiana, U.S.A.

Masashi Tanaka   Department of Gene Therapy, Gifu International Institute of Biotechnology, Kakamiganara, Gifu, Japan

Shugo Ueda   Translational Research Center, Kyoto University Hospital, Kyoto, Japan

Henk van den Berg   Netherlands Nutrition Centre, Den Haag, The Netherlands

Robin van den Berg   TNO Quality of Life, Zeist
The Netherlands

Wim Vanden Berghe   Laboratory of Eukaryotic Gene Expression
and Signal Transduction (LEGEST), Department of Molecular
Biology, University of Gent, Gent, Belgium

Linda Vermeulen   Laboratory of Eukaryotic Gene Expression
and Signal Transduction (LEGEST), Department of Molecular
Biology, University of Gent, Gent, Belgium

Marie Yeo   Genomic Research Center for Gastroenterology, Ajou
University School of Medicine, Suwon, South Korea

Junji Yodoi   Translational Research Center and Institute for
Virus Research, Kyoto University Hospital and Kyoto University,
Kyoto, Japan

Sang-Oh Yoon   Department of Biological Sciences, Korea
Advanced Institute of Science and Technology, Daejeon, South
Korea

Byung Pal Yu   Department of Physiology, The University of
Texas Health Science Center at San Antonio, San Antonio, Texas,
U.S.A.

Contents

1

Oxidants and Antioxidants: Mechanism of Action and Regulation of Gene Expression by Bioflavonoids

LESTER PACKER

Department of Molecular Pharmacology
and Toxicology, School of Pharmacy, University
of Southern California, Los Angeles,
California, U.S.A.

ABSTRACT

The identification of oxidant-sensitive pathways of cell signaling mandated that antioxidants regulate gene expression. Basically, three different mechanisms exist: First, by

International Symposium on: "Inflammatory and Oxidative Cell Death and Tissue Damage: Molecular Mechanisms, Biomarkers, and Chemoprevention."

affecting the cell's redox status; this is largely controlled by the balance between oxidants, free radicals, and the antioxidant network. Second, there is the activation of phase 2 enzymes, which strengthen antioxidant defense, and are also active in detoxification and chemoprevention. Then, there are the specific effects of antioxidants on key targets of the transciptome. In the 1990s, we turned our attention toward antioxidant regulation of oxidant-sensitive transcription factors, such as NfκB and dependent genes as cell adhesion molecule and iNOS gene expression. Our findings identified specific cell targets of gene regulation by bioflavonoid-rich botanical extracts as Ginkgo biloba EGb 761. Interestingly, this extract exhibits greater activity than any of their individual bioflavonoid components, thus, suggesting synergistic effects by their complex mixture of components. Global gene expression analysis with high-density oligonucleotide arrays is being used for studies of gene regulation and some recent findings will be reported for Ginkgo biloba extract EGb 761 which has been investigated in cell culture and animal model systems. In cell culture systems, oxidative stress linked affects on genes expression can readily be demonstrated. In vivo results reveal the presence of neuromodulatory substances in EGb 761.

Results in vivo indicate cell and tissue specific effects of antioxidants, which apart from their redox regulating effects in the system of antioxidant defense, would not have been predicted a priori from knowledge of their in vitro reaction mechanisms. Antioxidant- and micronutrient research will increasingly be reliant on genomic and proteonomic studies to elucidate the molecular basis of their health effects.

I. INTRODUCTION TO FREE RADICALS AND ANTIOXIDANTS

The chemical definition for free radicals (oxidants) is "atoms or groups of atoms with one or more unpaired electrons." This property makes them very unstable and highly reactive, trying to capture the needed electron from other compounds to gain stability. When the "attacked" molecule loses its electron, it becomes a free radical itself, beginning a chain reaction. Some free radicals, such as the toxic oxyradical species hydroxyl radical ($^\bullet$OH) or the less reactive superoxide radical ($O_2^{\bullet-}$), arise normally during mitochondrial oxidative metabolism. Also, the body's immune system cells purposefully generate free radicals to neutralize viruses and bacteria. Environmental factors such as pollution, radiation, cigarette smoke, and herbicides can also generate free radicals. Therefore, biological systems are continuously interacting with free radicals arising either from metabolism or from environmental sources, i.e., leading to a process called oxidation.

As stated above, free radicals have a precise chemical definition. However, the word "antioxidant" refers to a broad range of substances, which have the ability to neutralize free radicals by donating one of their own electrons, i.e., antioxidation. They act as scavengers, helping to prevent cell and tissue damage. Antioxidants display many different properties in biological systems. Consequently, it is difficult to find a precise description for the term "antioxidant" and many definitions abound.

In general, to be called antioxidant, a compound must be able to donate an electron and/or hydrogen atom and prevent or delay oxidation of an oxidizable substrate. They can act in different ways: by metal chelation (preventing free radical formation), scavenging free radicals, acting as a chain-breaking (stopping propagation of the free radicals), being part of the redox antioxidant network, and/or regulating gene expression.

The major natural antioxidants from dietary sources include vitamins C (ascorbate) and E, polyphenols and bioflavonoids (redox active substances that can be oxidized and reduced), and carotenoids, which act as free radical sinks. Other biofactors important in antioxidant defense are lipoic acid, coenzyme Q_{10}, and the various metals essential for the activity of antioxidant enzymes such as various nutritional metals (micronutrients) selenium, copper, zinc, manganese, and iron. The process of evolution has also equipped us with enzymes such as superoxide dismutases, catalases, glutathione, and thioredoxin peroxidasis to neutralize or prevent the formation of free radicals.

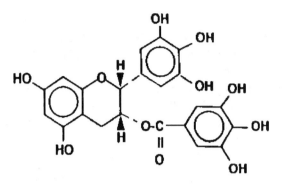

Figure 1 Molecular structure of epigallocatechin gallate, the major antioxidant substance in green tea.

When polyphenolic antioxidants, such as epigallocatechin gallate (Fig. 1), donate an electron or hydrogen atom to quench a free radical, they become free radicals. However, since the unpaired electron is delocalized by resonance around the aromatic ring, this confers greater stability and thus less reactivity to natural antioxidants. Thus, they are less dangerous than the free radicals they have destroyed. The property of resonance stability by the free radical form of antioxidants is also exhibited by vitamin E (four different forms of tocopherols and tocotrienols) and vitamin C.

II. MECHANISMS OF ANTIOXIDANT DEFENSE BY FLAVONOIDS

Three different modes of action are indicated by the available evidence.

1. The mechanism which has been most extensively investigated is metal chelation preventing free radical formation and secondly, the direct scavenging of oxygen and nitrogen free radicals such as hydroxyl, superoxide, and nitric oxide radicals. Most of these studies have been made in vitro or ex vivo systems. Evidence for such activity in body tissues is generally absent or inconclusive.

A pro-oxidant action of quercetin but not of rutin in vitro leading to hydrogen peroxide production has been reported (1). However, the pH dependence of this effect indicates that it is not likely to occur in vivo.

The following two mechanisms are likely for many of the protective effects of flavonoids in vivo.

2. A mechanism of great importance in antioxidant defense, detoxification, and cancer prevention is the induction of phase 2 enzymes. Evidence of this important in vivo response to bioflavonoids is from cell, animal, and human studies. Figure 2 illustrates the major pathways of phase 2 enzyme induction. Phase 2 enzyme induction results in synthesis of flavonol and other conjugation enzymes. Major flavonol conjugation enzymes are:

Glutathione as in GSH S-tranferases
Glucuronic acid as in UDP-glucuronosyltransferases
Methylation as with methyl transferases
Sulfation as sulfotranferases.

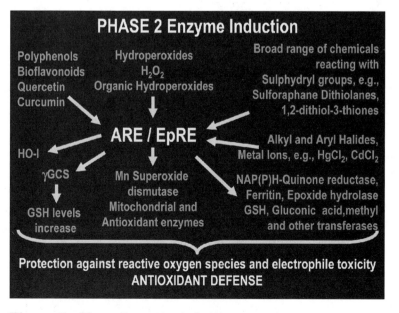

Figure 2 Phase 2 enzyme induction.

After absorption from the intestine, both Rutin and
quercetin are mainly present in vivo in conjugated
form. The biological activity of aglycones and conjugated forms is very different. Where studied, free flavonoids are generally more effective antioxidants and
thus more protective than the conjugated form (e.g.,
quercetin as compared with its conjugate quercetin-3-
glucoside) (2).

3. Another mechanism is action by flavonoids is on
direct molecular targets with high specificity. These
include effects on receptors, enzymes involved in cell
signaling (1992), transcription, and gene expression.
Some information for selected bioflavonoids has been
reported. Animal feeding studies with high-density
oligonucleotide arrays (gene chips) demonstrate the
presence of target proteins and enzymes (3).

The direct scavenging action against radical flavonols has
been extensively studied in in vitro systems which often are
not so relevant to their actions in vivo where little is known
about the concentration, distribution, and metabolism (4).

Mechanisms 2 and 3 are currently the subject of intense
investigation.

All cells possess elaborate antioxidant defense systems
that consist of interacting micronutrients, enzymes, and other
molecules. Oxidants such as hydroperoxides (H_2O_2, CH_3OOH,
ROOH), many natural antioxidant substances, and a broad
range of chemicals reacting with sulfydryl groups induce
phase 2 enzymes. Phase 2 enzymes protect against oxidants
and electrophilic toxicity, i.e., antioxidant defense, detoxification, and cancer prevention. This induction leads to a series of
coordinated increases in the gene expression and the enzymatic activity of antioxidant defense system.

Examples of phase 2 enzymes, proteins, and products
important in antioxidant defense include glutamate-
cysteine ligase, hemeoxygenase-1, Mn superoxide dismutase,
NAD(P) H:quinone reductase, dihyrodiol dehydrogenase, exp-
oxide hydrolase, leukotiene B4 dehydrogenase, aflatoxin B1

dehydrogenase, and the major intracellular iron binding protein ferritin. Upregulation of these systems result in the protection against reactive oxygen species and electrophile toxicity and therefore are very important in antioxidant defense, detoxification, and cancer prevention.

III. OXIDATIVE STRESS AND THE ANTIOXIDANT NETWORK

Oxidants and antioxidants must be kept in balance to minimize molecular, cellular, and tissue damage because if the balance is upset in favor of the former "oxidative stress" occurs. The term "oxidative stress" was first coined by Helmut Sies in 1986 Ref. 5 referring to the imbalance that arises when exposure to oxidants changes the normal redox status of major tissue antioxidants, especially glutathione, the cell's primary preventative antioxidant. Glutathione is usually present in tissues in millimolar amounts in the aqueous compartments of cells and their organelles. Under normal conditions, glutathione exists primarily in its reduced form (GSH). However, upon exposure to oxidants generated during flux through the respiratory chain (cytochrome P450), electron transport reactions metabolism of foreign compounds, ligand/receptor interaction, immune system activation, or exposure to environmental oxidants, GSH is oxidized to glutathione disulfide (GSSG), thus changing the redox status of the glutathione system. This change in the reduced/oxidized ratio of glutathione is an example of "oxidative stress." Recovery of the oxidative imbalance requires NADPH and NADH dependent reactions of reducing metabolism.

Oxidative stress often results in oxidative damage. This can arise from body metabolism, strenuous, and/or traumatic exercise, exposure to environmental stressors such as ultraviolet irradiation and cigarette smoke, as well as from infection (microorganisms, viruses, parasites, etc.), and of course during aging. Molecular markers of oxidative damage to lipids (e.g., isoprostanes, age pigment, or lipofuscin), proteins (e.g., carbonyl and nitrotyrosine derivatives of protein), and DNA (products of DNA fragmentation and oxidized bases such as 8-hydroxy-2-deoxyguanosine) accumulate during aging, and therefore tissues from aged individuals are more susceptible to disease. In addition, mild oxidative stress affects cell signaling pathways and thus gene expression.

All the redox-based antioxidants appear to interact with one another through the "antioxidant network" comprised of nonenzymatic and enzymatic reactions (Fig. 3). In this way, antioxidants are recycled or regenerated by biological reductants. For example, bioflavonoids and polyphenols interact with the vitamin C radical lengthening its lifetime.

Oxidative stress can weaken the strength of the entire antioxidant system. Redox active antioxidants from food or food supplements bolster antioxidant defenses and help in preventing oxidative damage. In addition to those, carotenoids, while not redox based antioxidants, also have an important effect upon bolstering antioxidant defense acting as "sinks"

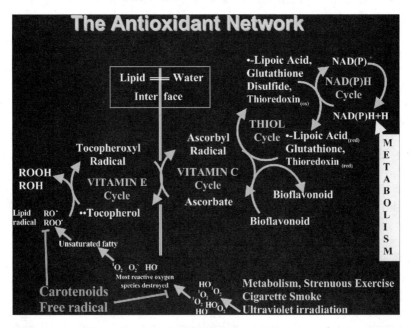

Figure 3 The redox antioxidant network.

for quenching free radicals such as lipid peroxyl radicals. Presence of a double bond system in their hydrocarbon chain enables them to take many hits in destroying free radicals.

IV. REDOX REGULATION OF GENE EXPRESSION

Antioxidants affect gene expression in two general ways, either by changing the redox status of the cell, or directly by specific interactions with molecular targets in a manner partially independent of their antioxidant/radical-scavenging ability.

Some cell signaling pathways are very sensitive to oxidants and changes in cellular redox status. Hence, free radicals and antioxidants must regulate gene expression. Activation of the plasma membrane NADPH oxidase, autooxidation of polyphenols or bioflavonoids, and mitochondrial metabolism can generate the formation of hydrogen peroxide (H_2O_2), sometimes referred to as a second messenger (Fig. 4). The presence of H_2O_2 in the cytosol changes the glutathione/glutathione disulfide ratio (important for cell signaling systems) and provides an oxidizing environment, which is needed for the binding of activated transcription factors to their consensus sequence on DNA. H_2O_2 also activates MAP kinase signaling cascade and transcription factors such as NfκB, and inhibits protein phosphatases, causing hyper phosphorylation of enzymes and proteins of cell regulatory processes.

Any or all of these events will modulate gene expression. Therefore, antioxidants are critical components of cell growth, development, differentiation, and function.

In addition, some antioxidants act on specific molecular targets that modulate cell gene expression. For example, protein kinase C is only inhibited by the alpha tocopherol form of vitamin E; protein tyrosine kinase is inhibited by the isoflavone genistein; and intracellular gap junctional communication is attenuated by carotenoids. Specific targeting of cell proteins by antioxidants is a new horizon in antioxidant research.

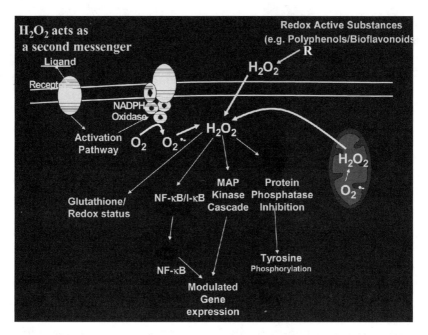

Figure 4 Some sources of H_2O_2 and its effects on cell signaling.

Selected references on oxidative stress, antioxidants, the antioxidant network, antioxidant and redox regulation of cell signaling, and gene expression and related topics are found in Refs. 6–16.

Herbal medicines and dietary supplements are widely used for health benefits but the molecular basis of their therapeutic potentials are often poorly defined. Ginkgo biloba extract EGb 761 is a highly standardized bioflavonoid-rich preparation whose components interact in a redox antioxidant network. Since free radicals and other oxidants are known to modulate important cell signaling systems, it can be predicted that flavonoid rich botanical extracts will exhibit gene modulatory activity. We hypothesized that in vitro and in vivo assays that allow quantitative analysis of gene expression profiles combined with targeted biochemical analysis could help to identify effects of phytochemicals. This hypothesis was tested by application of high-density oligonucleotide microarrays to define mRNA expression in cell cultures (20) and brain tissue after Ginkgo biloba extract (EGb 761) treatment (3).

Human bladder cancer cells (T-24) with the constitutively active, oncogenic G-protein, Ha-ras, was used as in vitro model to test the transcriptional response to EGb 761 (21,22). Analysis revealed a net activation of transcription. Functional classification of the affected mRNAs showed the largest changes in the abundance of mRNAs for intracellular vesicular transport, mitochondria, transcription, and antioxidants. The transcripts for hemeoxygenase-1, mitochondrial superoxide dismutase and the regulatory subunit of γ-glutamyl-cysteinyl ligase (formerly synthetase), and their encoded proteins were elevated (Figs. 5 and 6). Treatment also increased intracellular glutathione, the transcripts for DNA repair and synthesis, and decreased ^{3}H-thymidine incorporation (Fig. 7). These results demonstrate that EGb 761 initiates an adaptive transcriptional response that

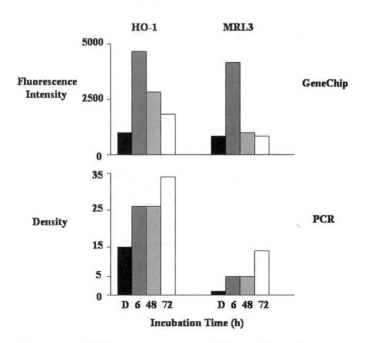

Figure 5 EGb 761 treatment of T-24 cells enhances expression of hemeoxygenase (17).

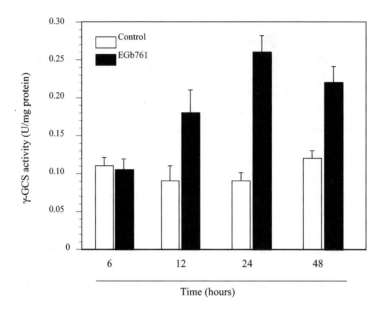

Figure 6   EGb 761 enhances γ-GCL activity (18).

augments the "antioxidant status" of the cells and inhibits
DNA damage (Fig. 8).

V.   NEUROACTIVE RESPONSES IN VIVO
   TO GINKGO BILOBA EGb 761

An important limitation of the in vitro assay of gene expression analysis of herbal extracts is that it does not account
for the processes that affect bioavailability and biotransformation. These biological processes of the alimentary tract
and the liver will modify herb components. Therefore, an in
vivo study was undertaken in normal mice to define the
potential in vivo transcriptional activity of EGb 761 (3). The
hypothesis was that a centrally active herbal extract should
change the gene expression profile in the brain. Survey of
the literature identified several reports that indicated central
effects of orally administered EGb 761 in rodents (23,24) and
in humans (25,26).

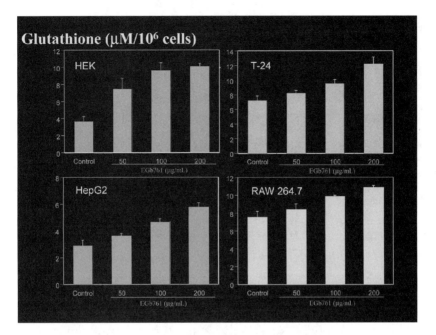

Figure 7 Glutathione levels are elevated after EGb 761 treatment of cultured cells (HEK, human keratinocytes; T-24, human bladder cells; HepG2, liver cells; and RAW, murine macrophages) (19).

Cortex and hippocampus were chosen as target neuroanatomical sites for gene expression analysis of ~12,000 mouse genes (Affymetrix, Murine Genome Mu74Av2). Mice were fed a diet supplemented with Ginkgo biloba extract as previously described (3) which was supplemented with 1 mg of EGb 761 per day. Plasma was obtained from blood samples and analyzed for flavonoids as previously described (27). Dietary supplementation for 30 days increased the presence of flavonoids, quercetin, and kempherol, in plasma of mice supplemented with EGb 761 (3). These data support previous observations that orally administered extract is bioavailable in rodents (28,29) and in humans (30,31). The extract was well tolerated as indicated by absence of any effects on the dietary intake or the whole body weight of mice.

Analysis of the expression of 12,000 mouse genes in cortex and hippocampus showed that the orally administered

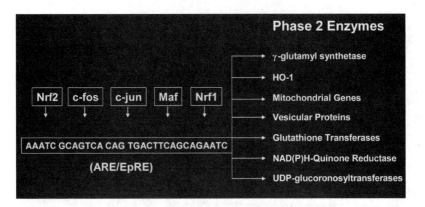

Figure 8 Antioxidant and electrophile response elements appear to be the common targets of transcription factors activated by Ginkgo biloba extract EGb 761 (18).

extract was centrally active (3). The expression of 43 genes and 13 genes was up regulated by at least twofold in cortex and hippocampus, respectively. A small number of genes such as those encoding growth hormone, prolactin, serum albumin LINE-1 repeat, and serine protease inhibitor were activated in both the brain regions. These observations are remarkable for three reasons. First, they provide molecular evidence for the action of a dietary supplement in the brain. Second, they show that EGb 761 has differential effects in the two brain regions that were selected for gene analysis. Third, the extract increased the abundance of the affected mRNAs suggesting increase in transcription although we cannot exclude the possibility of increased stability of mRNAs.

The heterogeneity in the responsiveness of the two neuroanatomical regions raises an important issue of targeting specific brain regions by dietary supplementation for desired molecular outcome and its effects on physiological, pathophysiological, and behavioral outcomes. Functional classification of the affected genes showed that Ginkgo biloba extract affected the expression of genes encoding transcription factors, ion-channels, growth factors and neuromodulators, synaptic vesicles and transport, cell surface, and protein kinases and phosphates.

The selective induction and transtheyretin transcript in hippocampus but not in cortex is noteworthy. Defects in hippocampal functions have been described in Alzheimer's disease (32,33). Transthyretin, a tetrameric protein important in the transport of vitamin A and an accessory protein for the transport of thyroxine, plays a role in the formation of amyloid products present in degenerating brain (34,35). The activation of transthyretin gene by EGb 761 suggests that the extract may regulate the metabolism of amyloid products in vivo. The mechanisms by which transthyretin may affect the formation of neuronal amyloid is poorly understood. However, a role forthyroxine may be implicated because it binds to trasnsthyretin and affects neuronal growth and development (36). A gene related to that of transthyretin, serum albumin gene with LINE-1 repeat, was induced (eightfold) in cortex and to a lesser magnitude (threefold) in hippocampus. These observations emphasize the differential effects of herbal extract in the brain.

EGb 761 also activated the transcription of genes that encode peptides important in cellular growth and function. These included growth hormone, prolactin, oxytosin neurophysin 1, placental growth factor, brain derived neurothropic factor, platelet derived growth factor, and insulin like growth factor receptor 1. Also noteworthy was the simultaneous induction of the genes for oligodendrocyte basic protein and proteolipid protein, both of which are essential for the biogenesis and assembly of myelin (37–39). Collectively, these observations provide some of the potential molecular targets through which the extract of Ginkgo biloba confers neuroprotective effects observed in rodent models of human stroke (40,41) and may contribute to regeneration of neurons.

The complex processes of learning and memory recruit multiple cellular types and molecular pathways that involve ion-channels (42,43). The expression of several genes encoding proteins of ion-channels was induced in the brain cortex and hippocampus after 30 days of dietary supplementation (3). These included the genes for chloride protein 3, calcium activated potassium channel, alpha subunit of $GABA_A$ recep-

Table 1   Effect of Ginkgo Biloba Extract (EGb 761), Ginkgolide A, B, and Bilobalide on GSH Levels in (HaCaT) Human Keratinocytes (18)

	GSH	
Control	3.32 ± 0.20	100
EGb 761 (100 µg/mL)	6.40 ± 0.26	193
Ginkgolide A 4 µg/mL	3.17 ± 0.34	95
Ginkgolide B 4 µg/mL	2.97 ± 0.33	89
Bilobalide 4 µg/mL	2.92 ± 0.20	88

tor, alpha-2 subunit of glutamate receptor, and type VII alpha polypeptide of voltage gated sodium channel. The induction of the ion-channel genes was accompanied by the activation of several genes encoding proteins in the signal transduction pathways. These included calmodulin 3, neuronal tyrosine threonine phosphatase 1, the beta subunit of phosphatidylinositol 4-phosphate 5-kinase, cAMP dependent regulatory subunit of protein kinase, magnesium dependent protein phosphatase 1B, protein tyrosine phosphatese IF2p, protein tyrosine phosphatase-1, serine/arginine rich protein specific kinase 2, Janus N-terminal kinase 2, Gz subunit of GTP binding protein and epsilon subunit of protein kinase C.

In summary, gene expression profiling of hippocampus and cortex from mice whose diets were supplemented with EGb 761 for 4 weeks showed significant increases in the transcripts encoding brain proteins. Many of these proteins play a vital role in neuronal and synaptic plasticity. The data reviewed here identify a molecular phenotype of EGb 761 actions. The behavioral phenotype of EGb 761 in health and in neurological disease remains to be defined. The data may offer some molecular correlates of behavioral changes observed in some studies (24,44). Future studies that combine the tools of functional genomics and behavioral analysis to study the effects Ginkgo biloba leaves are essential for critical and objective analysis of the in vivo effects of the dietary antioxidant supplements.

REFERENCES

1. Canada A, Giannella E, Nguyen T, Mason R. The production of reactive oxygen species by dietary flavonols. Free Radic Biol Med 1990; 9(5):441–449.

2. Shimoi K, Mochizuki M, Tomita I, Kaji K, Kuruto R, Nozawa R, Kumazawa S, Terao J, Nakayama T. Vascular permeability and functional activity of quercetin conjugates. 1st International Conference on Polyphenols and Health, Vichy, France, 2003.

3. Watanabe CM, Wolffram S, Ader P, Rimbach G, Packer L, Maguire JJ, Schultz PG, Gohil K. The in vivo neuromodulatory effects of the herbal medicine ginkgo biloba. Proc Natl Acad Sci USA 2001; 98(12):6577–6580.

4. Azzi A, Davies K, Kelly F. Free radical biology—terminology and critical thinking. FEBS Lett 2003; 1–3:3–6.

5. Sies H. Biochemistry of oxidative stress. Angew Chem Int Ed 1986; 25:1058–1071.

6. Sies H. Strategies of antioxidant defense. Eur J Biochem 1993; 215:213–219.

7. Packer L. Vitamin E is nature's master antioxidant. Sci Am, Sci Med 1994; 1:54–63.

8. Sen CK, Roy S, Packer L. Antioxidant and redox regulation of gene transcription. FASEB J 1996; 10:709–720.

9. Forman HJ, Cadenas E, eds. Oxidative Stress and Signal Transduction. New York: Chapman & Hall, 1997.

10. Montagnier L, Olivier R, Pasquier C, Eds Packer L and Cadenas E. Oxidative Stress in Cancer, AIDS and Neurodegenerative Diseases. The Oxidative Stress and Disease Series. New York: Marcel Dekker Inc, 1999.

11. Packer L, Yodoi J, Cadenas E, eds. Redox Regulation of Cell Signaling and its Clinical Application. The Oxidative Stress and Disease Series. New York: Marcel Dekker Inc, 1999.

12. Sen CK, Sies H, Baeurerle P, eds. Antioxidant and Redox Regulation of Genes. Academic Press, 1999.

13. Talalay P. Chemoprotection against cancer by induction of phase 2 enzymes. Biofactor 2000; 12(1–4):5–11.

14. Cadenas E, Packer L. Handbook of Antioxidants. In: Packer L, Cadena E, eds. The Oxidative Stress and Disease Series. New York: Marcel Dekker Inc, 2001.

15. Packer L, Traber M, Kraemer K, Frei B. The Antioxidant Vitamins C and E. Champaign, IL: AOCS Press, 2002.

16. Rice-Evans CA, Packer L. Flavonoids in Health and Disease. In: Packer L, Cadenas E, eds. Oxidative Stress and Disease. 2nd ed. Revised and expanded. New York, NY: Marcel Dekker, 2003:1–467.

17. Gohil K, Packer L. Ginkgo biloba extract and gene expression. In: Nesaretnam K, Packer L, eds. Micro Nutrients and Health: Molecular Biological Mechanisms. Champaign, IL: AOCS Press, 2001:217–224.

18. Rimbach G, Wolffram S, Watanabe CG, Packer L, Gohil K. Effect of Ginkgo biloba (EGb 761) on differential gene expression. Pharmacopsychiatry 2003; 36:S95–S99.

19. Rimbach G, Gohil K, Matsugo S, Moini H, Saliou C, Virgili F, Weber S, Packer L. Induction of glutathione synthesis in human keratinocytes in Ginkgo biloba. Biofactors 2001; 15:39–52.

20. Gohil K, Packer L. Global gene expression analysis identifies cell and tissue specific actions of ginkgo biloba extract, EGb 761. Cell Mol Biol 2002; 48:531–625.

21. Gohil K, Moy R, Farzin S, Maguire JJ, Packer L. mRNA expression profile of a human cancer cell line in response to Ginkgo biloba extract: induction of antioxidant response and the Golgi system. Free Radic Res 2000; 33:831–849.

22. Gohil K, Packer L. Bioflavonoid-rich botanical extracts show antioxidant and gene regulatory activity. Ann NYAS 2002; 957:1–8.

23. Hoyer S, Lannert H, Noldner M, Chatterjee S. Damaged neurol energy metabolism and behavior are improved by Ginkgo biloba extract (EGb 761). J Neural Transm 1999; 106: 1171–1188.

24. Stoll S, Scheuer K, Pohl O, Muller W. Ginkgo biloba extract (EGb 761) independently improves changes in passive avoidance learning and brain membrane fluidity in the aging mouse. Pharmacopsychiatry 1996; 29:144–149.

25. Cesarani A, Meloni F, Alpini D, Barozzi S, Verdorio L, Boscani P. Ginkgo biloba (EGb 761) in the treatment of equilibrium disorders. Adv Ther 1998; 15:291–304.

26. Itil T, Eralp E, Tsambis E, Itil K, Stein U. Central nervous system effects of Ginkgo biloba, a plant extract. Am J Ther 1996; 3:63–73.

27. Ader P, Wessmann A, Wolffram S. Bioavailability and metabolism of flavonol quercetin in the pig. Free Radic Biol Med 2000; 28:1059–1067.

28. Li C, Wong Y. The bioavailability of ginkgolides in Ginkgo biloba extracts. Planta Med 1997; 63:563–565.

29. Pietta P, Gardana C, Mauri P, Maffei-Facino R, Carini M. Identification of flavonoid metabolites after oral administration to rats of a Ginkgo biloba extract. J Chormatogr: B Biomed Appl 1995; 673:75–80.

30. Pietta P, Gardana C, Mauri P. Identification of Ginkgo Biloba flavonol metabolites after oral administration to humans. J Chormatogr: B Biomed Sci Appl 1997; 693:249–255.

31. Wojcicki J, Gawronska-Szklarz B, Bieganowski W, Patalan M, Smulski H, Samochowiec L, Zakzewski J. Comparative pharmacokinetics and bioavailability of flavonoid glycosides of Ginkgo Biloba after a single oral administration of three formulations to healthy volunteers. Mater Med Pol 1995; 27:141–146.

32. Savaskan E, Olivieri G, Meier F, Ravid R, Muller-Spahn F. Hippocampal estrogen beta-receptor immunoreactivity is increased in Alzheimer's disease. Brain Res 2001; 908:113–119.

33. Sencakova D, Graff-Radford N, Willis F, Lucas J, Parfitt F, Cha R, O'brien P, Peterson R, Jack C Jr. Hippocampal atropy correlates with clinical features of Alzheimer disease in African Americans. Arch Neurol 2001; 58:1593–1597.

34. Andersson K, Olofsson A, Nielsen E, Svehag S, Lundgren E. Only amyloidogenic intermediates of transthyretin induce apoptosis. Biochem Biophys Res Commun 2002; 294:309–314.

35. Koo E, Lansbury P Jr, Kelly J. Amyloid disease: abnormal protein aggregation in neurodegeneration. Proc Natl Acad Sci USA 1999; 96:9989–9990.

36. Hamilton J, Benson M. Transthyretin: a review from a structural perspective. Cell Mol Life Sci 2001; 58:1491–1521.

37. Banik N, Gohil K, Davison A. The action of snake venom, phospholipase A and trypsin on purified myelin in vitro. Biochem J 1976; 159:273–277.

38. Sabri M, Tremblay C, Banik N, Scott T, Gohil K, Davison A. Biochemical and morphological changes in the subcellular fractions during myelination of rat brain. Biochem Soc Trans 1975; 3:275–276.

39. Wahle S, Stoffel W. Cotranslational integgration of myelin proteolipid protein (PLP) into the membrane of endoplasmic reticulum: analysis of topology by glycosylation scanning and protease domain protection assay. Glia 1998; 24:226–235.

40. Clark W, Rinker L, Lessov N, Lowery S, Cipolla M. Efficacy of antioxidant therapies in transient focal ischemia in mice. Stroke 2001; 32:1000–1004.

41. Zhang W, Hayashi T, Kitagawa H, Sasaki C, Sakai K, Warita H, Wang J, Shiro Y, Uchida M, Abe K. Protective effect of ginkgo biloba extract on rat brain with transient middle cerebral artery occlusion. Neurol Res 2000; 22:517–521.

42. Alkon DL. Ionic conductance determinants of synaptic memory nets and their implications for Alzheimer's disease. J Neurosci Res 1999; 58:24–33.

43. Giese K, Peters M, Vernon J. Modulation of excitability as a learning and memory mechanism: a molecular genetic perspective. Physiol Behav 2001; 73:803–810.

44. Gajewski A, Hensch S. Ginkgo biloba and memory for a maze. Pyschol Rep 1999; 84:481–484.

2

Transcriptional Regulation of Cellular Antioxidant Defense Mechanisms

YOUNG-JOON SURH

Laboratory of Biochemistry and Molecular
Toxicology, National Research Laboratory,
College of Pharmacy, Seoul, National University,
Seoul, South Korea

Reactive oxygen species (ROS) are constantly generated in human body. Enzymatic and nonenzymatic antioxidants detoxify ROS and minimize damage to biomolecules. An imbalance between the production of ROS and cellular antioxidant capacity leads to a state of "oxidative stress" that contributes to the pathogenesis of a vast variety of clinical abnormalities (1–3). The susceptibility of the target organs or cells to oxidative injury depends largely on their capability to control protective ROS scavenging systems. The primary endogenous antioxidants are present normally at low levels in human tissues and are not timely induced when exposed to oxidative stress. However, the response of certain antioxidant enzymes, constituting the critical primary defense against exogenous oxidative stress, occurs rapidly in proportion to oxidant insult. This chapter focuses on molecular aspects of adaptive cytoprotection in response to oxidative stress.

I. TOXICOLOGICAL vs. PHYSIOLOGICAL FUNCTIONS OF ROS

During the normal aerobic metabolism, ROS are generated at low levels as unwanted by-products as a consequence of the transfer of a single electron. ROS, which include oxygen free radicals and their nonradical derivatives, play an integral role in maintaining and modulating a wide spectrum of vital physiological functions. In recent years, it has become more apparent that ROS are important mediators of intracellular signaling and redox regulation responsible for cellular homeostasis (4–6). In fact, some growth factors, cytokines, hormones, and neurotransmitters utilize ROS as secondary messengers in executing normal physiological processes (7). However, excessive production of ROS by exogenous redox chemicals, physical agents (e.g., ultraviolet and ionizing radiations), bacterial or viral infection, or under abnormal pathophysiologic conditions such as oxygen shortage (hypoxia) can be destructive. ROS not only induce direct damage to critical biomolecules, such as DNA, proteins, membrane lipids, and carbohydrates but also indirectly alter or dysregulate the cellular signaling events (Fig. 1). The pathophysiological consequences of such oxidative injury include cancer, neurodegenerative disorders, diabetes, rheumatoid arthritis, etc. Multiple lines of compelling evidence from laboratory and clinical studies support the involvement of ROS as a major cause of cellular injuries in a number of human diseases. ROS hence elicit a wide spectrum of toxicologic as well as physiologic responses.

Oxidative stress refers to the situation of a serious imbalance or mismatched redox equilibrium between production of ROS and the ability of cells to defend against them. Oxidative

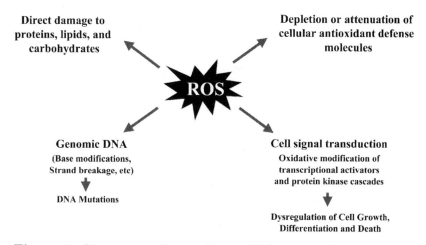

Figure 1 Direct vs. indirect effects of ROS.

stress thus occurs when the production of ROS increases, elimination of ROS or repair of oxidatively damaged macromolecules decreases, or both. Narrowly interpreted, the family of ROS consists of superoxide radical anion ($O_2{}^{\cdot-}$), hydroperoxyl radical ($HO_2{}^{\cdot}$), hydrogen peroxide (H_2O_2), and hydroxyl radical ($HO^{\cdot}$). Superoxide, its protonated form $HO_2{}^{\cdot}$, and $HO^{\cdot}$ are relatively short-lived, whereas H_2O_2 is comparatively stable and can cross cell membranes.

H_2O_2 is a representative ROS that is produced through auto-oxidation of redox xenobiotics as well as incomplete oxidation in the electron transport chain via dismutation of resultant superoxide by superoxide dismutase (SOD). H_2O_2 is recently considered to play a pivotal role as a messenger in intracellular signaling cascades (5,8,9). Though not reactive per se, H_2O_2 forms highly reactive hydroxyl radical by the Fenton reaction in the presence of transition metal ion. Hydroxyl radical reacts rapidly with almost every critical cellular macromolecules including DNA, lipids, and proteins, and thereby causes functional as well as structural alterations in these biomolecules. Enzymatic and nonenzymatic antioxidants detoxify H_2O_2 and other ROS and minimize damage to critical biomolecules.

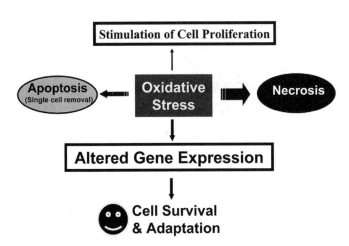

Figure 2 Differential effects of ROS.

Typically, low concentrations of ROS are mitogenic, and promote cell proliferation, while intermediate concentrations result in either temporary or permanent growth arrest, such as replicative senescence. Severe oxidative stress ultimately causes cell death via either apoptotic or necrotic mechanisms (Fig. 2).

II. ADAPTIVE RESPONSE TO OXIDATIVE INJURIES

During evolution, multicellular aerobic organisms have been adapted to tolerate and survive the oxidant burden evoked by endogenous oxidative metabolism in the cells or external insults including the chemicals, low-level radiation and microbial infection. The defense repertoire exists at all levels of the biological hierarchy—from the molecular and biochemical level to the cellular and tissue level, further extending to the organ level (10). The cellular and molecular adaptive response to oxidative stress involves elevated synthesis of low-molecular weight antioxidants, increased expression of antioxidant/detoxifying enzymes and other stress-inducible cytoprotective proteins aimed at reversing the oxidative imbalance or restoring the normal redox status, thereby achieving and maintaining the cellular homeostasis. Central to such adaptive response is the stimulation of specific signal transduction pathways mediating both transcriptional and post-translational modification of proteins needed to cope with oxidative stress.

III. CELLULAR ANTIOXIDANT DEFENSES

To minimize oxidative injury to major cellular components, thereby protecting cells under stress conditions, the multicellular organisms are endowed with distinct sets of integrated antioxidant defense mechanisms responsible for scavenging ROS or converting them to less reactive products that can be quenched by other antioxidants (11). These endogenous antioxidants comprise a co-operative network which employs a series of redox reactions. If there exists too much oxidant burden, ROS may overwhelm or attenuate/inactivate the antioxidant defense system. Consequences of oxidative stress hence depend on types of cells affected and nature of ROS involved, endogenous antioxidant status, co-operation and/or compensation between different antioxidant systems, and also the cellular capability to induce or potentiate the antioxidant defense system, all of which determine the competence of the cellular defense mechanism against oxidative insult. In general, the preservation of the redox status of the cell is vital for survival.

The response of cells to oxidative stress is very complex. The cellular antioxidant defense system is composed of enzymatic and nonenzymatic components, which can be classified into several categories (Table 1). These include: (i) metal chelators (e.g., metalothionein, ceruloplasmin, ferritin, transferrin, and deferoxamine) capable of preventing free radical formation by inhibiting metal catalyzed reactions such as Fenton reaction; (ii) low-molecular weight antioxidants (e.g., ascorbic acid, glutathione, tocopherols); (iii) enzymes synthesizing or regenerating the reduced forms of antioxidants, such as glutamate-cysteine ligase (GCL) and glutathione

Table 1   Cellular Antioxidant Defense System

Inhibiting ROS formation
Scavenging ROS or their precursors
Binding metal ions needed for catalysis of ROS generation
Upregulation of endogenous antioxidant enzymes
Maintenance of cellular thiol (e.g., GSH) levels

---

reductase (GR) (iv) ROS-interacting enzymes such as SOD, glutathione peroxidases (GPx), and catalases (CAT). Antioxidant defense mechanisms include prevention of formation of ROS, direct elimination or scavenging of ROS, or stimulation of the repair of damaged oxidative lesions. The levels and functions of individual antioxidant components should be maintained in a co-ordinated manner so that they can efficiently counteract oxidative burden.

Endogenous antioxidant systems are critically important in limiting ROS-mediated cellular damage. The severity and duration of oxidative stress are themselves important in regulating levels of both antioxidant enzymes and endogenous ROS scavengers. Besides classical low-molecular weight antioxidants, such as ascorbic acid, tocopherol, and GSH, thioredoxin has recently attracted increasing attention as a cellular protectant against oxidative injury (12,13). Thioredoxin proteins act as disulfide bond reductants and electron donors for thioredoxin peroxidases. Oxidized thioredoxin is reduced by thioredoxin reductases that require NADPH. Thioredoxin protects cells from oxidative injury.

Endogenous antioxidant enzymes, such as SOD, CAT GPx, thioredoxin reductase, and GCL function co-operatively to destroy excess ROS (Fig. 3), thereby rescuing cells from oxidative injuries (14). SOD inactivates superoxide anion by converting it to H_2O_2 which, in turn, is detoxified by CAT or GPx. While CAT reacts with H_2O_2 to form water and molecular oxygen, GPx detoxifies H_2O_2 in the presence of GSH, producing H_2O and oxidized glutathione (GSSG) which is recycled to GSH by GR. GSH can be replenished through de novo synthesis which is catalyzed by GCL. When a cellular GSH level is low, H_2O_2 can produce extremely reactive hydroxyl

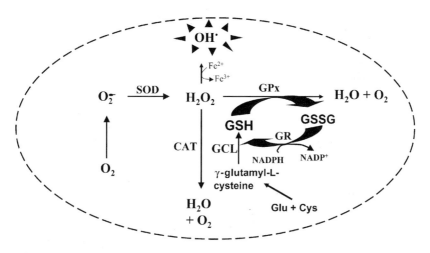

Figure 3 Representative cellular enzymes responsible for eliminating ROS or potentiating the antioxidant capacity. *Abbreviations*: SOD, superoxide dismutase; CAT, Catalase; GPx, glutathione peroxidase; GR, glutathione reductase; GCL, glutamate-cysteine ligase.

radical in the presence of transition metal ions via the Fenton reaction. If SOD is formed at a very high level, excessive amounts of H_2O_2 can be generated, provoking a hydrogen peroxide stress. Thus, it is desirable that SOD induction should be accompanied by timely induction of a gene coding for an H_2O_2-inactivating enzyme, such as CAT or GPx. Therefore, the response of cells to oxidative stress should be tightly regulated.

IV. TRANSCRIPTIONAL REGULATION OF CELLULAR ANTIOXIDANT AND DETOXIFYING ENZYMES

Oxidative stress provokes cellular responses which principally involve transcriptional activation of genes encoding proteins participating in the antioxidant defense. The eukaryotic cell contains a multitude of signal transduction pathways coupling pro-oxidative stimuli to the specific regulation of gene expression for adaptive responses. The signal transduction pathways that lead to the oxidative stress response are much less understood. In recent years, several components of the signal transduction cascades that sense oxidative injury have been identified in higher eukaryotes. These include the mitogen-activated protein (MAP) kinases and transcription factors that are subject to redox regulation. ROS can function as physiological mediators of transcriptional control of downstream genes encoding antioxidant enzymes and cytoprotective or stress-responsive proteins, which may lead to the adaptive responses to oxidative stress. Among the transcription factors involved in cellular defense/adaptation to oxidative stress response, two early response transcriptional complexes NF-κB and AP-1 have been considered to be most important because they can be redox regulated (15,16). The activation of NF-κB and AP-1 is influenced following exposure to chemical, physical and biological agents that can alter the cellular redox status.

Recent studies have identified other transcription factors distinct from NF-κB and AP-1. Of particular interest is Nrf2, a member of the "cap 'n' collar" (CNC) family of basic region/leucine zipper (bZIP) transcription factors, highly homologous to the NF-E2 transcription factors (17).

IV.A. NF-κB

The ubiquitous eukaryotic transcription factor NF-κB is known to regulate expression of numerous genes involved in stress responses and plays a crucial role in cell survival. ROS can serve as common mediators of NF-κB activation signals. The DNA binding and transcriptional activities of NF-κB were constitutively elevated in selected clones of PC12 cells resistant to oxidative stress induced by beta-amyloid peptide and H_2O_2 (18). In addition, PC12 cells rendered tolerant to H_2O_2- or beta-amyloid-induced oxidative stress by ectopic expression of antiapoptotic Bcl-2 exhibited constitutive activation of NF-κB (19,20). Conversely, NF-κB inhibitors, such as pyrrolidine dithiocarbamate and N-tosyl-L-phenylalanine chloromethyl ketone, or antisense oligonucleotide containing a specific κB binding sequence exacerbated the oxidative cell death (18–24). Likewise, suppression of transcriptional activity of NF-κB with dexamethasone or ectopic expression of a super-repressor mutant form of IκBα reversed the oxidative stress-resistant phenotype (18).

Recently, there has been increasing evidence supporting the role of NF-κB in the regulation of antiapoptotic gene expression and promotion of cell survival (25,26). The transcriptional regulation of CAT, SOD and GCL is mediated partially by NF-κB (Fig. 4). Sequence analysis of genes encoding mouse GPx and CAT revealed the existence of binding sites for NF-κB (23). The 5'-flanking region of SOD harbors the putative binding sites for NF-κB and other transcription factors including Sp-1 and AP-1, and a specific regulatory sequence called antioxidant-response element (ARE)/electrophile-response element (EpRE) (27,28). The promoter regions of both catalytic and regulatory subunits of GCL gene contain multiple potential regulatory elements, such as NF-κB, Sp-1, AP-1, AP-2, and ARE/EpRE (29,30). Taking into consideration all these findings, it is likely that NF-κB is a key component of signaling pathways mediating cellular adaptation to oxidative stress through augmentation of endogenous antioxidant capacity.

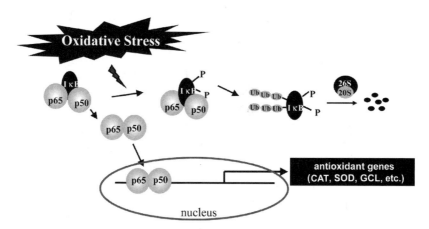

Figure 4 NF-κB signaling pathways leading to upregulation of antioxidant genes.

IV.B. AP-1

AP-1 is another redox-sensitive transcription factor that regulates expression of genes involved in adaptation, inflammation, immune responses, cell proliferation and differentiation, and tumor promotion. AP-1 consists of either homo- or heterodimers between members of Jun and Fos family proteins which interact via the heptad repeats and hydrophobic residues located in a leucine zipper domain. The activation of AP-1 through dimerization by means of leucine zippers is known to be mediated by protein kinase C (PKC)-catalyzed phosphorylation. Like NF-κB, AP-1 mediates pleiotropic effects of pro-oxidative stimuli in cellular signaling cascades although few target genes are identified. The 5′-flanking region of the human GPx encoding gene contains the DNA binding element for AP-1 (31). Cigarette smoke increases AP-1 DNA binding in human epithelial cells in vivo (32). Increased levels of ROS derived from cigarette smoke appear to alter the redox balance, which is sensed by the epithelial cells, leading to activation of AP-1 and downstream target gene expression.

The expression and/or activity of AP-1 can be induced under pro-oxidant conditions. However, the transcriptional activity of AP-1 can also be downregulated post-translationally by oxidant stimuli as evidenced by decreased DNA binding upon oxidation of critical cysteine residues of AP-1, which could be reversed by thiol reduction (33). In support of the latter notion, AP-1 DNA binding was enhanced in the presence of reduced thioredoxin (34) by apurinic/apyrimidinic endonuclease/redox factor-1 (APE/Ref-1), a multifunctional protein originally identified as endonuclease. Ref-1 is considered to mediate redox activation of AP-1 and other stress-inducible transcription factors including NF-κB, p53, and hypoxia-inducible factor 1 (35).

IV.C. Nrf2

The Nrf2 protein was first isolated by an expression cloning procedure using an oligonucleotide harboring the NF-E2 DNA binding motif as a probe to screen closely related proteins (36). NF-E2 is a dimeric protein originally considered to be involved in the regulation of globin gene expression in hematopoietic cells (37,38). NF-E2 activates gene transcription following binding to its consensus DNA binding motif 5'-TGCTGAGTCAC-3' as a heterodimer consisting of a 45-kDa and an 18-kDa subunit (39). The p45 subunit is a bZIP protein consisting of a transactivation domain within the N-terminal region and a basic DNA binding region/leucine zipper structure in the C-terminal region of the protein (37,38). The p18 subunit was subsequently identified as a member of the small Maf proteins containing the basic region/leucine zipper but lacking any apparent transactivation domain (39). Whereas expression of the p45 NF-E2 subunit is restricted to hematopoietic tissues (37,38), two other related bZIP family members, Nrf1 and Nrf2, are ubiquitously expressed in a wide range of tissues and cell types. The tissue distribution profile of these proteins combined with their DNA binding motif being similar in comparison to that of the ARE suggests that ARE-responsive genes may be regulated by Nrf1 and/or Nrf2 (40). Nrf1 is a widely expressed CNC-bZIP factor and together with small Maf proteins binds as heterodimers to the NF-E2/AP-1 site (41). Nrf1 was not studied much in comparison with Nrf2. It is an essential gene in mice, the targeted disruption of which results in embryonic lethality at around mid-gestation period.

Nrf2, as a basic leucine zipper transcriptional factor, has two functional termini. One is DNA binding C-terminus and the other is transactivating N-terminus. During oxidative stress or other types of toxic insult, Nrf2 heterodimerizes and binds to the ARE/EpRE sequence to activate transcription of genes coding for antioxidant or phase II detoxifying enzymes. A role of Nrf2 in the regulation of ARE-driven expression of antioxidant/phase II detoxification enzymes has been shown in studies utilizing Nrf2 (−/−) null mice (42–44). These mice have defects in detoxifying certain carcinogens and were more susceptible to chemically induced carcinogenesis and oxidative injury compared with the wild-type animals.

In cytosol, Nrf2 activity is controlled by an interaction between Nrf2 and a cytoskeleton-associated inhibitory protein called Kelch-like ECH associating protein 1 (Keap1) (45,46). Nrf2 dissociates from its cytoplasmic docking protein Keap1 and translocates to nuclei when cells are exposed to stimulus which involves oxidative stress. Nrf2, once bound to ARE, induces ARE-dependent gene expression. The actual mechanism of dissociation of Nrf2 from Keap1 remains unresolved, but it is considered to involve thiol modification/oxidation of the cysteine residues of Keap1 and/or phosphorylation of serine or threonine residues located in Nrf2 (47,48). Oxidative stress appears to activate several signaling molecules such as MAP kinases, PKC, or phosphatidylinositol 3-kinase (PI3K), which facilitates dissociation of Nrf2 from Keap1 and/or its nuclear translocation (49–56) as schematically illustrated in Fig. 5.

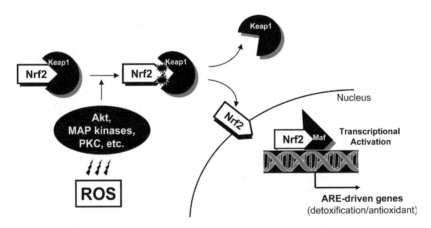

Figure 5 Activation of Nrf2 in response to oxidative stress. Reactive oxygen species can activate the distinct upstream kinases, such as MAP kinases, PKC, and PI3K-Akt, which can phosphorylate the specific Ser/Thr residues of Nrf2. This may facilitate the nuclear translocation of Nrf2 and subsequent ARE/EpRE binding. Alternatively, ROS can oxidize the thiol residues in the cysteine-rich Keap1, stimulating the dissociation of this inhibitory protein from Nrf2.

The phase II gene inducers, such as butylated hydroxy anisole, *tert*-butylhydroquinone, the green tea polyphenol (–)-epigallocatechin-3-gallate, carnosol, and the phenethylisothiocyanate sulforaphane, can activate the Nrf2 which dimerizes with small Mafs and binds to ARE/EpRE enhancers (56–59).

V. ARE AS A CRITICAL COMPONENT OF Nrf2-MEDIATED ANTIOXIDANT DEFENSE

ARE is a *cis*-acting regulatory element or enhancer sequence, which is found in promoter regions of genes coding phase II detoxification enzymes and antioxidant proteins. Okuda et al. (60) described an enhancer element which is similar to 12-*O*-tetradecanolylphorbol-13-acetate-responsive element or AP-1 site in the rat glutathione *S*-transferase-P gene. Subsequently, it was also found in the promoter regions of mouse glutathione *S*-transferase Ya (61), and human NAD(P) H:quinone oxidoreductase 1 (62) genes.

Through computer search of various databases, a number of genes containing an ARE-like sequence have been identified (63,64). The promoters of these genes have a common sequence called the core sequence (5′-TGACnnnGC-3′) and newly searched genes bear the same sequence. ARE contains the core sequence that resembles the half-site recognized by a member of the AP-1 and ATF/CREB families of transcription factors. All members of these two families of proteins belong to a class of transcription factors collectively named bZIP proteins, which can bind to DNA as part of a homodimer and/or heterodimer complex. It has been suggested that bZIP proteins may be involved in regulating ARE. Although many proteins including AP-1 have been suggested to regulate ARE, the underlying ARE activation mechanisms have been elucidated with the identification of Nrf2 (40,45).

VI. SUMMARY

There is increasing evidence that oxidative stress induced by ROS and their equivalents is implicated in a vast variety of human disorders. In response to pro-oxidant insult, cells tend to potentiate or fortify the antioxidant mechanisms to minimize the oxidative damage. This can often be achieved by upregulation of antioxidant enzymes or stress-response proteins. Transient changes in cellular oxidation–reduction status, communicated via a series of cellular redox-sensitive signaling circuitry, initiate de novo synthesis of a distinct set of antioxidant or phase II detoxification enzymes that are responsible for adaptive cellular responses to toxic environmental insults. Certain stress-activated protein kinases or other upstream signaling enzymes can activate the redox-sensitive transcription factors, such as NF-κB, AP-1, and Nrf2, thereby upregulating the synthesis of early response genes to tolerate or survive the subsequent oxidative injury. Pharmacological and dietary manipulation of cellular antioxidant capacity may provide an important strategy to cope with the oxidative stress-mediated ailments.

ACKNOWLEDGMENT

This work has been supported by the National Research Laboratory grant from the Ministry of Science and Technology, Republic of Korea.

REFERENCES

1. Barnham KJ, Masters CL, Bush AI. Neurodegenerative diseases and oxidative stress. Nature Rev Drug Discov 2004; 3: 205–214.

2. Ceconi C, Boraso A, Cargnoni A, Ferrari R. Oxidative stress in cardiovascular disease: myth or fact? Arch Biochem Biophys 2003; 420:217–221.

3. Arteel GE. Oxidants and antioxidants in alcohol-induced liver disease. Gastroenterology 2003; 124:778–790.

4. Sundaresan M, Yu ZX, Ferrans VJ, Irani K, Finkel T. Requirement for generation of H_2O_2 for platelet-derived growth factor signal transduction. Science 1995; 270:296–299.

5. Rhee SG. Redox signaling: hydrogen peroxide as intracellular messenger. Exp Mol Med 1999; 31:53–59.

6. Dalton TP, Shertzer HG, Puga A. Regulation of gene expression by reactive oxygen. Annu Rev Pharmacol Toxicol 1999; 39: 67–101.

7. Thannickal VJ. Reactive oxygen species in cell signaling. Am J Physiol Lung Cell Mol Physiol 2000; 279:L1005–L1028.

8. Saran M, Bros W. Oxygen radicals acting as chemical messengers: a hypothesis. Free Radic Res Commun 1989; 7:213–220.

9. Finkle T. Oxygen radicals and signaling. Curr Opin Cell Biol 1998; 10:248–253.

10. Trosko JE. Hierarchical and cybernetic nature of biologic systems and their relevance to homeostatic adaptation to low-level exposures to oxidative stress-inducing agents. Environ Health Perspect 1998; 106 (suppl 1):331–339.

11. Comhair SAA, Erzurum SC. Antioxidant responses to oxidant-mediated lung diseases. Am J Physiol Lung Cell Mol Physiol 2002; 283:L246–L255.

12. Hirota K, Nakamura H, Masutani H, Yodoi J. Thioredoxin superfamily and thioredoxin-inducing agents. Ann NY Acad Sci 2002; 957:189–199.

13. Norberg J, Arner ESJ. Reactive oxygen species, antioxidants, and the mammalian thioredoxin system. Free Radic Biol Med 2001; 31:1287–1312.

14. Curtin JF, Donovan M, Cotter TG. Regulation and measurement of oxidative stress in apoptosis. J Immunol Methods 2002; 265:49–72.

15. Gius D, Botero A, Shah S, Curry HA. Intracellular oxidation/reduction status in the regulation of transcription factors NF-κB and AP-1. Toxicol Lett 1999; 106:93–106.

16. Peng M, Huang L, Xie ZJ, Huang WH, Askari A. Oxidant-induced activations of nuclear factor-kappa B and activator protein-1 in cardiac myocytes. Cell Mol Biol Res 1995; 41:189–197.

17. Otterbein LE, Choi AMK. The saga of leucine zippers continues in response to oxidative stress. Am J Respir Cell Mol Biol 2002; 26:161–163.

18. Lezoualc'h F, Sagara Y, Holsboer F, Behl C. High constitutive NF-κB activity mediates resistance to oxidative stress in neuronal cells. Neuroscience 1998; 18:3224–3232.

19. Jang J-H, Surh Y-J. Bcl-2 attenuation of oxidative cell death is associated with upregulation of γ-glutamate-cysteine ligase via constitutive NF-κB activation. J Biol Chem 2004; 279: 38779–38786.

20. Jang J-H, Surh Y-J. Bcl-2 protects against beta-amyloid-induced oxidative PC12 cell death via potentiation of antioxidant capacity. Biochem Biophys Res Commun 2004; 320: 880–886.

21. Blum D, Torch S, Nissou M-F, Verna J-M. 6-Hydroxydopamine-induced nuclear factor-κB activation in PC12 cells. Biochem Pharmacol 2001; 62:473–481.

22. Mitsiades N, Mitsiades CS, Poulaski V, Chauhan D, Rishcardson P, Hideshima T, Munshi N, Treon SP, Anderson KC. Biologic sequelae of nuclear factor-κB blockade in multiple myeolma: therapeutic applications. Blood 2002; 99:4079–4086.

23. Zhou LZ, Johnson AP, Rando TA. NF-κB and AP-1 mediate transcriptional responses to oxidative stress in skeletal muscle cells. Free Radic Biol Med 2001; 31:1405–1416.

24. Taglialatela G, Robinson R, Peerz-Polo JR. Inhibition of nuclear factor κB (NF-κB) activity induces nerve growth factor-resistant apoptosis in PC12 cells. J Neurosci Res 1997; 47:155–162.

25. Bours V, Bentires-Alj M, Hellin AC, Viatour P, Robe P, Delhalle S, Benoit V, Merville MP. Nuclear factor-kappa B, cancer, and apoptosis. Biochem Pharmacol 2000; 60: 1085–1089.

26. Jang J-H, Surh Y-J. Potentiation of cellular antioxidant capacity by Bcl-2: implications for its antiapoptotic function. Biochem Pharmacol 2003; 66:1371–1379.

27. Kim HT, Kim YH, Nam JW, Lee HJ, Rho HM, Jung G. Study of 5'-flanking region of human Cu/Zn superoxide dismutase. Biochem Biophys Res Commun 1994; 201:1526–1533.

28. Jones PL, Kucera G, Gordon H, Boss JM. Cloning and characterization of the murine manganous superoxide dismutase-encoding gene. Gene 1995; 153:155–161.

29. Yang H, Wang J, Huang ZZ, Ou X, Lu SC. Cloning and characterization of the 5′-flanking region of the rat glutamate-cysteine ligase catalytic subunit. Biochem J 2001; 357: 447–455.

30. Mulcahy RT, Gipp JJ. Identification of a putative antioxidant response element in the 5′-flanking region of the human gamma-glutamylcysteine synthetase heavy subunit gene. Biochem Biophys Res Commun 1995; 209:227–233.

31. Yoshimura S, Suemizu H, Tanigushi Y, Arimori K, Kawabe N, Moriuchi T. The human plasma glutathione peroxidase-encoding gene: organization, sequence and localization to chromosome 5q32. Gene 1994; 145:293–297.

32. Rahman I, MacNee W. Lung glutathione and oxidative stress: implications in cigarette smoke-induced airway disease. Am J Physiol Lung Cell Mol Physiol 1999; 277:L1067–L1088.

33. Abate C, Patel L, Rauscher FJ III, Curran T. Redox regulation of fos and jun DNA-binding activity in vitro. Science 1990; 249: 1157–1161.

34. Hirota K, Matsui M, Iwata S, Nishiyama A, Mori K, Yodoi J. AP-1 transcriptional activity is regulated by a direct association between thioredoxin and Ref-1. Proc Natl Acad Sci 1997; 94:3633–3638.

35. Fritz G, Grosch S, Tomicic M, Kaina B. APE/ref-1 and the mammalian response to genotoxic stress. Toxicology 2003; 193: 67–78.

36. Moi P, Chan K, Asunis I, Cao A, Kan YW. (1994) Isolation of NF-E2 related factor 2 (Nrf2), a NF-E2-like basic leucine zipper transcriptional activator that binds to the tandem NF-E2/AP1 repeat of the β-globin locus control region. Proc Natl Acad Sci USA 1994; 91:9926–9930.

37. Andrews NC, Erdjument BH, Davidson MB, Tempst P, Orkin SH. Erythroid transcription factor NF-E2 is a haematopoietic-specific basic-leucine zipper protein. Nature 1993; 362: 722–728.

38. Ney PA, Andrews NC, Jane SM, Safer B, Purucker ME. Purification of the human NF-E2 complex: cDNA cloning of

the hematopoietic cell-specific subunit and evidence for an associated partner. Mol Cell Biol 193; 11:5604–5612.

39. Igarashi K, Kataoka K, Itoh K, Hayashi N, Nishizawa M. Regulation of transcription by dimerization of erythroid factor NF-E2 p45 with small Maf proteins. Nature 1994; 367:568–572.

40. Venugopal R, Jaiswal AK. Nrf1 and Nrf2 positively and c-Fos and Fra1 negatively regulate the human antioxidant response element-mediated expression of NAD(P)H:quinone oxidoreductase1 gene. Proc Natl Acad Sci USA 1996; 93:14960–14965.

41. Mandy K, Yuet WK, Jefferson YC. The CNC basic leucine zipper factor, Nrf1, is essential for cell survival in response to oxidative stress-inducing agents. J Biol Chem 1999; 274:37491–37498.

42. Chan K, Kan YW. Nrf2 is essential for protection against acute pulmonary injury in mice. Proc Natl Acad Sci USA 1999; 96: 12731–12736.

43. Cho H-Y, Jedlicka AE, Reddy SPM, Kensler TW, Yamamoto M, Zhang L-Y, Kleeberger SR. Role of NRF2 in protection against hyperoxic lung injury in mice. Am J Respir Cell Mol Biol 2002; 26:175–182.

44. Kwak MK, Wakabayashi N, Itoh K, Motohashi H, Yamamoto M, Kensler TW. Modulation of gene expression by cancer chemopreventive dithiolethiones through the Keap1-Nrf2 pathway. Identification of novel gene clusters for cell survival. J Biol Chem 2003; 278:8135–8145.

45. Itoh K, Wakabayashi N, Katoh Y. Keap1 represses nuclear activation of antioxidant responsive elements by Nrf2 through binding to the amino-terminal Neh2 domain. Genes Dev 1999; 13:76–86.

46. Itoh K, Wakabayashi N, Katoh Y, Ishii T, O'Connor T, Yamamoto M. Keap1 regulates both cytoplasmic-nuclear shuttling and degradation of Nrf2 in response to electrophiles. Genes Cells 2003; 8:379–391.

47. Wakabayashi N, Dinkova-Kostova AT, Holtzclaw WD, Kang MI, Kobayashi A, Yamamoto M, Kensler TW, Talalay P. Protection against electrophile and oxidant stress by induction of the phase 2 response: fate of cysteines of the Keap1 sensor modified by inducers. Proc Natl Acad Sci USA 2004; 101:2040–2045.

48. Surh Y-J. Chemoprevention with dietary phytochemicals. Nature Rev Cancer 2003; 3:768–780.

49. Alam J, Wicks C, Stewart D, Gong P, Touchard C, Otterbein S, Choi AMK, Burrow ME, Tou J. Mechanisms of heme oxygenase-1 gene activation by cadmium in MCF-7 mammary epithelial cells: role of p38 kinase and Nrf2 transcription factor. J Biol Chem 2000; 275:27694–27702.

50. Yu R, Chen C, Mo YY, Hebbar V, Owuor ED, Tan TH, Kong ANT. Activation of mitogen-activated protein kinase pathways induces antioxidant response element-mediated gene expression via a Nrf2-dependent mechanism. J Biol Chem 2000; 275: 39907–39913.

51. Lee JM, Hanson JM, Chu WA, Johnson JA. Phosphatidylinositol 3-kinase, not extracellular signal-regulated kinase, regulates activation of the antioxidant-responsive element in IMR-32 human neuroblastoma cells. J Biol Chem 276; 20011–20016.

52. Kang KW, Park EY, Kim SG. Activation of CCAAT/enhancer-binding protein β by 2'-amino-3'-methoxyflavone (PD98059) leads to the induction of glutathione S-transferase A2. Carcinogenesis 2003; 24:475–482.

53. Martin D, Rojo AI, Salinas M, Diaz R, Gallardo G, Alam J, De Galarreta CM, Cuadrado A. Regulation of heme oxygenase-1 expression through the phosphatidylinositol 3-kinase/Akt pathway and the Nrf2 transcription factor in response to the antioxidant phytochemical carnosol. J Biol Chem 2004; 279: 8919–8929.

54. Jaiswal AK. Nrf2 signaling in coordinated activation of antioxidant gene expression. Free Radic Biol Med 2004; 36: 1199–1207.

55. Cho HY, Reddy SP, Debiase A, Yamamoto M, Kleeberger SR. Gene expression profiling of NRF2-mediated protection against oxidative injury. Free Radic Biol Med 2005; 38:325–343.

56. Keum YS, Jeong WS, Kong AN. Chemoprevention by isothiocyanates and their underlying molecular signaling mechanisms. Mutat Res 2004; 555:191–202.

57. Chen C, Kong AN. Dietary chemopreventive compounds and ARE/EpRE signaling. Free Radic Biol Med 2004; 36:1505–1516.

58. Zhang DD, Hannink M. Distinct cysteine residues in Keap1 are required for Keap1-dependent ubiquitination of Nrf2 and for stabilization of Nrf2 by chemopreventive agents and oxidative stress. Mol Cell Biol 2003; 23:8137–8151.

59. Rushmore TH, Kong AN. Pharmacogenomics, regulation and signaling pathways of phase I and II drug metabolizing enzymes. Curr Drug Metab 2002; 3:481–490.

60. Okuda A, Imagawa M, Meada Y, Sakai M, Muramatsu M. Structural and functional analysis of an enhancer GPEI having a phorbol 12-*O*-tetradecanoate 13-acetate responsive element-like sequence found in the rat glutathione transferase P gene. J Biol Chem 1989; 264:16919–16926.

61. Friling RS, Bensimon A, Tichauer Y, Daniel V. Xenobiotic-inducible expression of murine glutathione *S*-transferase Ya subunit gene is controlled by an electrophile-responsive element. Proc Natl Acad Sci USA 1990; 87:6258–6262.

62. Li Y, Jaiswal AK. Regulation of human NAD(P)H:quinone oxidoreductase gene. Role of AP1 binding site contained within human antioxidant response element. J Biol Chem 1992; 267: 15097–15104.

63. Johnsen Q, Murphy P, Prydz H, Kolst AB. Interaction of the CNC-bZIP factor TCF11/LCR-F1/Nrf1 with MafG: binding-site selection and regulation of transcription. Nucleic Acids Res 1998; 26:512–520.

64. Wasserman WW, Fahl WE. Functional antioxidant responsive elements. Proc Natl Acad Sci USA 1997; 94:5361–5366.

3

Redox Regulation of Inflammatory Tissue Damage by Thioredoxin

SHUGO UEDA, HAJIME NAKAMURA, and TAKAYUKI NAKAMURA

Translational Research Center, Kyoto University Hospital, Kyoto, Japan

JUNJI YODOI

Translational Research Center and Institute for Virus Research, Kyoto University Hospital and Kyoto University, Kyoto, Japan

ABSTRACT

The thioredoxin (TRX) and glutathione (GSH) systems regulate the cellular redox (reduction/oxidation) status. Thioredoxin is a ubiquitous small multifunctional molecule that has a redox-active disulfide/dithiol within the conserved –Cys-Gly-Pro-Cys– sequence. It is a stress-inducible protein, and exerts intracellular and extracellular functions. Intracellularly, TRX scavenges reactive oxygen species (ROS) and protects cells against oxidative stress. It also regulates various signal transduction pathways, including proliferation, apoptosis, and gene expressions through the protein–protein or protein–nucleic acid interaction. It translocates from cytosol to nucleus upon stress and augments the DNA binding activity of several transcription factors. Extracellularly, TRX suppresses the inflammation by regulating the neutrophil activation and extravasation and exerts the cytoprotective activity. Thioredoxin-transgenic mice (TRX-Tg mice) show the milder tissue damage or injury than the control mice in several oxidative stress-associated disease models. Moreover, the administration of recombinant human TRX (rhTRX) ameliorates the acute lung injury (ALI) associated with neutrophil infiltration in mice. Therefore, aiming at the clinical application, the translational research project is in process using rhTRX as a therapeutic drug of ALI/acute respiratory distress syndrome (ARDS). Accumulating evidence suggests that administrating rhTRX can be a novel modality regulating cellular redox on the several severe diseases or complications related to inflammation and oxidative stress.

I. INTRODUCTION—REDOX CONTROL BY THIOREDOXIN SYSTEM

Reactive oxygen species (ROS) are generated by a variety of oxidative stress; internally by the respiratory chain in mitochondria or NADPH oxidases, and externally ultraviolet (UV), ionizing irradiation, virus infection, and drugs. They not only damage cells but also play an important role in intracellular signal transduction pathways, such as proliferation, gene activation, cell cycle arrest, and apoptosis (1–3). Cellular redox (reduction/oxidation) status is controlled by the thioredoxin (TRX) and glutathione (GSH) systems that scavenge ROS. The TRX system is composed of TRX reductase, TRX, and peroxiredoxin (4) (Fig. 1).

Thioredoxin is a small (12 kDa) ubiquitous protein, which has the thiol-mediated reducing activity at its conserved active site: –Cys-Gly-Pro-Cys–. It was first reported as a hydrogen donor for ribonucleotide reductase in *Escherichia*

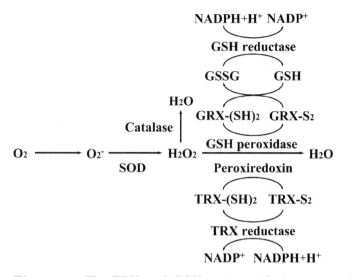

Figure 1 The TRX and GSH systems that scavenge ROS. GSH reductase, glutathione reductase; GRX-(SH)2, reduced glutaredoxin; GRX-S2, oxidized glutaredoxin; GSH peroxidase, glutathione peroxidase; TRX-(SH)2, reduced thioredoxin; TRX-S2, oxidized thioredoxin; SOD, superoxide dismutase.

coli in 1964 (5). Yodoi and colleagues cloned human TRX from the culture supernatant of human T-cell leukemia virus type I (HTLV-I)-transformed ATL2 cells as an adult T-cell leukemia-derived factor (ADF) in 1989 (6). It was also cloned as an autocrine growth factor that is produced by Epstein–Barr virus-transformed B cell line 3B6 (7). Thioredoxin reduces target protein with its active disulfide/dithiol, while TRX itself becomes oxidized. Oxidized TRX is converted to the reduced form by accepting hydrogen from NADPH via TRX reductase (Fig. 1).

Mammalian TRX is a stress-inducible protein, is secreted from cells, and protects cells against a variety of oxidative stress. Intracellularly, TRX interacts other molecules (protein or nucleotide) through the redox regulation of cysteine residues, and modulates several signal transduction pathways (Fig. 2).

Under physiological conditions, intracellular reducing environment is maintained by the disulfide/dithiol reducing activity of the both GSH and TRX systems. Glutathione is a

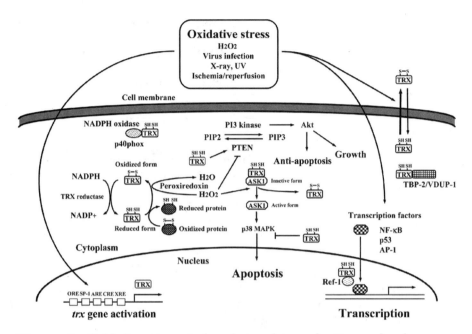

Figure 2   Oxidative stress-induced signal transduction and redox regulation by TRX.

cysteine-containing tripeptide (γ-glutamyl-cysteinyl-glycine), which is a major component of cytosolic antioxidant (millimolar concentration). Although the amount of TRX (micromolar concentration) is less than GSH, TRX has the much stronger activity of modulating signal transduction such as transcription factors binding to target DNA than GSH (4).

II.   TRX FAMILY

Recently, an increasing number of molecules are reported to share the similar active site: –Cys-Xxx-Yyy-Cys– (two cysteines separated by two other amino acid residues) and proposed to designate as members of TRX family.

   Mitochondria-specific thioredoxin (TRX-2) was cloned by Spyrou et al. (8). It has a mitochondrial translocation signal peptide at the N-terminus and the conserved active disulfide/dithiol. It is more resistant to oxidative stress and scavenges ROS generated in mitochondria, which are a major physiological source of ROS during respiration. We reported that TRX-2 forms a complex with cytochrome *c* in mitochondria and protects mitochondria-mediated apoptosis using TRX-2-deficient DT40 cells (9).

Glutaredoxin was originally identified as a GSH-dependent hydrogen donor for ribonucleotide reductase in a mutant *Escherichia coli* lacking TRX (10). It is a member of TRX family and catalyzes GSH-dependent disulfide reduction. Mammalian cells contain two isoforms of glutaredoxin. Glutaredoxin-1 is a cytosolic protein and glutaredoxin-2 is located in the mitochondria and nucleus. The active site sequence of glutaredoxin-2 is –Cys-Ser-Tyr-Cys–, which is different from –Cys-Pro-Tyr-Cys– sequence, the active site of cytosolic glutaredoxin (11,12).

Nucleoredoxin is predominantly located in the nucleus and has –Cys-Pro-Pro-Cys– sequence at the active site (13). Thioredoxin-related protein of 32 kDa (TRP32) having –Cys-Gly-Pro-Cys– (identical to TRX) is a cytosolic protein (14). Transmembrane thioredoxin-related protein (TMX) possesses a signal peptide at the N-terminus, followed by TRX-like domain with unique –Cys-Pro-Ala-Cys– sequence at the active site. The TMX-overexpressed cells showed resistance to the apoptosis induced by an endoplasmic reticulum (ER)–Golgi transport inhibitor, suggesting that TMX may relieve ER stress (15).

Protein disulfide isomerases (PDIs) catalyze the formation of disulfide bond of newly synthesized proteins in the ER. They possess two or three TRX-like –Cys-Gly-His-Cys– domains at the active site (16,17).

III. TRX SYSTEM AND MODULATION OF SIGNAL TRANSDUCTION (Fig. 2)

Thioredoxin is induced by a variety of oxidative stress caused by hydrogen peroxide, x-ray- or ultraviolet (UV) -irradiation, viral infection, ischemia-reperfusion, and drugs such as *cis*-diamminedichloroplatinum (II) and hemin. Hydrogen peroxide induces TRX gene transcription through an oxidative responsive element (18), and hemin induces the gene transcription by regulating NF-E2-related factor (Nrf2) through antioxidant responsive element (ARE) (19).

Thioredoxin translocates from the cytosol to the nucleus upon stress, such as UV, phorbol 12-myristate acetate (PMA), tumor necrosis factor-α (TNF-α), and an anticancer drug (20–22). In the nucleus, TRX enhances DNA binding of transcription factors such as NF-κB, AP-1, and p53 (21–24). Oxidative stress induces activation of NF-κB and antioxidant such as N-acetylcysteine (NAC) suppressed the activation (25). Cytoplasmic TRX suppresses the NF-κB signaling, whereas intranuclear TRX enhances the DNA-binding activity by reducing the key cysteine residue in NF-κB (21). In co-operation with redox factor-1 (Ref-1), TRX enhances the transcriptional activity of p53, "the guardian of genome," upregulates p53-dependent p21 expression, and affords cells to repair damaged DNA by inducing cell cycle G1 arrest (22).

Apoptosis signal-regulating kinase 1 (ASK1) was identified by Ichijo et al. (26) as a mitogen-activated protein (MAP) kinase that activates c-Jun N-terminal kinase (JNK) and p38 MAP kinase and induces stress-mediated apoptosis signal. Reduced TRX binds to ASK1 and inhibits the activity of ASK1. Upon oxidative stress, TRX is oxidized and dissociated from ASK1, resulting in the activation of ASK1 (27). In addition, TRX negatively regulates p38 MAP kinase activation (28). Therefore, the cytoprotective effect of TRX can be partly explained by the regulation of the activity of ASK1 or p38 MAP kinase.

The activity of several intracellular enzymes is regulated by TRX. Tumor suppressor PTEN, which is a protein tyrosine phosphatase and reverses the action of phosphoinositide 3-kinase, is inactivated by hydrogen peroxide, and oxidized PTEN is reduced and reactivated by TRX (29). The activity of caspase-3, an apoptosis inducer, is regulated by TRX-mediated redox state. Thioredoxin recovered the activity of caspase-3 that is inactivated by thiol-oxidant (30), suggesting that TRX shifts the cell death mode from necrosis to apoptosis induced by oxidative stress (3).

Thioredoxin-binding proteins (TBPs) were reported in addition to ASK1. We identified TBP-1 as p40phox, a phagocyte oxidase component, and TBP-2 as vitamin D3 upregulated protein 1 (VDUP1), using the yeast two-hybrid system (31,32). Intriguingly, TBP-2/VDUP1 negatively regulates the reducing activity of TRX (31). The expression of TBP-2/VDUP1 is downregulated in cancer by histone deacetylation, and an inhibitor of histone deacetylase caused cell cycle arrest (33). We recently reported that TBP-2/VDUP1 is correlated with interleukin (IL)-2 dependent growth in HTLV-I infected cell lines and that TBP-2 induces cell cycle G1 arrest by increasing p16 expression (34). The mutation of TBP-2/VDUP1 gene causes hyperlipidemia (35), although it needs to be clarified whether affected TRX-mediated cellular redox is involved in the onset of the disease.

IV. EXTRACELLULAR TRX

Thioredoxin is secreted from cells via a nonclassical pathway, since TRX does not possess signal sequence (36,37). Extracellular TRX has cytokine- or chemokine-like activity. Human TRX has growth-promoting effect on leukemic cells and hepatocellular carcinoma cells (7,38). Thioredoxin enhances migration of eosinophils (39) and shows the chemotactic activity on neutrophils, monocytes, and T lymphocytes (40,41). Because TRX does not increase intracellular Ca^{2+}, the chemotactic activity of TRX seems to differ from that of known chemokines whose receptors are coupled with G proteins (40).

Recently, we reported that the release of TRX is augmented by hydrogen peroxide and suppressed by the addition of NAC or recombinant human TRX (rhTRX) extracellularly (42,43) (Fig. 2). We reported that fluorescence-labeled or histidine-tagged rhTRX enters the cells within 1–3 days (42), suggesting that cytoprotective effect of extracellular TRX may partly be due to scavenging ROS intracellularly after entering cells as well as extracellularly coupled with peroxiredoxin IV (secretable type) (Figs. 2 and 3).

The serum or plasma level of TRX is regarded as a good marker of oxidative stress. The enzyme-linked immunosorbent assay (ELISA) system using anti-TRX monoclonal

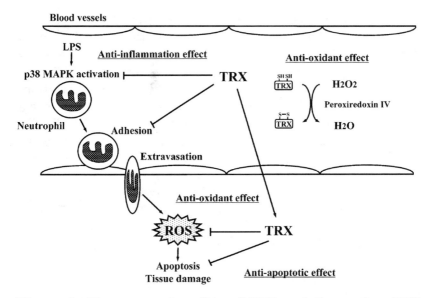

Figure 3 The suppressive effect of TRX on inflammation; TRX suppresses the activation, adhesion, and extravasation of neutrophils. LPS, lipopolysaccharide.

antibodies has been developed to measure the levels of TRX in serum/plasma or other body fluids (44). The serum/plasma levels of TRX in healthy persons are around 20–30 ng/mL. Elevation of serum/plasma TRX is reported in patients with HIV (43,44), hepatitis C virus infection (45), rheumatoid arthritis (46–48), heart failure (49,50), and surgical stress (51). The elevated levels of TRX seem to be correlated with the severity of the diseases, and the level can be a good indicator of the diagnosis or the therapy.

V. TRX-KNOCKOUT MOUSE

Analysis using knockout mice is an excellent method to evaluate the function of the target molecule. Mice with heterozygotes carrying disrupted TRX gene are viable, fertile, and appear normal. However, homozygous mutants die shortly after implantation (52). TRX-2-knockout mice are also embryonic lethal (53), which is consistent with our report that TRX-2-deficient DT40 cells undergo apoptosis (9). These results suggest that both TRX and TRX-2 are indispensable for cell viability and normal development of mouse embryo and that they do not compensate the function each other.

VI. TRX-TRANSGENIC MOUSE

Thioredoxin-transgenic mice overexpressed with human TRX under the control of β-actin promoter were generated to investigate the role of the protein in vivo. The expression levels of human TRX in TRX-Tg mice are several-fold higher than that of endogenous mouse TRX (54). Bone marrow cells from TRX-Tg mice are more resistant to UVC than cells from control mice. Thioredoxin-transgenic mice show the higher telomerase activity. These results support the longer life span of TRX-Tg mice than control mice (55). The embryos with TRX transgene show the resistance to oxidative stress and less developmental abnormalities induced by 25% oxygen than the control embryos (56).

There are several reports showing that TRX-Tg mice are resistant in several disease models. Using the brain infarction model, TRX transgene expression in mice reduced the focal ischemic injury after the middle cerebral artery occlusion with the decreased production of oxidized proteins (54). In the renal warm ischemia-reperfusion model, TRX-Tg mice show the resistance to the damage in the vulnerable thick ascending limb in the outer medulla of the kidney (57). Thioredoxin-transgenic mice are resistant to the bleomycin-induced lung injury (58), the adriamycin-induced cardiotoxicity (59), or the retinal photo-oxidative damage induced by the intense white fluorescent light (60). Moreover, overexpressed TRX attenuated the acute lethal hepatitis induced by drugs, such as thioacetamide or lipopolysaccharide (LPS) plus D-galactosamine, suggesting that TRX protects mainly mitochondria-mediated apoptosis (61).

Thioredoxin-transgenic mice are resistant to the influenza virus infection. Interestingly, pneumonia induced by the influenza infection is reduced in TRX-Tg mice without affecting the systemic immune responses (62). In addition to scavenging ROS, inhibiting extravasation of leukocytes is another rationale that accounts for the TRX-mediated protection of tissue damage (41). The combined administration of IL-18 and IL-2 induces lung-specific injury accompanied by interstitial infiltration of inflammatory cells (63). Thioredoxin-transgenic mice show the resistance to the IL-18/IL-2-induced lethal lung injury (58).

VII. THERAPEUTIC APPROACH USING RECOMBINANT HUMAN TRX— TRANSLATIONAL RESEARCH

In vitro, the extracellular administration of rhTRX protected cell death induced by TNF-α or anti-Fas antibody (64), or hydrogen peroxide (65). The administration of rhTRX also protected the cytotoxicity in PC12 cells induced by 1-methyl-4-phenylpyridinium ion (MPP+), an active metabolite of 1-methyl-4-phenyl-1,2,3,6-tetrahydropyridine (MPTP), which causes dopaminergic denervation and Parkinsonism in human (66).

In the lung ischemia-reperfusion model, the administration of rhTRX ameliorates the ischemia-reperfusion injury (67,68). The half-life of the intravenously injected rhTRX is around 1 hr (41). The intravenous administration of rhTRX protects the photic injury in retina (69), the cerebral ischemia-reperfusion injury (70), and the bleomycin- or IL-18/IL-2-induced lung fibrosis (58). Thus, both TRX-Tg mice and the rhTRX administration models provided the similar protective effect of TRX.

Fascinatingly, we have found that rhTRX enters into cells quite recently (42), suggesting that extracellularly administrated rhTRX can protect cells against oxidative stress after entering into target cells. Since redox-inactivated mutant rhTRX(C32S/C35S) does not enter the cells (42), some redox-sensitive target molecule may be involved in the entry of TRX.

Neutrophilic inflammation is characteristic of acute lung injury (ALI)/acute respiratory distress syndrome (ARDS). Activated neutrophils migrate and infiltrate into parenchyma, express proinflammatory cytokines, such as IL-1β, TNF-α, and IL-8, and finally induce respiratory burst. Especially, p38 MAP kinase plays an important role in the neutrophil activation, like the adhesion, chemotaxis, priming, oxygen release, and NF-κB activation. The inhibition of p38 MAP kinase blocked the LPS-induced adhesion, the NF-κB activation, and the synthesis of TNF-α, resulting in the neutrophil accumulation in the airspaces (71,72) (Fig. 3).

The administration of rhTRX inhibits the shedding of L-selectin, the adhesion to endothelial cells, the extravasation (chemotaxis), and the activation of p38 in neutrophils (41). Administrated rhTRX is expected to protect lung tissue injury by inhibiting the neutrophil-mediated inflammation in addition to protecting pulmonary cells by scavenging ROS (Fig. 3). The half-life of intraperitoneally administrated rhTRX in the lung and sera is 51.3 and 8.5 hr, respectively (58), indicating that rhTRX accumulates in the lung and that the lung should be the primary target organ protected by rhTRX. These results strongly suggest that TRX can be employed for the clinical therapeutics including interstitial lung disease and lung fibrosis.

We are developing rhTRX with the good manufacturing product (GMP) quality level for the clinical application. A clinical trial project of treating ALI/ARDS patients with rhTRX is in progress at the Translational Research Center, Kyoto University Hospital (Kyoto, Japan) since June 2003. The project is the preclinical research using animal models and the following translational clinical phase I/II studies to evaluate the therapeutic effect of rhTRX in patients with ALI/ARDS (http://www.kuhp.kyoto-u.ac.jp/~trx/index_e.html).

VIII. CONCLUDING REMARKS

In conclusion, TRX not only quenches ROS intracellularly or extracellularly as an antioxidant and controls various signal transduction pathways as a redox regulator but also inhibits the neutrophil extravasation as an anti-inflammatory drug. Evidence shows that TRX-Tg mice are resistant in several tissue injury models, and that the administration of rhTRX can ameliorate the tissue damage. Thus, since TRX is considered as a unique and promising target molecule regulating the inflammation, the translational research using rhTRX as a therapeutic drug of ALI/ARDS is in progress. In the near future, the clinical application of TRX may be expanded to other oxidative stress-associated diseases.

REFERENCES

1. Nakamura H, Nakamura K, Yodoi J. Redox regulation of cellular activation. Annu Rev Immunol 1997; 15:351–369.

2. Schreck R, Rieber P, Baeuerle PA. Reactive oxygen intermediates as apparently widely used messengers in the activation of the NF-kappa B transcription factor and HIV-1. EMBO J 1991; 10:2247–2258.

3. Ueda S, Masutani H, Nakamura H, Tanaka T, Ueno M, Yodoi J. Redox control of cell death. Antioxid Redox Signal 2002; 4: 405–414.

4. Holmgren A. Thioredoxin and glutaredoxin systems. J Biol Chem 1989; 264:13963–13966.

5. Holmgren A. Thioredoxin. Annu Rev Biochem 1985; 54:237–271.

6. Tagaya Y, Maeda Y, Mitsui A, Kondo N, Matsui H, Hamuro J, Brown N, Arai K, Yokota T, Wakasugi H, Yodoi J. ATL-derived factor (ADF), an IL-2 receptor/Tac inducer homologous to thioredoxin; possible involvement of dithiol-reduction in the IL-2 receptor induction. EMBO J 1989; 8:757–764.

7. Wakasugi N, Tagaya Y, Wakasugi H, Mitsui A, Maeda M, Yodoi J, Tursz T. Adult T-cell leukemia-derived factor/ thioredoxin, produced by both human T-lymphotropic virus type I- and Epstein–Barr virus-transformed lymphocytes, acts as an autocrine growth factor and synergizes with interleukin 1 and interleukin 2. Proc Natl Acad Sci USA 1990; 87:8282–8286.

8. Spyrou G, Enmark E, Miranda-Vizuete A, Gustafsson J. Cloning and expression of a novel mammalian thioredoxin. J Biol Chem 1997; 272:2936–2941.

9. Tanaka T, Hosoi F, Yamaguchi-Iwai Y, Nakamura H, Masutani H, Ueda S, Nishiyama A, Takeda S, Wada H, Spyrou G, Yodoi J. Thioredoxin-2 (TRX-2) is an essential gene regulating mitochondria-dependent apoptosis. EMBO J 2002; 21:1695–1703.

10. Holmgren A. Hydrogen donor system for *Escherichia coli* ribonucleoside-diphosphate reductase dependent upon glutathione. Proc Natl Acad Sci USA 1976; 73:2275–2279.

11. Lundberg M, Johansson C, Chandra J, Enoksson M, Jacobsson G, Ljung J, Johansson M, Holmgren A. Cloning and expression of a novel human glutaredoxin (Grx2) with mitochondrial and nuclear isoforms. J Biol Chem 2001; 276: 26269–26275.

12. Gladyshev VN, Liu A, Novoselov SV, Krysan K, Sun QA, Kryukov VM, Kryukov GV, Lou MF. Identification and characterization of a new mammalian glutaredoxin (thioltransferase), Grx2. J Biol Chem 2001; 276:30374–30380.

13. Kurooka H, Kato K, Minoguchi S, Takahashi Y, Ikeda J, Habu S, Osawa N, Buchberg AM, Moriwaki K, Shisa H, Honjo T. Cloning and characterization of the nucleoredoxin gene that encodes a novel nuclear protein related to thioredoxin. Genomics 1997; 39:331–339.

14. Lee KK, Murakawa M, Takahashi S, Tsubuki S, Kawashima S, Sakamaki K, Yonehara S. Purification, molecular cloning, and characterization of TRP32, a novel thioredoxin-related mammalian protein of 32 kDa. J Biol Chem 1998; 273:19160–19166.

15. Matsuo Y, Akiyama N, Nakamura H, Yodoi J, Noda M, Kizaka-Kondoh S. Identification of a novel thioredoxin-related transmembrane protein. J Biol Chem 2001; 276:10032–10038.

16. Freedman RB, Hirst TR, Tuite MF. Protein disulphide isomerase: building bridges in protein folding. Trends Biochem Sci 1994; 19:331–336.

17. Turano C, Coppari S, Altieri F, Ferraro A. Proteins of the PDI family: unpredicted non-ER locations and functions. J Cell Physiol 2002; 193:154–163.

18. Taniguchi Y, Taniguchi-Ueda Y, Mori K, Yodoi J. A novel promoter sequence is involved in the oxidative stress-induced expression of the adult T-cell leukemia-derived factor (ADF)/human thioredoxin (Trx) gene. Nucleic Acids Res 1996; 24:2746–2752.

19. Kim YC, Masutani H, Yamaguchi Y, Itoh K, Yamamoto M, Yodoi J. Hemin-induced activation of the thioredoxin gene by Nrf2. A differential regulation of the antioxidant responsive element by a switch of its binding factors. J Biol Chem 2001; 276:18399–18406.

20. Masutani H, Hirota K, Sasada T, Ueda-Taniguchi Y, Taniguchi Y, Sono H, Yodoi J. Transactivation of an inducible anti-oxidative stress protein, human thioredoxin by HTLV-I Tax. Immunol Lett 1996; 54:67–71.

21. Hirota K, Murata M, Sachi Y, Nakamura H, Takeuchi J, Mori K, Yodoi J. Distinct roles of thioredoxin in the cytoplasm and in the nucleus. A two-step mechanism of redox regulation of transcription factor NF-kappaB. J Biol Chem 1999; 274:27891–27897.

22. Ueno M, Masutani H, Arai RJ, Yamauchi A, Hirota K, Sakai T, Inamoto T, Yamaoka Y, Yodoi J, Nikaido T. Thioredoxin-dependent redox regulation of p53-mediated p21 activation. J Biol Chem 1999; 274:35809–35815.

23. Okamoto T, Ogiwara H, Hayashi T, Mitsui A, Kawabe T, Yodoi J. Human thioredoxin/adult T cell leukemia-derived factor activates the enhancer binding protein of human immunodeficiency virus type 1 by thiol redox control mechanism. Int Immunol 1992; 4:811–819.

24. Hirota K, Matsui M, Iwata S, Nishiyama A, Mori K, Yodoi J. AP-1 transcriptional activity is regulated by a direct association between thioredoxin and Ref-1. Proc Natl Acad Sci USA 1997; 94:3633–3638.

25. Meyer M, Schreck R, Baeuerle PA. H_2O_2 and antioxidants have opposite effects on activation of NF-kappa B and AP-1 in intact cells: AP-1 as secondary antioxidant-responsive factor. EMBO J 1993; 12:2005–2015.

26. Ichijo H, Nishida E, Irie K, ten Dijke P, Saitoh M, Moriguchi T, Takagi M, Matsumoto K, Miyazono K, Gotoh Y. Induction of apoptosis by ASK1, a mammalian MAPKKK that activates

SAPK/JNK and p38 signaling pathways. Science 1997; 275:90–94.

27. Saitoh M, Nishitoh H, Fujii M, Takeda K, Tobiume K, Sawada Y, Kawabata M, Miyazono K, Ichijo H. Mammalian thioredoxin is a direct inhibitor of apoptosis signal-regulating kinase (ASK) 1. EMBO J 1998; 17:2596–2606.

28. Hashimoto S, Matsumoto K, Gon Y, Furuichi S, Maruoka S, Takeshita I, Hirota K, Yodoi J, Horie T. Thioredoxin negatively regulates p38 MAP kinase activation and IL-6 production by tumor necrosis factor-alpha. Biochem Biophys Res Commun 1999; 258:443–447.

29. Lee SR, Yang KS, Kwon J, Lee C, Jeong W, Rhee SG. Reversible inactivation of the tumor suppressor PTEN by H_2O_2. J Biol Chem 2002; 277:20336–20342.

30. Ueda S, Nakamura H, Masutani H, Sasada T, Yonehara S, Takabayashi A, Yamaoka Y, Yodoi J. Redox regulation of caspase-3(-like) protease activity: regulatory roles of thioredoxin and cytochrome *c*. J Immunol 1998; 161:6689–6695.

31. Nishiyama A, Matsui M, Iwata S, Hirota K, Masutani H, Nakamura H, Takagi Y, Sono H, Gon Y, Yodoi J. Identification of thioredoxin-binding protein-2/vitamin D(3) up-regulated protein 1 as a negative regulator of thioredoxin function and expression. J Biol Chem 1999; 274:21645–21650.

32. Nishiyama A, Ohno T, Iwata S, Matsui M, Hirota K, Masutani H, Nakamura H, Yodoi J. Demonstration of the interaction of thioredoxin with p40phox, a phagocyte oxidase component, using a yeast two-hybrid system. Immunol Lett 1999; 68:155–159.

33. Butler LM, Zhou X, Xu WS, Scher HI, Rifkind RA, Marks PA, Richon VM. The histone deacetylase inhibitor SAHA arrests cancer cell growth, up-regulates thioredoxin-binding protein-2, and down-regulates thioredoxin. Proc Natl Acad Sci USA 2002; 99:11700–11705.

34. Nishinaka Y, Nishiyama A, Masutani H, Oka S, Ahsan KM, Nakayama Y, Ishii Y, Nakamura H, Maeda M, Yodoi J. Loss of thioredoxin binding protein-2/vitamin D3 up-regulated protein 1 (TBP-2/VDUP1) in HTLV-I-dependent

T-cell transformation: implications for ATL leukemogenesis. Cancer Res 2004; 64:1287–1292.

35. Bodnar JS, Chatterjee A, Castellani LW, Ross DA, Ohmen J, Cavalcoli J, Wu C, Dains KM, Catanese J, Chu M, Sheth SS, Charugundla K, Demant P, West DB, de Jong P, Lusis AJ. Positional cloning of the combined hyperlipidemia gene Hyplip1. Nat Genet 2002; 30:110–116.

36. Ericson ML, Horling J, Wendel-Hansen V, Holmgren A, Rosen A. Secretion of thioredoxin after in vitro activation of human B cells. Lymphokine Cytokine Res 1992; 11:201–207.

37. Rubartelli A, Bajetto A, Allavena G, Wollman E, Sitia R. Secretion of thioredoxin by normal and neoplastic cells through a leaderless secretory pathway. J Biol Chem 1992; 267: 24161–24164.

38. Nakamura H, Masutani H, Tagaya Y, Yamauchi A, Inamoto T, Nanbu Y, Fujii S, Ozawa K, Yodoi J. Expression and growth-promoting effect of adult T-cell leukemia-derived factor. A human thioredoxin homologue in hepatocellular carcinoma. Cancer 1992; 69:2091–2097.

39. Hori K, Hirashima M, Ueno M, Matsuda M, Waga S, Tsurufuji S, Yodoi J. Regulation of eosinophil migration by adult T cell leukemia-derived factor. J Immunol 1993; 151: 5624–5630.

40. Bertini R, Howard OM, Dong HF, Oppenheim JJ, Bizzarri C, Sergi R, Caselli G, Pagliei S, Romines B, Wilshire JA, Mengozzi M, Nakamura H, Yodoi J, Pekkari K, Gurunath R, Holmgren A, Herzenberg LA, Ghezzi P. Thioredoxin, a redox enzyme released in infection and inflammation, is a unique chemoattractant for neutrophils, monocytes, and T cells. J Exp Med 1999; 189:1783–1789.

41. Nakamura H, Herzenberg LA, Bai J, Araya S, Kondo N, Nishinaka Y, Yodoi J. Circulating thioredoxin suppresses lipopolysaccharide-induced neutrophil chemotaxis. Proc Natl Acad Sci USA 2001; 98:15143–15148.

42. Kondo N, Ishii Y, Kwon Y-W, Tanito M, Horita H, Nishinaka Y, Nakamura H, Yodoi J. Redox-sensing release of human thioredoxin from T lymphocytes with negative feedback loops. J Immunol 2004; 172:442–448.

43. Nakamura H, De Rosa SC, Yodoi J, Holmgren A, Ghezzi P, Herzenberg LA. Chronic elevation of plasma thioredoxin: inhibition of chemotaxis and curtailment of life expectancy in AIDS. Proc Natl Acad Sci USA 2001; 98:2688–2693.

44. Nakamura H, De Rosa S, Roederer M, Anderson MT, Dubs JG, Yodoi J, Holmgren A, Herzenberg LA. Elevation of plasma thioredoxin levels in HIV-infected individuals. Int Immunol 1996; 8:603–611.

45. Sumida Y, Nakashima T, Yoh T, Nakajima Y, Ishikawa H, Mitsuyoshi H, Sakamoto Y, Okanoue T, Kashima K, Nakamura H, Yodoi J. Serum thioredoxin levels as an indicator of oxidative stress in patients with hepatitis C virus infection. J Hepatol 2000; 33:616–622.

46. Yoshida S, Katoh T, Tetsuka T, Uno K, Matsui N, Okamoto T. Involvement of thioredoxin in rheumatoid arthritis: its costimulatory roles in the TNF-alpha-induced production of IL-6 and IL-8 from cultured synovial fibroblasts. J Immunol 1999; 163:351–358.

47. Maurice MM, Nakamura H, Gringhuis S, Okamoto T, Yoshida S, Kullmann F, Lechner S, van der Voort EA, Leow A, Versendaal J, Muller-Ladner U, Yodoi J, Tak PP, Breedveld FC, Verweij CL. Expression of the thioredoxin– thioredoxin reductase system in the inflamed joints of patients with rheumatoid arthritis. Arthritis Rheum 1999; 42: 2430–2439.

48. Jikimoto T, Nishikubo Y, Koshiba M, Kanagawa S, Morinobu S, Morinobu A, Saura R, Mizuno K, Kondo S, Toyokuni S, Nakamura H, Yodoi J, Kumagai S. Thioredoxin as a biomarker for oxidative stress in patients with rheumatoid arthritis. Mol Immunol 2002; 38:765–772.

49. Kishimoto C, Shioji K, Nakamura H, Nakayama Y, Yodoi J, Sasayama S. Serum thioredoxin (TRX) levels in patients with heart failure. Jpn Circ J 2001; 65:491–494.

50. Shioji K, Nakamura H, Masutani H, Yodoi J. Redox regulation by thioredoxin in cardiovascular diseases. Antioxid Redox Signal 2003; 5:795–802.

51. Nakamura H, Vaage J, Valen G, Padilla CA, Bjornstedt M, Holmgren A. Measurements of plasma glutaredoxin and

thioredoxin in healthy volunteers and during open-heart surgery. Free Radic Biol Med 1998; 24:1176–1186.

52. Matsui M, Oshima M, Oshima H, Takaku K, Maruyama T, Yodoi J, Taketo MM. Early embryonic lethality caused by targeted disruption of the mouse thioredoxin gene. Dev Biol 1996; 178:179–185.

53. Nonn L, Williams RR, Erickson RP, Powis G. The absence of mitochondrial thioredoxin 2 causes massive apoptosis, exencephaly, and early embryonic lethality in homozygous mice. Mol Cell Biol 2003; 23:916–922.

54. Takagi Y, Mitsui A, Nishiyama A, Nozaki K, Sono H, Gon Y, Hashimoto N, Yodoi J. Overexpression of thioredoxin in transgenic mice attenuates focal ischemic brain damage. Proc Natl Acad Sci USA 1999; 96:4131–4136.

55. Mitsui A, Hamuro J, Nakamura H, Kondo N, Hirabayashi Y, Ishizaki-Koizumi S, Hirakawa T, Inoue T, Yodoi J. Overexpression of human thioredoxin in transgenic mice controls oxidative stress and life span. Antioxid Redox Signal 2002; 4:693–696.

56. Kobayashi-Miura M, Nakamura H, Yodoi J, Shiota K. Thioredoxin, an anti-oxidant protein, protects mouse embryos from oxidative stress-induced developmental anomalies. Free Radic Res 2002; 36:949–956.

57. Kasuno K, Nakamura H, Ono T, Muso E, Yodoi J. Protective roles of thioredoxin, a redox-regulating protein, in renal ischemia/reperfusion injury. Kidney Int 2003; 64:1273–1282.

58. Hoshino T, Nakamura H, Okamoto M, Kato S, Araya S, Nomiyama K, Oizumi K, Young HA, Aizawa H, Yodoi J. Redox-active protein thioredoxin prevents proinflammatory cytokine- or bleomycin-induced lung injury. Am J Respir Crit Care Med 2003; 168:1075–1083.

59. Shioji K, Kishimoto C, Nakamura H, Masutani H, Yuan Z, Oka S, Yodoi J. Overexpression of thioredoxin-1 in transgenic mice attenuates adriamycin-induced cardiotoxicity. Circulation 2002; 106:1403–1409.

60. Tanito M, Masutani H, Nakamura H, Oka S, Ohira A, Yodoi J. Attenuation of retinal photooxidative damage in thioredoxin transgenic mice. Neurosci Lett 2002; 326:142–146.

61. Okuyama H, Nakamura H, Shimahara Y, Araya S, Kawada N, Yamaoka Y, Yodoi J. Overexpression of thioredoxin prevents acute hepatitis caused by thioacetamide or lipopolysaccharide in mice. Hepatology 2003; 37:1015–1025.

62. Nakamura H, Tamura S, Watanabe I, Iwasaki T, Yodoi J. Enhanced resistancy of thioredoxin-transgenic mice against influenza virus-induced pneumonia. Immunol Lett 2002; 82:165–170.

63. Okamoto M, Kato S, Oizumi K, Kinoshita M, Inoue Y, Hoshino K, Akira S, McKenzie AN, Young HA, Hoshino T. Interleukin 18 (IL-18) in synergy with IL-2 induces lethal lung injury in mice: a potential role for cytokines, chemokines, and natural killer cells in the pathogenesis of interstitial pneumonia. Blood 2002; 99:1289–1298.

64. Matsuda M, Masutani H, Nakamura H, Miyajima S, Yamauchi A, Yonehara S, Uchida A, Irimajiri K, Horiuchi A, Yodoi J. Protective activity of adult T cell leukemia-derived factor (ADF) against tumor necrosis factor-dependent cytotoxicity on U937 cells. J Immunol 1991; 147:3837–3841.

65. Nakamura H, Matsuda M, Furuke K, Kitaoka Y, Iwata S, Toda K, Inamoto T, Yamaoka Y, Ozawa K, Yodoi J. Adult T cell leukemia-derived factor/human thioredoxin protects endothelial F-2 cell injury caused by activated neutrophils or hydrogen peroxide. Immunol Lett 1994; 42:75–80.

66. Bai J, Nakamura H, Hattori I, Tanito M, Yodoi J. Thioredoxin suppresses 1-methyl-4-phenylpyridinium-induced neurotoxicity in rat PC12 cells. Neurosci Lett 2002; 321:81–84.

67. Yokomise H, Fukuse T, Hirata T, Ohkubo K, Go T, Muro K, Yagi K, Inui K, Hitomi S, Mitsui A, Hirakawa T, Yodoi J, Wada H. Effect of recombinant human adult T cell leukemia-derived factor on rat lung reperfusion injury. Respiration 1994; 61:99–104.

68. Fukuse T, Hirata T, Yokomise H, Hasegawa S, Inui K, Mitsui A, Hirakawa T, Hitomi S, Yodoi J, Wada H. Attenuation of ischaemia reperfusion injury by human thioredoxin. Thorax 1995; 50:387–391.

69. Tanito M, Masutani H, Nakamura H, Ohira A, Yodoi J. Cytoprotective effect of thioredoxin against retinal photic injury in mice. Invest Ophthalmol Vis Sci 2002; 43:1162–1167.

70. Hattori I, Takagi Y, Nakamura H, Nozaki K, Bai J, Kondo N, Sugino T, Nishimura M, Hashimoto N, Yodoi J. Intravenous administration of thioredoxin decreases brain damage following transient focal cerebral ischemia in mice. Antioxid Redox Signal 2004; 6:81–87.

71. Nick JA, Young SK, Brown KK, Avdi NJ, Arndt PG, Suratt BT, Janes MS, Henson PM, Worthen GS. Role of p38 mitogen-activated protein kinase in a murine model of pulmonary inflammation. J Immunol 2000; 164:2151–2159.

72. Nick JA, Avdi NJ, Young SK, Lehman LA, McDonald PP, Frasch SC, Billstrom MA, Henson PM, Johnson GL, Worthen GS. Selective activation and functional significance of p38 alpha mitogen-activated protein kinase in lipopolysaccharide-stimulated neutrophils. J Clin Invest 1999; 103: 851–858.

4

Regulation of NF-κB-Driven Inflammatory Genes by Phytoestrogens

WIM VANDEN BERGHE, NATHALIE
DIJSSELBLOEM, AN DE NAEYER, LINDA
VERMEULEN, and GUY HAEGEMAN
Laboratory of Eukaryotic Gene Expression and
Signal Transduction (LEGEST), Department
of Molecular Biology, University of Gent,
Gent, Belgium

I. INTRODUCTION

The inflammatory response is a highly regulated physiological process that is critically important to homeostasis. The pleiotropic cytokine interleukin-6 (IL6) affects inflammatory reactions, hematopoiesis, bone metabolism, reproduction (spermatogenesis, menstrual cycle), and ageing. Aberrant IL6 gene expression has been associated with multiple myeloma, neoplasia, rheumatoid arthritis, bowel disease, psoriasis, obesity, coronary heart disease, Alzheimer's disease, reduced longevity, and postmenopausal osteoporosis (1,2). Serum IL6 levels are currently considered a diagnostic marker for tumor progression and prognosis in various types of cancer (renal cell carcinoma, breast, lung, ovarian, and gut cancer) (3,4). Briefly, the IL6 promoter behaves as a sophisticated biosensor for environmental stress, thus controlling immunological homeostasis (5). Interleukin-6 is normally tightly regulated and expressed at low levels, except during infection, trauma, ageing, or other stress conditions (5). Among several factors that can downregulate IL6 gene expression are estrogen and testosterone hormones. After menopause or andropause, IL6 levels are elevated, even in the absence of infection, trauma, or stress (6–8). Natural menopause is associated with a rapid decline in circulating sex hormones and, apart from the loss of reproductive function, this can lead to unpleasant symptoms such as hot flushes and vaginal dryness, with a long-term increased risk of bone loss, cardiovascular disease, and neurological disorders (dementia). There is no unifying mechanism that would be able to explain all consequences of menopause on the metabolism of organs as diverse as bone, blood vessels, or adipose tissue. However, menopause triggered changes in the activity of proinflammatory cytokines are beginning to emerge as a common theme that may have a significant impact on the function of all of these tissues (2,8). It has been proposed that the age-associated increase in IL6 accounts for certain of the phenotypic changes of advanced age including cardiovascular disease, osteoporosis, arthritis, type 2 diabetes, certain cancers, lymphoproliferative disorders, multiple myeloma, periodontal disease, frailty, and chronic inflammatory disease (2). The functional interaction, or "crosstalk," between the estrogen receptor (ER) and the proinflammatory transcription factor, nuclear factor (NF)-κB, as demonstrated in vitro, has been suggested to play a key role in estrogen prevention of those age-related conditions in vivo, besides modulation of nitric oxide antioxidative effects, plasma membrane actions, and changes in immune cell functions (2,6–11). In this respect, restoring hormonal imbalance by hormone replacement therapy (HRT) (either by synthetic or plant derived hormone preparations) in the ageing population is an attractive therapeutic option, although the risks/benefits of HRT are still an area of hot debate (12–18). Because the average life expectancy for women in Western countries exceeds 80 year and women thus spend more than a third of their lifetime in postmenopause, the possible implications of estrogen deficiency on the rates of cardiovascular disease and osteoporosis are of enormous public health importance. Currently, there is a renewed interest in naturally occurring phytoestrogens as potential alternatives to synthetic "selective estrogen receptor modulators" (SERMs), currently applied in HRT. Traditionally, plant extracts have been used to treat various diseases and such therapies are still continuing (19). Various phytotherapeuticals with a claimed hormonal activity are recommended for prevention of discomforts related to a disturbed hormonal balance (8,12,14,15,20–22). Many hope that phytoestrogens can exert the cardioprotective, antiosteoporotic, and other beneficial effects of the estrogens used in HRT in postmenopausal women without adversely affecting the risk of thrombosis and the incidence of breast and uterine cancers. Despite their putative health benefits, it is clear that we need to know much more about the molecular mechanisms, safety, and efficacy of phytochemicals before they can be generally applied to postmenopausal women as an alternative treatment to estrogens for HRT (17,23–26).

II. (PHYTO)ESTROGENS

Thousands of chemical structures have been identified in plant foods. Given the wide range of botanical species and plant parts from which phytochemicals are derived, they can contribute a significant variety and complexity to the human diet (27). In the past, the medicinal uses of spices and herbs were often indistinguishable from their culinary uses, and for good reason: people have recognized for centuries both the inherent value, as well as the potential toxicity, of phytochemicals in relation to human health. Plants have the capacity to synthesize a diverse array of chemicals, and understanding how phytochemicals function in plants may further increase our understanding of the mechanisms by which they benefit humans. In plants, these compounds function to attract beneficial and repel harmful organisms, serve as photoprotectants, and to respond to environmental changes. In humans, they can have complementary and overlapping actions, including antioxidant effects, modulation of detoxification enzymes, reduction of inflammation, modulation of steroid metabolism, and antibacterial and antiviral effects. Embracing a cuisine rich in spice, as well as in fruit and vegetables, may further enhance the chemopreventive capacity of one's diet (12,20,22,27–30).

It has been well established that cancer rates differ strikingly in various populations. Hormone-related cancers of breast, ovary, endometrium, and prostate have been reported to vary by as much as 5- to 20-fold between populations, and migrant studies indicate that the difference is largely attributed to environmental factors rather than genetics (31–33). The highest incidences of these cancers are typically observed in populations with Western lifestyles that include relatively high fat, meat-based, and low-fiber diets, whereas the lowest rates are typically observed in Asian populations with Eastern lifestyles that include plant-based diets with a high content of phytoestrogens (34–39). Migrants from Asian to Western countries, who maintain their traditional diet, do not show increased risk for these diseases, whereas an increased risk accompanies a change towards a Westernized diet (31,40). Much of the evidence is based on the differences in consumption of soy products as the major source of isoflavones in different areas of the world. The consumption of soy products is estimated to be highest in particular Japanese populations, with levels in the diet up to 200 mg/day. Throughout Asia, the consumption of legumes is estimated to supply 25–45 mg total isoflavones in the diet each day, compared with Western countries with a consumption of less than 5 mg/day (41).

During the last years, there has been a growing interest in dietary natural plant estrogens (phytoestrogens), particularly those found in soy products, as a potential alternative to the synthetic estrogens in HRT (24,36,42,43). Together with lignans, coumestans, flavones, and flavanones, isoflavones belong to the larger group of nonsteroidal phytoestrogens. Interest in phytoestrogens has been fueled by observational studies showing a lower incidence of menopausal symptoms, osteoporosis, cardiovascular disease, and breast and endometrial cancers in Asian women who have a diet rich in soy products. Consistent with epidemiological studies are the findings that soy phytoestrogens prevent mammary tumors and bone loss in rodents and atherosclerosis of coronary arteries in monkeys. Soy protein relieves hot flashes in postmenopausal women and attenuates bone loss in the lumbar spine of perimenopausal women. Furthermore, a high intake of dietary phytoestrogens is associated with a lower incidence of cancers of the colon, breast, and prostate. Isoflavones and other phytoestrogens have been considered to exert anticarcinogenic actions, mainly through antiestrogenic, antiaromatase, or antiproliferative mechanisms (33,44,45). Soy seems to protect against breast cancer if consumed throughout life, particularly before and during adolescence. Whether the phytoestrogens are responsible for the protection is not known; it is more likely that the soybean products or grain–fiber complexes are protective in their entirety. Many postmenopausal women are taking phytoestrogens in an effort to alleviate menopausal symptoms without increasing their risk of developing breast cancer. Moreover, many women with a history of breast cancer take phytoestrogens to control menopausal symptoms because estrogens are contraindicated.

Natural, synthetic, and environmental estrogens have numerous effects on the development and physiology of mammals. Estrogen is primarily known for its role in the development and functioning of the female reproductive system (20,46–48). However, roles for estrogen in male fertility, bone, the circulatory, cardiovascular, and immune system have been established by clinical observations regarding sex differences in pathologies, as well as observations following menopause or castration, or from ER knockout studies. Estrogens display intriguing tissue-selective action that is of great biomedical importance in the development of optimal therapeutics for the prevention and treatment of breast cancer, for menopausal hormone replacement, and for fertility regulation. In recent years, it has become apparent through the use of ER agonists and antagonists that the biological actions of estrogens are multifaceted. Estrogens and antiestrogens mediate their effects through diverse molecular mechanisms (49). The predominant biological effects of estradiol are mediated through two distinct intracellular ERs, ERα and ERβ, each encoded by a unique gene but possessing the hallmark modular structure of functional domains characteristic of the steroid/thyroid hormone superfamily of nuclear receptors (47–51). There is a hypervariable N-terminal domain that contributes to the transactivation function, a highly conserved central domain responsible for specific DNA binding, dimerization, and nuclear localization, and a C-terminal domain involved in ligand binding and ligand-dependent transactivation (47–49,52). Recent investigations have revealed cellular/molecular mechanisms of ER signaling at multiple levels: (1) classical ligand-dependent; (2) ligand-independent; (3) DNA binding-independent; and (4) cell–surface (nongenomic) signaling (typically, leading to rapid events, which are initiated at the plasma membrane and result in the activation of intracellular signaling pathways within seconds to minutes). Certain compounds that act via the ER, now referred to as SERMs, can demonstrate remarkable differences in activity in the various estrogen target tissues, functioning as agonists in some tissues but as antagonists in others. Recent advances elucidating the tripartite nature of the biochemical and molecular actions of estrogens provide a good basis for understanding these tissue-selective actions (53–55).

Estrogens are used in HRT to prevent hot flashes, urogenital atrophy, and osteoporosis in postmenopausal women. Hormone replacement therapy also may prevent heart disease, Alzheimer's disease, and colon cancer. Unfortunately, HRT has not lived up to its potential to improve the health of women, because estrogens have been associated with an increased incidence of breast and endometrial cancer (18).

This relationship has hampered compliance with HRT severely and has sparked an intense pursuit for SERMs that have a safer profile. Recently, raloxifene has been approved for the prevention and treatment of osteoporosis. Raloxifene is classified as a SERM because it exhibits agonist activity in some tissues such as the bone, and acts as an antagonist in other tissues including the breast. Although these effects are extremely desirable, raloxifene also increases hot flashes, is weaker than estrogens at increasing bone mineral density, and does not improve cognitive function or prevent hip fracture. Thus, the quest for superior SERMs to be used in HRT continues to be intense (17). Isoflavones are widespread in the plant kingdom, as they are predominantly found in leguminous plants and are especially abundant in soy. Lignans exist as minor constituents of building blocks in the formation of lignin in plant cell walls. They are found widely in cereals, legumes, fruit, and vegetables, with exceptionally high concentrations in flaxseed.

The isoflavones, genistein, daidzein, and biochanin A, which are abundant in soybeans and widely available as herbal tablets, are especially popular among postmenopausal women (17). The structural similarities between these substances, endogenous mammalian estrogens, and potent synthetic estrogens have attracted a lot of attention (42,56). Numerous studies have revealed that the biological activity of 17β-estradiol greatly depends on the presence of at least two hydroxyl groups, one located in the A-ring of the steroid nucleus and the other at C (17) (see Fig. 1). Unlike 17β-estradiol, phytoestrogens are not steroids, but they possess hydroxyl groups (present or introduced by hydroxylation) that can be positioned in a stereochemical alignment resembling that of 17β-estradiol. The distances between two hydroxyl groups in isoflavones, mammalian lignans, and 17β-estradiol are comparable. It appears that this feature is an essential factor for strong binding to the ER. Phytoestrogens are also structurally related to the antiestrogen tamoxifen, which is widely used in the treatment of breast cancer (57).

Another possibility deserving consideration is that some of the phytoestrogen actions may be attributable to properties

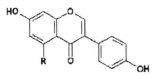

Daidzein: R = H
Genistein: R = OH

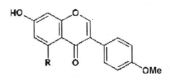

Formononetin: R = H
Biochanin A: R = OH

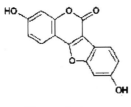

Coumestrol

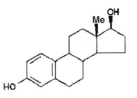

17β-estradiol

Figure 1   Different structures of (phyto)estrogens.

that do not involve hormone receptors such as antioxidative effects, interference with particular enzymes, protein synthesis, calcium transport, Na^+/K^+ adenosine triphosphatase, growth factor action, lipid oxidation. Besides its role in immune function, its activities may affect cell proliferation, angiogenesis, and cell differentiation, as well. In this respect, genistein is also known as a broad specificity tyrosine kinase inhibitor, blocks topoisomerase I and II activity and inhibits phosphodiesterases (30,42,58–65). Although preliminary results suggest phytoestrogens may inhibit the transcription factor NF-κB, which is strongly linked to inflammatory and immune responses and is associated with oncogenesis in certain models of cancer, the identification of molecular and cellular targets of chemopreventive phytochemicals is still incomplete (29).

III.   NF-κB

The onset of inflammatory gene expression is driven by the transcription factor NF-κB, of which the transcriptional activity is regulated at multiple levels (66,67). Transcriptional regulators of the NF-κB/IκB family promote the expression of over 100 target genes, the majority of which participate in the host immune response (68). Gene knockout and other studies not only establish roles for NF-κB in the ontogeny of the immune system, but also demonstrate that NF-κB participates at multiple steps during oncogenesis and regulation of programmed cell death (69,70). In addition, the involvement of the ubiquitous transcription factor NF-κB in the pathogenesis of the inflammatory response has been well documented by experiments both in vitro and in vivo (70–75).

Nuclear factor-κB transcription factors appear as homo- or heterodimers composed of members of the NF-κB/Rel family (66,67,70,76). They bind selectively to the κB consensus sequence GGGRNNYYCC (R = purine, Y = pyrimidine, N = any base), which is found in the promoter of a large variety of genes (77). The five mammalian NF-κB/Rel proteins generate more than 12 dimers recognizing 9–11 nucleotide NF-κB sites. Each dimer selectively regulates a few target promoters; however, several genes are redundantly induced by more than one dimer. Whether this property simply generates redundancy in target gene activation or underlies more complex regulatory mechanisms is an open issue (78–82). Proteins belonging to the NF-κB/Rel family share an N-terminal Rel homology domain and can be divided into two classes based on their C-terminal sequences (77,83). In vertebrates, one group includes RelA (p65), RelB, and cRel. These proteins contain one or more transactivation domains in their C-terminal end. Members of the second class (p50 and p52), which do not exert transactivation functions, are produced cotranslationally or by limited proteolysis from larger precursor proteins (i.e., p105 and p100, respectively). The C-terminal domain of these precursors consists of multiple copies of ankyrin repeats. Nuclear factor-κB activity is tightly regulated by members of the IκB family, which also contain ankyrin repeats. In most cell types, NF-κB proteins are sequestered in the cytoplasm in an inactive form through their noncovalent association with the inhibitor IκB (84). This subcellular localization results from efficient masking of the nuclear localization signal of NF-κB by IκBβ and IκBε (85,86). Complexes bound by IκBα continuously shuttle between the cytoplasm and the nucleus. A highly efficient nuclear export of these complexes ensures very low levels of nuclear NF-κB in the uninduced state (87,88).

The agents activating NF-κB include proinflammatory cytokines, such as tumor necrosis factor (TNF)α and IL1, phorbol esters, lipopolysacharide (LPS) and antigens. These stimuli activate through complex signaling cascades that are integrated by activation of an IκB-kinase complex (IKK) (67), consisting of two catalytic subunits (IKKα and IKKβ) and a regulatory subunit IKKγ (reviewed in Ref. 89). Phosphorylation of the IκB molecules at two conserved N-terminal Ser residues by IKKβ, subsequently leads to ubiquitination and degradation of the inhibitor and liberates the p65- and cRel-containing NF-κB dimers, which are now able to migrate to the nucleus and to activate NF-κB-regulated gene expression genes (67,90–92). Release of p52/RelB (93,94) as well as p50/RelB dimers (95) occurs through a "noncanonical" pathway requiring the NF-κB-inducing kinase (NIK) (94), which phosphorylates and activates IKKα (96,97). In turn, IKKα phosphorylates p100, thus directing its polyubiquitination and processing (93). This pathway is induced in response to a subset of stimuli such as BAFF, CD40 ligand, and LTβ-R triggering (98–100), and it has much slower activation kinetics than the canonical one (97).

Tumor necrosis factor α-induced NF-κB activation is one of the best characterized signaling pathway (101–106). After binding to its membrane-bound 55-kDa receptor, TNF initiates cell death mechanisms and signaling pathways leading to activation of the transcription factor NF-κB and of several MAPK pathways (p38, ERK and Jun N-terminal kinase) effecting gene expression (107,108). Oligomerization of the type I receptor by TNFα triggers the assembly of signaling molecules on the intracellular domain of the receptor, including TNFα receptor-associated factor (TRAF) -2, TRAF-5, and RIP (14–18,109–111). The IKK complex becomes recruited to TNFα receptor-associated proteins, where it is activated and the subsequent phosphorylation of IκB occurs. In that way, regulated IKK activity may be also cell type- and time-dependent (67,92,97,112). In addition to stimulus-induced nuclear translocation of NF-κB, several lines of evidence suggest that stimulus-induced phosphorylation of the p65 subunit plays a key role in the transcriptional activation after the nuclear translocation (113–119). Several candidate kinases that phosphorylate each serine residue have been identified (113–119), such as protein kinase A and mitogen- and stress-activated protein kinase-1 (MSK1) for Ser276, PKCζ for Ser311, casein kinase II for Ser529, and IKKα/β and the phosphatidylinositol 3-kinase (PI3K)-Akt for Ser536. It has been shown that phosphorylation at Ser276 is critical for p65 NF-κB activation by regulating the interaction with coactivators p300/CBP or the histone deacetylase HDAC1. Collectively, all this evidence further supports the importance of p65 phosphorylation in the nuclear function of NF-κB.

One important aspect of nuclear regulation of NF-κB activity relies on its interactions with enzymatic cofactor complexes (120–124). Upon comparing the function and subunit specificity of NF-κB driven genes with the sequence of the κB DNA-binding site in various knockout cells of Rel family members, little correlation was found, indicating that NF-κB subunit specificity for endogenous promoters is not solely encoded by the κB site sequence itself. Multiple protein–DNA complexes may be able to assemble on a given promoter depending on the cell-type and stimulus (125–127). Transcriptional synergy (128) between adjacent transcription factors and within enhanceosomes (129), as well as specific coactivator requirements (130), have been observed in nonchromosomal experimental systems and have long been thought to play a role in generating specificity through combinatorial control. Differential specificity of NF-κB protein family members in interactions with contextual transcriptional factors may thus account for family member-specific requirements for gene activation. Chromatin has also been shown to control NF-κB/Rel accessibility in vivo (79–81,118,131–134) and may in fact do so in a manner that is specific for a subset of RHD protein dimers. Furthermore, some genes require chromatin reorganization for gene activation (135–138), and this may be dependent on protein–protein interactions specific to a particular family member. Proinflammatory cytokines and microbial products stimulate transcription of several genes involved in the inflammatory and immunological response through the coordinate induction of multiple signaling pathways, including the three major MAPK pathways—ERKs, p38 and c-Jun NH2-terminal kinases (JNKs)—and IKK (66,67,139). A major function of these kinase pathways is to control eukaryotic gene expression programs in response to extracellular signals by phosphorylating transcription factors. However, it is now becoming clear that transcriptional regulation in response to IKK/MAPK signaling is more complex as these kinases can also target coactivators and corepressors and affect nucleosomal structure by inducing histone modifications. Furthermore, multiple inputs into individual promoters can be elicited by targeting different components of the same coregulatory complex ("enhanceosomes"), and/or nucleosome components, or by triggering different events on the same transcription factor, which contributes to establishment of dynamic gene regulatory networks (126,140). Proinflammatory stimuli induce the rapid and transient translocation of NF-κB to the nucleus, where it activates transcription from several genes, including those encoding inflammatory cytokines and chemokines, adhesion molecules, and cytoprotective proteins with different kinetics potencies. Assuming that NF-κB dimers present in the nucleus are in excess, with respect to the immediately available binding sites, most will remain unbound and will scan the chromatin for newly exposed binding sites (79–81,118,131,132,135,136,141). The appearance of new docking sites reflects active promoter modifications. In some cases, the MAPK and/or IKK pathway(s), at least in part through the induction of H3 phosphorylation and phosphoacetylation, may orchestrate the modifications that lead to enhanced accessibility of NF-κB sites. An alternative and not mutually exclusive possibility is that H3 phosphorylation and/or phosphoacetylation may permit or enhance binding to chromatin of a transcription factor required for cooperative recruitment of NF-κB to a subset of target genes. Thus, it is possible that the NF-κB sites contained in some promoters are not strong NF-κB binding sites (due either to sequence context relative to the flanking sequences or to physical constraints imposed by intrinsic DNA bending or chromatin organization). Efficient NF-κB recruitment to these promoters may either require corecruitment of an NF-κB partner, and/or stimulated by H3 phosphorylation or phosphoacetylation. As a final result, promoters that require active modification to efficiently recruit NF-κB will be activated if two conditions are fulfilled: stimulation activates above a critical threshold the pathway(s) required to induce promoter modification (for example, p38) and NF-κB dimers persist in the nucleus long enough to come into contact with the newly available binding sites. Nuclear factor-κB dimers have extremely potent transcriptional activation domains; enhancing their recruitment to target genes in a promoter-specific manner represents a very efficient control strategy.

IV. INDUCTION MECHANISMS OF NF-κB-DRIVEN GENE INDUCTION

Previously, we have explored how the multiresponsive IL6 promoter is modulated by the proinflammatory cytokine TNF and what the underlying mechanisms for promoter stimulation might be (5,114). Briefly, the regulation of expression of the IL6 gene is adapted to its key function, namely a systemic alarm signal that recruits diverse host defence mechanisms in order to limit tissue injury. The IL6 promoter behaves as a sophisticated biosensor for environmental stress, surveys immunological homeostasis, and is induced by a plethora of chemical or physiological compounds, including bacterial endotoxins, viruses (HIV, human T-cell leukemia virus), or inflammatory cytokines such as TNF and IL1. Characterization of the human IL6 promoter revealed a highly conserved control region of 300 bp upstream of the transcriptional initiation site that contains most, if not all, of the elements necessary for its induction by a variety of stimuli commonly associated with acute inflammatory or proliferative states. Electrophoretic mobility shift assays, as well as promoter deletion and point mutation analysis, revealed the presence and functional involvement of an NF-κB-binding element between positions −73 and −63, a multiple response element consisting of CRE followed by a binding site for the CCAAT enhancer-binding protein (C/EBPβ

β or NFIL6) between -173 and -145, and an activator protein-1 (AP1) site located between -283 and -277 (124,142,143). We found that, in addition to TNF-induced cytoplasmic NF-κB activation and nuclear DNA binding, the TNF-activated p38 and ERK MAPK pathways contribute to transcriptional activation of the IL6 promoter by modulating the transactivation capacity of the NF-κB p65 subunit. Recently, we found that phosphorylation of the p65 NF-κB subunit is a prerequisite for gene expression in response to TNF, and point to Ser276 in the RHD of p65 as the crucial residue for regulation of the nuclear transactivation capacity. More particularly, phosphorylation of Ser276 is an essential element for engagement of nuclear cofactors and their associated HAT activity, necessary to elaborate gene expression. Whether this engagement relies on conformational changes of CBP/HDAC upon interaction with NF-κB (144,145) and/or is the result of phosphorylations of NF-κB and CBP/HDAC, e.g., by TNF induction, is not clear at present (116,146,147). Interestingly, reconstitution of p65$^{-/-}$ MEFs with various serine to alanine substituted mutants of p65 recently was found to completely rescue TNF-induced IL6 production with S529A, S536A, and S529A/S536A S536A mutants, whereas the S276A mutant had an impaired ability to rescue this response (148). We identified the dually regulated nuclear kinase MSK1 as a candidate for phosphorylation of Ser276 (118). This kinase, which acts downstream of p38 and ERK MAPKs, is indeed activated by TNF, associates with p65 in a stimulus-dependent manner and specifically phosphorylates the serine residue at position 276, thus leading to its positioning at NF-κB-containing promoter sections and selective stimulation of particular NF-κB-driven genes. Subsequent to MSK1-mediated phosphorylation of p65, a conformational switch may allow engagement of CBP to create a transcriptionally competent enhanceosome. Although deacetylase inhibitors have been shown to prolong phosphorylation of p65 (149,150), it is not clear yet how transcription factor phosphorylation is linked further to acetylation.

Whether MSK1 is the only nuclear p65 kinase is uncertain. It should be noted that besides MSK1 and MSK2, H89 inhibits several protein kinases, such as PKA, RSK, etc.

(151). Furthermore, reduction of p65 phosphorylation in MSK1$^{-/-}$/MSK2$^{-/-}$ MEFs in response to TNF is not complete (data not shown) and the transactivation of Gal4-p65 is not entirely knocked out in MSK1$^{-/-}$/MSK2$^{-/-}$ cells (118). Several growth factor and stress signals have been shown to promote phosphorylation of CREB at Ser133, with comparable stoichiometry and kinetics. Coincident with this wide profile of inducibility, CREB is a substrate for various cellular kinases including pp90rsk, PKA, PKC, Akt, MSK1, MSK2, MAPKAPK-2, and Ca2^{+}/calmodulin kinases II and IV (152). The question can be raised whether p65 Ser276, considering its high homology with the CREB Ser133 motif, displays a similar promiscuity for different kinases and how the transcriptional system centered at p65 can distinguish between various stimuli to produce different gene responses (153); however, the ability of two stimuli to activate distinct genetic programs through a common component has now been noted in eukaryotes and yeast (152,154). Zhong et al. (155) pointed to a PKAc subunit present in the cytoplasmic NF-κB/IκB complex, which is able to phosphorylate a subfraction of the cytoplasmic p65 molecules at Ser276 in response to LPS. Whether the same occurs upon stimulation with TNF has not been described. Alternatively, another kinase(s) may also contribute to specific Ser276 phosphorylation of p65. Depending on starvation conditions, cellular confluence and senescence, varying background phosphorylation intensities have been observed, which may explain conflicting results of inducible vs. constitutive p65 Ser 276 phosphorylation levels: moreover, persistent activation of the Ras-MAPK pathway in oncogene-transformed or embryonal cells has regularly been found to result in elevated basal phosphorylation levels of histone H3 and NF-κB p65 (156–158). Furthermore, whether multiple kinases are present simultaneously at the enhanceosome (static model) (5,114) or act subsequently at different stages (dynamic model) (159–161) and/or at different locations (different subnuclear structures) (162) to ensure transcription needs to be further investigated. So far, the lack of specific inhibitors without cross-reactivity does not allow the relative contribution of each kinase in NF-κB-driven gene expression to be untangled easily; therefore, comparison of PKA-, MSK- or RSK-defective MEFs might possibly reveal unique or overlapping signaling functions. Interestingly, whereas p65 S276A mutant mice are lethal (S. Ghosh, personal communication), $MSK1^{-/-}/MSK2^{-/-}$ and $RSK2^{-/-}$ mice are viable (163,164). This may suggest a more selective involvement of these kinases in particular subsets of genes. To what extent the core promoter architecture, other MSK1 targets such as CREB and histone H3, and other (as yet unidentified) mechanistic factors also contribute to determine a specific gene expression pattern of NF-κB-driven genes, needs to be explored further.

Although multiple kinases may target the same NF-κB p65 Ser276 residue, a specific biological response may also depend on the complete pattern of modifications (such as acetylation, ubiquitylation, methylation, or SUMOylation) present in surrounding transcription factor or chromatin domains at a particular moment (135,152,153); by analogy with the proposed histone (165–167) and cofactor code (168,169), the complete set of p65 modifications may also determine a unique transcription factor code (170–173). Alternatively, specificity may also originate from variable activation kinetics of the kinases (114,154,174,175). Part of the specificity and kinetics may also be determined by histone modifications; differences in NF-κB responses (early vs. late genes) have now been correlated with differences in H4 acetylation and H3 phosphorylation, as well as with methylation patterns at particular promoters (79–81). Of special interest is the direct interaction of transcription factor transactivation domains with histones H3/H4, which may restrict signaling effects to promoters with selected nucleosome settings and a particular transcription factor/histone code (137,138,141,176–178). Taken together, we propose a model (see Fig. 2) in which cytoplasmic NF-κB activation is followed by nuclear phosphorylation of the p65 subunit at Ser276. This involves effective recruitment of the dually MAPK-activated, nuclear kinase MSK1 into the enhanceosome, followed by recruitment of CBP/p300 and accompanying phosphoacetylation of the chromatin environment, e.g., histone H3.

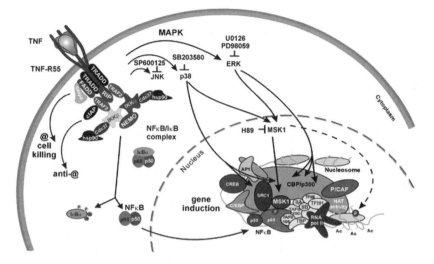

Figure 2 Dual activation scheme for IL6 gene expression in response to TNF. In this model, the IL6 promoter recruits CBP which, upon phosphorylation of NF-κB p65, renders the NF-κB/CBP CBP complex transcriptionally competent.

V. NEGATIVE MODULATION OF NF-κB-DRIVEN GENE INDUCTION BY PHYTOESTROGENS

Several nutrients and non-nutritive phytochemicals are being evaluated in intervention trials for their potential as health benefiting compounds (28,179–181). Despite significant advances in our molecular understanding of various diseases, little is known about the mechanism of action of most phytochemicals (12,29,30,64,182). The chemopreventive effects that most dietary phytochemicals exert are likely to be the sum of several distinct mechanisms. A wide array of phenolic substances, particularly those present in edible and medicinal plants, have been reported to retain antioxidative and antiinflammatory properties which appear to contribute to their chemopreventive or chemoprotective activity. Various studies demonstrate that flavonoids suppress NF-κB activation. As NF-κB is a transcription factor governing the expression of genes involved in the immune response, embryo or cell lineage development, cell apoptosis, cell cycle progression, inflammation, and oncogenesis, considerable attention has been paid to the upstream signaling pathways that lead to the activation of NF-κB. Many of these signaling molecules can serve as potential pharmaceutical targets for the specific inhibition of NF-κB activation leading to interruption of disease processes. Since many of the signal molecules in this pathway relay more than one of the upstream signals to downstream targets, it has been suggested that the transmission of signals involves a network, rather than a linear sequence in the activation of NF-κB. Thus, the detailed elucidation of the upstream signaling molecules involved in NF-κB activation and evaluation of their sensitivity to particular phytochemicals are currently a hot topic as this may allow the design of new chemoprotective drugs which selectively modulate NF-κB activity in various pathophysiological conditions. Disruption or deregulation of intracellular signaling cascades often leads to pathologies, and it is therefore important to identify these molecules in the signaling network that can be affected by individual chemopreventive phytochemicals to allow for a better assessment of their underlying mechanisms (12,29,30,64,182).

In this respect, we and others are investigating in more detail how phytoestrogens can specifically modulate NF-κB driven gene expression (i.e., IL6, IL8, iNOS, COX2, ...). Multiple lines of evidence indicate that IKK, PI3K, MAPK, and hormones are key elements of the intracellular-signaling cascades regulating NF-κB activity, and various levels of crosstalk have already been described (2,8,114,183,184). For example, breast cancers often progress from a hormone-dependent, nonmetastatic, anti-estrogen-sensitive phenotype to a hormone-independent, antiestrogen- and chemotherapy-resistant phenotype with highly invasive and metastatic growth properties (185–195). This progression is usually accompanied by transition from inducible to constitutive receptor tyrosine kinase (RTK)/MAPK signaling which affects ER function (becomes hormone ligand-independent), NF-κB regulation (high turnover of its inhibitor IκB results in constitutive NF-κB activity), or outgrowth of ER-negative cancer cells. The potential mechanisms for either intrinsic or acquired endocrine resistance are still poorly comprehended, but they clearly include ER-coregulatory proteins and cross-talk between the ER, NF-κB, growth factors, and kinase networks. Soy isoflavones are believed to contribute to the putative breast- and prostate-cancer-preventive activity of soy by abrogating NF-κB/DNA binding, as a consequence of reduced IκB phosphorylation, suppressed Akt activity, and inhibition of nuclear translocation of NF-κB. Besides some effects on DNA-binding, inhibition of the nuclear transactivation capacity of NF-κB has been observed too, independent of IKK activity.

In the past few years, evidence has emerged that phytoestrogen binding hormone receptors, such as ER, AR, PR, AhR, PPAR, interact with NF-κB, and transcriptionally modulate each other (9–11,196–203). So far, the mode, according to which phytoestrogenic compounds mediate their estrogenic effects, has been studied in most detail. Assumptions range from mimicking normal estrogenic action to competitive inhibitory effects, which may block normal estrogenic action. Structural studies clearly reveal changes in ER-conformation upon binding of classic estrogens as compared to phytoestrogens, which may already suggest ligand selective (cofactor) dependent activities (48,49,52,196,204–211). Interestingly, phytoestrogens seem to preferentially mediate their effects via ERβ, whereas classical estrogen acts via both receptors ERα and ERβ (212,213). As such, phytoestrogens may act as natural SERMs that elicit distinct clinical effects from estrogens used for hormone replacement by selectively recruiting coregulatory proteins to ERβ that trigger transcriptional pathways (53–55,212,214). Interestingly, ERβ was found to be more potent in NF-κB transrepression than ERα and may suggest distinct modulatory effects of synthetic estrogens vs. phytochemicals at the level of NF-κB gene regulation. As crosstalk of phytoestrogens with PPAR, AhR, PR, vitamin D, ERR responses has been reported as well, the complexity of their hormone activities is further augmented (9,64,196–200,215–217). At another level, involvement of tyrosine phosphorylation (218,219) in TNF (101,220–222), Toll like receptors (TLR) (223) and growth factor signaling (EGF, her/neu) (32,33,224–229) has been demonstrated at the receptor level, which may be sensible to inhibition by phytoestrogens (62,230,231), and decrease further downstream-signaling cascades such as Akt, IKK, and MAPK signaling pathways (29,33,182,184,232–234).

Finally, it has been postulated that reactive oxygen species (ROS) may act as second messengers leading to NF-κB activation, whereas antioxidants may block this activity (235). Whether this mechanism holds through, has recently become a matter of hot debate, since antioxidants were found to lower ligand-receptor affinity (i.e., TNF/TNFR) and in this way, lower the magnitude of receptor signaling (236–238).

Considering the pleiotropic activities of flavonoids, i.e., as hormone ligands (i.e., ER, AR, AhR, PPAR, ESR, ...), tyrosine kinase inhibitors, or antioxidants, it will be interesting to evaluate what activity predominates in its NF-κB modulatory activities in a gene- or cell-specific context, as compared to effects of classical estrogens (12,29,64). Further structure function analysis of phytochemicals may allow to define core structure elements, which are responsible to fulfill part of their subactivities.

In many cases, the chemopreventive effects of dietary chemopreventives in cultured cells or tissues are only achievable at supraphysiological concentrations (such concentrations might not be attained when the phytochemicals are administered as part of diet) since metabolization and/or chaperone protection of these compounds may be less efficient in tissue culture setups (12,27,56,239). As phenolic phytochemicals are often present as glycosides or are converted to other conjugated forms after absorption, it would be of high interest to compare relative activities of each metabolite in modulation of NF-κB activity. Both pharmacokinetic properties and bioavailability are key problems in investigating the dietary prevention of cancer and should be assessed carefully before undertaking intervention trials with dietary supplements. The development and use of chemopreventive agents for intervention trials involve many scientific disciplines. With the advances in techniques to assess single nucleotide polymorphisms (SNPs), we are now more aware of the specific genes that can directly and indirectly contribute to individual differences in the susceptibility to carcinogenesis and/or drug metabolization pathways (240). When high-risk groups are identified, practitioners might be able to recommend specific dietary supplements that can modulate or restore the cellular signaling events which are likely to be disrupted in these individuals. The term "nutragenomics" has been coined, and much attention is being focused on this relatively new area of research (28,240). Tailored supplementation with designer foods that consist of chemopreventive phytochemicals—each having their own distinct anticancer mechanisms—will be available in the near future. These should be developed in line with advances in the genetic and molecular epidemiology ofcarcinogenesis.

VI. CONCLUSION

In the past decade, nutrition research has undergone an important shift in focus from epidemiology and physiology to molecular biology and genetics. This is a result of a growing realization that the effects of nutrition on health and disease cannot be understood without a profound understanding of how nutrients act at the molecular level: micronutrients and macronutrients can be potent dietary signals that influence the metabolic programming of cells and have an important role in the control of homeostasis and/or disease (28,240). Phytoestrogens are plant-derived molecules with estrogen-like action. Numerous reports exist on the potential beneficial role of nutritional phytoestrogens in human health (cancer chemoprevention, relief of postmenopausal symptoms, osteoporosis amelioration) (13–15,18–20). Herbal extracts are frequently used in traditional Chinese medicine for the treatment of hormonal disturbances or as health benefiting supplements (19). Often these preparations are formulated as complex mixtures, without any knowledge about the nature of the active components. Therefore, not all of these herbs and dietary supplements are risk-free. It is clear that we need to know much more about the molecular mechanisms, safety, and efficacy of bioactive natural compounds before they can be generally recommended as health-benefiting food supplements (22).

Because of its central role in various physiopathological processes (i.e., inflammatory diseases, bone diseases, oncogenesis, and neurodegeneration), NF-κB has become the first target for innovative, anti-inflammatory, and/or antitumoral therapies in various pharmaceutical companies in Europe and the United States (241). The identification of new regulatory target molecules involved in NF-κB signaling, which are useful for pharmacological applications, has been a hot topic for many years. The number of newly formulated herbal products is growing rapidly and the Holy grail of pharmaceutical companies today has rather shifted from exploration of synthetic compound libraries to the identification of compounds in medicinal plant extracts with potential NF-κB inhibitory activity and/or inhibitory effects on targets of the inflammatory /apoptotic cascade (242). These (non) steroidal modulators may serve as lead compounds for developing new pharmaceuticals to be used in the treatment of inflammatory diseases, hormonal disturbances, cancer, or conditions associated with an increased activity of the transcription factor NF-κB.

Our observation that nuclear NF-κB activity is susceptible to various important modulations adds a new perspective to therapeutical applications, as several pharmaceutical companies which currently develop selective NF-κB inhibitors, focus on the cytoplasmic regulatory event in which NF-κB is released from its physiologic inhibitor IκB (243). As NF-κB is also a crucial component of the immune system, the latter strategy has the important disadvantage that blocking IκB degradation is a very drastic event with detrimental side effects. In this respect, innovative anti-inflammatory strategies may focus on less aggressive therapies, which interfere with NF-κB activity without affecting its DNA binding. To this purpose, the main scientific interest may shift from the cytoplasm to the NF-κB/chromatin interface in the nucleus. Furthermore, future drug compounds should interfere selec-

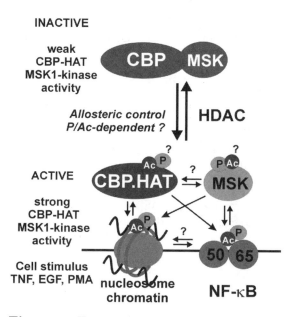

Figure 3 Working model illustrating relation between NF-κB and chromatin regulation.

tively with interactions of NF-κB with particular cofactors or defined nucleosomes, which could restrict drug activity to a limited set of disease-related genes, leaving homeostatic gene expression unaffected. Unraveling the relationship between chromatin and NF-κB regulation will definitely shed a new light on the design of innovative anti-inflammatory and/or anticancer drugs (79–81,174). A working model displaying this relationship is presented in Fig. 3. As chromatin-embedded promoter enhanceosomes nowadays behave like sophisticated protein modules receptive to various signals, future therapies may benefit from combined structural (selective ligand) and signaling (selective inhibitors) approaches to establish effective but harmless treatments (113,118,173,178,244,245).

REFERENCES

1. Akira S, Taga T, Kishimoto T. Interleukin-6 in biology and medicine. Adv Immunol 1993; 54:1–78.

2. Erschler WB, Keller, ET. Age-associated increased interleukin-6 gene expression, late-life diseases, and frailty. Annu Rev Med 2000; 51:245–270.

3. Trikha M, Corringham R, Klein B, Rossi JF. Targeted anti-interleukin-6 monoclonal antibody therapy for cancer: a review of the rationale and clinical evidence. Clin Cancer Res 2003; 9:4653–4665.

4. Zhang GJ, Adachi I. Serum interleukin-6 levels correlate to tumor progression and prognosis in metastatic breast carcinoma. Anticancer Res 1999; 19:1427–1432.

5. Vanden Berghe W, Vermeulen L, De Wilde G, De Bosscher K, Boone E, Haegeman G. Signal transduction by tumor necrosis factor and gene regulation of the inflammatory cytokine interleukin-6. Biochem Pharmacol 2000; 60:1185–1195.

6. Burger HG, Dudley EC, Robertson DM, Dennerstein L. Hormonal changes in the menopause transition. Recent Prog Horm Res 2002; 57:257–275.

7. Kiecolt-Glaser JK, Preacher KJ, MacCallum RC, Atkinson C, Malarkey WB, Glaser R. Chronic stress and age-related increases in the proinflammatory cytokine IL-6. Proc Natl Acad Sci USA 2003; 100:9090–9095.

8. Pfeilschifter J, Koditz R, Pfohl M, Schatz H. Changes in proinflammatory cytokine activity after menopause. Endocr Rev 2002; 23:90–119.

9. Evans MJ, Eckert A, Lai K, Adelman SJ, Harnish DC. Reciprocal antagonism between estrogen receptor and NF-kappaB activity in vivo. Circ Res 2001; 89:823–830.

10. Galien R, Garcia T. Estrogen receptor impairs interleukin-6 expression by preventing protein binding on the NF-kappaB site. Nucleic Acids Res 1997; 25:2424–2429.

11. Ray P, Ghosh SK, Zhang DH, Ray A. Repression of interleukin-6 gene expression by 17 beta-estradiol: inhibition of the DNA-binding activity of the transcription factors NF-IL6 and NF-kappa B by the estrogen receptor. FEBS Lett 1997; 409:79–85.

12. Bolego C, Poli A, Cignarella A, Paoletti R. Phytoestrogens: pharmacological and therapeutic perspectives. Curr Drug Targets 2003; 4:77–87.

13. Wuttke W, Jarry H, Westphalen S, Christoffel V, Seidlova-Wuttke D. Phytoestrogens for hormone replacement therapy? J Steroid Biochem Mol Biol 2002; 83:133–47.

14. Wells G, Tugwell P, Shea B, et al. Meta-analyses of therapies for postmenopausal osteoporosis. V. Meta-analysis of the efficacy of hormone replacement therapy in treating and preventing osteoporosis in postmenopausal women. Endocr Rev 2002; 23:529–539.

15. Marsden J. The menopause, hormone replacement therapy and breast cancer. J Steroid Biochem Mol Biol 2002; 83:123–132.

16. Guyatt GH, Cranney A, Griffith L, et al. Summary of meta-analyses of therapies for postmenopausal osteoporosis and the relationship between bone density and fractures. Endocrinol Metab Clin North Am 2002; 31:659–679,xii.

17. Kronenberg F, Fugh-Berman A. Complementary and alternative medicine for menopausal symptoms: a review of randomized, controlled trials. Ann Intern Med 2002; 137:805–813.

18. Beral V. Breast cancer and hormone-replacement therapy in the Million Women Study. Lancet 2003; 362:419–427.

19. Mann J. Natural products in cancer chemotherapy: past, present and future. Nat Rev Cancer 2002; 2:143–148.

20. Wuttke W, Jarry H, Becker T, et al. Phytoestrogens: endocrine disrupters or replacement for hormone replacement therapy? Maturitas 2003; 44(suppl 1):S9–S20.

21. Cranney A, Guyatt G, Griffith L, Wells G, Tugwell P, Rosen C. Meta-analyses of therapies for postmenopausal osteoporosis. IX: Summary of meta-analyses of therapies for postmenopausal osteoporosis. Endocr Rev 2002; 23:570–578.

22. Albertazzi P, Purdie D. The nature and utility of the phytoestrogens: a review of the evidence. Maturitas 2002; 42:173–185.

23. Ewies AA. Phytoestrogens in the management of the menopause: up-to-date. Obstet Gynecol Surv 2002; 57: 306–313.

24. Knight DC, Eden JA. A review of the clinical effects of phytoestrogens. Obstet Gynecol 1996; 87:897–904.

25. Stark A, Madar Z. Phytoestrogens: a review of recent findings. J Pediatr Endocrinol Metab 2002; 15:561–572.

26. Russell L, Hicks GS, Low AK, Shepherd JM, Brown CA. Phytoestrogens: a viable option? Am J Med Sci 2002; 324:185–188.

27. Dixon RA, Steele CL. Flavonoids and isoflavonoids—a gold mine for metabolic engineering. Trends Plant Sci 1999; 4:394–400.

28. Muller M, Kersten S. Nutrigenomics: goals and strategies. Nat Rev Genet 2003; 4:315–322.

29. Surh YJ. Cancer chemoprevention with dietary phytochemicals. Nat Rev Cancer 2003; 3:768–780.

30. Ososki AL, Kennelly EJ. Phytoestrogens: a review of the present state of research. Phytother Res 2003; 17:845–869.

31. Ziegler RG, Hoover RN, Pike MC, et al. Migration patterns and breast cancer risk in Asian–American women. J Natl Cancer Inst 1993; 85:1819–1827.

32. Bange J, Zwick E, Ullrich A. Molecular targets for breast cancer therapy and prevention. Nat Med 2001; 7:548–552.

33. Ali S, Coombes C. Endocrine-responsive breast cancer and strategies for combating resistance. Nat Rev Mol Cell Biol 2002; 2:101–115.

34. Parkin DM. Cancers of the breast, endometrium and ovary: geographic correlations. Eur J Cancer Clin Oncol 1989; 25:1917–1925.

35. Birt DF, Hendrich S, Wang W. Dietary agents in cancer prevention: flavonoids and isoflavonoids. Pharmacol Ther 2001; 90:157–177.

36. Kris-Etherton PM, Hecker KD, Bonanome A, et al. Bioactive compounds in foods: their role in the prevention of cardiovas-

cular disease and cancer. Am J Med 2002; 113(suppl 9B): 71S–88S.

37. Adlercreutz H. Phyto-oestrogens and cancer. Lancet Oncol 2002; 3:364–373.

38. Valsta LM, Kilkkinen A, Mazur W, et al. Phyto-oestrogen database of foods and average intake in Finland. Br J Nutr 2003; 89:822–830.

39. Brownson DM, Azios NG, Fuqua BK, Dharmawardhane SF, Mabry TJ. Flavonoid effects relevant to cancer. J Nutr 2002; 132:3482S–3489S.

40. Knight DC, Eden JA. Phytoestrogens—a short review. Maturitas 1995; 22:167–175.

41. Coward L, Barnes NC, Setchell KD, Barnes S. The isoflavones genistein and daidzein in soybean foods from American and Asian diets. J Agric Food Chem 1993; 41:1961–1967.

42. Dixon RA, Ferreira D. Genistein. Phytochemistry 2002; 60:205–211.

43. Adlercreutz H, Mazur W. Phyto-oestrogens and Western diseases. Ann Med 1997; 29:95–120.

44. Rosenberg Zand RS, Jenkins DJ, Diamandis EP. Flavonoids and steroid hormone-dependent cancers. J Chromatogr B Analyt Technol Biomed Life Sci 2002; 777:219–232.

45. Powles TJ. Anti-oestrogenic chemoprevention of breast cancer—the need to progress. Eur J Cancer 2003; 39:572–579.

46. Hall JM, Couse JF, Korach KS. The multifaceted mechanisms of estradiol and estrogen receptor signaling. J Biol Chem 2001; 276:36869–36872.

47. Moggs JG, Orphanides G. Estrogen receptors: orchestrators of pleiotropic cellular responses. EMBO Rep 2001; 2:775–781.

48. Weihua Z, Andersson S, Cheng G, Simpson ER, Warner M, Gustafsson JA. Update on estrogen signaling. FEBS Lett 2003; 546:17–24.

49. Gustafsson JA. What pharmacologists can learn from recent advances in estrogen signalling. Trends Pharmacol Sci 2003; 24:479–485.

50. Segars JH, Driggers PH. Estrogen action and cytoplasmic signaling cascades. Part I: membrane-associated signaling complexes. Trends Endocrinol Metab 2002; 13:349–354.

51. Driggers PH, Segars JH. Estrogen action and cytoplasmic signaling pathways. Part II: the role of growth factors and phosphorylation in estrogen signaling. Trends Endocrinol Metab 2002; 13:422–427.

52. Metivier R, Stark A, Flouriot G, et al. A dynamic structural model for estrogen receptor-alpha activation by ligands, emphasizing the role of interactions between distant A and E domains. Mol Cell 2002; 10:1019–1032.

53. Sun J, Meyers MJ, Fink BE, Rajendran R, Katzenellenbogen JA, Katzenellenbogen BS. Novel ligands that function as selective estrogens or antiestrogens for estrogen receptor-alpha or estrogen receptor-beta. Endocrinology 1999; 140:800–804.

54. Shiau AK, Barstad D, Radek JT, et al. Structural characterization of a subtype-selective ligand reveals a novel mode of estrogen receptor antagonism. Nat Struct Biol 2002; 9:359–364.

55. Katzenellenbogen BS, Katzenellenbogen JA. Biomedicine. Defining the "S" in SERMs. Science 2002; 295:2380–2381.

56. Havsteen BH. The biochemistry and medical significance of the flavonoids. Pharmacol Ther 2002; 96:67–202.

57. Jordan VC. Tamoxifen: a most unlikely pioneering medicine. Nat Rev Drug Discov 2003; 2:205–213.

58. Matsukawa Y, Marui N, Sakai T, et al. Genistein arrests cell cycle progression at G2-M. Cancer Res 1993; 53:1328–1331.

59. Agarwal R. Cell signaling and regulators of cell cycle as molecular targets for prostate cancer prevention by dietary agents. Biochem Pharmacol 2000; 60:1051–1059.

60. Casagrande F, Darbon JM. Effects of structurally related flavonoids on cell cycle progression of human melanoma cells: regulation of cyclin-dependent kinases CDK2 and CDK1. Biochem Pharmacol 2001; 61:1205–1215.

61. Kim MH, Gutierrez AM, Goldfarb RH. Different mechanisms of soy isoflavones in cell cycle regulation and inhibition of invasion. Anticancer Res 2002; 22:3811–3817.

62. Akiyama T, Ishida J, Nakagawa S, et al. Genistein, a specific inhibitor of tyrosine-specific protein kinases. J Biol Chem 1987; 262:5592–5595.

63. Ye R, Bodero A, Zhou BB, Khanna KK, Lavin MF, Lees-Miller SP. The plant isoflavenoid genistein activates p53 and Chk2 in an ATM-dependent manner. J Biol Chem 2001; 276: 4828–4833.

64. Cos P, De Bruyne T, Apers S, Vanden Berghe D, Pieters L, Vlietinck AJ. Phytoestrogens: recent developments. Planta Med 2003; 69:589–599.

65. Nichols MR, Morimoto BH. Differential inhibition of multiple cAMP phosphodiesterase isozymes by isoflavones and tyrphostins. Mol Pharmacol 2000; 57:738–745.

66. Richmond A. Nf-kappab, chemokine gene transcription and tumour growth. Nat Rev Immunol 2002; 2:664–674.

67. Li Q, Verma IM. NF-kappaB regulation in the immune system. Nat Rev Immunol 2002; 2:725–734.

68. Ghosh S, May MJ, Kopp EB. NF-kappaB and rel proteins: evolutionarily conserved mediators of immune responses. Annu Rev Immunol 1998; 16:225–260.

69. Rayet B, Gelinas C. Aberrant rel/nfkb genes and activity in human cancer. Oncogene 1999; 18:6938–6947.

70. Perkins ND. The Rel/NF-kappaB family: friend and foe. Trends Biochem Sci 2000; 25:434–440.

71. Baeuerle PA, Baichwal VR. NF-κB as a frequent target for immunosuppressive and anti-inflammatory molecules. Adv Immunol 1997; 65:111–137.

72. Silverman N, Maniatis T. NF-kappaB signaling pathways in mammalian and insect innate immunity. Genes Dev 2001; 15:2321–2342.

73. Baldwin AS Jr. Series introduction: the transcription factor NF-kappaB and human disease. J Clin Invest 2001; 107:3–6.

74. Baldwin AS. Control of oncogenesis and cancer therapy resistance by the transcription factor NF-kappaB. J Clin Invest 2001; 107:241–246.

75. Tak PP, Firestein GS. NF-kappaB: a key role in inflammatory diseases. J Clin Invest 2001; 107:7–11.

76. Karin M, Delhase M. The I kappa B kinase (IKK) and NF-kappa B: key elements of proinflammatory signalling. Semin Immunol 2000; 12:85–98.

77. Pahl HL. Activators and target genes of Rel/NF-kappaB transcription factors. Oncogene 1999; 18:6853–6866.

78. Saccani S, Pantano S, Natoli G. Modulation of NF-kappaB activity by exchange of dimers. Mol Cell 2003; 11:1563–1574.

79. Saccani S, Pantano S, Natoli G. p38-Dependent marking of inflammatory genes for increased NF-kappa B recruitment. Nat Immunol 2002; 3:69–75.

80. Saccani S, Pantano S, Natoli G. Two waves of nuclear factor kappaB recruitment to target promoters. J Exp Med 2001; 193:1351–1359.

81. Saccani S, Natoli G. Dynamic changes in histone H3 Lys 9 methylation occurring at tightly regulated inducible inflammatory genes. Genes Dev 2002; 16:2219–2224.

82. Zhou A, Scoggin S, Gaynor RB, Williams NS. Identification of NF-kappa B-regulated genes induced by TNFalpha utilizing expression profiling and RNA interference. Oncogene 2003; 22:2054–2064.

83. Gilmore TD. The Rel/NF-kappaB signal transduction pathway: introduction. Oncogene 1999; 18:6842–6844.

84. Sen R, Baltimore D. Inducibility of kappa immunoglobulin enhancer-binding protein Nf-kappa B by a posttranslational mechanism. Cell 1986; 47:921–928.

85. Malek S, Chen Y, Huxford T, Ghosh G. IkappaBbeta, but not IkappaBalpha, functions as a classical cytoplasmic inhibitor of NF-kappaB dimers by masking both NF-kappaB nuclear localization sequences in resting cells. J Biol Chem 2001; 276:45225–45235.

86. Tam WF, Sen R. IkappaB family members function by different mechanisms. J Biol Chem 2001; 276:7701–7704.

87. Huang TT, Kudo N, Yoshida M, Miyamoto S. A nuclear export signal in the N-terminal regulatory domain of Ikappa-

Balpha controls cytoplasmic localization of inactive NF-kappaB/IkappaBalpha complexes. Proc Natl Acad Sci USA 2000; 97:1014–1019.

88. Johnson C, Van Antwerp D, Hope TJ. An N-terminal nuclear export signal is required for the nucleocytoplasmic shuttling of IkappaBalpha. EMBO J 1999; 18:6682–6693.

89. Israel A. A regulator branches out. Nature 2003; 423:596–597.

90. Zandi E, Chen Y, Karin M. Direct phosphorylation of IkappaB by IKKalpha and IKKbeta: discrimination between free and NF-kappaB-bound substrate. Science 1998; 281:1360–1363.

91. Karin M, Lin A. NF-kappaB at the crossroads of life and death. Nat Immunol 2002; 3:221–227.

92. Ghosh S, Karin M. Missing pieces in the NF-kappaB puzzle. Cell 2002; 109(suppl):S81–S96.

93. Senftleben U, Cao Y, Xiao G, et al. Activation by IKKalpha of a second, evolutionary conserved, NF-kappa B signaling pathway. Science 2001; 293:1495–1499.

94. Xiao G, Harhaj EW, Sun SC. NF-kappaB-inducing kinase regulates the processing of NF-kappaB2 p100. Mol Cell 2001; 7:401–409.

95. Muller JR, Siebenlist U. Lymphotoxin beta receptor induces sequential activation of distinct NF-kappa B factors via separate signaling pathways. J Biol Chem 2003; 278: 12006–12012.

96. Regnier CH, Song HY, Gao X, Goeddel DV, Cao Z, Rothe M. Identification and characterization of an IkappaB kinase. Cell 1997; 90:373–383.

97. Pomerantz JL, Baltimore D. Two pathways to NF-kappaB. Mol Cell 2002; 10:693–695.

98. Dejardin E, Droin NM, Delhase M, et al. The lymphotoxin-beta receptor induces different patterns of gene expression via two NF-kappaB pathways. Immunity 2002; 17:525–535.

99. Coope HJ, Atkinson PG, Huhse B, et al. CD40 regulates the processing of NF-kappaB2 p100 to p52. EMBO J 2002; 21:5375–5385.

100. Claudio E, Brown K, Park S, Wang H, Siebenlist U. BAFF-induced NEMO-independent processing of NF-kappa B2 in maturing B cells. Nat Immunol 2002; 3:958–965.

101. Aggarwal BB. Signalling pathways of the TNF superfamily: a double-edged sword. Nat Rev Immunol 2003; 3:745–756.

102. Fiers W, Beyaert R, Boone E, et al. TNF-induced intracellular signaling leading to gene induction or to cytotoxicity by necrosis or by apoptosis. J Inflamm 1995; 47:67–75.

103. Wallach D, Varfolomeev EE, Malinin NL, Goltsev YV, Kovalenko AV, Boldin MP. Tumor necrosis factor receptor and Fas signaling mechanisms. Annu Rev Immunol 1999; 17:331–367.

104. Chen G, Goeddel V. TNF Connections Map. http://stke.sciencemag.org/cgi/cm/CMP_7107, 2002.

105. Chen G, Cao P, Goeddel DV. TNF-induced recruitment and activation of the IKK complex require Cdc37 and Hsp90. Mol Cell 2002; 9:401–410.

106. Chen G, Goeddel DV. TNF-R1 signaling: a beautiful pathway. Science 2002; 296:1634–1635.

107. Heyninck K, De Valck D, Vanden Berghe W, et al. The zinc finger protein A20 inhibits TNF-induced NF-kappaB-dependent gene expression by interfering with an RIP- or TRAF2-mediated transactivation signal and directly binds to a novel NF-kappaB- inhibiting protein ABIN. J Cell Biol 1999; 145:1471–1482.

108. Zhang SQ, Kovalenko A, Cantarella G, Wallach D. Recruitment of the IKK signalosome to the p55 TNF receptor: RIP and A20 bind to NEMO (IKKgamma) upon receptor stimulation. Immunity 2000; 12:301–311.

109. Sakurai H, Suzuki S, Kawasaki N, et al. Tumor necrosis factor-alpha-induced IKK phosphorylation of NF-kappaB p65 on serine 536 is mediated through the TRAF2, TRAF5, and TAK1 signaling pathway. J Biol Chem 2003; 278: 36916–36923.

110. Vandevoorde V, Haegeman G, Fiers W. Induced expression of trimerized intracellular domains of the human tumor

necrosis factor (TNF) p55 receptor elicits TNF effects. J Cell Biol 1997; 137:1627–1638.

111. Boone E, Vandevoorde V, De Wilde G, Haegeman G. Activation of p42/p44 mitogen-activated protein kinases (MAPK) and p38 MAPK by tumor necrosis factor (TNF) is mediated through the death domain of the 55-kDa TNF receptor. FEBS Lett 1998; 441:275–280.

112. Schmidt C, Peng B, Li Z, et al. Mechanisms of proinflammatory cytokine-induced biphasic NF-kappaB activation. Mol Cell 2003; 12:1287–1300.

113. Schmitz ML, Bacher S, Kracht M. I kappa B-independent control of NF-kappa B activity by modulatory phosphorylations. Trends Biochem Sci 2001; 26:186–190.

114. Vanden Berghe W, De Bosscher K, Vermeulen L, De Wilde G, Haegeman G. Induction and Repression of NF-kB-driven Inflammatory Genes. Ernst Schering Res Found Workshop 2002; 40:233–278.

115. Vermeulen L, De Wilde G, Notebaert S, Vanden Berghe W, Haegeman G. Regulation of the transcriptional activity of the nuclear factor-kappaB p65 subunit. Biochem Pharmacol 2002; 64:963–970.

116. Zhong H, May MJ, Jimi E, Ghosh S. The phosphorylation status of nuclear NF-kappa B determines its association with CBP/p300 or HDAC-1. Mol Cell 2002; 9:625–636.

117. Duran A, Diaz-Meco MT, Moscat J. Essential role of RelA Ser311 phosphorylation by zetaPKC in NF-kappaB transcriptional activation. EMBO J 2003; 22:3910–3918.

118. Vermeulen L, De Wilde G, Damme PV, Vanden Berghe W, Haegeman G. Transcriptional activation of the NF-kappaB p65 subunit by mitogen- and stress-activated protein kinase-1 (MSK1). EMBO J 2003; 22:1313–1324.

119. Drier EA, Huang LH, Steward R. Nuclear import of the Drosophila Rel protein Dorsal is regulated by phosphorylation. Genes Dev 1999; 13:556–568.

120. Torchia J, Glass C, Rosenfeld MG. Co-activators and co-repressors in the integration of transcriptional responses. Curr Opin Cell Biol 1998; 10:373–383.

121. Kuo MH, Allis CD. Roles of histone acetyltransferases and deacetylases in gene regulation. Bioessays 1998; 20:615–626.

122. Berger SL. Gene activation by histone and factor acetyltransferases. Curr Opin Cell Biol 1999; 11:336–341.

123. Sheppard KA, Rose DW, Haque ZK, et al. Transcriptional activation by NF-kappaB requires multiple coactivators. Mol Cell Biol 1999; 19:6367–6378.

124. Vanden Berghe W, De Bosscher K, Boone E, Plaisance S, Haegeman G. The nuclear factor-kappaB engages CBP/p300 and histone acetyltransferase activity for transcriptional activation of the interleukin-6 gene promoter. J Biol Chem 1999; 274:32091–32098.

125. Hoffmann A, Leung TH, Baltimore D. Genetic analysis of NF-kappaB/Rel transcription factors defines functional specificities. EMBO J 2003; 22:5530–5539.

126. Hoffmann A, Levchenko A, Scott ML, Baltimore D. The IkappaB-NF-kappaB signaling module: temporal control and selective gene activation. Science 2002; 298:1241–1245.

127. Falvo JV, Uglialoro AM, Brinkman BM, et al. Stimulus-specific assembly of enhancer complexes on the tumor necrosis factor alpha gene promoter. Mol Cell Biol 2000; 20:2239–2247.

128. Lin YS, Carey M, Ptashne M, Green MR. How different eukaryotic transcriptional activators can cooperate promiscuously. Nature 1990; 345:359–361.

129. Thanos D, Maniatis T. Virus induction of human IFN beta gene expression requires the assembly of an enhanceosome. Cell 1995; 83:1091–1100.

130. Merika M, Williams AJ, Chen G, Collins T, Thanos D. Recruitment of CBP/p300 by the IFN beta enhanceosome is required for synergistic activation of transcription. Mol Cell 1998; 1:277–287.

131. Anest V, Hanson J, Cogswell P, Steinbrecher K, Strahl B, Baldwin A. A nucleosomal function for IkB kinase-a in NF-kB-dependent gene expression. Nature 2003; 423:659–663.

132. Yamamoto Y, Verma U, Prajapati S, Kwak Y, Gaynor R. Histone H3 phosphorylation by IKK-a is critical for cytokine-induced gene expression. Nature 2003; 423:655–659.

133. Smale ST, Fisher AG. Chromatin structure and gene regulation in the immune system. Annu Rev Immunol 2002; 20:427–462.

134. Smale ST. Core promoters: active contributors to combinatorial gene regulation. Genes Dev 2001; 15:2503–2508.

135. Fischle W, Wang Y, Allis CD. Binary switches and modification cassettes in histone biology and beyond. Nature 2003; 425:475–479.

136. Narlikar GJ, Fan HY, Kingston RE. Cooperation between complexes that regulate chromatin structure and transcription. Cell 2002; 108:475–487.

137. Lomvardas S, Thanos D. Modifying gene expression programs by altering core promoter chromatin architecture. Cell 2002; 110:261–271.

138. Lomvardas S, Thanos D. Opening chromatin. Mol Cell 2002; 9:209–211.

139. Dong C, Davis RJ, Flavell RA. MAP kinases in the immune response. Annu Rev Immunol 2002; 20:55–72.

140. Tian B, Brasier AR. Identification of a nuclear factor kappa B-dependent gene network. Recent Prog Horm Res 2003; 58:95–130.

141. Martone R, Euskirchen G, Bertone P, et al. Distribution of NF-kappaB-binding sites across human chromosome 22. Proc Natl Acad Sci USA 2003; 100:12247–12252.

142. Vanden Berghe W, Plaisance S, Boone E, et al. p38 and extracellular signal-regulated kinase mitogen-activated protein kinase pathways are required for nuclear factor-kappaB p65 transactivation mediated by tumor necrosis factor. J Biol Chem 1998; 273:3285–3290.

143. Beyaert R, Cuenda A, Vanden Berghe W, Plaisance S, Lee JC, Haegeman G, Cohen P, Fiers W. The p38/RK mitogen-activated protein kinase pathway regulates interleukine-6

synthesis in response to tumour necrosis factor. EMBO J 1996; 15:1914–1923.

144. Canettieri G, Morantte I, Guzman E, et al. Attenuation of a phosphorylation-dependent activator by an HDAC-PP1 complex. Nat Struct Biol 2003; 10:175–181.

145. Parker D, Jhala US, Radhakrishnan I, et al. Analysis of an activator:coactivator complex reveals an essential role for secondary structure in transcriptional activation. Mol Cell 1998; 2:353–359.

146. Kovacs KA, Steinmann M, Magistretti PJ, Halfon O, Cardinaux JR. CCAAT/enhancer-binding protein family members recruit the coactivator CREB-binding protein and trigger its phosphorylation. J Biol Chem 2003; 278:36959–36965.

147. Li QJ, Yang SH, Maeda Y, Sladek FM, Sharrocks AD, Martins-Green M. MAP kinase phosphorylation-dependent activation of Elk-1 leads to activation of the co-activator p300. EMBO J 2003; 22:281–291.

148. Okazaki T, Sakon S, Sasazuki T, et al. Phosphorylation of serine 276 is essential for p65 NF-kappaB subunit-dependent cellular responses. Biochem Biophys Res Commun 2003; 300:807–812.

149. Ashburner BP, Westerheide SD, Baldwin AS Jr. The p65 (RelA) subunit of NF-kappaB interacts with the histone deacetylase (HDAC) corepressors HDAC1 and HDAC2 to negatively regulate gene expression. Mol Cell Biol 2001; 21: 7065–7077.

150. Chen LF, Mu Y, Greene WC. Acetylation of RelA at discrete sites regulates distinct nuclear functions of NF-kappaB. EMBO J 2002; 21:6539–6548.

151. Davies SP, Reddy H, Caivano M, Cohen P. Specificity and mechanism of action of some commonly used protein kinase inhibitors. Biochem J 2000; 351:95–105.

152. Mayr BM, Canettieri G, Montminy MR. Distinct effects of cAMP and mitogenic signals on CREB-binding protein recruitment impart specificity to target gene activation via CREB. Proc Natl Acad Sci USA 2001; 98:10936–10941.

153. Deisseroth K, Tsien RW. Dynamic multiphosphorylation passwords for activity-dependent gene expression. Neuron 2002; 34:179–182.

154. Murphy LO, Smith S, Chen RH, Fingar DC, Blenis J. Molecular interpretation of ERK signal duration by immediate early gene products. Nat Cell Biol 2002; 4:556–564.

155. Zhong H, SuYang H, Erdjument-Bromage H, Tempst P, Ghosh S. The transcriptional activity of NF-kappaB is regulated by the IkappaB-associated PKAc subunit through a cyclic AMP-independent mechanism. Cell 1997; 89: 413–424.

156. Anrather J, Csizmadia V, Soares MP, Winkler H. Regulation of NF-kappaB RelA phosphorylation and transcriptional activity by p21(ras) and protein kinase Czeta in primary endothelial cells. J Biol Chem 1999; 274:13594–13603.

157. Bradbury EM. Chromatin structure and dynamics: state-of-the-art. Mol Cell 2002; 10:13–19.

158. Strelkov IS, Davie JR. Ser-10 phosphorylation of histone H3 and immediate early gene expression in oncogene-transformed mouse fibroblasts. Cancer Res 2002; 62:75–78.

159. Freeman BC, Yamamoto KR. Disassembly of transcriptional regulatory complexes by molecular chaperones. Science 2002; 296:2232–2235.

160. Freeman BC, Yamamoto KR. Continuous recycling: a mechanism for modulatory signal transduction. Trends Biochem Sci 2001; 26:285–290.

161. Morimoto R. Dynamic remodeling of transcription complexes by molecular chaperones. Cell 2002; 110:281.

162. Hager GL, Elbi C, Becker M. Protein dynamics in the nuclear compartment. Curr Opin Genet Dev 2002; 12:137–141.

163. Dufresne SD, Bjorbaek C, El-Haschimi K, et al. Altered extracellular signal-regulated kinase signaling and glycogen metabolism in skeletal muscle from p90 ribosomal S6 kinase 2 knockout mice. Mol Cell Biol 2001; 21:81–87.

164. Wiggin GR, Soloaga A, Foster JM, Murray-Tait V, Cohen P, Arthur JS. MSK1 and MSK2 are required for the

mitogen- and stress-induced phosphorylation of CREB and ATF1 in fibroblasts. Mol Cell Biol 2002; 22:2871–2881.

165. Strahl BD, Allis CD. The language of covalent histone modifications. Nature 2000; 403:41–45.

166. Jenuwein T, Allis CD. Translating the histone code. Science 2001; 293:1074–1080.

167. Turner BM. Cellular memory and the histone code. Cell 2002; 111:285–291.

168. Rosenfeld MG, Glass CK. Coregulator codes of transcriptional regulation by nuclear receptors. J Biol Chem 2001; 276:36865–36868.

169. Gamble MJ, Freedman LP. A coactivator code for transcription. Trends Biochem Sci 2002; 27:165–167.

170. Lipford JR, Deshaies RJ. Diverse roles for ubiquitin-dependent proteolysis in transcriptional activation. Nat Cell Biol 2003; 5:845–850.

171. Tansey WP. Transcriptional activation: risky business. Genes Dev 2001; 15:1045–1050.

172. Wang Z, Frederick J, Garabedian MJ. Deciphering the phosphorylation "Code" of the glucocorticoid receptor in vivo. J Biol Chem 2002; 277:26573–26580.

173. Holmberg CI, Tran SE, Eriksson JE, Sistonen L. Multisite phosphorylation provides sophisticated regulation of transcription factors. Trends Biochem Sci 2002; 27:619–627.

174. Merienne K, Pannetier S, Harel-Bellan A, Sassone-Corsi P. Mitogen-regulated RSK2-CBP interaction controls their kinase and acetylase activities. Mol Cell Biol 2001; 21: 7089–7096.

175. Hazzalin CA, Mahadevan LC. MAPK-regulated transcription: a continuously variable gene switch? Nat Rev Mol Cell Biol 2002; 3:30–40.

176. Agalioti T, Chen G, Thanos D. Deciphering the transcriptional histone acetylation code for a human gene. Cell 2002; 111:381–392.

177. Cirillo LA, Lin FR, Cuesta I, Friedman D, Jarnik M, Zaret KS. Opening of compacted chromatin by early developmental transcription factors HNF3 (FoxA) and GATA-4. Mol Cell 2002; 9:279–289.

178. Schreiber SL, Bernstein BE. Signaling network model of chromatin. Cell 2002; 111:771–778.

179. Messina MJ. Legumes and soybeans: overview of their nutritional profiles and health effects. Am J Clin Nutr 1999; 70: 439S–450S.

180. Cassidy A. Potential risks and benefits of phytoestrogen-rich diets. Int J Vitam Nutr Res 2003; 73:120–126.

181. Hughes CL, Dhiman TR. Dietary compounds in relation to dietary diversity and human health. J Med Food 2002; 5: 51–68.

182. Surh YJ, Chun KS, Cha HH, et al. Molecular mechanisms underlying chemopreventive activities of anti-inflammatory phytochemicals: down-regulation of COX-2 and iNOS through suppression of NF-kappa B activation. Mutat Res 2001; 480–481:243–268.

183. De Bosscher K, Vanden Berghe W, Haegeman G. The interplay between the glucocorticoid receptor and nuclear factor-kappaB or activator protein-1: molecular mechanisms for gene repression. Endocr Rev 2003; 24:488–522.

184. Yang SH, Sharrocks AD, Whitmarsh AJ. Transcriptional regulation by the MAP kinase signaling cascades. Gene 2003; 320:3–21.

185. Nakshatri H, Bhat-Nakshatri P, Martin DA, Goulet RJ Jr, Sledge GW Jr. Constitutive activation of NF-kappaB during progression of breast cancer to hormone-independent growth. Mol Cell Biol 1997; 17:3629–3639.

186. Pratt MA, Bishop TE, White D, et al. Estrogen withdrawal-induced NF-kappaB activity and bcl-3 expression in breast cancer cells: roles in growth and hormone independence. Mol Cell Biol 2003; 23:6887–6900.

187. Garg AK, Hortobagyi GN, Aggarwal BB, Sahin AA, Buchholz TA. Nuclear factor-kappaB as a predictor of

treatment response in breast cancer. Curr Opin Oncol 2003; 15:405–411.

188. Biswas DK, Cruz AP, Gansberger E, Pardee AB. Epidermal growth factor-induced nuclear factor kappa B activation: a major pathway of cell-cycle progression in estrogen-receptor negative breast cancer cells. Proc Natl Acad Sci USA 2000; 97:8542–8547.

189. Biswas DK, Dai SC, Cruz A, Weiser B, Graner E, Pardee AB. The nuclear factor kappa B (NF-kappa B): a potential therapeutic target for estrogen receptor negative breast cancers. Proc Natl Acad Sci USA 2001; 98:10386–10391.

190. Biswas DK, Cruz A, Pettit N, Mutter GL, Pardee AB. A therapeutic target for hormone-independent estrogen receptor-positive breast cancers. Mol Med 2001; 7:59–67.

191. Gong L, Li Y, Nedeljkovic-Kurepa A, Sarkar FH. Inactivation of NF-kappaB by genistein is mediated via Akt signaling pathway in breast cancer cells. Oncogene 2003; 22: 4702–4709.

192. Romieu-Mourez R, Landesman-Bollag E, Seldin DC, Sonenshein GE. Protein kinase CK2 promotes aberrant activation of nuclear factor-kappaB, transformed phenotype, and survival of breast cancer cells. Cancer Res 2002; 62: 6770–6778.

193. Patel NM, Nozaki S, Shortle NH, et al. Paclitaxel sensitivity of breast cancer cells with constitutively active NF-kappaB is enhanced by IkappaBalpha super-repressor and parthenolide. Oncogene 2000; 19:4159–4169.

194. Sovak MA, Bellas RE, Kim DW, et al. Aberrant nuclear factor-kappaB/Rel expression and the pathogenesis of breast cancer. J Clin Invest 1997; 100:2952–2960.

195. Bhat-Nakshatri P, Sweeney CJ, Nakshatri H. Identification of signal transduction pathways involved in constitutive NF-kappaB activation in breast cancer cells. Oncogene 2002; 21:2066–2078.

196. Benassayag C, Perrot-Applanat M, Ferre F. Phytoestrogens as modulators of steroid action in target cells. J Chromatogr B Analyt Technol Biomed Life Sci 2002; 777:233–248.

197. Beck V, Unterrieder E, Krenn L, Kubelka W, Jungbauer A. Comparison of hormonal activity (estrogen, androgen and progestin) of standardized plant extracts for large scale use in hormone replacement therapy. J Steroid Biochem Mol Biol 2003; 84:259–268.

198. Palvimo JJ, Reinikainen P, Ikonen T, Kallio PJ, Moilanen A, Janne OA. Mutual transcriptional interference between RelA and androgen receptor. J Biol Chem 1996; 271:24151–24156.

199. Tian Y, Ke S, Denison MS, Rabson AB, Gallo MA. Ah receptor and NF-kappaB interactions, a potential mechanism for dioxin toxicity. J Biol Chem 1999; 274:510–515.

200. Tian Y, Rabson AB, Gallo MA. Ah receptor and NF-kappaB interactions: mechanisms and physiological implications. Chem Biol Interact 2002; 141:97–115.

201. Wissink S, van Heerde EC, Schmitz ML, et al. Distinct domains of the RelA NF-kappaB subunit are required for negative cross-talk and direct interaction with the glucocorticoid receptor. J Biol Chem 1997; 272:22278–22284.

202. Kalkhoven E, Wissink S, van der Saag PT, van der Burg B. Negative interaction between the RelA(p65) subunit of NF-kappaB and the progesterone receptor. J Biol Chem 1996; 271:6217–6224.

203. Delerive P, De Bosscher K, Besnard S, et al. Peroxisome proliferator-activated receptor alpha negatively regulates the vascular inflammatory gene response by negative cross-talk with transcription factors NF-kappaB and AP-1. J Biol Chem 1999; 274:32048–32054.

204. Jordan VC. The secrets of selective estrogen receptor modulation: cell-specific coregulation. Cancer Cell 2002; 1:215–217.

205. Levenson AS, Svoboda KM, Pease KM, et al. Gene expression profiles with activation of the estrogen receptor alpha-selective estrogen receptor modulator complex in breast cancer cells expressing wild-type estrogen receptor. Cancer Res 2002; 62:4419–4426.

206. Jordan VC, Schafer JM, Levenson AS, et al. Molecular classification of estrogens. Cancer Res 2001; 61:6619–6623.

207. Gustafsson JA. Therapeutic potential of selective estrogen receptor modulators. Curr Opin Chem Biol 1998; 2:508–511.

208. Dobrzycka K, Townson S, Oesterreich S. Estrogen receptor corepressors—a role in human breast cancer? Endoct. Relat. Cancer 2003; 10:517–536.

209. Fujita N, Jaye DL, Kajita M, Geigerman C, Moreno CS, Wade PA. MTA3, a Mi-2/NuRD complex subunit, regulates an invasive growth pathway in breast cancer. Cell 2003; 113:207–219.

210. Mazumdar A, Wang RA, Mishra SK, et al. Transcriptional repression of oestrogen receptor by metastasis-associated protein 1 corepressor. Nat Cell Biol 2001; 3:30–37.

211. Pike AC, Brzozowski AM, Hubbard RE. A structural biologist's view of the estrogen receptor. J Steroid Biochem Mol Biol 2000; 74:261–268.

212. An J, Tzagarakis-Foster C, Scharschmidt TC, Lomri N, Leitman DC. Estrogen receptor beta-selective transcriptional activity and recruitment of coregulators by phytoestrogens. J Biol Chem 2001; 276:17808–17814.

213. Kuiper GG, Lemmen JG, Carlsson B, et al. Interaction of estrogenic chemicals and phytoestrogens with estrogen receptor beta. Endocrinology 1998; 139:4252–4263.

214. Enmark E, Gustafsson JA. Newly discovered estrogen receptor. New therapeutic possibilities in postmenopausal symptoms, osteoporosis, cancer of the breast and prostate. Lakartidningen 1998; 95:1945–1949.

215. Rosselli M, Reinhart K, Imthurn B, Keller PJ, Dubey RK. Cellular and biochemical mechanisms by which environmental estrogens influence reproductive function. Hum Reprod Update 2000; 6:332–350.

216. Dang ZC, Audinot V, Papapoulos SE, Boutin JA, Lowik CW. Peroxisome proliferator-activated receptor gamma (PPAR-gamma) as a molecular target for the soy phytoestrogen genistein. J Biol Chem 2003; 278:962–967.

217. Suetsugi M, Su L, Karlsberg K, Yuan YC, Chen S. Flavone and isoflavone phytoestrogens are agonists of estrogen-related receptors. Mol Cancer Res 2003; 1:981–991.

218. Larner AC, Finbloom DS. Protein tyrosine phosphorylation as a mechanism which regulates cytokine activation of early response genes. Biochim Biophys Acta 1995; 1266: 278–287.

219. Schlessinger J. Cell signaling by receptor tyrosine kinases. Cell 2000; 103:211–225.

220. Huang WC, Chen JJ, Inoue H, Chen CC. Tyrosine phosphorylation of I-kappa B kinase alpha/beta by protein kinase C-dependent c-Src activation is involved in TNF-alpha-induced cyclooxygenase-2 expression. J Immunol 2003; 170: 4767–4775.

221. Huang WC, Chen JJ, Chen CC. c-Src-dependent tyrosine phosphorylation of IKKbeta is involved in tumor necrosis factor-alpha-induced intercellular adhesion molecule-1 expression. J Biol Chem 2003; 278:9944–9952.

222. Darnay BG, Aggarwal BB. Inhibition of protein tyrosine phosphatases causes phosphorylation of tyrosine-331 in the p60 TNF receptor and inactivates the receptor-associated kinase. FEBS Lett 1997; 410:361–367.

223. Chen LY, Zuraw BL, Zhao M, Liu FT, Huang S, Pan ZK. Involvement of protein tyrosine kinase in Toll-like receptor 4-mediated NF-kappa B activation in human peripheral blood monocytes. Am J Physiol Lung Cell Mol Physiol 2003; 284:L607–L613.

224. Hirota K, Murata M, Itoh T, Yodoi J, Fukuda K. Redox-sensitive transactivation of epidermal growth factor receptor by tumor necrosis factor confers the NF-kappa B activation. J Biol Chem 2001; 276:25953–25958.

225. Schiff R, Massarweh S, Shou J, Osborne CK. Breast cancer endocrine resistance: how growth factor signaling and estrogen receptor coregulators modulate response. Clin Cancer Res 2003; 9:447S–454S.

226. Hou MF, Lin SB, Yuan SS, et al. The clinical significance between activation of nuclear factor kappa B transcription factor and overexpression of HER-2/neu oncoprotein in Taiwanese patients with breast cancer. Clin Chim Acta 2003; 334:137–144.

227. Zwick E, Bange J, Ullrich A. Receptor tyrosine kinase signalling as a target for cancer intervention strategies. Endocr Relat Cancer 2001; 8:161–173.

228. Burgess AW, Cho HS, Eigenbrot C, et al. An open-and-shut case? Recent insights into the activation of EGF/ErbB receptors. Mol Cell 2003; 12:541–552.

229. Dancey JE, Freidlin B. Targeting epidermal growth factor receptor—are we missing the mark? Lancet 2003; 362:62–64.

230. Lee AV, Schiff R, Cui X, et al. New mechanisms of signal transduction inhibitor action: receptor tyrosine kinase down-regulation and blockade of signal transactivation. Clin Cancer Res 2003; 9:516S–523S.

231. Morin MJ. From oncogene to drug: development of small molecule tyrosine kinase inhibitors as anti-tumor and anti-angiogenic agents. Oncogene 2000; 19:6574–6583.

232. Devin A, Lin Y, Liu ZG. The role of the death-domain kinase RIP in tumour-necrosis-factor-induced activation of mitogen-activated protein kinases. EMBO Rep 2003; 4:623–627.

233. Chang F, Steelman LS, Lee JT, et al. Signal transduction mediated by the Ras/Raf/MEK/ERK pathway from cytokine receptors to transcription factors: potential targeting for therapeutic intervention. Leukemia 2003; 17:1263–1293.

234. Sebolt-Leopold JS. Development of anticancer drugs targeting the MAP kinase pathway. Oncogene 2000; 19:6594–6599.

235. Cotelle N. Role of flavonoids in oxidative stress. Curr Top Med Chem 2001; 1:569–590.

236. Hayakawa M, Miyashita H, Sakamoto I, et al. Evidence that reactive oxygen species do not mediate NF-kappaB activation. EMBO J 2003; 22:3356–3366.

237. Sakon S, Xue X, Takekawa M, et al. NF-kappaB inhibits TNF-induced accumulation of ROS that mediate prolonged MAPK activation and necrotic cell death. EMBO J 2003; 22:3898–3909.

238. Brash DE, Havre PA. New careers for antioxidants. Proc Natl Acad Sci USA 2002; 99:13969–13971.

239. Hendrich S. Bioavailability of isoflavones. J Chromatogr B Analyt Technol Biomed Life Sci 2002; 777:203–210.

240. Orzechowski A, Ostaszewski P, Jank M, Berwid SJ. Bioactive substances of plant origin in food—impact on genomics. Reprod Nutr Dev 2002; 42:461–477.

241. Zambrowicz BP, Sands AT. Knockouts model the 100 best-selling drugs—will they model the next 100? Nat Rev Drug Discov 2003; 2:38–51.

242. Heinrich M. Ethnobotany and natural products: the search for new molecules, new treatments of old diseases or a better understanding of indigenous cultures? Curr Top Med Chem 2003; 3:141–154.

243. O'Neill L. Inhibiting NF-kB. Trends Immunol 2001; 22:478.

244. Michael LF, Asahara H, Shulman AI, Kraus WL, Montminy M. The phosphorylation status of a cyclic AMP-responsive activator is modulated via a chromatin-dependent mechanism. Mol Cell Biol 2000; 20:1596–1603.

245. Goriely S, Demonte D, Nizet S, et al. Human IL-12(p35) gene activation involves selective remodeling of a single nucleosome within a region of the promoter containing critical Sp1-binding sites. Blood 2003; 101:4894–4902.

5

Nuclear Factor-κB as Target for Chemoprevention

BHARAT B. AGGARWAL

Cytokine Research Section, Department
of Bioimmunotherapy, The University of Texas
M. D. Anderson Cancer Center, Houston,
Texas, U.S.A.

ABBREVIATIONS

NF-κB, nuclear factor kappa B; IκB, inhibitor of NF-κB; CAPE, caffeic acid phenethyl ester; PBIT; S,S'-l,4-Phenylene-bis (1,2-ethanediyl) bis-isothiourea; PDTC, pyrrolidine dithocarbamate.

ABSTRACT

The process of tumorigenesis requires cellular transformation, hyperproliferation, invasion, angiogenesis, and metastasis. Several genes that mediate these processes are regulated by the transcription factor NF-κB. The latter is activated by various carcinogens, inflammatory agents, and tumor promoters. The NF-κB, a transcription factor, is present normally in the cytoplasm as an inactive heterotrimer consisting of p50, p65, and IκBα subunits. When activated, NF-B migrates to the nucleus translocates to the as a p50-p65 heterodimer. This factor regulates the expression of various genes that control apoptosis, viral replication, tumorigenesis, various autoimmune diseases, and inflammation. The NF-κB has been linked to the development of carcinogenesis for several reasons. First, various carcinogens and tumor promoters have been shown to activate NF-κB. Second, activation of NF-κB has been shown to block apoptosis and promote proliferation. Third, the tumor microenvironment can induce NF-κB activation. Fourth, constitutive expression of NF-κB is frequently found in tumor cells. Fifth, NF-κB activation induces resistance to chemotherapeutic agents. Sixth, several genes involved in tumor initiation, promotion, and metastasis are regulated by NF-κB. Seventh, various chemopreventive agents have been found to downregulate the NF-κB activation. All these observations suggest that NF-κB could mediate tumorigenesis and thus can be used as a target for chemoprevention and for the treatment of cancer. Agents, which suppress NF-κB activation, can suppress the expression of genes involved in carcinogenesis and tumorigenesis in vivo.

I. CARCINOGENESIS/TUMORIGENESIS

The process of tumorigenesis is a process that requires cellular transformation, hyperproliferation, invasion, angiogenesis, and metastasis. This process is activated by various carcinogens (such as cigarette smoke), inflammatory agents (such as TNF and H_2O_2), and tumor promoters (such as phorbol ester and okadaic acid) (1). Although initially identified as an anticancer agent (2), TNF has now been shown to be involved in cellular transformation (3), tumor promotion (4), and induction of metastasis (5–7). In agreement with these observations, mice deficient in TNF have been shown to be resistant to skin carcinogenesis (8). For several tumors,

TNF has been shown to be a growth factor (9,10). Like phorbol ester, TNF mediates these effects in part through activation of a protein kinase C pathway (11). Similar to TNF, other inflammatory cytokines have also been implicated in tumorigenesis (12,13). Thus, agents that can suppress the expression of TNF and other inflammatory agents have chemopreventive potential (14,15). Most carcinogens, inflammatory agents, and tumor promoters including cigarette smoke, phorbol ester, okadaic acid, H_2O_2, and TNF, have been shown to activate the transcription factor NF-κB.

II. CIGARETTE SMOKE AND CANCER

Cigarette smoke(CS) is a major cause of cancers of the lung, larynx, oral cavity and pharynx, esophagus, pancreas, kidney, and bladder (16). Worldwide, one in seven or 15% (1.1 million new cases per year) of all cancer cases are attributable to CS, 25% in men and 4% in women. Recent estimates indicate that CS causes approximately 80–90% of lung cancer in the United States (17). Smoking during pregnancy and passive exposure to CS may increase the risk of cancer for children and adults (18–20). These estimates do not include the disease resulting from smokeless tobacco (taken orally or as snuff), which is a substantial cause of cancer mortality, particularly on the Indian subcontinent (21).

Tobacco smoke is a complex mixture containing at least 40 different carcinogens, which mediate tumor initiation and promotion. These carcinogens include nitrosamine, polycyclic aromatic hydrocarbons (PAH), aromatic amines, unsaturated aldehydes (e.g., crotonaldehyde), and some phenolic compounds (acrolein). The most potent carcinogenic agent contained in CS is the nitrosamine 4-(methylnitrosoamino) - l-(3-pyridyl) -l-butanone (NNK); formed by nitrosation of nicotine, it is thought to be an important etiological factor in tobacco–smoke related human cancers (22). The NNK is a site-specific carcinogen in that, irrespective of the route of administration, NNK has remarkable specificity for the lung (23). Because side-stream smoke often contains higher amounts of NNK than mainstream smoke, passive exposure to CS has been suggested to be quite harmful (22). An enzyme 11 beta-hydroxysteroid dehydrogenase 1(11 beta-HSD1), which is involved in metabolism of endogenous steroids, is also responsible for the metabolism of NNK. Thus inhibition of 11 beta-HSD1 can increase the circulating levels of NNK by impairing its metabolism. Ethanol has been shown to be a potent inhibitor of 11 beta-HSD (24) and thus may increase the risk of lung cancer for active or passive smokers. An alcohol consumption and cigarette smoking have also been shown to increases the frequency of p53, a tumor suppressor gene, mutation in lung cancer (25).

Cigarette smoke has been shown to induce aryl hydrocarbon hydroxylase (AHH) activity, an activator of respiratory tract carcinogens of the PAH (e.g., benzo[a] pyrene) group (26), in human pulmonary macrophages (27) and in patients with smoking-associated malignant cancers (28). It has been postulated that individuals with high activity of oxidative enzymes (cytochrome P-450 enzymes) or a low activity of detoxifying enzymes (e.g., glutathione s-transferase and epoxide hydroxylase) may be at increased risk for cancer caused by CS (29). Low intake of dietary constituents with antioxidant properties such as β carotene, vitamin C, and vitamin E further increases the cancer risk in smokers (30).

Lung tumors from nonsmokers exhibit elevated NAD(P) H:(quinone-acceptor) oxidoreductase (QAO) activity compared to normal tissue, but tumors from smokers show increases in tumor QAO (31). This could influence the response of these tumors to quinone drugs (commonly used to treat cancer) or toxic agents that are metabolized by QAO. Quinone anticancer drugs are activated to alkylating species by reduction to hydroquinone. Metabolism by QAO is responsible for the formation of alkylating species from doxorubicin (32) and other cytotoxic drugs (33).

Another possible mechanism by which CS can cause cancer involves the effects of PAH on the p53 gene. For instance, exposure of cells to benzo(a)pyrene adducts can induce the same mutation in p53 as is found in 60% of all lung cancers (34). Also exposure of cells to PAH and its metabolites results in a rapid accumulation of the p53 gene product (35,36) through activation of a transcription factor, NF-κB (37).

III. EFFECT OF CIGARETTE SMOKE ON PULMONARY INFLAMMATION

Experimental epidemiological and clinical evidence indicates that CS is a primary risk factor for chronic obstructive pulmonary disease (COPD), which includes chronic bronchitis and emphysema. These two conditions result from obstruction of airflow and usually coexist. An increased proteolytic activity in the lung due to an imbalance between proteases, especially elastase and alfa1-protease inhibitor (alfa-I-PI, an antielastase), has been suggested as a primary cause for COPD caused by CS. This occurs for three reasons. First, CS causes the generation of chemotactic factors (such as chemokines) (38), which recruit inflammatory cells (such as neutrophils and macrophages) to the lung, and these cells release proteolytic enzymes. Second, free radicals present in CS can either inactivate alfa-I-PI by oxidation of an active site methionyl residue present in the protein sequence or damage macromolecules to make them more susceptible to proteolysis. Third, components in CS can suppress elastin synthesis by inhibiting the cross-linking enzyme lysyl oxidase. Thus neutrophil recruitment, inactivation of protease inhibitors, and depressed tissue repair are considered responsible for the pathogenesis of CS-induced emphysema, although, only one in six smokers develop extensive COPD.

The inhalation of CS also results in inflammation of the pulmonary epithelia. Reactive oxygen intermediates (ROIs) are some of the most important effector molecules of acute inflammation. The inflammatory cell response to CS has been studied extensively either in cells harvested by bronchoalveolar lavage from cigarette smokers or smoke-exposed animals or in macrophages exposed to CS in vitro. Alveolar macrophages lavaged from smokers have increased oxidative metabolism compared to those in nonsmokers, and this leads to increased apoptosis of fibroblasts, which could be prevented by oxidant scavenging agents. Thus oxidants generated by alveolar macrophages from smokers may facilitate tissue destruction (39).

IV. OXIDATIVE DAMAGE BY CIGARETTE SMOKE

Cigarette smoke has been implicated as major risk factor in COPD such as chronic bronchitis and emphysema, in chemical carcinogenesis, and in atherosclerotic arterial diseases. The mechanisms of the adverse biological effects of CS appear, in part, to include oxidative damage to essential biological consti- tuents. The CS increases the number of phagocytes in the blood and lungs (40), decreases plasma levels of high-density lipopro- teins (HDL) (41), and induces lipid peroxidation of LDL (42). Several plasma proteins have been shown to undergo modifica- tion by exposure to CS (43,44). In CS-bubbled buffers, H_2O_2 and hydroxyl radical were generated from aqueous extracts of tar (45,46). A superoxide radical was an intermediate in these reactions. Superoxide formed from CS impairs active oxygen generation from neutrophils.

V. COMPOSITION OF CIGARETTE SMOKE

Cigarette smoke is a complex mixture consisting of tarry particles of respirable size suspended in a mixture of organic and inorganic gases and containing more than 4000 chemical compounds. Inhaled mainstream, exhaled mainstream, and sidestream CS differ in composition. The CS contains two classes of free radicals, one in the gas phase and another in tar. The gas phase radicals consist of inorganic radicals (e.g., nitric oxide, NO) as well as organic radicals such as carbon- and oxygen-centered radicals. Nitric oxide is slowly oxidized to NO_2. It is estimated that there are approximately 10^{17} organic radicals per puff in gas phase smoke (Ref.46 and refer- ence therein). Gas phase smoke is unstable and inactivates alfa-I-PI. In contrast, tar radicals in the particulate phase are stable indefinitely and contain as many as 10^{18} free radicals per gram, the major ones being quinone–hydroquinone complex. This complex is an active redox system capable of reducing molecular oxygen to produce superoxide, eventually leading to H_2O_2 and OH radicals. Tar also chelates metals, such as iron, that catalyze the decomposition of H_2O_2. An aqueous suspension of tar produces hydroxyl radicals and has been shown to cleave DNA. Many smokers have switched from high- to low-tar cigarettes. Though low tar cigarettes may expose the lungs to lower levels of carcinogens, they produce a higher burden of oxidants. Nicotine is the most important smoke component present in the blood of smokers, and it has a half-life of 2 hr. Nicotine affects the respiratory, cardiovascular, central nervous, and the endocrine systems. Another significant component of CS is Cd compounds, which have a long half-life, accumulate in the lungs, and induce acute inflammatory reactions in the lung and increased lung epithelial permeability.

VI. WHAT IS NF-κB?

The NF-κB represents a group of five proteins namely c-Rel, RelA (p65), Rel B, NF-κBl (p50 and pl05), and NF-κB2 (p52) (16). The NF-κB proteins are regulated by inhibitors of the IκB family, which includes IκBα, IκBβ, IκBε, IκBγ, Bcl-3, pl00, and pl05 (47). In an inactive state, NF-κB is present in the cytoplasm as a heterotrimer consisting of p50, p65, and IκBα subunits. In response to an activation signal, the IκBα subunit is phosphorylated at serine residues 32 and 36, ubiquitinated at lysine residues 21 and 22 and degraded through the proteosomal pathway, thus exposing the nuclear localization signals on the p50-p65 heterodimer. The p65 is then phosphorylated, leading to nuclear translocation and binding to a specific sequence in DNA, which in turn results in gene transcription. The phosphorylation of κBα is catalyzed by the IKK. The IKK consists of three subunits IKK-α, IKK-β, and IKK-γ (also called NEMO) (for references see Ref. 48). Gene deletion studies have indicated that IKK-β is essential for NF-kB activation by most agents (49). The kinase that induces the phosphorylation of p65 is controversial, but

IKK-β, protein kinase C, and protein kinase A have been implicated (17–19).

VII. RELEVANCE OF NF-κB TO CIGARETTE SMOKING

There are several reasons to believe NF-κB is a good target by which to examine CS-induced lung cancer development and its chemoprevention. First, benzo[a]pyrene, a component of CS, has recently been shown to activate NF-κB in lung adenocarcinoma cells (37) and in vascular smooth muscle cells (50). Second, CS is also a potent source of ROIs (44–46), which are required for NF-κB activation (47). Our laboratory and others have shown that antioxidants and overexpression of cells with antioxidant enzymes such as Mn superoxide dismutase or with γ-glutamylcysteinyl synthase (51–53) block NF-κB activation. Third, NF-κB activation has been implicated in chemical carcinogenesis and tumorigenesis (54,55). Fourth, CS has been shown to induce NF-κB-regulated chemokine genes in bronchial epithelium (Ref. 38 and references therein). Lastly our laboratory and others have shown that most chemopreventive agents suppress NF-κB activation (56–60).

VIII. WHY NF-κB IS IMPORTANT FOR CANCER?

The NF-κB has been shown to regulate the expression of a number of genes whose products are involved in tumorigenesis (20,21). These include antiapoptosis genes [e.g., cIAP, suvivin, TRAF, bcl-2, and bcl-xl) COX2; MMP-9; genes encoding adhesion molecules, chemokines, inflammatory cytokines and iNOS; and cell cycle regulatory genes (e.g., cyclin Dl) (22).] Thus, agents that can suppress NF-κB activation have the potential to suppress carcinogenesis and have therapeutic potential (21,23). The therapeutic role of phytochemicals in prevention and treatment of cancer has been indicated (24–26). Thus, plant-derived phytochemicals that could suppress NF-κB activation by various carcinogens have been shown (Table 1).

Table 1 Chemopreventive Agents That Block NF-κB

Chemopreventive agent	Source	Chemical name
Curcumin	*Curcuma longa*	(E,E)-1,7-Bis(4-hydroxy-3-methoxy-phenyl)-1,6-heptadiene-3,5-dione; Diferuloylmethane
Retinoids	Vitamin A	N-(4-hydroxyphenyl) retinamide; All-trans retinoc acid
Capsaicin	*Capsicum* sp.	N-(4-hydroxy-3-methoxybenzyl)-8-methylnon-trans-6-enamide
Vitamin E	Plants	α-tocopherol, ; 5,7,8-trimethyltocol
Quercetin	*Rhododendron cinnabarinum*	3,3′,4′,5,7-pentahydroxyflavone
Dihydroxy vitamin D3	Fish liver oils	1α,25-dihydroxycholecalciferol
Resveratrol	*Polygonum cuspidatum*	Trans-3,5,4′-trihydroxystilbene
Silymarin	*Silybum marianum*.	Silybin
Lapachone	Bignoniaceous plants e.g. Lapacha tree	β-Lapachone
Sulindac	Synthetic	(Z)—Fluoro-2-methyl-1-[[4-(methyl-sulfinyl) phenyl]methylene]-1H-indene-3-acetic acid
Celecoxib	Synthetic	1,2-diarylpyrrole
Tea polyphenols	Green or Black Tea	(−) Epigallocatechin 3-gallate; Theaflavin
Sulforaphane	Cruciferous vegitables eg. Broccoli	1-isothiocyanato-(4R)-(methylsulfinyl) butane
Asprin	Synthetic	2-(Acetyloxy)benzoic acid
Caffeic acid phenethyl ester (CAPE)	Honeybee hives	3,4-dihydroxycinnamic acid phenethyl ester

(Continued)

Table 1 Chemopreventive Agents That Block NF-κB (*Continued*)

Chemopreventive agent	Source	Chemical name
S,S′-1,4-Phenylene-bis (1,2-ethanediyl) bis-isothiourea (PBIT)	Synthetic	S,S′-1,4-phenylene-bis(1,2-ethanediyl) bis-isothiourea (PBIT)
Pyrrolidine dithocarbamate (PDTC)	Synthetic	Pyrrolidine dithocarbamate
Anethole	*Pimpinella anisum*	p-propenylanisole
Oleandrin	*Nerium oleander*	(3β,5β,16β)-16-(Acetyloxy)-3-[(2,6-di-deoxy-3-Omethyl-αarabino-hexopyranosyl)oxyl]-14-hydroxycard-20(22)-enolide
Wortmannin	*Penicillium wormanni*	[1S-(1α,6bα, 9aβ, 11α, 11bβ)]-11-(Acetyloxy)-1,6b,7,8,9a,10,11,11b-octahydro-1-(methoxymethyl)-9a, 11b-dimethyl-3H-furo[4,3,2-de]indeno[4,5-h]-2-benzopyran-3,6,9-trione
Emodin	*Aloe barbandensis, polygonum cuspidatum*	1,3,8-Trihydroxy-6-methyl-9,10-anthracenedione
Ursolic acid	*Rosemarinus officinalis, Eriobotrya japonica, Calluna vulgaris, Ocimum sanctum,* and *Eugenia jumbolana*	(3β-hydroxy-urs-12-en-28-oic acid)
Betulinic acid	*Tryphyllum peltaum, Ancistrocladus heyneaus, Zizyphus joazeiro, Diospyoros leucomelas, Tetracera boliviana,* and *Syzygium formosanum)*	3β-Hydroxy-20(29)-lupaene-28-oic acid

IX. CHEMOPREVENTIVE AGENTS INHIBIT NF-κB ACTIVATION

Several agents that suppress carcinogenesis have been shown to block NF-κB activation. These include curcumin, green tea polyphenols, silymarin, and resveratrol (Fig. 1). Curcumin is a polyphenol (diferuloylmethane) derived from the roots of *Curcuma longa*, and it inhibits both tumor initiation induced by BP and 7,12 dimethylbenz(a)anthracene and phorbol ester-induced tumor promotion (61–63). Both B[a]P and phorbol esters are potent activators of NF-κB. Curcumin has also been shown to suppress the expression of several genes involved in carcinogenesis including COX 2, lipooxygenases, and iNOS (64–67), also known to require NF-κB activation. Additionally, our laboratory has shown that curcumin blocks the TNF-induced expression of ICAM-1, VCAM-1, and ELAM-1, all

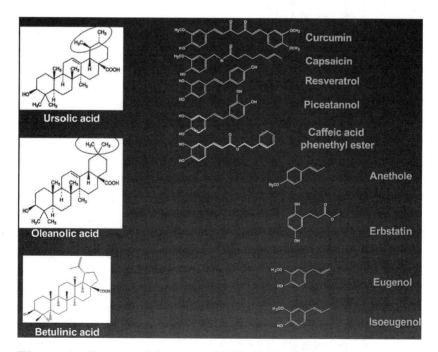

Figure 1 Structural features of NF-κB blockers.

NF-κB-regulated genes in endothelial cells, and needed for tumor metastasis (68). Our laboratory has also shown that curcumin suppresses the NF-κB activation induced by various tumor promoters in different cell types (56). Similarly, Silymarin, a flavonoid isolated from milk thistle (artichoks), has been demonstrated to suppress carcinogenesis (69), and we have shown that this compound also inhibits NF-κB activation through blocking the phosphorylation and degradation of IκB (59). Resveratrol, derived primarily from grapes and peanuts, exhibits chemopreventive activity by inhibiting cellular events associated with tumor initiation, promotion, and progression (70). Our laboratory and others showed that resveratrol also blocks NF-κB activation and NF-κB-regulated expression of monocyte chemoattractant protein (MCP)-l (60,71). Thus, several of these examples suggest that suppression of NF-κB activation correlates with chemoprevention.

The epidemiological evidences also indicate that certain cancers (e.g., breast, prostate, colon, and lung) are more prevalent in the developed countries than in the developing countries. It is most likely because of differences in dietary constituents (16,17). We propose that there are constituents of the every-day diet that regulate the activity of certain transcription factors such as NF-κB that plays a critical role in carcinogenesis.

X. CONCLUSION

Evidence presented above suggests that activation of NF-κB can lead to tumor cell proliferation, invasion, angiogenesis, and metastasis. Thus suppression of NF-κB in cancer cells may provide an additional target for prevention of cancer. The NF-κB blockers can also be considered for the therapy of cancer, perhaps in combination with chemotherapeutic agents or gamma irradiation.

ACKNOWLEDGMENTS

This research was supported by grants from the Clayton Foundation, National Institute of Health (1P01 CA91844-1) and by the Specialized Program Of Research Excellence grant from the National Institues of Health (of BBA).

REFERENCES

1. Kelloff G IV. Perspectives on cancer–chemoprevention research and drug development. Adv Cancer Res 2000; 78:199–334.

2. Aggarwal BB. Tumour necrosis factors receptor associated signaling molecules and their role in activation of apoptosis, JNK and NF-kappaB. Ann Rheum Dis 2000; 59(suppl 1):i6–16.

3. Komori A, Yatsunami J, Suganuma M, Okabe S, Abe S, Sakai A, Sasaki K, Fujiki H. Tumor necrosis factor acts as a tumor promoter in BALB/3T3 cell transformation. Cancer Res 1993; 53:1982–1985.

4. Suganuma M, Okabe S, Marino MW, Sakai A, Sueoka E, Fujiki H. Essential role of tumor necrosis factor alpha (TNF-alpha) in tumor promotion as revealed by TNF-alpha-deficient mice. Cancer Res 1999; 59:4516–4518.

5. Orosz P, Echtenacher B, Falk W, Ruschoff J, Weber D, Mannel DN. Enhancement of experimental metastasis by tumor necrosis factor. J Exp Med 1993; 177:1391–1398.

6. Hafner M, Orosz P, Kruger A, Mannel DN. TNF promotes metastasis by impairing natural killer cell activity. Int J Cancer 1996; 66:388–392.

7. Orosz P, Kruger A, Hubbe M, Ruschoff J, Von Hoegen P, Mannel DN. Promotion of experimental liver metastasis by tumor necrosis factor. Int J Cancer 1995; 60:867–871.

8. Moore RJ, Owens DM, Stamp G, Arnott C, Burke F, East N, Holdsworth H, Turner L, Rollins B, Pasparakis M, Kollias G, Balkwill F. Mice deficient in tumor necrosis factor-alpha are resistant to skin carcinogenesis. Nat Med 1999; 5:828–831.

9. Aggarwal BB, Schwarz L, Hogan ME, Rando RF. Triple helix-forming oligodeoxyribonucleotides targeted to the human tumor necrosis factor (TNF) gene inhibit TNF production and block the TNF-dependent growth of human glioblastoma tumor cells. Cancer Res 1996; 56:5156–5164.

10. Giri DK, Aggarwal BB. Constitutive activation of NF-kappaB causes resistance to apoptosis in human cutaneous T cell lymphoma HuT-78 cells. Autocrine role of tumor necrosis factor and reactive oxygen intermediates. J Biol Chem 1998; 273: 14008–14014.

11. Arnott CH, Scott KA, Moore RJ, Hewer A, Phillips DH, Parker P, Balkwill FR, Owens DM. Tumour necrosis factor-alpha mediates tumour promotion via a PKC alpha- and AP-1-dependent pathway. Oncogene 2002; 22:4728–4738.

12. Suganuma M, Okabe S, Kurusu M, Iida N, Ohshima S, Saeki Y, Kishimoto T, Fujiki H. Discrete roles of cytokines, TNF-alpha, IL-1, IL-6 in tumor promotion and cell transformation. Int J Oncol 2002; 20:131–136.

13. Hehlgans T, Stoelcker B, Stopfer P, Muller P, Cernaianu G, Guba M, Steinbauer M, Nedospasov SA, Pfeffer K, Mannel DN. Lymphotoxin–beta receptor immune interaction promotes tumor growth by inducing angiogenesis. Cancer Res 2002; 62: 4034–4040.

14. Suganuma M, Okabe S, Sueoka E, Iida N, Komori A, Kim SJ, Fujiki H. A new process of cancer prevention mediated through inhibition of tumor necrosis factor alpha expression. Cancer Res 1996; 56:3711–3715.

15. Sueoka N, Sueoka E, Okabe S, Fujiki H. Anti-cancer effects of morphine through inhibition of tumour necrosis factor-alpha release and mRNA expression. Carcinogenesis 1996; 17: 2337–2341.

16. Parkin DM. Cancer in developing countries. Cancer Surv 1994; 19–20:519–561.

17. Wingo PA, Ries LA, Giovino GA, Miller DS, Rosenberg HM, Shopland DR, Thun MJ, Edwards BK. Annual report to the nation on the status of cancer, 1973–1996, with a special section on lung cancer and tobacco smoking. J Natl Cancer Inst 1999; 1(8):675–690.

18. Stjernfeldt M, Ludvigsson J, Berglund K, Lindsten J. Maternal smoking during pregnancy and the risk of childhood cancer. Lancet 1986; 2(8508):687–688.

19. Sandier DP, Wilcox AJ, Everson RB. Cumulative effects of lifetime passive smoking on cancer risk. Lancet 1985; 1(8424): 312–315.

20. Saracci R, Riboli E. Passive smoking and lung cancer: current evidence and ongoing studies at the International Agency for Research on Cancer. Mutat Res 1989; 222(2):117–127.

21. Bhisey RA, Ramchandani AG, D'Souza AV, Borges AM, Notani PN. Long-term carcinogenicity of pan masala in Swiss mice. Int J Cancer 1999; 83(5):679–684.

22. Hecht SS. Tobacco smoke carcinogens and lung cancer. J Natl Cancer Inst 1999; 91(14):1194–1210.

23. Castonguay A, Pepin P, Stoner GD. Lung tumorigenicity of NNK given orally to A/J mice: its application to chemopreventive efficacy studies. Exp Lung Res 1991; 17(2):485–499.

24. Riddle MC, McDaniel PA. Acute reduction of renal 11 beta-hydroxysteroid dehydrogenase activity by several antinatriuretic stimuli. Metab Clin Exp 1993; 42(10):1370–1374.

25. Ahrendt SA, Chow JT, Yang SC, Wu L, Zhang MJ, Jen J, Sidransky D. Alcohol consumption and cigarette smoking increase the frequency of p53 mutations in non-small cell lung cancer. Cancer Res 2000; 60(12):3155–3159.

26. Kouri RE, McKinney CE, Slomiany DJ, Snodgrass DR, Wray NP, McLemore TL. Positive correlation between high aryl hydrocarbon hydroxylase activity and primary lung cancer as analyzed in cryopreserved lymphocytes. Cancer Res 1982; 42(12): 5030–5037.

27. McLemore TL, Martin RR, Toppell KL, Busbee DL, Cantrell ET. Comparison of aryl hydrocarbon hydroxylase induction in cultured blood lymphocytes and pulmonary macrophages. J Clin Invest 1977; 60(5):1017–1024.

28. Korsgaard R, Trell E, Simonsson BG, Stiksa G, Janzon L, Hood B, Oldbring J. Aryl hydrocarbon hydroxylase induction levels in patients with malignant tumors associated with smoking. J Cancer Res Clin Oncol 1984; 108(3):286–289.

29. Mace K, Bowman ED, Vautravers P, Shields PG, Harris CC, Pfeifer AM. Characterisation of xenobiotic-metabolising enzyme

expression in human bronchial mucosa and peripheral lung tissues. Europ J Cancer 1998; 34(6):914–920.

30. Cade JE, Margetts BM. Relationship between diet and smoking—is the diet of smokers different? J Epidemiol Community Health 1991; 45(4):270–27.

31. Schlager JJ, Powis G. Cytosolic NAD(P)H:(quinone-acceptor) oxidoreductase in human normal and tumor tissue: effects of cigarette smoking and alcohol. Intern J Cancer 1990; 45:403–409.

32. Sinha BK, Katki AG, Batist G, Cowan KH, Myers CE. Adriamycin-stimulated hydroxyl radical formation in human breast tumor cells. Biochem Pharmacol 1987; 36(6):793–796.

33. Talcott RE, Levin VA. Glutathione-dependent denitrosation of *N,N'*-bis(2-chloroethyl)*N*-nitrosourea (BCNU): nitrite release catalyzed by mouse liver cytosol in vitro. Drug Metab Disposition 1983; 11(2):175–176.

34. Denissenko MF, Pao A, Tang M. Pfeifer GP. Preferential formation of benzo[a]pyrene adducts at lung cancer mutational hotspots in P53. Science 1996; 274(5286):430–432.

35. Hellin AC, Calmant P, Gielen J, Bours V, Merville MP. Nuclear factor—kappaB-dependent regulation of p53 gene expression induced by daunomycin genotoxic drug. Oncogene 1998; 16(9):1187–1195.

36. Venkatachalam S, Denissenko M, Wani AA. Modulation of (+/−)-anti-BPDE mediated p53 accumulation by inhibitors of protein kinase C and poly(ADP-ribose) polymerase. Oncogene 1997; 14(7):801–809.

37. Pei XH, Nakanishi Y, Takayama K, Bai F, Hara N. Benzo[a]-pyrene activates the human p53 gene through induction of nuclear factor kappaB activity. J Biol Chem 1999; 274(49): 35240–35246.

38. Nishikawa M, Kakemizu N, Ito T, Kudo M, Kaneko T, Suzuki M, Udaka N, Ikeda H, Okubo T. Superoxide mediates cigarette smoke-induced infiltration of neutrophils into the airways through nuclear factor-kappaB activation and IL-8 mRNA expression in guinea pigs in vivo. Am J Respir Cell Mol Biol 1999; 20(2):189–198.

39. Repine JE, Bast A, Lankhorst L. Oxidative stress in chronic obstructive pulmonary disease. Oxidative Stress Study Group [Rev] [296 refs]. Amer J Respir Crit Care Med 1997; 156(2 Pt l): 341–357.

40. Cantin A, Crystal RG. Oxidants, antioxidants and the pathogenesis of emphysema [Rev] [77 Refs.]. Euro J Respir Dis Suppl 1985; 139:7–17.

41. Nesje LA, Mjos OD. Plasma HDL cholesterol and the subclasses HDL2 and HDL3 in smokers and non-smokers. Artery 1985; 13(1):7–18.

42. Frei B, Forte TM, Ames BN, Cross CE. Gas phase oxidants of cigarette smoke lipid induce peroxidation and changes in lipoprotein properties in human blood plasma. Protective effects of ascorbic acid. Biochem J 1991; 277(Pt 1):133–138.

43. Reznick AZ, Cross CE, Hu ML, Suzuki YJ, Khwaja S, Safadi A, Motchnik PA, Packer L, Halliwell B. Modification of plasma proteins by cigarette smoke as measured by protein carbonyl formation. Biochem J 1992; 286(Pt 2):607–611.

44. Cross CE, O'Neill CA, Reznick AZ, Hu ML, Marcocci L, Packer L, Frei B. Cigarette smoke oxidation of human plasma constituents. Ann NY Acad Sci 1993; 686:72–89 [Discussion 89–90].

45. Cosgrove JP, Borish ET, Church DF, Pryor WA. The metal-mediated formation of hydroxyl radical by aqueous extracts of cigarette tar. Biochem Biophys Res Comm 1985; 132:390–396.

46. Pryor WA. Cigarette smoke radicals and the role of free radicals in chemical carcinogenicity [Rev] [110 Refs.]. Environ Health Perspect 1997; 105(suppl 4):875–882.

47. Baeuerle PA, Baichwal VR. NF-kappa B as a frequent target for immunosuppressive and anti-inflammatory molecules [Rev] [105 Refs.]. Adv Immunol 1997; 65:111–137.

48. Karin M. The beginning of the end: IkappaB kinase (IKK) and NF-κB activation. J Biol Chem 1999; 274(39):27339–27342.

49. Pahl HL. Activators and target genes of Rel/NF-κB transcription factors. Oncogene 1999; 18:6853–6866.

50. Yan Z, Subbaramaiah K, Camilli T, Zhang F, Tanabe T, McCaffrey TA, Dannenberg AJ, Weksler BB. Benzo[a]pyrene

induces the transcription of cyclooxygenase-2 in vascular smooth muscle cells. Evidence for the involvement of extracellular signal-regulated kinase and NF-κB. J Biol Chem 2000; 275:4949–4955.

51. Manna S, Zhang HJ, Yan T, Oberley LW, Aggarwal BB. Overexpression of Mn-superoxide dismutase suppresses TNF-induced apoptosis and activation of nuclear transcription factor-κB and activated protein-1. J Biol Chem 1998; 273: 13245–13254.

52. Manna SK, Kuo MT, Aggarwal BB. Overexpression of γ-glutamylcysteine synthetase abolishes tumor necrosis factor-induced apoptosis and activation of nuclear transcription factor-kappa B and activator protein-1. Oncogene 1999; 18:4371–4382.

53. Schmidt KN, Amstad P, Cerutti P, Baeuerle PA. The roles of hydrogen peroxide and superoxide as messengers in the activation of transcription factor NF-kappa B. Chemistry Biol 1995; 2(1):13–22.

54. Sharma HW, Narayanan R. The NF-κB transcription factor in oncogenesis. Anticancer Res 1996; 16:589–596.

55. Waddick KG, Uckun FM. Innovative treatment programs against cancer: II.Nuclear factor-kappaB (NF-kappaB) as a molecular target. Biochem Pharmacol 1999; 57:9–17.

56. Singh S, Aggarwal BB. Activation of transcription factor NF-κB is suppressed by curcumin (Diferulolylmethane). J Biol Chem 1995; 270:24995–25000.

57. Natarajan K, Singh S, Burke TR Jr, Grunberger D, Aggarwal BB. Caffeic acid phenethyl ester (CAPE) is a potent and specific inhibitor of activation of nuclear transcription factor NF-κB. Proc Natl Acad Sci USA 1996; 93:9090–9095.

58. Kumar K, Dhawan S, Aggarwal BB. Emodin (3-methyl-1,6, 8-trihydroxyanthraquinone) inhibits the TNF-induced NF-κB activation, IκB degradation and expression cell surface adhesion protein in human vascular endothelial cells. Oncogene 1998; 17:913–918.

59. Manna SK, Mukhopadhyay A, Van NT, Aggarwal BB. Silymarin suppresses TNF-induced activation of nuclear tran-

scription factor-κB, c-Jun N-terminal kinase and apoptosis. J Immunol 1999; 163:6800–6809.

60. Manna SK, Mukhopadhyay A, Aggarwal BB. Resveratrol suppresses TNF-induced activation of nuclear transcription factors NF-κB, activator protein-1, and apoptosis: potential role of reactive oxygen intermediates and lipid peroxidation. J Immunol 2000; 164:6509–6519.

61. Huang MT, Wang ZY, Georgiadis CA, Laskin JD, Conney AH. Inhibitory effects of curcumin on tumor initiation by benzo[a]-pyrene and 7,12-dimethylbenz[a]anthracene. Carcinogenesis 1992; 13:2183–2186.

62. Huang MT, Smart RC, Wong C-Q, Conney AH. Inhibitory effect of curcumin, chlorogenic acid, caffeic acid, and ferulic acid on tumor promotion in mouse skin by 12-O-tetradecanoylphorbol-13-acetate. Cancer Res 1988; 48:5941–5946.

63. Conney AM, Lysz T, Ferraro T, Abidi TF, Manchand PS, Laskin JD, Huang MT. Inhibitory effect of curcumin and some related dietary compounds on tumor promotion and arachidonic acid metabolism in mouse skin. Adv Enzyme Regul 1991; 31:385–396.

64. Plummer SM, Holloway KA, Manson MM, Munks RJ, Kaptein A, Farrow S, Howells L. Inhibition of cyclo-oxygenase 2 expression in colon cells by the chemopreventive agent curcumin involves inhibition of NF-kappaB activation via the NIK/IKK signalling complex. Oncogene 1999; 18(44):6013–6020.

65. Zhang F, Altorki NK, Mestre JR, Subbaramaiah K, Dannenberg AJ. Curcumin inhibits cyclooxygenase-2 transcription in bile acid- and phorbol ester-treated human gastrointestinal epithelial cells. Carcinogenesis 1999; 20(3):445–451.

66. Began G, Sudharshan E, Appu Rao AG. Inhibition of lipoxygenase 1 by phosphatidylcholine micelles-bound curcumin. Lipids 1998; 33(12):1223–1228.

67. Chan MM, Huang HI, Fenton MR, Fong D. In vivo inhibition of nitric oxide synthase gene expression by curcumin, a cancer preventive natural product with anti-inflammatory properties. Biochem Pharmacol 1998; 55:1955–1962.

68. Kumar A, Dhawan S, Hardegen NJ, Aggarwal BB. Curcumin (Diferuloylmethane) inhibition of tumor necrosis factor (TNF)-mediated adhesion of monocytes to endothelial cells by suppression of cell surface expression of adhesion molecules and of nuclear factor-kappaB activation. Biochem Pharmacol 1998; 55(6):775–783.

69. Lahiri-Chatterjee M, Katiyar SK, Mohari RR, Agarwal R. A flavonoid antioxidant, silymarin, affords exceptionally high protection against tumor promotion in the SENCAR mouse skin tumorigenesis model. Cancer Res 1999; 59:622–632.

70. Jang M, Cai L, Udeani GO, Slowing KV, Thomas CF, Beecher CW, Fong HH, Farnsworth NR, Kinghorn AD, Mehta RG, Moon RC, Pezzuto JM. Cancer chemopreventive activity of resveratrol, a natural product derived from grapes. Science 1997:275–218.

71. Holmes-McNary M, Baldwin AS Jr. Chemopreventive properties of trans-resveratrol are associated with inhibition of activation of the IkappaB kinase. Cancer Res 2000; 60(13): 3477–3483.

6

Heme Oxygenase-1: An Emerging Major Cellular Defense Against Cigarette Smoke-Induced Oxidative Stress and Cytotoxicity

THOMAS MÜLLER

Philip Morris Research
Laboratories GmbH,
Köln, Germany

ABSTRACT

Cigarette smoke (CS) harbors a strong oxidative stress potential, which broadly impacts exposed cells. A special feature of the cellular response to CS corresponds to the strong transcriptional upregulation of antioxidant genes, among which, as evaluated by cDNA microarray analysis, the gene encoding heme oxygenase-1 (HO-1) was found to be the most strongly expressed gene, both in vitro and in vivo. Exposure of

murine 3T3 fibroblast cells with strongly impaired HO-1
expression to otherwise subcytotoxic concentrations of
aqueous extracts of CS resulted in a significant increase in
the extent of cell death, suggesting that CS-induced HO-1
expression is potentially cytoprotective in this system. Cell
death induced by CS in Swiss 3T3 cells with attenuated
HO-1 expression showed specific apoptotic traits, as indicated
by the induction of *fasL*, followed by a significant increase in
the number of Annexin V-positive/propidium iodide-negative
cells, which was partly sensitive to caspase-8 inhibition. The
apoptotic signal was traced to Jun-N-terminal kinase (JNK)
and p38 mitogen-activated protein kinase (MAPK) signal
transduction pathways, as inhibition of either pathway
resulted in a significant decrease in the degree of apoptosis
in CS-exposed cells with attenuated HO-1 expression. In contrast to wild-type 3T3 fibroblasts, a substantial fraction of
human myeloid U937 cells undergo apoptosis when exposed
to similarly low concentrations of CS. Similar to CS-exposed
HO-1-deprived 3T3 cells, apoptosis in CS-treated U937 cells
appears to be executed in an autocrine manner as indicated
by *fasL* expression and activation of caspase-8. Although CS-
exposed U937 cells also strongly express HO-1, the underlying
kinetics significantly differs from that seen in 3T3 cells, particularly by a later onset of HO-1 expression. Notably, after 6 hr
of CS exposure, U937 cells undergoing apoptosis are completely devoid of HO-1 protein, whereas the fraction of viable cells
efficiently expresses HO-1. Altogether, these results strongly
support the general concept that under stressing conditions,
HO-1 plays a central antiapoptotic, cytoprotective role.

I. INTRODUCTION

Cigarette smoke (CS) harbors a strong oxidative stress potential, which eventually may result in chronic inflammation,
and therefore is believed to be involved, at least in part, in
the development of CS-related diseases, such as cancer,
cardiovascular disease, and chronic obstructive pulmonary
disease (1). Although the underlying molecular mechanisms

of disease development lack a detailed characterization, previous studies of exposed cells in culture have demonstrated that the water-soluble fraction of CS [smoke-bubbled phosphate-buffered saline (PBS)] triggers a specific stress response characterized, e.g., by distinct transient morphological reactions, such as membrane ruffling; the upregulation of known stress-responsive genes, such as *heme oxygenase-1* (*HO-1*) (2,3), c-fos (4), and c-myc (5) and induction of cell cycle arrest (4). These specific cellular responses may be provoked by CS-induced cellular lesions, such as DNA damage, e.g., DNA single-strand breaks (6), peroxidation of lipids in the cellular membrane (7,8), and the strong transient depletion of the intracellular glutathione (GSH) content (9,10). Previous investigations of CS constituents potentially implicated in cell damage revealed that hydroxyl radicals produced by Fenton chemistry are involved mainly in inducing DNA single-strand breaks (6) while CS-related aldehydes, such as formaldehyde, acetaldehyde, and acrolein, were identified as efficient depletors of the intracellular GSH content (9). The latter effect obviously facilitates the interference of peroxynitrite, which is formed in aqueous solutions of CS from superoxide and nitric oxide, with crucial signaling components, thus leading to the expression of c-fos (11).

HO-1 represents a major target of CS-induced stress but is also strongly induced under other diverse stressing conditions (for review, see Refs. 12, 13). Heme oxygenase-1 catalyzes the initial and rate-limiting steps in heme catabolism, leading to the formation of biliverdin and CO, as well as to the release of iron (Fe^{2+}) cations. In the presence of biliverdin reductase, biliverdin is converted to bilirubin (14). As both biliverdin and bilirubin are efficient antioxidants (15,16), the "HO-1 pathway" represents a prime defense tool in protecting the cell from stress-dependent adverse effects induced mainly by reactive oxygen and/or reactive nitrogen species from different sources (1,17–22). Moreover, the parallel expression of *ferritin* is considered to be an integral part of this antioxidant pathway in protecting the cell from oxidative stress due to the HO-1-dependent release of "free" iron (23). Mechanistically, the expression of HO-1 is controlled by a complex system consisting of a promoter with a proximal enhancer and two distal enhancers, which are distributed over more than 10 kb upstream of the transcriptional start site (reviewed in Ref. 24). This ternary system harbors a whole array of *cis* regulatory motifs addressing transcription factors known to be involved in oxidative stress and inflammatory responses, such as nuclear factor-κB (NF-κB), activating protein-1 (AP-1), NF-E2-related factor 2 (Nrf2), and heat shock factor-1 (HSF-1).

In addition to its ability to provide the cell with potent antioxidants, HO-1 expression has been linked with an increased resistance to cytotoxicity and a reduced inflammatory response, as exemplified by an HO-1-linked reduced extent of cell death in cells exposed to TNF-α (25), H_2O_2 (20), and peroxynitrite (21), while transgenic mice overexpressing HO-1 are protected against hypoxia-induced inflammatory cell infiltration (22). Recently, the cytoprotective potential of HO-1 has been mechanistically attributed to the release of CO during heme oxygenation (26,27).

II. HO-1 IS THE MOST STRONGLY INDUCED ANTIOXIDANT GENE IN CS-EXPOSED CELLS, IN VITRO AND IN VIVO

cDNA microarray experiments, performed to study CS-induced differential gene expression on a comprehensive basis, identified HO-1 as the strongest expressed antioxidant gene, both in vitro (28) and in vivo (29). Kinetic studies in Swiss 3T3 fibroblasts exposed to subcytotoxic concentrations of aqueous extracts of CS showed an immediate strong increase in HO-1 expression resulting in a maximal 84-fold induction of this gene in comparison with untreated cells after 8 hr of exposure (28). As discussed in the section on Introduction, in order to prevent oxidative damage from the release of "free" iron as the result of heme degradation, transcriptional activation of HO-1 in CS-exposed cells was paralleled in an obvious, co-ordinated manner by the expression of *ferritin*. Apart from the paramount induction of HO-1, further results showed the transcriptional activation of other stress-related genes, such as those encoding metallothionein I and II, heat shock proteins, and stress-related transcription factors, such as JunD and C/EBP-β, and more notably, the induction of genes described as mediators of an inflammatory/immune-regulatory response, such as *st2*, *kc*, and *id3* (28).

During a recent follow-up study addressing gene expression studies of CS-exposed cells in vivo, Sprague–Dawley rats were exposed to CS either once or repeatedly over 3 weeks, with and without a 20 hr recovery period before sacrifice (29). Similar to what was observed with CS-exposed cultured cells, HO-1 was generally the most strongly induced antioxidant gene in vivo. However, when differential gene expression in CS-exposed tissues of the respiratory tract (i.e., nasal epithelium, tracheal epithelium, and lung tissue, using micro-arrays carrying more than 2000 cDNA probes) was quantitatively compared, it became obvious that the strength of HO-1 expression was significantly different between the various tissues and the parameters: in tissue samples obtained from rats exposed once for 3 hr, HO-1 was found to be induced more than 50-fold in the nasal epithelium, approximately 10-fold in the tracheal epithelium, but only approximately 3-fold in lung tissue. As this effect was, in principle, also observed for Phase II-related genes, such as NAD(P)H:quinone oxireductase and γ-glutamyl-cysteine-synthetase, this phenomenon is a strong indication that there is a deposition gradient of (stress) gene-inducing CS compounds from the upper to the lower respiratory tract. Surprisingly, the apparent deposition gradient is obviously effectively cleared within 20 hr, even in rats repeatedly exposed for 3 weeks, as evidenced by the fact that the differential gene expression seen for HO-1 and nearly all other genes in CS-exposed rats without recovery was not seen in exposed rats with 20 hr recovery (29).

After 3 weeks of repeated exposure without recovery, expression of HO-1 and Phase II-related genes followed basically the same pattern as in rats which were exposed once, but the expression rates were significantly decreased (2- to 5-fold) when compared with the corresponding samples from a single exposure. Because HO-1 and all Phase II-related genes, found to be differentially expressed by CS, are efficiently downregulated after 20 hr recovery, regardless of whether there was a single 3 hr exposure or repeated exposures over 3 weeks, this effect clearly points to an adaptive response (29). Like HO-1, Phase II-related genes are also transcriptionally regulated, at least in part, by the transcription factor Nrf2 (30,31). Therefore, in mechanistic terms, the phenomenon of an adaptive response in this class of genes may be interpreted by a steadily decreasing sensitivity of Nrf2 to become activated by CS-dependent stressors over prolonged exposure periods. Consequently, this effect may gradually compromise the ability of the respiratory tract cells to adequately respond to CS-dependent (oxidative) stress during chronic exposure, and may therefore be critically involved in CS-dependent carcinogenesis or other diseases.

Notably, in contrast to Phase II-related genes, genes encoding Phase I-metabolizing enzymes, such as cytochrome P4501A1 and aldehyde dehydrogenase type 3, are obviously not expressed along a deposition gradient in CS-exposed rats as shown by equally high expression rates for these genes in nasal and lung tissues. Nor is the expression of Phase I-related genes subject to an adaptation process, as similar rates of induction were detected after a single exposure and after 3 weeks of repeated exposure, generally indicating that CS induces Phase I- and Phase II-related genes in completely different ways regarding both the inducing principles and the inducing mechanism (29).

III. CS-EXPOSED FIBROBLASTS WITH IMPAIRED HO-1 INDUCIBILITY EXHIBIT INCREASED CYTOTOXICITY

In order to elucidate the meaning of the strong HO-1 expression by CS in vitro and in vivo, 3T3 fibroblasts were inhibited in their ability to fully express this gene either by the stable expression of HO-1 antisense (AS) RNA in NIH3T3 cells or by treating Swiss 3T3 cells with sodium vanadate (SV; $100\,\mu M$), which was found to be a potent inhibitor of HO-1 expression in CS-treated cells (32). Expression of HO-1

AS-RNA in NIH3T3 cells resulted in a general, significant attenuation of HO-1 transcription; i.e., in comparison with wild-type cells, HO-1 protein was reduced by nearly 90% in untreated (control) cells, while after 6 hr of exposure to smoke-bubbled PBS, the reduction was still more than 50%. Swiss 3T3 cells pretreated with SV (15 min prior to the addition of smoke-bubbled PBS) were not affected in their background expression of HO-1, but virtually did not show any increase in HO-1 when exposed to CS. The decrease in HO-1 expression in CS-exposed HO-1-AS-RNA-expressing NIH3T3 cells and SV-pretreated Swiss 3T3 cells could be confirmed on the enzymatic level, as demonstrated by the significantly reduced ability of CS-exposed SV-pretreated and HO-1-AS-RNA-expressing cells to convert hemin to bilirubin.

Investigations of the viability of CS-exposed HO-1-AS-RNA-transfected NIH3T3 or SV-pretreated Swiss 3T3 cells revealed that cells inhibited in HO-1 expression, at otherwise subcytotoxic concentrations of CS (0.03 puffs/mL medium), are vulnerable to cytotoxic processes, which become obvious after 6–8 hr of exposure. Morphologically, this effect is indicated by light microscopically detectable changes, such as rounding-up of the cells resulting in loss of contact with neighboring cells. Interestingly, in HO-1-deficient Swiss 3T3 cells, this effect is first visible in small clusters of cells and gradually spreads over the whole monolayer. In contrast, although CS-exposed NIH3T3 wild-type cells and Swiss 3T3 cells not pretreated with SV show slight morphological signs of stress, they remain viable and are subject to cell cycle arrest as described previously (4). Altogether, these results clearly show that murine 3T3 fibroblasts exposed to subcytotoxic concentrations of aqueous extracts of CS are protected from cell death in an HO-1-dependent manner.

IV. SWISS 3T3 CELLS WITH COMPROMISED HO-1 INDUCIBILITY UNDERGO APOPTOSIS WHEN EXPOSED TO CS

NIH3T3 cells are obviously equipped with an aberrant pathway of apoptosis as evidenced, e.g., by a striking robust

resistance to Fas ligand (FasL)-induced apoptosis while, paradoxically, specific inhibition of caspase-8 sensitizes these cells
to TNF-induced cell death (33). Therefore, HO-1-AS-RNA-
expressing NIH3T3 cells are not suitable to elucidate whether
CS-exposed cells with attenuated HO-1 inducibility are subject to necrosis or apoptosis. Nevertheless, it should be
stressed that irrespective of the obvious limitations in apoptosis induction, cells of this cell line are prone to CS-induced cell
death when compromised in HO-1 expression.

Swiss 3T3 cells, however, upon exposure to exogenous
soluble FasL, undergo the regular path of apoptosis involving
the activation of caspase-8 and subsequently caspase-3,
finally resulting in DNA-laddering (32) and, accordingly, were
used after pretreatment with SV to evaluate the type of cell
death induced by CS in cells with impaired HO-1 inducibility.
In technical terms, discrimination between apoptosis and
necrosis was accomplished by quantitatively determining
the binding of Annexin V to externalized phosphatidylserine
and the uptake of propidium iodide using flow-cytometric
methods. Using this approach, Swiss 3T3 cells pretreated
with SV and subsequently exposed to smoke-bubbled PBS
showed a significant increase in the fraction of apoptotic cells
(16–20%) sensitive to caspase-8 inhibition as could be deduced
from experiments conducted in the presence of the caspase-8-
specific inhibitor z-IETD.fmk. In contrast, the amount of
apoptotic cells in untreated (control) cells was below 3% and
remained unchanged when the cells were exposed to SV or
CS alone. The number of necrotic cells, which varied between
15% and 20% irrespective of treatment, is thought to result
mainly from the detachment procedure (32).

CS-dependent apoptosis involving caspase-8 activation
suggests an autocrine mechanism, which is initiated by the
expression of *fasL*, followed by the interaction of FasL protein
with its receptor, and subsequent execution of the apoptotic
process. In fact, such an autocrine mechanism of apoptosis
has been described for genotoxic physical and chemical stresses, such as those induced by UV-irradiation or protein synthesis inhibition (34,35). When CS-exposed Swiss 3T3 cells were
monitored for *fasL* expression using semiquantitative reverse

transcriptase (RT)–PCR, an up to 6-fold transient induction of this gene was detected. Maximal rates of expression were observed after 4 hr of exposure, which in kinetic terms is consistent with the first appearance of apoptotic cells after 6–8 hr in CS-exposed SV-pretreated cell samples (32). In summary, these results provide conclusive evidence that CS-exposed Swiss 3T3 cells with impaired HO-1 expression undergo apoptosis in an Fas/FasL/caspase-8-dependent autocrine manner.

V. THE CS-DEPENDENT APOPTOTIC SIGNAL IS TRANSDUCED BY JNK AND p38 MAPK PATHWAYS

Data from the literature describing stress-induced apoptosis by an autocrine mechanism in T-cells link stress-mediated *fasL* expression, with the prior activation of the Jun-N-terminal kinase (JNK) signaling pathway when cells were exposed to genotoxic physical stresses (34) and with the activation of the p38 mitogen-activated protein kinase (p38 MAPK) signaling pathway when the cells were challenged with inhibitors of protein synthesis (35). In turn, both pathways are supposed to convey their signal to the transcription factor AP-1, specifically composed of ATF-2 and c-Jun proteins, eventually leading to the transcriptional activation of *fasL* (36).

Investigation of CS-exposed Swiss 3T3 cells for activated JNK and p38 MAPK signal transduction pathways by screening the cells for phosphorylated JNK and p38 MAPK proteins revealed a strong activation of both pathways. Activated JNK and p38 MAPK proteins were first seen after 30–60 min, but were still clearly detectable after 8 hr of exposure. Considering that CS has been shown to exert both genotoxic effects (6) and protein synthesis inhibitory effects (2), the activation of both pathways by CS is conceivable. In addition to activated JNK and p38 MAPK signaling, increased protein binding to model nucleotides containing an AP-1/TPA-responsive element (TRE) consensus binding site was observed in protein extracts from CS-exposed cells, while virtually no increase was observed in the amount of pre-existing DNA protein complexes when an AP-1/cAMP response element (CRE)-specific oligonucleotide, which is more efficiently recognized by c-Jun/ATF-2 heterodimers (37), was used for analysis. However, the trans-activation activity of AP-1, composed of c-Jun/ATF-2, is mainly determined by phosphorylation rather than by a considerable effect on the DNA-binding activity (37). In fact, the specific activation of c-Jun/ATF-2 as a potential trans-activator of *fasL* expression in CS-exposed cells was readily observed when the cells were monitored for phosphorylated c-Jun and ATF-2. As revealed by Western blotting, both activated proteins produced strong signals, which were first detectable approximately 30 min after the start of exposure.

Finally, the crucial involvement of the JNK as well as the p38 MAPK signaling pathways in CS-induced apoptosis were demonstrated when SV-pretreated Swiss 3T3 cells were quantified for apoptosis formation in the presence of specific inhibitors of either JNK (SP600125) or p38 MAPK (SB 203580): a significant ($\sim$60%) reduction in the amount of apoptotic cells was discernable for either inhibitor, clearly indicating that apoptosis of CS-exposed Swiss 3T3 cells with impaired HO-1 expression is dependent on activation of JNK and p38 MAPK stress signal transduction (32).

VI. ABSENCE OF HO-1 EXPRESSION IS ASSOCIATED WITH APOPTOSIS INDUCTION IN CS-EXPOSED U937 CELLS

It has been reported that in other cell lines, e.g., U937 cells, a certain fraction of the cells are prone to CS-induced cytotoxicity when exposed to smoke-bubbled PBS concentrations at which wild-type 3T3 cells are completely resistant to cytotoxicity (i.e., <0.04 puffs/mL smoke-bubbled PBS) (38). Similar to Swiss 3T3 cells hampered in HO-1 expression, CS-treated wild-type U937 cells start to upregulate *fasL*, first seen after 2 hr of exposure, whereas after 6 hr of exposure considerable apoptosis-specific DNA laddering, which is sensitive to caspase-8 inhibition but insensitive to caspase-9 inhibition, is observed in these cells. In addition, cleaved procaspase-8 products are detectable in these cells, clearly indicating the apoptotic nature of cell death induced by CS in U937 cells, which, in parallel to smoke-treated SV-pretreated Swiss 3T3 cells, appears to be executed in a Fas/FasL/caspase-8-dependent manner. Regarding the obvious HO-1-dependent suppression of CS-dependent cell death in wild-type murine 3T3 fibroblasts, these data immediately raise the question of whether the apoptotic fraction of CS-exposed U937 cells is not protected by an increased HO-1 activity. Experiments addressing this issue showed that although CS-exposed U937 cells, like smoke-exposed 3T3 cells, strongly upregulate HO-1 in a time-dependent manner, there are some remarkable differences between the expression characteristics of HO-1 in CS-treated 3T3 cells and U937 cells: (a) as evaluated by Northern and Western blotting, U937 cells, unlike 3T3 cells, do not show any detectable HO-1 background expression; (b) in CS-exposed U937 cells, HO-1 mRNA is first detectable after 2 hr of exposure, whereas an immediate increase in this mRNA species is observed in 3T3 cells upon smoke exposure; which (c) in kinetic terms is fully reflected on the protein level, as indicated by a first detectable CS-dependent increase in HO-1 protein in 3T3 cells as early as 2 hr after the start of exposure, whereas in CS-exposed U937 cells, HO-1 is first visible after 4–8 hr of exposure (32).

The obvious different kinetics of HO-1 expression with regard to CS-exposed 3T3 cells, together with the observation that at lower smoke doses only a minor portion of cells undergo apoptosis, raises the question of whether HO-1 expression in CS-exposed U937 cells is associated to any extent with death and survival decisions in these cells. In order to address this issue, apoptotic and viable cells were screened for the presence of HO-1 protein using a dual-labeling flow-cytometric approach. Six hours after the start of exposure to smoke-bubbled PBS, the vast majority of viable (Annexin V-negative) cells showed efficient binding of the HO-1 label demonstrating significant HO-1 expression, whereas, in marked contrast, a fraction of apoptotic (Annexin V-positive) cells did not show any affinity to the HO-1 label

(32). This result clearly indicates that, at least in the early phase of apoptosis onset, survival in U937 cells strictly correlates with HO-1 expression. Altogether, these data further emphasize the apoptosis-suppressive potential of HO-1 seen with CS-exposed 3T3 cells compromised in HO-1 expression. At the same time, they also raise the question of why a certain fraction of smoke-exposed U937 cells are insensitive to CS-induced HO-1 expression and therefore prone to apoptosis.

VII. CONCLUDING REMARKS

Experimental data described here provide evidence for a cytoprotective role of HO-1 in cells exposed to CS, thus adding to several other publications that demonstrate an anticytotoxic role of HO-1 under several apparently unrelated procytotoxic conditions (1,17–22,25). Previous reports (26,27) have hypothesized that it might be primarily the carbon monoxide (CO) released during the catalytic action of HO-1 that is involved in the cytoprotective and anti-inflammatory effects exerted by HO-1. In mechanistic terms, HO-1-dependent CO should evoke an antiapoptotic signal by activating cell survival-related genes in a p38 MAPK/NF-κB-dependent manner. According to this line of evidence, CS-induced HO-1 expression could inhibit cell death via the release of CO during heme oxygenation followed by p38 MAPK activation, as described here, and the subsequent induction of NF-κB and cell survival genes, as previously shown for smoke-bubbled PBS-treated cells (5). On the other hand, CS itself contains large amounts of CO, which, according to the suggested cytoprotective mechanism (26,27), should inhibit apoptosis formation independently of HO-1-delivered CO. Moreover, a cytoprotective role of activated p38 MAPK is compatible neither with the data reported here nor with apoptosis formation in anisomycin-treated cells (35). Thus, more experimentation is needed in order to elucidate the potential role of HO-1-delivered CO and/or other HO-1 enzymatic products in HO-1-induced cell survival.

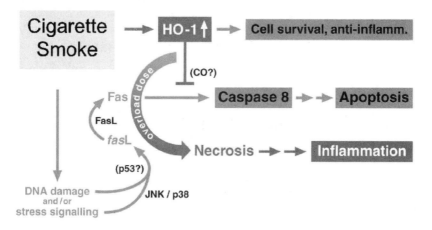

Figure 1 CS-induced expression of HO-1 protects smoke-exposed cells from an autocrine type of apoptotic cell death triggered by the activation of the JNK and p38 MAPK stress signal transduction pathways. Overloading of the cytoprotective "HO-1 pathway" may result in necrotic cell death and the onset of inflammation. For further details, see text.

With regard to CS-related adverse physiological effects, the data reported here (schematically summarized in Fig. 1) indicate that up to a certain dose of CS, cytotoxic and inflammatory effects tend to be preventable by HO-1 enzymatic action. However, if the HO-1-dependent cytoprotective system is compromised, e.g., by chronic inhalation of CS considerably exceeding a putative critical dose, a significant increase in the number of necrotic cells can be assumed, which may be a potential source for the onset of inflammatory processes. In fact, we observed that exposure of SV-pretreated Swiss 3T3 cells and U937 cells to considerably higher concentrations of smoke-bubbled PBS resulted in a significant decrease in the number of apoptotic cells, while the amount of necrotic cells increased dramatically (32). Therefore, pathologies caused by CS, such as chronic obstructive pulmonary disease, cardiovascular disease, and cancer, may all arise from inflammatory processes due to necrotic cell death resulting from uncompensated cellular stress.

ACKNOWLEDGMENT

The author thanks L. Conroy for expert editorial support.

REFERENCES

1. Rahman I. Oxidative stress, chromatin remodeling and gene transcription in inflammation and chronic lung diseases. J Biochem Mol Biol 2003; 36:95–109.

2. Müller T, Gebel S. Heme oxygenase expression in Swiss 3T3 cells following exposure to aqueous cigarette smoke fractions. Carcinogenesis 1994; 15:67–72.

3. Favatier F, Polla B. Tobacco-smoke inducible human haem oxygenase-1 gene expression: role of distinct transcription factors and reactive oxygen intermediates. Biochem J 2001; 353:475–482.

4. Müller T. Expression of c-*fos* in quiescent Swiss 3T3 cells exposed to aqueous cigarette smoke fractions. Cancer Res 1995; 55:1927–1932.

5. Gebel S, Müller T. The activity of NF-κB in Swiss 3T3 cells exposed to aqueous extracts of cigarette smoke is dependent on thioredoxin. Toxicol Sci 2001; 59:75–81.

6. Nakayama T, Kaneko M, Kodama M, Nagata C. Cigarette smoke induces DNA strand breaks in human cells. Nature 1985; 314:462–464.

7. Frei B, Forte TM, Ames BN, Cross CE. Gas phase oxidants of cigarette smoke induce lipid peroxidation and changes in lipoprotein properties in human blood plasma. Biochem J 1991; 277:133–138.

8. Niki E, Minamisawa S, Oikawa M, Komuro E. Membrane damage from lipid oxidation induced by free radicals and cigarette smoke. Ann NY Acad Sci 1993; 686:29–38.

9. Müller T, Gebel S. The cellular stress response induced by aqueous extracts of cigarette smoke is critically dependent on the intracellular glutathione concentration. Carcinogenesis 1998; 19:797–801.

10. Bilimoria MH, Ecobichon DJ. Protective antioxidant mechanisms in rat and guinea pig tissues challenged by acute exposure to cigarette smoke. Toxicology 1992; 72:131–144.

11. Müller T, Haussmann H-J, Schepers G. Evidence for peroxynitrite as an oxidative stress-inducing compound of aqueous extracts of cigarette smoke. Carcinogenesis 1997; 18:295–301.

12. Otterbein LE, Choi AMK. Heme oxygenase: colors of defense against cellular stress. Am J Lung Cell Mol Physiol 2000; 279: L1029–L1037.

13. Morse D, Choi AM. Heme oxygenase-1: the "emerging molecule" has arrived. Am J Respir Cell Mol Biol 2002; 27:8–16.

14. Tenhunen R, Marver HS, Schmid R. The enzymatic conversion of heme to bilirubin by microsomal heme oxygenase. Proc Natl Acad Sci USA 1968; 61:748–755.

15. Maines MD. Heme oxygenase: function, multiplicity, regulatory mechanisms, and clinical applications. FASEB J 1988; 2: 2557–2568.

16. Stocker RY, Yamamoto Y, McDonagh AF, Glazer AN, Ames BN. Bilirubin is an antioxidant of possible physiological importance. Science 1987; 235:1043–1046.

17. Applegate LA, Luscher P, Tyrrell RM. Induction of heme oxygenase: a general response to oxidant stress in cultured mammalian cells. Cancer Res 1991; 51:974–978.

18. Lee PJ, Alam J, Wiegand GW, Choi AM. Overexpression of heme oxygenase-1 in human pulmonary epithelial cells results in cell growth arrest and increased resistance to hyperoxia. Proc Natl Acad Sci USA 1996; 93:10393–10398.

19. Dennery PA, Sridhar KJ, Lee CS, Wong HE, Shokoohi V, Rodgers PA, Spitz DR. Heme oxygenase-mediated resistance to oxygen toxicity in hamster fibroblasts. J Biol Chem 1997; 272:14937–14942.

20. Yamada N, Yamaya M, Okinaga S, Lie R, Suzuki T, Nakayama K, Takeda KA, Yamaguchi T, Itoyama Y, Sekizawa K, Sasaki H. Protective effects of heme oxygenase-1 against oxidant-induced injury in the cultured human tracheal epithelium. Am J Respir Cell Mol Biol 1999; 21:428–435.

21. Foresti R, Motterlini R. The heme oxygenase pathway and its control of cellular homeostasis. Free Rad Res 1999; 31: 459–475.

22. Minamino T, Christou H, Hsieh C-H, Liu Y, Dhawan V, Abraham NG, Perrella MA, Mitsialis SA, Kourembanas S. Targeted expression of heme oxygenase-1 prevents the pulmonary inflammatory and vascular responses to hypoxia. Proc Natl Acad Sci USA 2001; 98:8798–8803.

23. Tyrrell R. Redox regulation and oxidant activation of heme oxygenase-1. Free Rad Res 1999; 31:335–340.

24. Choi AMK, Alam J. Heme oxygenase-1: function, regulation, and implication of a novel stress-inducible protein in oxidant-induced lung injury. Am J Respir Cell Mol Biol 1996; 15:9–19.

25. Petrache I, Otterbein LE, Alam J, Wiegand GW, Choi AMK. Heme oxygenase-1 inhibits TNF-α-induced apoptosis in cultured fibroblasts. Am J Physiol Lung Cell Mol Physiol 2000; 278:L312–L319.

26. Brouard S, Otterbein LE, Anrather J, Tobiasch E, Bach FH, Choi AMK, Soares MP. Carbon monoxide generated by heme oxygenase 1 suppresses endothelial cell apoptosis. J Exp Med 2000; 192:1015–1026.

27. Brouard S, Berberat PO, Tobiasch E, Seldon MP, Bach FH, Soares MP. Heme oxygenase-1-derived carbon monoxide requires the activation of transcription factor NF-κB to protect endothelial cells from tumor necrosis factor-α-mediated apoptosis. J Biol Chem 2000; 277:17950–17961.

28. Bosio A, Knörr C, Janssen U, Gebel S, Haussmann H-J, Müller T. Kinetics of gene expression profiling in Swiss 3T3 cells exposed to aqueous extracts of cigarette smoke. Carcinogenesis 2002; 23:741–748.

29. Gebel S, Gerstmayer B, Bosio A, Haussmann H-J, Van Miert E, Müller T. Gene expression profiling in respiratory tissues from the rats exposed mainstream cigarette smoke. Carcinogenesis, 2004; 25:169–178.

30. Alam J, Stewart D, Touchard C, Boinapally S, Choi AMK. Nrf2, a cap'n'collar transcription factor, regulates induction

of the heme oxygenase-1 gene. J Biol Chem 1999; 274: 26071–26078.

31. Nguyen T, Sherratt PJ, Pickett CB. Regulatory mechanisms controlling gene expression mediated by the antioxidant response element. Annu Rev Pharmacol Toxicol 2003; 43: 233–260.

32. Gebel S, Knörr C, Kindt R, Müller T. Heme oxygenase-1 expression protects cells from cell death induced by aqueous extracts of cigarette smoke. Manuscript in preparation.

33. Lüschen S, Ussat S, Scherer G, Kabelitz D, Adam-Klages S. Sensitization to death receptor cytotoxicity by inhibition of Fas-associated death domain protein (FADD)/caspase signaling: requirement of cell cycle progression. J Biol Chem 2000; 275:24670–24678.

34. Faris M, Kokot N, Latinis K, Kasibhatla S, Green DR, Koretzky GA, Nel A. The c-Jun N-terminal kinase cascade plays a role in stress-induced apoptosis in Jurkat cells by up-regulating Fas ligand expression. J Immunol 1998; 160: 134–144.

35. Hsu S-C, Gavrilin MA, Tsai M-H, Han J, Lai M-Z. p38 mitogen-activated protein kinase is involved in Fas ligand expression. J Biol Chem 1999; 274:25769–25776.

36. Faris M, Latinis KM, Kempiak SJ, Koretzky GA, Nel A. Stress-induced Fas ligand expression in T cells is mediated through MEK kinase 1-regulated response element in the Fas ligand promoter. Mol Cell Biol 1998; 18:5414–5424.

37. Karin M. The regulation of AP-1 activity by mitogen-activated protein kinases. J Biol Chem 1995; 270:16483–16486.

38. Vayssier M, Banzet N, Francois D, Bellmann K, Polla BS. Tobacco smoke induces both apoptosis and necrosis in mammalian cells: differential effects of HSP70. Am J Physiol Lung Cell Mol Physiol 1998; 19:L771–L779.

7

Biphasic Induction of HO-1 in Macrophages Treated with Lipopolysaccharide: Role of HO-1 Induction in Cell Survival from Oxidative Stress

KLAOKWAN SRISOOK and YOUNG-NAM CHA

Department of Pharmacology and Toxicology,
College of Medicine, Inha University,
Inchon, South Korea

I. INTRODUCTION

The aerobic mice, rats, and human are being infected constantly by air-borne pathogens like bacteria and yet, they maintain health by destroying the infecting pathogen using their inflammatory immune system. Outer surface of bacteria contains lipopolysaccharide (LPS) and when these aerobic animals are infected with bacteria or administered with

LPS, inflammatory responses similar to that seen in adult respiratory distress syndrome (ARDS) are observed in the lung tissue (1). Inflammatory responses seen in ARDS are caused in part by the oxidative stress induced by LPS. During an acute inflammatory response induced by bacterial infection or administrations of LPS or mitogens, there is an early influx of inflammatory phagocytic cells like polymorphonuclear cells (PMNs) and monocytes, which differentiate into macrophages (2). At the inflammatory site, all these phagocytic cells undergo oxidative burst and produce abundance of oxygen metabolites like superoxide radical (O_2^-), hydrogen peroxide (H_2O_2) and hydroxyl radical ($HO^\bullet$). These reactive oxygen species (ROS), in turn, damage DNA, proteins, and lipid membranes and cause cytotoxicity, not only to invading pathogens but also to phagocytic cells themselves and to surrounding host cells (3).

Even in the normal host cells and tissues, small amount of O_2^- is formed by one-electron reduction of molecular oxygen in mitochondria during respiratory electron transfer processes and also by the NADPH oxidases localized in plasma membrane during oxidative burst (4). Concentration of O_2^- present in a cell is determined by the balance between rates of its production catalyzed by several heme-containing enzymes and its clearance processed by various radical-scavenger biomolecules and antioxidant enzymes. Initially, O_2^- is converted to H_2O_2 by superoxide dismutases (SODs) localized both in cytoplasm and mitochondria. In biological tissues, particularly in the presence of transition metals like heme-iron, H_2O_2 can also be converted into highly reactive $HO^\bullet$ by the heme-iron catalyzed Fenton chemistry (5). In most cases, however, due to lack of free heme-iron in normal cells, H_2O_2 is converted back to H_2O and O_2 by enzymes like catalase or glutathione peroxidase (GPX). In the GPX reaction, reduced glutathione (GSH) is oxidized to glutathione disulfide (GSSG), lowering GSH/GSSG ratio and altering intracellular redox balance toward oxidation, thus, causing oxidative stress (6). Glutathione disulfide can be reconverted back to GSH by glutathione reductase (GR) in an NADPH-consuming process. Finally, NADPH is replenished by activation of glucose metabolism, and thus, when cells are deprived of glucose, cells are not able to cope even with a normal level of oxidative stress (7,8). If the initial amount of ROS production is small, pre-existing reserve antioxidant responses (e.g., catalase, GPX, GSH, GR, and NADPH) may be sufficient to clear the ROS, allowing the cells to survive without severe loss of redox balance.

In response to stronger oxidative stresses such as those caused by exposure to LPS or bacterial infection, however, macrophages induce expression of several genes whose products exhibit antioxidant activity. A major mechanism for oxidatively stressed cells to maintain redox homeostasis is based on ROS-mediated initiation of signal cascades that lead to activation of several redox-sensitive transcription factors involved in increased synthesis of antioxidant enzymes (9). These antioxidant enzymes, in turn, produce GSH, NADPH, or nitric oxide radical (NO). In the LPS-treated macrophages, in response to the overproduced O_2^-, synthesis of inducible NOS (iNOS) increases rapidly for an abundant and continuous production of NO (10). These O_2^- and NO radicals, although they cannot react directly with GSH, can oxidize GSH after undergoing intracellular redox reactions. Alternatively, the NO can react with O_2^- by radical–radical interaction forming peroxynitrite ($ONOO^-$), thus, clearing and scavenging O_2^- in a diffusion-limited rate (11). However, product of this reaction, $ONOO^-$, is a strong oxidizing agent, a reactive nitrogen species (RNS) which oxidizes GSH rapidly to GSSG and depletes intracellular store of GSH (Fig. 1).

In response to this depletion of GSH or lowered thiol/disulfide ratio, redox-sensitive transcription factors like activator protein-1 (AP-1), nuclear factor κ-B (NF-κB), and NF-E2-related factor (Nrf2) are activated and expressions of iNOS and heme oxygenase-1 (HO-1) are enhanced. Thus, in animal lung tissues exposed to LPS, high level iNOS expression together with abundant NO production as well as high level HO-1 protein with elevated HO activity are detected transiently, particularly in the inflammatory alveolar macrophage cells (1). This increased HO activity is involved in oxidative degradation of toxic heme and generation of bilirubin and carbon monoxide antioxidant. In particular, CO is known

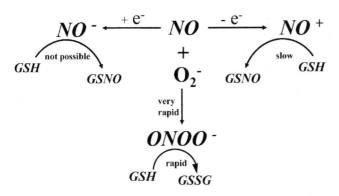

Figure 1 Reactivity of NO metabolites with reduced glutathione. NO does not react directly with GSH unless it is oxidized to NO^+ or combines with O_2^- to generate $ONOO^-$. NO reacts with O_2^- at $10^{11} M/sec$. $ONOO^-$ is generated at $10^{11} M/sec$ and oxidizes GSH to GSSG at $10^6 M/sec$. NO^+ reacts with GSH to generate GSNO at $10^3 M/sec$.

to bind the heme-iron contained in heme-enzymes catalyzing the production of O_2^- and NO radicals, blocking further production of these radicals (12,13). Thus, induction of HO-1 observed in the LPS-stimulated macrophages may function not only in degrading the toxic heme but also in producing CO, which inhibits additional production of O_2^- and NO radicals. These adaptive changes, like the activation of redox-sensitive transcription factors, induction of iNOS and HO-1, which occur in response to the oxidative stress initiated by LPS-derived oxidative burst in macrophages, appear to proceed in a highly co-ordinated and sequential manner (Fig. 2).

By employing the cultured murine macrophage cells (RAW 264.7 cell line) stimulated with LPS, we demonstrate in the present study that HO-1 expression is induced in a biphasic manner, the initial phase of HO-1 induction by ROS-derived oxidative stress and the second phase induction by RNS-derived nitrosative stress (Fig. 3). Based on the results obtained with LPS-stimulated macrophages, this review explores the role of HO-1 induction in inhibiting the cytotoxicity caused by ROS and RNS by degrading the cytotoxic heme while generating the cytoprotective CO (Fig. 4). Eventually,

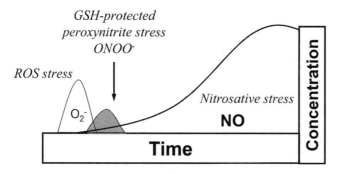

Figure 2 Relative time scale and magnitudes of O_2^- derived ROS stress, GSH protected $ONOO^-$ stress, and NO-derived nitrosative stress.

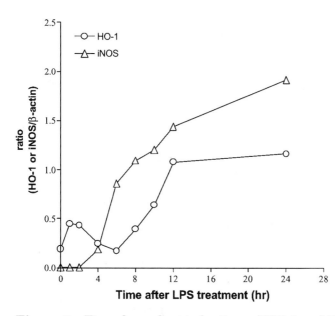

Figure 3 Time-dependent inductions of HO-1 and iNOS in macrophages stimulated with LPS. RAW 264.7 cells were treated with LPS (1 μg/mL) and harvested at the indicated times. Harvested cells were homogenized and homogenates were analyzed to determine the contents of HO-1 and iNOS employing immunoblot analysis. Graph shows the results of densitometric analysis on the time-dependent accumulation of HO-1 and iNOS proteins, which have been normalized to β-actin contents.

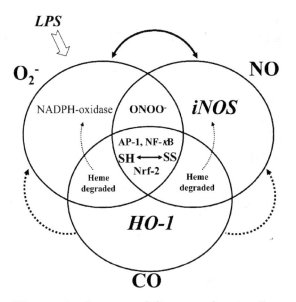

Figure 4   Conceptual diagram showing "crosstalks" between LPS-derived sequential overproductions of O_2^-, NO, and CO catalyzed respectively by NADPH oxidase, iNOS, and HO-1 mediated by reversible changes of intracellular thiol redox states. Activations of redox-sensitive transcription factors like AP-1, NF-κB and Nrf2 are involved in the induction of HO-1. With induction of HO-1, and with elevated HO activity, heme contained in NADPH oxidase and iNOS is degraded and with enhanced production of HO-derived CO, further generation of O_2^-, and NO is inhibited.

this will allow the LPS-stimulated macrophages to regain redox homeostasis and to survive.

II.   PROTECTIVE EFFECT OF NO AGAINST CYTOTOXICITY OF ROS ONLY IN THE PRESENCE OF GSH

Nearly two decades of biological research on free radicals has indicated that they play key roles both in physiology and pathophysiology. Reactive oxygen species derived from O_2^- and peroxide radicals can oxidize several key molecules in cells and cause numerous diseases (5). However, living cell also responds and utilizes the ROS in several mechanisms to protect itself from ROS-derived cellular injury (3). One of the immediate mechanisms of defense against ROS is to scavenge the ROS directly by overproducing NO.

The free radical NO has been described to have both physiological and pathological roles; it is essential not only for maintaining homeostasis (14,15), but it also has the ability to cause damage by reacting with specific cellular targets through formation of RNS (16,17). Further complexity is introduced by studies showing that NO can protect against oxidative injury mediated by ROS (11,17,18), while other reports suggest that NO augments the ROS-mediated toxicity (19–21).

Although the H_2O_2 generated from O_2^- by SOD is a relatively stable nonradical ROS compound, it mediates oxidation of biological molecules and causes oxidative tissue damage. While NO does not react directly with H_2O_2 (22), NO can protect cells against the toxicity caused by H_2O_2 (17,22–25). In support of this, NO produced rapidly from the Ca^{++}-activated eNOS of endothelial cells can protect vascular smooth muscle cells against the H_2O_2-derived damage (26). Lung fibroblasts exposed to H_2O_2 are known to exhibit marked cytotoxicity. However, addition of NONOates, a class of compounds that release NO in a controlled manner over the duration of H_2O_2 exposure (27), resulted in a protection against the cytotoxicity caused by H_2O_2 (22). Addition of these NO donors either before or after exposing the lung fibroblasts to H_2O_2 did not result in protection. Similar observations have been made with neuronal (22), hepatoma (28), and endothelial cells (17,29,30). Thus, NO protects cells against the toxicity caused by H_2O_2.

Freeman and coworkers have investigated the effect of NO on O_2^--derived lipid peroxidation caused by addition of xanthine oxidase (XO), an O_2^- producing condition, and found that NO acts as an inhibitor of lipid peroxidation by scavenging or clearing O_2^- (31). However, a strong oxidizing agent peroxynitrite ($ONOO^-$) is produced under this condition. Treatment of cells with a bolus of $ONOO^-$ resulted in deaths for both bacteria and mammalian cells (22,23). Conversely, treatment of cells with a combination of XO (generator of O_2^-) and slow-release NO donors such as the NONOates, believed to generate $ONOO^-$ in situ, was not necessarily toxic to various mammalian cell types. Thus, presence of NO appeared to protect cells against the toxicity caused by O_2^- derived from XO activity, perhaps by generating $ONOO^-$, which can be destroyed by GSH (17,22–25). These results suggested that in situ formation of $ONOO^-$ is not necessarily toxic to mammalian cells as long as there is ample intracellular store of GSH or is replenished rapidly (24). These results further suggested that there may be a distinct difference between exposing cells with a bolus of $ONOO^-$ at millimolar concentrations, as opposed to simultaneous intracellular generation of NO and O_2^- producing $ONOO^-$ in micromolar ranges in the presence of GSH.

Stimulated aerobic cells produce NO, small amounts by immediate Ca^{++}-dependent activation eNOS or nNOS and continuous large amounts by Ca^{++}-independent iNOS, perhaps to scavenge the O_2^-. Thus, it appears that oxidatively stressed aerobic cells utilize NO radical for an immediate clearance of O_2^- radical at the expense of forming $ONOO^-$, which consumes and oxidizes cellular GSH. This suggested that NO can provide cytoprotective effects against oxidative insults in mammalian cells only when abundant GSH is present in the cell. Although the $ONOO^-$, being generated by the radical–radical interaction between NO and O_2^-, is a potent oxidizing agent and may lower intracellular thiol/disulfide ratio, the NO-derived conversion of the GSH-unreactive O_2^- to the GSH-detoxifiable $ONOO^-$ may serve to control or abolish the toxicity of ROS in vivo. This may constitute the "GSH-protected $ONOO^-$ stress." This ROS-scavenging antioxidant effect provided by NO may be important in minimizing tissue injury caused by the ROS-dependent processes, only as long as the oxidatively stressed cell can synthesize GSH and replenish the reducing equivalent NADPH from glucose (Figs. 1 and 2).

III. OXIDATIVE STRESS AND INDUCTIONS OF iNOS AND HO-1

In macrophages stimulated with LPS, abundant ROS is produced immediately by the oxidative burst catalyzed by

NADPH oxidase, a heme-containing enzyme complex present in the plasma membrane. Reactive oxygen species causes oxidation and depletion of intracellular GSH, triggering the oxidative-stress signaling either by activating protein kinases involved in phosphorylating the signal-cascade proteins or by inhibiting the protein tyrosine phosphatases involved in the dephosphorylation of activated signal-cascade proteins (6). In response to these stress signals, cytoplasmic redox-sensitive transcription factors like AP-1, Nf-κB, and Nrf2 are activated. These activated transcription factors then translocate into nucleus, bind to enhancer sequences and enhance the transcription and translation of antioxidant enzymes which, among many, include the heme-containing iNOS and the non-heme HO-1, the stress-inducible enzymes, respectively, involved in overproduction of NO and CO (32–34).

The newly upregulated iNOS produces abundant NO and this NO is transformed into $ONOO^-$ upon scavenging the O_2^- overproduced by the LPS-driven oxidative burst. Alternatively, NO can also be transformed into nitrogen dioxide radical ($NO_2^\bullet$) upon binding O_2 in aerobic conditions. $NO_2^\bullet$ is a highly reactive nitrogen species which can oxidize GSH and cause additional nitrosative stress. Additionally, NO can also be transformed either to NO^+ or NO^- after undergoing intracellular redox metabolism. All these NO-derived redox metabolites, $ONOO^-$ and $NO_2^\bullet$ will cause oxidation GSH to GSSG and will lower the intracellular ratio of GSH/GSSG. This will activate the Nrf2, a redox-sensitive transcription factor known to be involved in the induction of HO-1 (35). Thus, continued overproductions of O_2^- and NO, caused respectively by the LPS-driven oxidative burst and by the upregulated iNOS, will lead to a marked and sustained activation of Nrf2 and upregulation of HO-1 expression to increase the HO activity. As the result of this increased HO activity, heme will be degraded at an accelerated rate while producing abundant bilirubin and CO. These products of HO activity, bilirubin can clear both the ROS and RNS and CO can limit further productions of O_2^- and NO, which are catalyzed respectively by NADPH oxidase and iNOS, the heme-containing enzymes (Fig. 4).

Expression of the heme-degrading and the CO-producing HO-1 protein is enhanced rapidly by exposure to heme, LPS, UVA irradiation, and various other conditions that promote oxidative stress (36,37). Heme oxygenase-1 expression is also induced by the excessive NO produced in situ by elevated iNOS and other conditions that promote nitrosative stress (38–40). Furthermore, the induction of HO-1, occurring in response to the overproduced O_2^- resulting from UVA irradiation or to the overproduced NO resulting from LPS-derived induction of iNOS, is enhanced even further by prior depletion of intracellular GSH level (36,37). Thus, it appeared that overproduction of O_2^- and NO catalyzed by the heme-containing enzymes is a common response to stressful condition in aerobic cells. Aerobic cells then utilize these overproduced radicals to deplete the GSH level and to activate redox-sensitive transcription factors. Aerobic cells then use the activated transcription factors to enhance the expression of HO-1. Finally, the aerobic cells utilize the enhanced HO-1 activity not only to destroy the heme but also utilize the product of heme degradation, namely CO, to inactivate the heme-containing enzymes. Elevated HO activity resulting from induction of HO-1 expression is involved in accelerated degradation of the radical-generating heme while enhancing the production of bilirubin and CO, respectively the antioxidant and inhibitor of radical-producing heme enzyme activities. Of these products of HO activity, bilirubin is known to be a strong antioxidant detoxifying both ROS and RNS (41–43) and CO is known to bind the heme-iron contained in NADPH oxidase and iNOS, inhibiting further generation of O_2^- and NO (12,13). Thus, induction of HO-1 expression mediated by depletion of GSH in oxidatively stressed aerobic cells appears to be a universal, effective, and efficient mechanism in inhibiting further production of O_2^- and NO radicals.

In addition to CO, the NO, which is overproduced by the newly induced iNOS, is also known to bind the heme-iron contained in iNOS itself as well as in NADPH oxidase, again inhibiting further productions of both NO and O_2^- (44). This suggested that there is co-operative and sequential interaction between the overproduced NO and CO, not only in clearing the

ROS but also in limiting the iNOS from further production of NO, all leading to abolish the LPS-initiated concomitant cytotoxic effects of ROS and RNS (Fig. 4). Thus, induction of HO-1 initiated originally by the O_2^- overproduced in LPS-stimulated macrophages serves also as an inducible defense mechanism against oxidative stress not only by removing the ROS-generating toxic heme but also by generating the antioxidant bilirubin and the CO to inhibit the production of O_2^- and NO. Therefore, sustained induction of HO-1 mRNA and its inducibility both by overproduction of O_2^- and NO observed in many mammalian tissues has been rendered as a useful marker of oxidative and nitrosative stresses (45). Thus, upregulation of HO-1 gene expression in response to oxidative and nitrosative stresses is currently one of the most actively studied models of cell survival mechanism in aerobic cells.

IV. HEME OXYGENASE

Heme contained in hemoglobin is involved in delivering O_2 to tissues, and also the heme contained in variety of heme enzymes catalyzes oxidative reactions and electron transfer processes involved in generating O_2^- and NO in cells. Heme can carry out these numerous physiological functions only when it is associated with various proteins. Free heme or protein unbound heme is toxic and it dose not occur in normal cells; it is deposited in tissues only under pathological conditions. Heme is a tetrapyrrole porphyrin molecule joined with Fe^{++} and this conjugation gives the specific color and fluorescence of various iron-porphyrins. Porphyrin molecule by itself does not bind O_2 and does not participate in oxidative reactions or carry out electron transfer processes; these properties are endowed only when the porphyrin is bound with divalent transition metals (i.e., Fe, Mg, Cu, Zn, Sn, and Co) (46). Protoheme is ferro-protoporphyrin (Fe^{++}-state) and it is readily oxidized to ferri-protoporphyrin (Fe^{+++}-state), which then undergoes reduction back to protoheme in the presence of O_2 and reducing agents. This generates O_2^- in aerobic cells. Because the free heme not bound to proteins can participate in Fenton or Haber–Weiss reactions and produces highly reactive hydroxyl radical ($HO^{\bullet}$), it is highly toxic by altering protein structures and by causing lipid peroxidation. To protect from this toxicity of free heme, aerobic organisms have developed highly efficient means to compartmentalize and regulate heme synthesis, to sequester heme into proteins, to store and to transport as well as to degrade.

Heme molecule can be released from heme proteins by the ROS-driven structural alteration, fragmentation, and enhanced proteolytic digestion (47). Released heme undergoes oxidative degradation by heme oxygenases (HOs) contained in endoplasmic reticulum membranes. Heme is the natural substrate of HO. In the course of HO catalyzed heme oxidation reaction, NADPH is utilized both for activation of O_2 and reduction of heme-iron substrate from the HO-undegradable Fe^{+++} form to the HO-degradable Fe^{++}-state (48). Thus, initial input of an electron from NADPH starts the multistep process of HO reaction, oxidatively cleaving the methene–carbon bridges in protoporphyrin tetrapyrrole substrate, which is holding the Fe^{++}. After an initial release of CO and Fe^{++}, biliverdin is formed. Thus, the microsomal HO enzyme system requires concerted activity of microsomal NADPH-P450 reductase, which transfers an electron from NADPH to the heme substrate and also to reduce the O_2 for utilization in oxidative cleavage of heme that has entered heme pocket in the HO enzyme protein (see below). For each molecule of heme oxidized, three molecules of O_2 and three molecules of NADPH are utilized. Also, for each mole of O_2 used in the HO reaction, one mole of H_2O is produced and thus, HO is a microsomal-mixed function oxidase (46).

There are at least three distinct isoforms of HO, the (heme-substrate) inducible HO-1 and the constitutively expressed noninducible HO-2 and HO-3 (49). These HO isoforms are the products of different genes and they have wide differences in amino acid composition (50–52). Despite major differences between HO-1 and HO-2, there is a conserved 24 amino acid segment both in HO-1 and HO-2 (53). This conserved 24 amino acid segment is hydrophobic and forms a pocket that binds the hydrophobic heme substrate, thus constituting the heme pocket (54). This heme pocket does not recognize the metal portion of metalloporphyrins, but has specificity for the side chain of porphyrin ring. Thus, some of the nonphysiological metalloporphyrins like Zn- or Sn-protoporphyrins, which have the same porphyrin side chain as the Fe-protoporphyrin (heme), can compete with heme binding to the pocket and can inhibit the HO activity (55). Both HO-1 and HO-2 proteins are anchored to endoplasmic reticulum by another hydrophobic amino acid sequences present at the carboxyl terminal of these HO proteins and this hydrophobic region is not involved in the catalytic activity (56).

In recent years, physiological role of HO has changed from a simple task of heme degradation to the biological effects of its activity, namely, biliverdin (bilirubin) and CO. During the late 1980s, Ames and coworkers demonstrated that both biliverdin and bilirubin have strong antioxidant properties detoxifying ROS and RNS (42). Also, in the early 1990s, Marks et al. (57) and Snyder and coworkers (58) have suggested that CO mediates vasorelaxation and neurotransmission, respectively. Therefore, all products of HO activity are biologically active; CO is a signal molecule activating guanyl cyclase (GC) to enhance the generation of cGMP (59,60), Fe^{++} downregulates the expression of many genes including that of iNOS (61), and bilirubin (biliverdin) is a potent antioxidant (42,43,62,63). Thus, transgenic mice overexpressing HO-1 and HO-2 were found to have lower activity of lipid peroxidation (64), perhaps due to enhanced degradation of heme and also to the overproduction of bilirubin, the antioxidant. In addition, these transgenic animals were found to have altered behavioral patterns, perhaps due to abundant production of CO, a retrograde neurotransmitter in brain (65).

Aside from these functions of the products of HO activity, HO system plays a key role in maintaining cellular redox homeostasis by eliminating the toxic-free heme. Thus, HO system protects cells from deleterious effects of free-heme molecule, which is known to be the most effective promoter causing formation of reactive OH• and lipid peroxidation (66). Furthermore, HO is involved not only in the catabolic pathway of disposing toxic heme, but also in the anabolic pathway of producing bioactive molecules like bilirubin pigments and CO, which bring, respectively, the antioxidant and the cGMP-elevating physiologic effects. Thereafter, it was also learned that HO activity could be enhanced not only by heme, the native substrate, but also by various other nonheme agents like endotoxin (LPS), heavy metals, cytokines, mitogens, and hormones (67–74). Indeed, studies have revealed that following induction of HO-1 expression in vascular tissues, there is an increased production of CO and cGMP (75,76). It has also been shown that upregulation of HO-1 expression and consequent overproduction of intracellular bilirubin is associated with protection against the peroxynitrite (ONOO$^-$)-mediated apoptosis (77), oxidant-dependent microvascular leukocyte adhesion (78), and postischemic myocardial dysfunction (79). In support of this, cells treated with hemin (oxidized heme) had elevated HO-1 expression and bilirubin production and these hemin-treated cells were found to be highly resistant to the cytotoxicity caused by stronger or additional oxidants. High resistance to cytotoxicity was observed only when the cells are producing the bile pigment actively. This strongly implicated that activation or induction of HO-1 pathway overproducing bilirubin provides cytoprotection against the toxicity caused by oxidative stress (62,80). Thereafter, this increase of HO activity was found to be due to de novo synthesis of HO-1 protein and subsequently, the list of HO-1 inducers has been expanded to include many of the heme and nonheme compounds as well as stressful nonspecific physical stimuli. Intense efforts have been made to define the unifying mechanism to explain the induction of HO-1, which is caused by chemically and structurally unrelated diverse agents.

In 1991, Tyrrell and coworkers (81) noted that many known inducers of HO-1 could promote cellular oxidative stress and that induction of HO-1 represented a general cellular response to oxidative stress. Cellular oxidative stress, resulting either from increased production of ROS or from decreased levels of intracellular reductants, appeared to be the common effector mechanism for various inducers of

HO-1 expression. They concluded that induction of HO-1 "reflects a powerful mechanism by which the pro-oxidant state of cells can be transiently reduced in order to avoid cellular damage during a sustained oxidative stress" (82). In support of this, Nath et al. (83) demonstrated that prior induction of HO-1 protected rats from renal failure and mortality resulting from glycerol-induced oxidative stress called rhabdomyolysis; and the opposite effect was observed upon inhibition of HO-1 activity. Subsequently, many investigators have confirmed the protective function of HO-1 induction in preventing the oxidative injury caused by heme or nonheme insults (62,70,84–86).

V. CO-OPERATIVE INTERACTION BETWEEN iNOS AND HO-1

Biological function of HO system, the nonheme enzymes that oxidize heme to generate CO and the function of NOS system, the heme-containing enzymes that use heme to produce NO, is related to heme. HO degrades heme, but NOS needs heme. However, there are certain similarities and differences between the HO and NOS systems. As for the similarities, both the HO and NOS enzyme systems have constitutive and inducible isoforms. Expression of the inducible form of NOS (iNOS) is enhanced by some of the oxidative stimuli such as bacterial endotoxins (LPS), cytokines, and ROS, which also induce HO-1 (38). Also, expression of both HO-1 and iNOS can be induced in many of the same organs and tissues subjected to stressful conditions, and their induction involves gene activation and de novo synthesis of HO-1 and iNOS enzyme proteins. During inflammation, both iNOS and HO-1 are induced highly in macrophages, neutrophils, and mononuclear cells.

Extending the similarities, products of HO and NOS systems, respectively the CO and NO, are both gaseous ligands that bind to the heme-iron contained in hemoglobin, NADPH oxidase, NOSs, cytochromes and various other hemoprotein enzymes which catalyze several essential biological processes like O_2^- and NO production, oxygen transport, electron transfer, energy production, and biotransformation of xenobiotics. The Fe^{++} form of heme, but not the Fe^{+++} form of heme, contained in these hemoproteins reacts with π-bonding diatomic gaseous ligands like O_2, CO, and NO at physiological pH (46). However, the relative affinity of heme for binding the O_2, CO, and NO differs; CO has stronger affinity for hemoglobin than O_2 and thus, CO can displace O_2 readily from heme. However, NO binds more firmly to hemoglobin heme than CO and thus, NO will replace both O_2 and CO from heme (87,88). Alternatively, the NO bound to hemoglobin will undergo oxidation or reduction and will be released, thus delivering NO to tissues.

There are some differences between the HO and NOS systems. As mentioned above, products of HO and NOS systems, respectively CO and NO, share affinity for the heme-iron and they resemble one another in activating the heme-containing GC toward markedly increased production of cGMP (89). Guanyl cyclase is a dimeric hemoprotein and each dimer contains two hemes. Ignarro et al. (90) have demonstrated that NO binds to the heme-iron present in GC and increase its V_{max} for enhanced conversion of GTP to cGMP. Furthermore, it was demonstrated that CO generated in vivo by HO activity also binds to the heme in GC to stimulate the production of cGMP, however, much less effectively (only 1/30) than that enhanced by the NO binding (89,91). Furthermore, unlike NO, which is a free radical causing damage to invading pathogens as well as to surrounding host cells, CO is not a free radical and cannot kill invading pathogens or produce tissue damage. In fact, the ability of an organism to mount the O_2^-- and NO-producing inflammatory immune response to invading pathogen is suppressed when the background HO-1 activity is increased. This may be linked to the HO-derived overproduction of CO, which can inhibit the NADPH oxidase catalyzed oxidative burst. Carbon monoxide also binds to the heme-iron contained in iNOS and can inhibit NO production. Thus, CO can inhibit additional production of both O_2^- and NO. Thus, unlike the NO, which is proinflammatory and cytotoxic, CO provides both anti-inflammatory and cytoprotective activity (83,92,93).

In this connection, Willis et al. (92) observed that HO activity in various tissues is increased markedly in response to inflammatory stimuli and the tissues with elevated HO activity is protected from both oxidative and nitrosative injuries. Conversely, when the elevated HO activity is inhibited by administration of Zn-protoporphyrin (Zn-PP), inflammatory response is enhanced further and oxidative or nitrosative injury is potentiated. This observation suggested that elevated HO activity plays a significant role in suppressing host-cell injury caused by excessive inflammatory immune response. Thus, unlike the iNOS, HO-1 may play a key role in protecting host cells against toxicity caused by activation of inflammatory responses, such as those caused by excessive and prolonged induction of iNOS.

Heme oxygenase-1 is also known as heat-shock protein-32 (HSP32) and is induced not only by the heme itself (native physiological substrate) but also by all kinds of stimuli and agents that cause oxidative stress and pathological conditions (e.g., heat shock, ischemia, GSH oxidation and depletion, irradiation, glucose deprivation, hyperoxia or hypoxia, cellular transformation, and disease states) (36,45,80,94–98). Markedly increased level of HO-1 mRNA and protein was detected in the stressed cardiovascular and gastrointestinal systems as well as in the activated immune cells. Specifically, using an ischemia-reperfusion model, oxidative stress applied to kidney, heart, and aorta was shown to cause prominent and sustained increase of HO-1 mRNA and protein expression (94). Also, during the first weeks of life, when the liver of newborn is actively engaged in removing or degrading the fetal hemoglobin, there is a surge of HO activity and this is due mainly to marked increase of HO-1 expression (99,100). However, the hepatic expression of HO-1 declines to an undetectable level when removal of fetal hemoglobin is completed in few weeks. In the normal adult liver, HO-2 is the predominant form expressed constitutively. However, when the adult liver is stressed by an administration of heavy metal (e.g., $CdCl_2$), hematin (derived from hemolyzed RBC), bromobenzene (a free radical generator), or phenylhydrazine (a GSH-depleting drug), there is massive induction of HO-1 mRNA (up to 100-fold). This induction of HO-1 occurs without any increase of HO-2 expression (101).

The physiological sources of heme, to be used both as the native substrate of HO and also as the inducer of HO-1 expression, includes hemoglobin in RBC, myoglobin in muscles, cytochromes involved in microsomal drug oxidation (P-450), and mitochondrial respiratory chain (cytochrome *c*) as well as the newly synthesized free heme (i.e., before incorporation into apohemoproteins). There are several reasons why such heme molecules may become available for oxidative degradation by HO. (a) Free heme is being synthesized constantly for ready incorporation into apohemoproteins and the excess of this newly synthesized heme not incorporated into heme proteins is available for degradation. This may be catalyzed by HO-2, the constitutively expressed HO isoform (102). (b) Under oxidative stress, demand for additional synthesis of heme enzymes which generate ROS is halted and hence, more of the unincorporated newly synthesized heme may become available to be used not only as the substrate for degradation by HO but also as the inducer for enhancing expression of HO-1. (c) Most importantly, under the ROS-producing oxidative stress conditions, heme-containing proteins undergo denaturation, fragmentation, and enhanced proteolysis, all promoted by the ROS produced by heme enzymes themselves (6). This releases heme. This liberated heme then undergoes degradation by the HO system and also enhances the expression of HO-1 (103,104). Thus, induction of HO-1 may serve as a primary cellular defense in oxidatively stressed cells to degrade the pro-oxidant heme liberated from denatured hemoproteins and at the same time, to produce the biologically active antioxidant components like bile pigments and CO.

Under oxidatively stressed conditions, expression of iNOS is also enhanced and large amount of NO is produced. This NO, being produced continuously and abundantly by the newly induced iNOS, can act as an effective scavenger of ROS and also as an additional source for further stress. The overproduced NO derived from iNOS can move freely in and out of the cell and can bind to the heme-iron present in iNOS itself and other hemoproteins. Thus, overproduced NO can inhibit not only the additional production of NO catalyzed by iNOS, but also the energy production catalyzed by mitochondrial respiratory cytochromes and the O_2^- production catalyzed by NADPH oxidases (105). Also, as with the binding of O_2^-, the binding of NO to these heme-containing enzymes can also promote destabilization, fragmentation, and proteolysis by proteasomes. Thus, NO can also promote the release of heme from iNOS and NADPH oxidase, and this released heme can undergo degradation and removal by the HO-1 activity, which is enhanced by the released heme (106). This would create a situation in which the CO production increases at the expense of NO production. This CO can move freely in and out of the cell and bind to the heme-iron present in the remaining and functional iNOS. As the result, iNOS activity is inhibited and additional production of NO is blocked. With such "double-punch" inhibitions on NO production (first by NO itself and second by CO), further production of NO will be "turned off," allowing the cells to survive from both oxidative and nitrosative stresses.

In this manner, the CO-producing HO activity and the NO-producing NOS activity are intimately linked in protecting cells, initially against the oxidative injury caused by O_2^- and subsequently against the nitrosative stress caused by NO. Inhibiting the elevated HO activity by exposing oxidatively stressed tissues to Zn-PP has been demonstrated to enhance the synthesis of iNOS protein to an even higher level and to enhance the rate of NO production even further, causing markedly increased cytotoxicity (107). Then, a question arises as to why should there be such an inhibitory regulation on the NO-producing iNOS induction by the CO-producing HO-1 induction. NO and the NO-derived RNS (like $ONOO^-$ and NO^+) can damage cells by interacting with DNA, –SH groups, aromatic amino acids, and transition metals such as iron in heme-containing proteins and iron–sulfur centers (108,109). Thus, if the iNOS-derived continuous overproduction of NO is not stopped, cells with elevated iNOS activity will be damaged. Therefore, continuing overproduction of NO and the resulting RNS-derived toxicity must be checked by another biological means; current evidence suggests that induction of HO-1 can fulfill this role both by degrading the heme in iNOS and by generating the CO at the same time. Thus, induction of HO-1 expression in aerobic cells plays very effective bifunctional role in cellular defense against the toxicity caused by enhanced expression of iNOS and continuing overproduction of NO, which occurs in response to the oxidative stress initiated by LPS (Fig. 4).

Induction of HO-1 is likely to modulate the overproduction of NO in many ways and several of these modulation mechanisms reflect the fact that iNOS is a hemoprotein. These include the following. (a) Elevated HO-1 activity would accelerate the degradation of newly synthesized heme and would impair de novo synthesis of functional iNOS by limiting the availability of heme needed for incorporation into newly synthesized apo-iNOS protein (active site of iNOS requires two heme molecules) (110). (b) When the NO overproduced by the iNOS itself binds to the heme contained in the iNOS protein, iNOS activity is inhibited (111). In addition, the NO-bound iNOS undergoes destabilization and enhanced proteolytic digestion by proteasomes and releases heme (47). This liberated heme can now be degraded to produce and release CO by the elevated HO-1 activity. (c) This CO moves freely within the cell and binds to the heme-iron present in the remaining iNOS, which may still be functioning. This CO binding to the iNOS-heme would inhibit further production of NO by blocking the iNOS activity. In fact, CO has been reported to bind iNOS and to inhibit NO production (12,13). (d) Iron released from the degraded heme can inhibit additional transcription of iNOS gene and can suppress additional synthesis of iNOS protein (61). Based on these inhibitory effects of HO-1 induction on iNOS protein and activity, proposed mechanisms involved in this inhibitory interaction between HO-1 and iNOS is presented schematically (Fig. 5).

Conversely, NO has been shown to modulate the HO activity, both to inhibit as well as to activate the HO activity. Willis et al. (93) demonstrated that HO-2 activity in various tissues is inhibited by an in vivo administration of NO-donors. However, Motterlini et al. (112) have demonstrated that HO-1

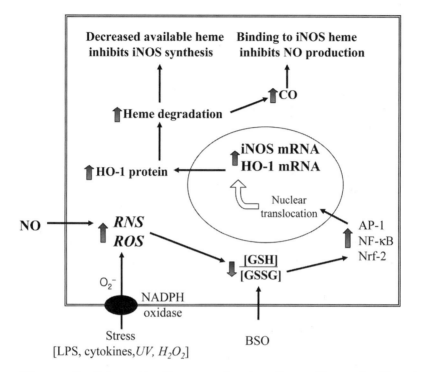

Figure 5 Schematic diagram showing the pathway leading to iNOS and HO-1 induction. Upon induction of HO-1, further synthesis of iNOS is inhibited due to lack of heme available to be incorporated into the newly synthesized apo-iNOS protein and further production of NO by iNOS is inhibited by the CO generated from the degraded heme.

activity in endothelial cells is increased by exposure to NO donors. This apparent discrepancy may reflect the chemical reactivity of NO with the cysteine-SH groups present in HO-2, but not in HO-1 (113). NO, by being a free radical, can bind to the –SH groups present in HO-2 which has three cysteine residues (114), and can inhibit the HO-2 activity (113). Furthermore, NO can also inhibit both HO-1 and HO-2 activities by binding to the heme substrate with high affinity and prevent the required O_2-binding, a process required for the heme substrate to be oxidatively degraded by both HO isoforms (47). This may explain the observation made by

Willis et al. (93). At the same time, Motterlini's observation
(112) can be explained by the fact that NO, being a free
radical and serving to oxidize GSH, induces the expression
of HO-1 via activation of redox-sensitive transcription factors.
Majority of stimuli that upregulate the expression of HO-1 do
so by lowering cellular sulfhydryl–disulfide (SH/SS) redox
ratio, which triggers the activation of redox-sensitive transcription factors involved in HO-1 induction. In the promoter
sequence of HO-1 gene, there are AP-1, Nrf2, and NF-κB
responsive elements (nucleotide sequences) (1,32,67,115).

VI. LPS-DERIVED DEPLETION OF GSH
AND SEQUENTIAL INDUCTION
OF iNOS AND HO-1

As mentioned earlier, HO-1 is an inducible stress-response
protein, the expression of which can be increased markedly
in eukaryotes by its own substrate heme and a wide variety
of nonheme compounds as well as nonspecific physical stimuli
that cause a transient oxidation of cellular GSH and lowers
the GSH/GSSG ratio (82). Importance of HO-1 induction in
physiology and in disease is emphasized because of the functional role played by HO-1, not only in degrading the toxic
heme but also in producing CO and bilirubin, particularly
in conditions associated with moderate or severe inflammatory stresses in which the expression of iNOS is enhanced
and NO is overproduced (1). Recent evidences show that NO
activates the HO-1 expression in a redox-dependent manner
(116). Thus, in this section, based on the results obtained
recently in our laboratory, we will focus on the biological relevance of GSH oxidation caused by the overproduced O_2^- and
NO that lead to upregulation of HO-1 expression.

The fact that UV radiation and many other oxidizing
agents deplete cellular GSH and induce both iNOS and HO-
1 expression potently in human fibroblasts (117) provided
the basis for an emerging hypothesis indicating that aerobic
cells respond to the GSH-depleting oxidative stress by combined activation of NO- and CO production. This appears to

be a ubiquitous response occurring in aerobic cells undergoing oxidative stress. Since then, several comprehensive reviews have provided detailed information on molecular regulation of the CO generating HO-1 induction in response to agents and conditions that are known to promote ROS and RNS production (1,118,119). From these and other studies, it has been well established that a prerequisite for such increase in HO-1 expression both in in vitro and in vivo systems is the lowered cellular redox balance and that oxidation of GSH or depletion of intracellular GSH level appeared to be the common denominator (120) (Fig. 6).

Cellular glutathione exists both in its reduced (GSH) and oxidized (GSSG) forms and these undergo reversible oxidation

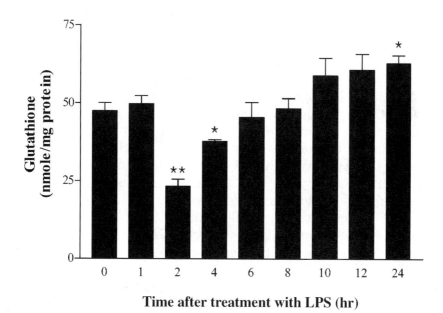

Time after treatment with LPS (hr)

Figure 6 Time-dependent changes in total glutathione level in LPS-stimulated macrophages. Cells suspended in DMEM containing 10% FBS were treated with LPS (1 μg/mL) and were harvested at the indicated times. Cells were treated with 5-SSA and solubilized, and the acidic supernatants were analyzed to determine the total glutathione (GSH + GSSG). Graph shows mean ± SEM obtained from three separate experiments.

and reduction. Maintaining their respective ratio generally at $>10:1$ with ample concentrations of each plays an essential role in keeping the intracellular redox in a normal balanced state. Oxidation and depletion of intracellular GSH occurs in conditions of moderate or severe oxidative stress and when excessive, cells are more susceptible to oxidative injury (121). In response to the stresses caused by overproduced O_2^- and NO, cellular GSH is oxidized to GSSG and GSH/GSSG ratio is lowered. Also, as the GSSG can leak out easily, cellular level of total GSH decreases. Then, a question arises as to how the O_2^- or NO could oxidize GSH to GSSG. It is well known that both O_2^- and NO do not react or oxidize GSH directly. However, their reaction product $ONOO^-$ reacts directly with GSH and oxidizes it to GSSG rapidly (Fig. 1). In support of this, in macrophages exposed to SIN-1, which releases O_2^- and NO simultaneously and generates $ONOO^-$, intracellular level of GSH becomes rapidly and severely depleted and HO-1 expression was increased markedly (122). However, when the macrophages were exposed to both SIN-1 and SOD together, the SIN-1-mediated GSH depletion and HO-1 induction was nearly abolished. This indicated that, in the presence of SOD or upon elimination of O_2^-, remaining NO alone being generated from SIN-1 does not appear to lower intracellular level of GSH and does not induce HO-1 expression by itself (123). From these results, it can be deduced that reactions of NO with O_2^- or generation of $ONOO^-$ are essential for the depletion of GSH. This may have been responsible for the GSH-protected $ONOO^-$ stress (Fig. 2) and also for the early GSH depletion observed soon after the LPS stimulation (Fig. 6).

The signaling gaseous molecule NO radical is generated by a family of NOSs, and NO plays an essential regulatory role in a variety of physiological and pathophysiological processes in cardiovascular, nervous, gastrointestinal, and immune systems (15). The distinctive biological activities evoked by NO radical are due to its uncharged nature. Having no charge, NO radical can move freely in and out of the cell, but it can also undergo ready oxidation and reduction either by losing or gaining an electron in the cell, respectively forming nitrosonium cation (NO^+) or nitroxyl anion (NO^-) (108). These NO metabolites can react with wide range of structural and functional targets in the cell. One of the fundamental aspects of NO radical is its ability to lose an electron forming NO^+, which then reacts with sulfhydryl centers in proteins and produces *S*-nitrosyl derivatives or RSNO. This chemical process, known as *S*-nitrosation, has been suggested to serve as cellular tool to stabilize and to preserve NO by holding the NO bound in cysteine residues in proteins (124). Formation of low-molecular weight RSNOs like *S*-nitroso-glutathione (GSNO) and *S*-nitroso-cysteine (CSNO) by reactions respectively with free GSH and cysteine may also play significant role in lowering the level of free NO within the cell. In this regard, intracellular concentration of GSH becomes an important determinant for the reactivity and fate of NO, because this cysteine-containing tripeptide (GSH) is abundant (5–10 mM) in most cells and thus, may function as an effective buffer and storage for excess NO at the expense of its oxidation and depletion.

Thus, in macrophages exposed to LPS undergoing oxidative burst and overproducing O_2^-, there is an immediate (2 hr) dramatic decrease of intracellular GSH level (Fig. 6). In response to such depletion of GSH, several redox-sensitive transcription factors like AP-1, NF-κB, and Nrf2 are activated and transcription of many antioxidant enzyme genes are activated. These include, most notably, the genes involved in GSH biosynthesis, NO production, and CO production (121). Thus, the γ-glutamylcysteine synthetase, an enzyme involved in GSH-biosynthesis, is induced rapidly in the LPS-stimulated macrophages. To support this increased rate of GSH biosynthesis, expression of cystine transporter system (X_c^-), required for an increased supply of cysteine, is also enhanced rapidly (125,126). Thus, in macrophages stimulated with LPS, a rapid monophasic upregulation of xCT mRNA (1 hr) was observed (Fig. 7A).

Also, along with rapid overexpression of xCT (a component protein in X_c^- system), rapid upregulations of both iNOS mRNA (2 hr) and HO-1 mRNA (1 hr) transcription occur as well. However, the upregulation of HO-1 mRNA occurred in

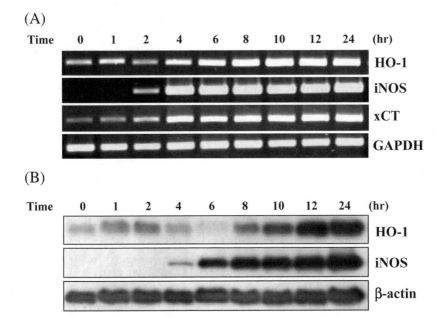

Figure 7 Time-dependent inductions of HO-1, iNOS, and xCT in LPS-stimulated macrophages. Cells were treated with LPS (1 µg/ml) and harvested at the indicated time points. (A) Total RNA was extracted and subjected to RT-PCR. (B) Harvested cells were homogenized and homogenates were analyzed to determine the contents of HO-1 and iNOS employing immunoblot analysis.

a biphasic manner, initially at 1 hr and returning to basal level before being increased again at 4 hr (Fig. 7A). In accordance with this biphasic increases in HO-1 mRNA expression, HO-1 protein levels were also elevated in a biphasic manner, first at 1–2 hr and then at 8–24 hr (Fig. 7B). While the increased iNOS mRNA expression became evident beginning at 2 hr, the increased iNOS protein level (Western blotting) began to be visible at 4 hr and significant overproduction of NO began to be observed starting at 6 hr (Griess reaction) (Fig. 8). This indicated that the NO overproduced by LPS-driven elevation of iNOS was responsible for the second-phase upregulation of HO-1 expression (Fig. 8). This also suggested further that initial depletion of GSH caused by the O_2^- over-produced from LPS-driven oxidative burst is responsible for

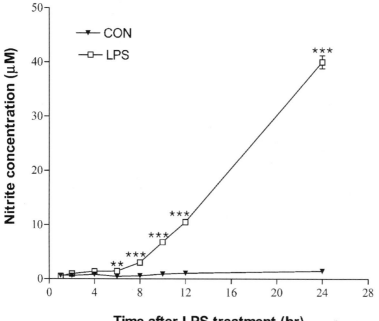

Figure 8 Time-dependent accumulation of nitrite in the culture medium of LPS-stimulated macrophages. Accumulated concentrations of nitrite present in the conditioned media of LPS-stimulated cells were determined at the indicated sampling times.

the small initial induction of HO-1 (Fig. 3). Collectively, this indicated that there is a direct link between the decrease of GSH level and rapid upregulations of both iNOS and HO-1, leading to overproduction of NO and CO.

In further support, when the iNOS-derived overproduction of NO was inhibited with either L-NAME or AG (Fig. 9A and B), the LPS-derived induction of HO-1 was decreased (Fig. 10A and B). Conversely, in macrophages treated only with exogenous NO donors like spermine NONOate in the absence of LPS stimulation, a marked depletion of total GSH level was observed and this was accompanied with a marked increase in HO-1 expression (Fig. 11). Thus, depletion of GSH caused by NO appears to be essential in enhancing the expression of HO-1 and the extent of NO-derived initial

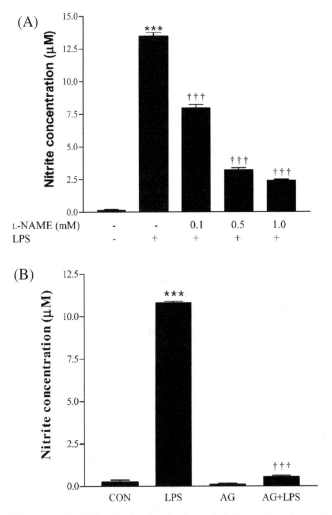

Figure 9 Effect of L-NAME and AG pretreatment on nitrite production in LPS-stimulated macrophages. (A) Cells were pretreated
with varying doses of L-NAME (0.1, 0.5, 1.0 mM) at 30 min before
addition of LPS (1 μg/ml). Cells were harvested at 12 hr after the
addition of LPS. Accumulated nitrite concentrations present in
the conditioned media were determined. (B) Cells were pretreated
with AG (0.5 mM) at 30 min before addition of LPS (1 μg/ml). Cells
were harvested at 12 hr after the addition of LPS. Accumulated
nitrite concentrations present in the conditioned media were
determined.

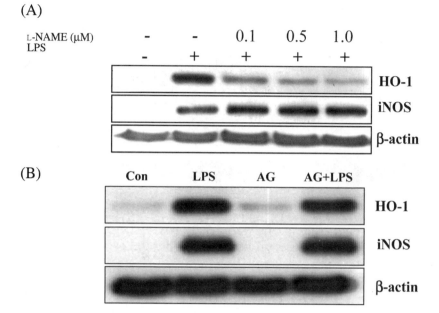

Figure 10 Effect of L-NAME and AG pretreatment on the LPS-derived induction of HO-1 and iNOS. (A) Cells were pretreated with varying doses of L-NAME (0.1, 0.5, 1.0 mM) at 30 min before addition of LPS (1 µg/ml). Cells were harvested at 12 hr after the addition of LPS. Accumulated nitrite concentrations present in the conditioned media were determined. Harvested cells were homogenized and the accumulated contents of HO-1 and iNOS were determined. (B) Cells were pretreated with AG (0.5 mM) at 30 min before addition of LPS (1 µg/ml). Cells were harvested at 12 hr after the addition of LPS. Accumulated nitrite concentrations present in the conditioned media were determined. Harvested cells were homogenized and the accumulated contents of HO-1 and iNOS were determined.

GSH depletion is critical. In support of this, in macrophages with depleted GSH level achieved only by treatment with buthionine sulfoximine (BSO), an inhibitor of GSH biosynthesis, expression of HO-1 was increased (Fig. 12A and B). Furthermore, when these GSH-depleted cells were treated with LPS, an immediate, nearly continuous, and marked monophasic induction of HO-1 expression was observed along

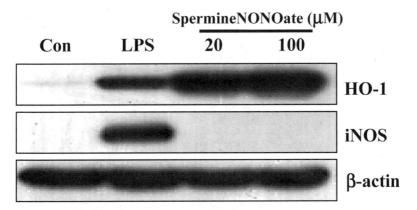

Figure 11 Effect of spermine NONOate, an NO donor, on HO-1 and iNOS expression. Cells were treated with spermine NONOate (20, 100 µM) and harvested at 12 hr. Homogenates were analyzed to determine the contents of HO-1 and iNOS proteins employing immunoblot analysis.

with somewhat reduced induction of iNOS and depressed overproduction of NO (Figs. 13 and 14). The level of HO-1 induction observed under this condition was almost twofold greater than that observed in the LPS-treated cells having normal level of GSH (Fig. 13). This indicated that depletion of GSH combined with overproduction of NO potentiates the HO-1 induction in a synergistic or additive manner. Conversely, when LPS was given to macrophages with elevated GSH level achieved by pretreatment with *N*-acetylcysteine (NAC), HO-1 induction was nearly abolished (Fig. 12A and B). This suggested that thiols like NAC and GSH antagonize the effects of O_2^- and NO in enhancing the expression of HO-1 and also that depletion of GSH is essential for both the O_2^-- and NO-derived induction of HO-1 expression.

Further support for the hypothesis, suggesting that depletion of GSH is the main cause of inducing HO-1 expression, came from direct comparison between agents that release NO, NO^+, or NO^-. Their intrinsic ability to deplete GSH level to various degrees was positively correlated to their ability to increase HO activity in endothelial cells (40,123). In this study, they also showed that HO-1 inducing effect

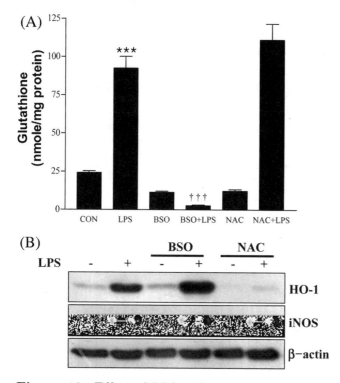

Figure 12 Effect of BSO (inhibitor of GSH synthesis) and NAC (precursor of GSH synthesis) on LPS-derived changes in cellular total GSH levels and in the induction of HO-1 and iNOS expression. Cells were incubated for 12 hr separately with LPS (1 μg/ml), BSO (100 μM), and NAC (2.5 mM) alone or in combinations of LPS and BSO or LPS and NAC. (A) Harvested cells were rinsed with 5-SSA and solubilized. Acidic supernatants were analyzed to determine the contents of total glutathione (GSH + GSSG). (B) Harvested cells were homogenized and homogenates were analyzed to determine the contents of HO-1 and iNOS proteins by employing immunoblot analysis.

produced either by oxidative or nitrosative stresses could be modified by several factors that modulate intracellular GSH level. Upon addition of thiols like NAC, which is utilized in the biosynthesis of GSH, HO-1 induction caused by NO donors was suppressed and HO activity was diminished significantly (40,123). This suggested that activation of specific

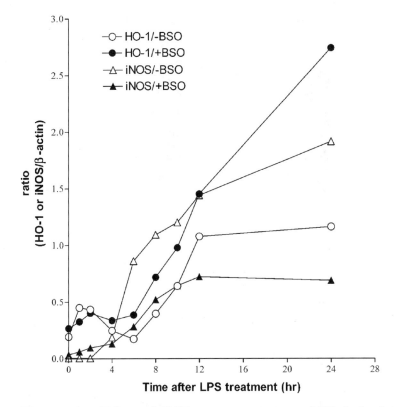

Figure 13 Effect of BSO pretreatment on LPS-derived time-dependent inductions of HO-1 and iNOS. Cells were pretreated with BSO (100 μM) for 4 hr before addition of LPS (1 μg/ml) and harvested at the indicated times after the LPS exposure. Harvested cells were homogenized and the homogenates were analyzed to determine the contents of HO-1 and iNOS proteins by employing immunoblot analysis. Graph shows the results of densitometric analysis on the time-dependent accumulations of HO-1 and iNOS, which have been normalized to the β-actin contents at each time point samples.

intracellular targets, possibly the redox-sensitive transcription factors specifically involved in the induction of HO-1 expression, is inhibited by GSH or conversely, the involved transcription factors are activated by oxidation and depletion of GSH (123). Others have also demonstrated such a

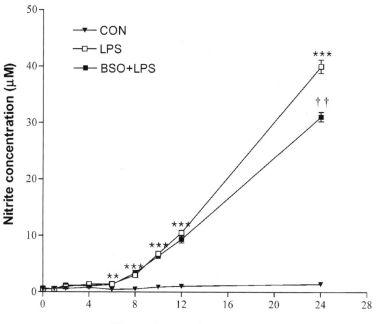

Figure 14 Suppression of LPS-derived overproduction of NO in macrophages with super-induced HO-1 expression. Cells were pre-treated with BSO (100 μM) for 4 hr before addition of LPS (1 μg/mL) and, at various times after the LPS expression, conditioned media were collected to determine the accumulated concentrations of nitrite. Along with super-induction of HO-1 and overproduction of CO but with suppression of iNOS expression, iNOS-derived NO production is inhibited.

causative association between NO-derived depletion of GSH levels and upregulations of HO-1 expression in variety of cells and tissues. These include murine macrophages (33), endothelial cells (39), rat brain (95), Chinese hamster ovary cells (127), human fibroblasts (128), mouse liver (129), and rat cardiomyocytes (130). Combined, these observations suggested that HO-1 could be induced by NO alone via its ability to oxidize and deplete the intracellular GSH level.

When intracellular levels of NO reach a critical threshold, in analogy with the oxidative stress caused by excessive production of O_2^-, metabolites of NO (RNS) react with –SH centers in proteins and protein functions may become altered as well. This phenomenon, caused by uncontrolled alteration of protein function by NO, has been termed as "nitrosative stress" (131). Thus, it appears that cells respond to both oxidative and nitrosative stresses, which oxidize and deplete GSH, in a similar manner. Oxidation and depletion of GSH leads to initiation of redox signaling by modulation of –SH centers in signal-cascade proteins, activating the redox-sensitive transcription factors, either directly or indirectly. This enhances de novo synthesis of antioxidant defense enzymes, most notably the expression of HO-1. The enhanced HO-1 activity, by virtue of its ability to eliminate the toxic heme while producing the antioxidant bilirubin and CO, can protect aerobic mammalian cells against the toxicity of $ONOO^-$, the combined product of O_2^- and NO, using bilirubin and also inhibit additional production of O_2^- and NO using CO. Therefore, by increasing the expression of HO-1, cells can adapt and resist against both oxidative and nitrosative insults by employing their own redox-regulated autoprotective mechanism. This autoprotective "crosstalks" between O_2^-, NO, and CO appears to be mediated by oxidation of cellular GSH or the decreased GSH/GSSG ratio (Fig. 4).

Recent evidence indicated that redox metabolites of NO (not the NO itself) could react directly with selective cysteine residues in the redox-sensitive transcription factors (120). This may lead to their functional activation and cause a pronounced overexpression of HO-1. However, specific transcription factors implicated in the induction of HO-1 gene transcription remain to be fully characterized. In view of the high inducibility of HO-1 observed in aerobic cells following exposure to NO and RNS (120), future work needs to address the molecular mechanisms involved in the upregulation of HO-1 expression caused by nitrosative stress. Specifically, they are: (a) chemical modification(s) within the target redox-sensitive transcription factor proteins, which is specifically responsible for the NO-mediated activation of HO-1 gene transcription, (b) transcriptionally activated regulator sequences of HO-1 gene which are sensitively modulated by the NO-derived changes of cellular redox status, and (c) biological significance of this response in protecting cells against nitrosative stress. Notable among many involved redox-sensitive transcription factors are the Fos/Jun (AP-1), NF-κB, and the recently identified Nrf2 proteins (1,35,132). Data show that NO can activate all of these redox-sensitive transcription factors and they are all known to contain cysteine residues (133). In many circumstances, activation of these transcription factors depends on the reversibility of cysteine nitrosation (134–136). Studies performed on smooth muscle cells and macrophages revealed that levels of HO-1 transcript increase by exposure to nitrosating agents like the NO-donor, spermine NONOate, and the increased HO-1 mRNA level appeared to be associated with enhanced DNA binding of AP-1 and Nrf2 (137). Once activated by lowered intracellular GSH/GSSG ratio or by depletion of GSH, these transcription factors can enhance the transcription of HO-1 gene by binding to specific DNA sequences present in the promoter and distal enhancer regions of HO-1 gene (137).

VII. CONCLUSION

The biological significance of a biphasic HO-1 upregulation induced by LPS stimulation, first by O_2^- and then by NO and NO-related species, remains to be fully elucidated. However, the striking parallelism between inductions of iNOS and HO-1 expression in sequence in the oxidatively stressed cells points to a functional role of HO-1 induction in counteracting against initial oxidative stress and also against the secondary nitrosative stress caused by the iNOS-derived overproduction of NO (Fig. 3). This hypothesis is supported by circumstantial evidences showing that some natural compounds enhancing the HO-1 activity can protect cells against both ROS- and RNS-derived injuries and that induction of HO-1 could participate directly in the restoration of cellular redox homeostasis through downregulation and inhibition of both NADPH oxidase and iNOS functions. The fact that HO-1 does not contain cysteine may be indicative of its evolutionary importance as a physiological protein, which is insensitive to inactivation either by oxidative and nitrosative stresses (113). Future studies on molecular mechanisms involved in the O_2^-- and NO-derived upregulation of HO-1 expression will enable scientists to characterize the structural and functional targets of ROS and RNS that are essential in promoting cellular adaptation to stressful conditions.

REFERENCES

1. Choi AM, Alam J. Heme oxygenase-1: function, regulation, and implication of a novel stress-inducible protein in oxidant-induced lung injury. Am J Respir Cell Mol Biol 1996; 15:9–19.

2. Abbas AK, Lichtman AH, Pober JS. Cellular and Molecular Immunology. 4th ed. Philadelphia: W.B. Saunders, 2000.

3. Pierre-Jacques P, Emmanuelle S, Olivier N, Fradelizi D. Auto-protective redox buffering systems in stimulated macrophages. BMC Immunol 2002; 3:3.

4. Babior BM. NADPH oxidase: an update. Blood 1999; 93: 1464–1476.

5. Halliwell B, Gutteridge JMC. Free Radicals in Biology and Medicine. 3rd ed. New York: Oxford Science Publications, 1999.

6. Droge W. Free radicals in the physiological control of cell function. Physiol Rev 2002; 82:47–95.

7. Lord-Fontaine S, Averill-Bates DA. Heat shock inactivates cellular antioxidant defenses against hydrogen peroxide: protection by glucose. Free Radic Biol Med 2002; 32:752–765.

8. Papadopoulos MC, Koumenis IL, Dugan LL, Giffard RG. Vulnerability to glucose deprivation injury correlates with glutathione levels in astrocytes. Brain Res 1997; 748: 151–156.

9. Forman HJ, Torres M. Redox signaling in macrophages. Mol Aspects Med 2001; 22:189–216.

10. Jacobs AT, Ignarro LJ. Lipopolysaccharide-induced expression of interferon-beta mediates the timing of inducible nitric-oxide synthase induction in RAW 264.7 macrophages. J Biol Chem 2001; 276:47950–47957.

11. Wink DA, Mitchell JB. Chemical biology of nitric oxide: insights into regulatory, cytotoxic, and cytoprotective mechanisms of nitric oxide. Free Radic Biol Med 1998; 25: 434–456.

12. White KA, Marletta MA. Nitric oxide synthase is a cytochrome P-450 type hemoprotein. Biochemistry 1992; 31:6627–6631.

13. McMillan K, Bredt DS, Hirsch DJ, Snyder SH, Clark JE, Masters BSS. Cloned, expressed rat cerebellar nitric oxide synthase contains stoichiometric amounts of heme, which binds carbon monoxide. Proc Natl Acad Sci USA 1992; 89:11141–11145.

14. Ignarro LJ. Biosynthesis and metabolism of endothelium-derived nitric oxide. Annu Rev Pharmacol Toxicol 1990; 30:535–560.

15. Moncada S, Palmer RM, Higgs EA. Nitric oxide: physiology, pathophysiology, and pharmacology. Pharmacol Rev 1991; 43:109–142.

16. Gross SS, Wolin MS. Nitric oxide: pathophysiological mechanisms. Annu Rev Physiol 1995; 57:737–769.

17. Gupta MP, Evanoff V, Hart CM. Nitric oxide attenuates hydrogen peroxide-mediated injury to porcine pulmonary artery endothelial cells. Am J Physiol 1997; 272:L1133–L1141.

18. Kelley EE, Wagner BA, Buettner GR, Burns CP. Nitric oxide inhibits iron-induced lipid peroxidation in HL-60 cells. Arch Biochem Biophys 1999; 370:97–104.

19. Pacelli R, Wink DA, Cook JA, Krishna MC, DeGraff W, Friedman N, Tsokos M, Samuni A, Mitchell JB. Nitric oxide potentiates hydrogen peroxide-induced killing of *Escherichia coli*. J Exp Med 1995; 182:1469–1479.

20. Hata Y, Ota S, Hiraishi H, Terano A, Ivey KJ. Nitric oxide enhances cytotoxicity of cultured rabbit gastric mucosal cells induced by hydrogen peroxide. Biochim Biophys Acta 1996; 1290:257–260.

21. Yamada M, Momose K, Richelson E, Yamada M. Sodium nitroprusside-induced apoptotic cellular death via production of hydrogen peroxide in murine neuroblastoma N1E-115 cells. J Pharmacol Toxicol Methods 1996; 35:11–17.

22. Wink DA, Hanbauer I, Krishna MC, DeGraff W, Gamson J, Mitchell JB. Nitric oxide protects against cellular damage and cytotoxicity from reactive oxygen species. Proc Natl Acad Sci USA 1993; 90:9813–9817.

23. Wink DA, Cook JA, Krishna MC, Hanbauer I, Degraff W, Gamson J, Mitchell JB. Nitric oxide protects against alkyl peroxide-mediated cytotoxicity: further insights into the role nitric oxide plays in oxidative stress. Arch Biochem Biophys 1995; 319:402–407.

24. Wink DA, Cook JA, Pacelli R, DeGraff W, Gamson J, Liebmann J, Krishna MC, Mitchell JB. The effect of various nitric oxide-donor agents on hydrogen peroxide-mediated toxicity: a direct correlation between nitric oxide formation and protection. Arch Biochem Biophys 1996; 331:241–248.

25. Kim YM, Bergonia H, Lancaster JR Jr. Nitrogen oxide-induced autoprotection in isolated rat hepatocytes. FEBS Lett 1995; 374:228–232.

26. Linas SL, Repine JE. Endothelial cells protect vascular smooth muscle cells from H_2O_2 attack. Am J Physiol 1997; 272:F767–F773.

27. Keefer LK, Nims RW, Davies KM, Wink DA. "NONOates" (1-substituted diazen-1-ium-1,2-diolates) as nitric oxide donors: convenient nitric oxide dosage forms. Methods Enzymol 1996; 268:281–293.

28. Wink DA, Hanbauer I, Laval F, Cook JA, Krishna MC, Mitchell JB. Nitric oxide protects against the cytotoxic effects of reactive oxygen species. Ann NY Acad Sci 1994; 738:265–278.

29. Chang J, Rao NV, Markewitz BA, Hoidal JR, Michael JR. Nitric oxide donor prevents hydrogen peroxide-mediated endothelial cell injury. Am J Physiol 1996; 270:L931–L940.

30. Degnim AC, Morrow SE, Ku J, Zar HA, Nakayama DK. Nitric oxide inhibits peroxide-mediated endothelial toxicity. J Surg Res 1998; 75:127–134.

31. Rubbo H, Parthasarathy S, Barnes S, Kirk M, Kalyanaraman B, Freeman BA. Nitric oxide inhibition of lipoxygenase-dependent liposome and low-density lipoprotein oxidation: termination of radical chain propagation reactions and formation of nitrogen-containing oxidized lipid derivatives. Arch Biochem Biophys 1995; 324:15–25.

32. Ishii T, Itoh K, Takahashi S, Sato H, Yanagawa T, Katoh Y, Bannai S, Yamamoto M. Transcription factor Nrf2 coordinately regulates a group of oxidative stress-inducible genes in macrophages. J Biol Chem 2000; 275:16023–16029.

33. Camhi SL, Alam J, Wiegand GW, Chin BY, Choi AMK. Transcriptional activation of the HO-1 gene by lipopolysaccharide is mediated by 5′ distal enhancers: role of reactive oxygen intermediates and AP-1. Am J Respir Cell Mol Biol 1998; 18:226–234.

34. Kim YM, Lee BS, Yi KY, Paik SG. Upstream NF-kappaB site is required for the maximal expression of mouse inducible nitric oxide synthase gene in interferon-gamma plus lipopolysaccharide-induced RAW 264.7 macrophages. Biochem Biophys Res Commun 1997; 236:655–660.

35. Alam J, Stewart D, Touchard C, Boinapally S, Choi AM, Cook JL. Nrf2, a Cap'n'Collar transcription factor, regulates induction of the heme oxygenase-1 gene. J Biol Chem 1999; 274:26071–26078.

36. Ossola JO, Tomaro ML. Heme oxygenase induction by UVA radiation. A response to oxidative stress in rat liver. Int J Biochem Cell Biol 1998; 30:285–292.

37. Basu-Modak S, Tyrrell RM. Singlet oxygen: a primary effector in the ultraviolet A/near-visible light induction of the human heme oxygenase gene. Cancer Res 993; 53:4505–4510.

38. Kitamura Y, Furukawa M, Matsuoka Y, Tooyama I, Kimura H, Nomura Y, Taniguchi T. In vitro and in vivo induction of heme oxygenase-1 in rat glial cells: possible involvement of nitric oxide production from inducible nitric oxide synthase. Glia 1998; 22:138–148.

39. Motterlini R, Foresti R, Bassi R, Calabrese V, Clark JE, Green CJ. Endothelial heme oxygenase-1 induction by hypoxia. Modulation by inducible nitric-oxide synthase and S-nitrosothiols. J Biol Chem 2000; 275:13613–13620.

40. Naughton P, Foresti R, Bains SK, Hoque M, Green CJ, Motterlini R. Induction of heme oxygenase 1 by nitrosative stress. A role for nitroxyl anion. J Biol Chem 2002; 277:40666–40674.

41. Stocker R, Yamamoto Y, McDonagh AF, Glazer AN, Ames BN. Bilirubin is an antioxidant of possible physiological importance. Science 1987; 235:1043–1046.

42. Stocker R, McDonagh AF, Glazer AN, Ames BN. Antioxidant activities of bile pigments: biliverdin and bilirubin. Methods Enzymol 1990; 186:301–309.

43. Mancuso C, Bonsignore A, Di Stasio E, Mordente A, Motterlini R. Bilirubin and S-nitrosothiols interaction: evidence for a possible role of bilirubin as a scavenger of nitric oxide. Biochem Pharmacol 2003; 66:2355–2363.

44. Abu-Soud HM, Wang J, Rousseau DL, Fukuto JM, Ignarro LJ, Stuehr DJ. Neuronal nitric oxide synthase self-inactivates by forming a ferrous–nitrosyl complex during aerobic catalysis. J Biol Chem 1995; 270:22997–23006.

45. Trekli MC, Riss G, Goralczyk R, Tyrrell RM. Beta-carotene suppresses UVA-induced HO-1 gene expression in cultured FEK4. Free Radic Biol Med 2003; 34:456–464.

46. Maines MD. The heme oxygenase system: a regulator of second messenger gases. Annu Rev Pharmacol Toxicol 1997; 37:517–554.

47. Davies KJ. Protein damage and degradation by oxygen radicals. I. General aspects. J Biol Chem 1987; 262: 9895–9901.

48. Liu Y, Ortiz de Montellano PR. Reaction intermediates and single turnover rate constants for the oxidation of heme by human heme oxygenase-1. J Biol Chem 2000; 275:5297–5307.

49. Elbirt KK, Bonkovsky HL. Heme oxygenase: recent advances in understanding its regulation and role. Proc Assoc Am Phys 1999; 111:438–447.

50. Cruse I, Maines MD. Evidence suggesting that the two forms of heme oxygenase are products of different genes. J Biol Chem 1988; 263:3348–3353.

51. Shibahara S, Yoshizawa M, Suzuki H, Takeda K, Meguro K, Endo K. Functional analysis of cDNAs for two types of human heme oxygenase and evidence for their separate regulation. J Biochem 1993; 113:214–218.

52. McCoubrey WK Jr, Huang TJ, Maines MD. Isolation and characterization of a cDNA from the rat brain that encodes hemoprotein heme oxygenase-3. Eur J Biochem 1997; 247:725–732.

53. Rotenberg MO, Maines MD. Characterization of a cDNA-encoding rabbit brain heme oxygenase-2 and identification of a conserved domain among mammalian heme oxygenase isozymes: possible heme-binding site? Arch Biochem Biophys 1991; 290:336–344.

54. Schacter BA, Cripps V, Troxler RF, Offner GD. Structural studies on bovine spleen heme oxygenase. Immunological and structural diversity among mammalian heme oxygenase enzymes. Arch Biochem Biophys 1990; 282:404–412.

55. Maines MD. Zinc protoporphyrin is a selective inhibitor of heme oxygenase activity in the neonatal rat. Biochim Biophys Acta 1981; 673:339–350.

56. McCoubrey WK Jr, Maines MD. Domains of rat heme oxyge-nase-2: the amino terminus and histidine 151 are required for heme oxidation. Arch Biochem Biophys 1993; 302:402–408.

57. Marks GS, Brien JF, Nakatsu K, McLaughlin BE. Does carbon monoxide have a physiological function? Trends Pharmacol Sci 1991; 12:185–188.

58. Verma A, Hirsch DJ, Glatt CE, Ronnett GV, Snyder SH. Carbon monoxide: a putative neural messenger. Science 1993; 259:381–384.

59. Brune B, Ullrich V. Inhibition of platelet aggregation by carbon monoxide is mediated by activation of guanylate cyclase. Mol Pharmacol 1987; 32:497–504.

60. Utz J, Ullrich V. Carbon monoxide relaxes ilial smooth mus-cle through activation of guanylate cyclase. Biochem Pharma-col 1991; 41:1195–2001.

61. Weiss G, Werner-Felmayer G, Werner ER, Grunewald K, Wachter H, Hentze MW. Iron regulates nitric oxide synthase

activity by controlling nuclear transcription. J Exp Med 1994; 180:969–976.

62. Clark JE, Foresti R, Green CJ, Motterlini R. Dynamics of haem oxygenase-1 expression and bilirubin production in cellular protection against oxidative stress. Biochem J 2000; 348:615–619.

63. Kaur H, Hughes MN, Green CJ, Naughton P, Foresti R, Motterlini R. Interaction of bilirubin and biliverdin with reactive nitrogen species. FEBS Lett 2003; 543:113–119.

64. Panahian N, Yoshiura M, Maines MD. Overexpression of heme oxygenase-1 is neuroprotective in a model of permanent middle cerebral artery occlusion in transgenic mice. J Neurochem 1999; 72:1187–1203.

65. Maines MD, Polevoda B, Coban T, Johnson K, Stoliar S, Huang TJ, Panahian N, Cory-Slechta DA, McCoubrey WK Jr. Neuronal overexpression of heme oxygenase-1 correlates with an attenuated exploratory behavior and causes an increase in neuronal NADPH diaphorase staining. J Neurochem 1998; 70:2057–2069.

66. Wagener FA, Volk HD, Willis D, Abraham NG, Soares MP, Adema GJ, Figdor CG. Different faces of the heme–heme oxygenase system in inflammation. Pharmacol Rev 2003; 55:551–571.

67. Camhi SL, Alam J, Otterbein L, Sylvester SL, Choi AMK. Induction of heme oxygenase-1 gene expression by lipopolysaccharide is mediated by AP-1 activation. Am J Respir Cell Mol Biol 1995; 13:387–398.

68. Carraway MS, Ghio AJ, Taylor JL, Piantadosi CA. Induction of ferritin and heme oxygenase-1 by endotoxin in the lung. Am J Physiol 1998; 275:L583–L592.

69. Sardana MK, Sassa S, Kappas A. Metal ion-mediated regulation of heme oxygenase induction in cultured avian liver cells. J Biol Chem 1982; 257:4806–4811.

70. Laniado-Schwartman M, Abraham NG, Conners M, Dunn MW, Levere RD, Kappas A. Heme oxygenase induction with attenuation of experimentally induced corneal inflammation. Biochem Pharmacol 1997; 53:1069–1075.

71. Cantoni L, Rossi C, Rizzardini M, Gadina M, Ghezzi P. Interleukin-1 and tumour necrosis factor induce hepatic haem oxygenase. Feedback regulation by glucocorticoids. Biochem J 1991; 279:891–894.

72. Rizzardini M, Terao M, Falciani F, Cantoni L. Cytokine induction of haem oxygenase mRNA in mouse liver. Interleukin 1 transcriptionally activates the haem oxygenase gene. Biochem J 1993; 290:343–347.

73. Rossi A, Santoro MG. Induction by prostaglandin A1 of haem oxygenase in myoblastic cells: an effect independent of expression of the 70 kDa heat shock protein. Biochem J 1995; 308:455–463.

74. Zhuang H, Pin S, Li X, Dore S. Regulation of heme oxygenase expression by cyclopentenone prostaglandins. Exp Biol Med 2003; 228:499–505.

75. Sammut IA, Foresti R, Clark JE, Exon DJ, Vesely MJ, Sarathchandra P, Green CJ, Motterlini R. Carbon monoxide is a major contributor to the regulation of vascular tone in aortas expressing high levels of haem oxygenase-1. Br J Pharmacol 1998; 125:1437–1444.

76. Wakabayashi Y, Takamiya R, Mizuki A, Kyokane T, Goda N, Yamaguchi T, Takeoka S, Tsuchida E, Suematsu M, Ishimura Y. Carbon monoxide overproduced by heme oxygenase-1 causes a reduction of vascular resistance in perfused rat liver. Am J Physiol 1999; 277:G1088–G1096.

77. Foresti R, Sarathchandra P, Clark JE, Green CJ, Motterlini R. Peroxynitrite induces haem oxygenase-1 in vascular endothelial cells: a link to apoptosis. Biochem J 1999; 339:729–736.

78. Hayashi S, Takamiya R, Yamaguchi T, Matsumoto K, Tojo SJ, Tamatani T, Kitajima M, Makino N, Ishimura Y, Suematsu M. Induction of heme oxygenase-1 suppresses venular leukocyte adhesion elicited by oxidative stress: role of bilirubin generated by the enzyme. Circ Res 1999; 85:663–671.

79. Clark JE, Foresti R, Sarathchandra P, Kaur H, Green CJ, Motterlini R. Heme oxygenase-1-derived bilirubin ameliorates

postischemic myocardial dysfunction. Am J Physiol Heart Circ Physiol 2000; 278:H643–H651.

80. Foresti R, Goatly H, Green CJ, Motterlini R. Role of heme oxygenase-1 in hypoxia-reoxygenation: requirement of substrate heme to promote cardioprotection. Am J Physiol Heart Circ Physiol 2001; 281:H1976–H1984.

81. Applegate LA, Luscher P, Tyrrell RM. Induction of heme oxygenase: a general response to oxidant stress in cultured mammalian cells. Cancer Res 1991; 51:974–978.

82. Alam J. Heme oxygenase-1: past, present, and future. Antioxid Redox Signal 2002; 4:559–562.

83. Nath KA, Balla G, Vercellotti GM, Balla J, Jacob HS, Levitt MD, Rosenberg ME. Induction of heme oxygenase is a rapid, protective response in rhabdomyolysis in the rat. J Clin Invest 1992; 90:267–270.

84. Nath KA, Haggard JJ, Croatt AJ, Grande JP, Poss KD, Alam J. The indispensability of heme oxygenase-1 in protecting against acute heme protein-induced toxicity in vivo. Am J Pathol 2000; 156:1485–1488.

85. Wang WP, Guo X, Koo MW, Wong BC, Lam SK, Ye YN, Cho CH. Protective role of heme oxygenase-1 on trinitrobenzene sulfonic acid-induced colitis in rats. Am J Physiol Gastrointest Liver Physiol 2001; 281:G586–G594.

86. Ohta K, Kikuchi T, Arai S, Yoshida N, Sato A, Yoshimura N. Protective role of heme oxygenase-1 against endotoxin-induced uveitis in rats. Exp Eye Res 2003; 77:665–673.

87. Sharma VS, Ranney HM. The dissociation of NO from nitrosylhemoglobin. J Biol Chem 1978; 253:6467–6472.

88. Motterlini R, Vandegriff KD, Winslow RM. Hemoglobin–nitric oxide interaction and its implications. Transfus Med Rev 1996; 10:77–84.

89. Stone JR, Marletta MA. Soluble guanylate cyclase from bovine lung: activation with nitric oxide and carbon monoxide and spectral characterization of the ferrous and ferric states. Biochemistry 1994; 33:5636–5640.

90. Ignarro LJ, Ballot B, Wood KS. Regulation of soluble guanylate cyclase activity by porphyrins and metalloporphyrins. J Biol Chem 1984; 259:6201–6207.

91. Kharitonov VG, Sharma VS, Pilz RB, Magde D, Koesling D. Basis of guanylate cyclase activation by carbon monoxide. Proc Natl Acad Sci USA 1995; 92:2568–2571.

92. Willis D, Moore AR, Frederick R, Willoughby DA. Heme oxygenase: a novel target for the modulation of the inflammatory response. Nat Med 1996; 2:87–90.

93. Willis D, Tomlinson A, Frederick R, Paul-Clark MJ, Willoughby DA. Modulation of heme oxygenase activity in rat brain and spleen by inhibitors and donors of nitric oxide. Biochem Biophys Res Commun 1995; 214:1152–1156.

94. Raju VS, Maines MD. Renal ischemia/reperfusion upregulates heme oxygenase-1 (HSP32) expression and increases cGMP in rat heart. J Pharmacol Exp Ther 1996; 277:1814–1822.

95. Ewing JF, Maines MD. Glutathione depletion induces heme oxygenase-1 (HSP32) mRNA and protein in rat brain. J Neurochem 1993; 60:1512–1519.

96. Takahashi S, Takahashi Y, Yoshimi T, Miura T. Oxygen tension regulates heme oxygenase-1 gene expression in mammalian cell lines. Cell Biochem Funct 1998; 16:183–193.

97. Lu R, Peng J, Xiao L, Deng HW, Li YJ. Heme oxygenase-1 pathway is involved in delayed protection induced by heat stress against cardiac ischemia-reperfusion injury. Int J Cardiol 2002; 82:133–140.

98. Chang SH, Barbosa-Tessmann I, Chen C, Kilberg MS, Agarwal A. Glucose deprivation induces heme oxygenase-1 gene expression by a pathway independent of the unfolded protein response. J Biol Chem 2002; 277:1933–1940.

99. Lin JH, Villalon P, Nelson JC, Abraham NG. Expression of rat liver heme oxygenase gene during development. Arch Biochem Biophys 1989; 270:623–629.

100. Sun Y, Maines MD. Heme oxygenase-2 mRNA: developmental expression in the rat liver and response to cobalt chloride. Arch Biochem Biophys 1990; 282:340–345.

101. Maines MD, Trakshel GM, Kutty RK. Characterization of two constitutive forms of rat liver microsomal heme oxygenase. Only one molecular species of the enzyme is inducible. J Biol Chem 1986; 261:411–419.

102. Maines MD, Kappas A. The induction of heme oxidation in various tissues by trace metals: evidence for the catabolism of endogenous heme by hepatic heme oxygenase. Ann Clin Res 1976; 8(suppl 17):39–46.

103. Maines MD, Kappas A. The degradative effects of porphyrins and heme compounds on components of the microsomal mixed function oxidase system. J Biol Chem 1975; 250:2363–2369.

104. Trakshel GM, Sluss PM, Maines MD. Comparative effects of tin- and zinc-protoporphyrin on steroidogenesis: tin-protoporphyrin is a potent inhibitor of cytochrome P-450-dependent activities in the rat adrenals. Pediatr Res 1992; 31:196–201.

105. Karupiah G, Harris N. Inhibition of viral replication by nitric oxide and its reversal by ferrous sulfate and tricarboxylic acid cycle metabolites. J Exp Med 1995; 181:2171–2179.

106. Kim YM, Bergonia HA, Muller C, Pitt BR, Watkins WD, Lancaster JR Jr. Loss and degradation of enzyme-bound heme induced by cellular nitric oxide synthesis. J Biol Chem 1995; 270:5710–5713.

107. Chakder S, Rathi S, Ma XL, Rattan S. Heme oxygenase inhibitor zinc protoporphyrin IX causes an activation of nitric oxide synthase in the rabbit internal anal sphincter. J Pharmacol Exp Ther 1996; 277:1376–1382.

108. Stamler JS, Singel DJ, Loscalzo J. Biochemistry of nitric oxide and its redox-activated forms. Science 1992; 258:1898–1902.

109. Beckman JS, Beckman TW, Chen J, Marshall PA, Freeman BA. Apparent hydroxyl radical production by peroxynitrite: implications for endothelial injury from nitric oxide and superoxide. Proc Natl Acad Sci USA 1990; 87:1620–1624.

110. Xie QW, Leung M, Fuortes M, Sassa S, Nathan C. Complementation analysis of mutants of nitric oxide synthase reveals that the active site requires two hemes. Proc Natl Acad Sci USA 1996; 93:4891–4896.

111. Colasanti M, Persichini T, Menegazzi M, Mariotto S, Giordano E, Caldarera CM, Sogos V, Lauro GM, Suzuki H. Induction of nitric oxide synthase mRNA expression. Suppression by exogenous nitric oxide. J Biol Chem 1995; 270:26731–26733.

112. Motterlini R, Foresti R, Intaglietta M, Winslow RM. NO-mediated activation of heme oxygenase: endogenous cytoprotection against oxidative stress to endothelium. Am J Physiol 1996; 270:H107–H114.

113. Ding Y, McCoubrey WK Jr, Maines MD. Interaction of heme oxygenase-2 with nitric oxide donors. Is the oxygenase an intracellular 'sink' for NO? Eur J Biochem 1999; 264:854–861.

114. Rotenberg MO, Maines MD. Isolation, characterization, and expression in *Escherichia coli* of a cDNA encoding rat heme oxygenase-2. J Biol Chem 1990; 265:7501–7506.

115. Menegazzi M, Guerriero C, Carcereri de Prati A, Cardinale C, Suzuki H, Armato U. TPA and cycloheximide modulate the activation of NF-kappa B and the induction and stability of nitric oxide synthase transcript in primary neonatal rat hepatocytes. FEBS Lett 1996; 379:279–285.

116. Motterlini R, Green CJ, Foresti R. Regulation of heme oxygenase-1 by redox signals involving nitric oxide. Antioxid Redox Signal 2002; 4:615–624.

117. Keyse SM, Tyrrell RM. Both near ultraviolet radiation and the oxidizing agent hydrogen peroxide induce a 32-kDa stress protein in normal human skin fibroblasts. J Biol Chem 1987; 262:14821–14825.

118. Tyrrell R. Redox regulation and oxidant activation of heme oxygenase-1. Free Radic Res 1999; 31:335–340.

119. Ryter SW, Otterbein LE, Morse D, Choi AM. Heme oxygenase/carbon monoxide signaling pathways: regulation and functional significance. Mol Cell Biochem 2002; 234–235:249–263.

120. Foresti R, Motterlini R. The heme oxygenase pathway and its interaction with nitric oxide in the control of cellular homeostasis. Free Radic Res 1999; 31:459–475.

121. Sies H. Glutathione and its role in cellular functions. Free Radic Biol Med 1999; 27:916–921.

122. Foresti R, Sarathchandra P, Clark JE, Green CJ, Motterlini R. Peroxynitrite induces haem oxygenase-1 in vascular endothelial cells: a link to apoptosis. Biochem J 1999; 339:729–736.

123. Foresti R, Clark JE, Green CJ, Motterlini R. Thiol compounds interact with nitric oxide in regulating heme oxygenase-1 induction in endothelial cells. Involvement of superoxide and peroxynitrite anions. J Biol Chem 1997; 272:18411–18417.

124. Stamler JS, Jaraki O, Osborne J, Simon DI, Keaney J, Vita J, Singel D, Valeri CR, Loscalzo J. Nitric oxide circulates in mammalian plasma primarily as an S-nitroso adduct of serum albumin. Proc Natl Acad Sci USA 1992; 89:7674–7677.

125. Moellering D, McAndrew J, Patel RP, Cornwell T, Lincoln T, Cao X, Messina JL, Forman HJ, Jo H, Darley-Usmar VM. Nitric oxide-dependent induction of glutathione synthesis through increased expression of gamma-glutamylcysteine synthetase. Arch Biochem Biophys 1998; 358:74–82.

126. Sato H, Kuriyama-Matsumura K, Hashimoto T, Sasaki H, Wang H, Ishii T, Mann GE, Bannai S. Effect of oxygen on induction of cystine transporter by bacterial lipopolysaccharide in mouse peritoneal macrophages. J Biol Chem 2001; 276:10407–10412.

127. Saunders EL, Maines MD, Meredith MJ, Freeman ML. Enhancement of heme oxygenase-1 synthesis by glutathione depletion in Chinese hamster ovary cells. Arch Biochem Biophys 1991; 288:368–373.

128. Lautier D, Luscher P, Tyrrell RM. Endogenous glutathione levels modulate both constitutive and UVA radiation/hydrogen peroxide inducible expression of the human heme oxygenase gene. Carcinogenesis 1992; 13:227–232.

129. Rizzardini M, Carelli M, Cabello Porras MR, Cantoni L. Mechanisms of endotoxin-induced haem oxygenase mRNA accumulation in mouse liver: synergism by glutathione depletion and protection by N-acetylcysteine. Biochem J 1994; 304:477–483.

130. Hoshida S, Nishida M, Yamashita N, Igarashi J, Aoki K, Hori M, Kuzuya T, Tada M. Heme oxygenase-1 expression and its

relation to oxidative stress during primary culture of cardio-myocytes. J Mol Cell Cardiol 1996; 28:1845–1855.

131. Hausladen A, Privalle CT, Keng T, DeAngelo J, Stamler JS. Nitrosative stress: activation of the transcription factor OxyR. Cell 1996; 86:719–729.

132. Alam J, Wicks C, Stewart D, Gong P, Touchard C, Otterbein S, Choi AM, Burow ME, Tou J. Mechanism of heme oxyge-nase-1 gene activation by cadmium in MCF-7 mammary epithelial cells. Role of p38 kinase and Nrf2 transcription factor. J Biol Chem 2000; 275:27694–27702.

133. Abate C, Patel L, Rauscher FJ, Curran T. Redox regulation of fos and jun DNA-binding activity in vitro. Science 1990; 249:1157–1161.

134. Nikitovic D, Holmgren A, Spyrou G. Inhibition of AP-1 DNA binding by nitric oxide involving conserved cysteine residues in Jun and Fos. Biochem Biophys Res Commun 1998; 242:109–112.

135. Peng HB, Libby P, Liao JK. Induction and stabilization of I kappa B alpha by nitric oxide mediates inhibition of NF-kappa B. J Biol Chem 1995; 270:14214–14219.

136. Xie QW, Kashiwabara Y, Nathan C. Role of transcription factor NF-kappa B/Rel in induction of nitric oxide synthase. J Biol Chem 1994; 269:4705–4708.

137. Hartsfield CL, Alam J, Cook JL, Choi AM. Regulation of heme oxygenase-1 gene expression in vascular smooth muscle cells by nitric oxide. Am J Physiol 1997; 273: L980–L988.

8

The Role of Gα12 Proteins in the PKC and JNK-Dependent Induction of NOS by Thrombin

SANG GEON KIM, SO YEON CHOI, MIN KYUNG CHO, and KEON WOOK KANG

National Research Laboratory, College of Pharmacy and Research Institute of Pharmaceutical Sciences, Seoul National University, Seoul, Korea

CHANG HO LEE

Department of Pharmacology and Institute of Biomedical Science, College of Medicine, Hanyang University, Seoul, Korea

ABBREVIATIONS

ATIII, antithrombin III; ERK, extracellular signal-regulated kinase; Gα12/13QL, Gα12/13 activated mutants; Gα12/13W, /13W, wild type Gα12/13; I-κBα, inhibitor-κBα; iNOS, inducible nitric oxide synthase; JNK, c-Jun N-terminal kinase; LPS, lipoplysaccharide; MAP kinase, mitogen-activated protein kinase; NF-κB, nuclear factor-κB; PKC, protein kinase C.

ABSTRACT

An imbalance between thrombin and antithrombin III contributes to vascular dysfunction, which can be attributed to excess production of inflammatory mediators including nitric oxide (NO) and prostaglandins (PGs). In view of the importance of the thrombin-activated coagulation pathway, excess NO and PGs as the culminating factors in uncontrolled vascular responses, this study investigated the effects of thrombin on the induction of inducible nitric oxide synthase (iNOS) and COX-2 in macrophages. Thrombin induced iNOS protein in the Raw264.7 cells, which was inhibited by a thrombin inhibitor, LB30057. Thrombin increased nuclear factor-κB (NF-κB) DNA binding, whose band was supershifted with anti-p65 and anti-p50 antibodies. Thrombin elicited the phosphorylation and degradation of inhibitor-κBα (I-κBα) prior to the nuclear translocation of p65. The NF-κB-mediated iNOS induction was stimulated by the overexpression of activated mutants of Gα12/13 (Gα12/13QL). Protein kinase C (PKC) depletion inhibited I-κBα degradation, NF-κB activation, and iNOS induction by thrombin or the iNOS induction by Gα12/13QL. c-Jun N-terminal kinase (JNK), p38 kinase, and ERK were all activated by thrombin. JNK inhibition by the stable transfection with a dominant negative mutant of JNK1 [JNK1(−)] completely suppressed the NF-κB-mediated iNOS induction by thrombin. Conversely, the inhibition of p38 kinase enhanced the expression of iNOS. In addition, JNK and p38 kinase oppositely controlled the NF-κB-mediated iNOS induction by Gα12/13QL. Hence, iNOS induction by thrombin was regulated by the opposed functions of JNK and p38 kinase downstream of Gα12/13. In the JNK1(−) cells, thrombin did not increase either the NF-κB binding activity or the I-κBα degradation despite I-κBα phosphorylation. Thrombin induced COX-2 in macrophages with increase in C/EBP DNA binding activity. CREB binding to the CREB binding oligonucleotide was also enhanced by thrombin treatment. Either PKC or JNK1 inhibition decreased COX-2 induction by thrombin, as was observed with the iNOS expression. These results demonstrated that thrombin induces iNOS in macrophages via

Gα12 and Gα13, which leads to NF-κB activation involving the PKC-dependent phosphorylation of I-κBα and the JNK-dependent degradation of phosphorylated I-κBα. COX-2 induction by thrombin was also dependent on PKC and JNK.

Expression of inducible nitric oxide synthase (iNOS) and production of nitric oxide (NO) greatly affect inflammatory processes (1,2). The cascade of inflammatory and clotting reactions induces the development of disseminated intravascular coagulation, and the microparticles cause extreme generation of thrombin (3). It has been shown that the level of thrombin and antithrombin complexes was higher in patients with sepsis than in healthy control subjects (4). In sepsis, the level of antithrombin III (ATIII), an endogenous coagulation inhibitor, is lowered as a result of complex formation with multiple activated clotting factors (5,6). Because of the consumption of ATIII in severe sepsis (7), the plasma ATIII level is low in septic patients. It is highly likely that activation of prothrombin to thrombin by coagulation pathway would increase the level of unbound free thrombin in these patients. In view of the imbalance between thrombin and antithrombin in septic patients as a result of exhaustion of antithrombin, we investigated the effect of thrombin on the production of NO as a potential major factor for septic shock. We explored the signaling pathways responsible for the induction of iNOS by thrombin. Thrombin induces iNOS through inhibitor-κBα (I-κBα) phosphorylation and subsequent nuclear factor-κB (NF-κB) activation under the pathways of protein kinase C (PKC) and c-Jun N-terminal kinase (JNK) in macrophage cells. Thrombin exerts mitogenic proliferation through the receptors coupled with Gα12 and Gα13 proteins belonging to the Gα12 subfamily (8,9). We examined whether thrombin activated NF-κB through the pathway involving Gα12/13. In view of the uncertainty of PKC linkage to the Gα12 subfamily, we assessed whether PKC was associated with the Gα12 family for the activation of NF-κB by thrombin. Gα12/13 proteins are implicated in Rho-dependent cytoskeletal shape change and activation of JNK (8,10). We further found that JNK was involved in the degradation of phosphorylated I-κBα downstream of Gα12/13.

I. INDUCTION OF iNOS BY THROMBIN

Thrombin, a serine protease of the trypsin family, is a key enzyme of the blood clotting system. Thrombin converts fibrinogen to fibrin and participates in the regulation of numerous physiological and pathological processes. Generation of too much thrombin leads to thrombosis, which is the major cause of morbidity and mortality. Thrombin potentiates both of IFN-γ and TNF-α-induced NO production in C6 glioma cells (11). Thrombin stimulates proliferation of smooth muscle cells and vascular disturbances through NF-κB activation (12).

We assessed the effect of thrombin on NO production in macrophages. Thrombin (10 units/mL, 12 hr) increased the production of NO by 4- to 5-fold. The extent of iNOS induction by thrombin was comparable to that by lipopolysaccharide (LPS). Although thrombin at 1 unit/mL (i.e., equivalent to the plasma concentration observed in healthy control animals) (13) minimally affected the expression of iNOS, thrombin at the concentrations of ≥ 10 units/mL notably induced the protein. Hence, the concentration used in this study is appropriate to assess the role of thrombin in sepsis. Treatment of Raw264.7 cells with LB30057, a direct thrombin inhibitor, completely blocked the induction of iNOS by thrombin, verifying that thrombin per se induces iNOS presumably through the thrombin receptor.

We showed that thrombin per se induces iNOS in macrophages. The concentration of 10 units/mL markedly increased the expression of iNOS, whereas that of 1 unit/mL showed minimal effect. The threshold effect may reflect the septic pathological situation, in which the level of unbound activated thrombin is increased as a result of the conversion of prothrombin to thrombin and the reciprocal consumption of ATIII by sepsis.

II. ACTIVATION OF NF-κB TRANSCRIPTION
FACTOR

The cis-acting elements identified on the promoter region of murine iNOS gene include NF-κB and AP-1 binding sites

(Fig. 1). The expression of iNOS is controlled primarily by the transcription factor NF-κB (14). To determine whether the induction of iNOS by thrombin was mediated with NF-κB activation, nuclear extracts prepared from cells treated with thrombin for 0.5–12 hr were probed with the radiolabeled NF-κB consensus oligonucloetide. NF-κB was activated by thrombin with the band intensity of slow migrating p65/p50 complex being increased from 30 min to 12 hr. Supershift analysis was carried out with anti-p65 and anti-p50 antibodies to confirm whether the retarded band consisted of p65 and p50 proteins. A 20-fold excess of NF-κB probe abolished the band retardation. Either anti-p65 or anti-p50 antibody supershifted the retarded band. Addition of both anti-p65 and anti-p50 antibodies also caused supershift with reduction in the band intensity of p65/p50 complex. These data indicate that p65 and p50 proteins were the components actively binding to the NF-κB binding site. Specificity of the thrombin effect on NF-κB activation was verified by LB30057.

Because p65 was the major component of NF-κB activated by thrombin, we determined translocation of p65 into the nucleus. Immunocytochemistry showed that p65 protein was located predominantly in the cytoplasm of control cells. In contrast, p65 protein moved into the nucleus after treatment with thrombin. Proteolytic degradation of I-κBα subunit precedes translocation of NF-κB to the nucleus. Studies were extended to determine whether NF-κB activation by thrombin resulted from the phosphorylation and degradation of I-κBα. Phosphorylation of I-κBα preceded I-κBα degradation. Thrombin increased phosphorylation of I-κBα at 15 min. The level of I-κBα was subsequently decreased from 30 min to 1 hr. Thus,

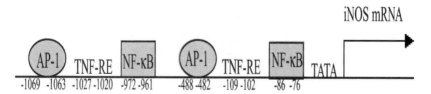

Figure 1   The promoter region of the murine iNOS gene.

thrombin activates NF-κB through the degradation of I-κBα following its phosphorylation.

III. NOS INDUCTION AND NF-κB ACTIVATION BY Gα12/13

Thrombin binding to its receptor stimulates guanine nucleotide exchange of Gα12 subfamily proteins (8). NF-κB as a pleiotropic regulator of many genes (e.g., iNOS) is involved in immune and inflammatory responses (14). We were interested in whether Gα12/13 subunits were responsible for NF-κB-mediated iNOS induction. The induction of iNOS and excess NO production by thrombin is mediated with NF-κB activation. In this study, thrombin activated p65/p50 NF-κB DNA binding complex and induced nuclear translocation of p65 protein. The role of the Gα12 family members, downstream of the thrombin signaling in NF-κB-mediated iNOS induction, was verified by the experiments using activated mutant of Gα12 or Gα13. It has been shown that thrombin induces stress fiber assembly via the Gα12 or Gα13-coupled receptor activation (8).

Expression of Gα12QL or Gα13QL also notably increased NO production in the cells. To confirm that thrombin induced iNOS in Raw264.7 cells through the pathway of Gα12/13, cells were transfected with the plasmid of Gα12W, Gα13W, Gα12QL, or Gα13QL (Gα12- or Gα13-activated mutant) and then treated with thrombin. Thrombin induced iNOS in cells transfected with the plasmid of Gα12/13W, which was comparable to that by Gα12QL or Gα13QL (Fig. 2). Next, we determined whether activated mutants of Gα12/13 stimulated binding of p65/p50 complex to NF-κB DNA consensus oligonucleotide. Gα12QL or Gα13QL increased the band intensity of p65/p50 complex.

IV. PKC-MEDIATED NF-κB ACTIVATION DOWNSTREAM OF Gα12/13

Thrombin induces cell differentiation via the PKC-dependent pathway (15). The potential role of PKC in NF-κB-mediated

iNOS ▶

Gα12W Gα12W   Gα12QL Gα13W Gα13W   Gα13QL
     +Thrombin   +Thrombin

Figure 2 Induction of iNOS by thrombin in cells transfected with the plasmid of wild-type Gα12 (Gα12W) or Gα13 (Gα13W), or by activated mutant of Gα12 (Gα12QL) or Gα13 (Gα13QL). Cells transfected with Gα12W, Gα13W, Gα12QL, or Gα13QL were exposed to thrombin for 12 hr (10 units/mL). iNOS protein and nitrite production were monitored 12 hr after 3 hr transfection of cells with the plasmid expressing Gα12QL or Gα13QL.

iNOS induction by thrombin was assessed in PKC-depleted cells. Pretreatment of cells with PMA completely inhibited both the induction of iNOS and the increase in NO production by thrombin. PKC depletion also inhibited NF-κB activation, I-κBα phosphorylation, and I-κBα degradation by thrombin, indicating that PKC plays a role in I-κBα phosphorylation by thrombin.

In the case of the pathways of Gα12 protein-coupled receptors, PKC-dependent phosphorylation is linked to the activation of Gα12/13 subunits (16). As an approach to determine whether PKC controlled the induction of iNOS downstream of Gα12/13 in Raw264.7 cells, the expression of iNOS was monitored in PKC-depleted cells transfected with Gα12/13QL. PKC depletion completely inhibited the iNOS induction by Gα12/13QL.

Thrombin differentiates normal lung fibroblasts to a myofibroblast phenotype via its receptor in a PKC-dependent pathway (15). Macrophage activation by external stimuli causes phosphorylation and degradation of I-κBα. Activation of I-κB kinase by LPS is dependent on PKC and extracellular signal-regulated kinase (ERK) (17). PKC was involved in NF-κB-mediated iNOS induction via phosphorylation and degradation of I-κBα. PKC depletion prevented the induction of iNOS by activated mutants of Gα12/13, raising the conclusion that the PKC pathway functions downstream of the activation of Gα12/13.

V. THE ROLE OF MAP KINASES IN iNOS INDUCTION

In view of the extensive studies on the mitogen-activated protein (MAP) kinase pathways for the iNOS induction (18), we were interested in whether MAP kinases contributed to the enzyme induction by thrombin. Thrombin activated all three MAP kinases, including JNK, p38 kinase, and ERK1/2, in Raw264.7 cells. The role of each of these MAP kinases in the induction of iNOS was assessed using cells stably transfected with dominant negative mutants or chemical inhibitors. Inhibition of JNK by stable transfection with a dominant negative mutant of JNK1 [JNK1(−)] completely suppressed the iNOS induction by thrombin. Conversely, inhibition of p38 kinase by SB203580 (10 μM) enhanced the enzyme expression. PD98059 (50 μM) failed to affect the iNOS expression. Hence, the induction of iNOS by thrombin was regulated by the distinct and opposed functions of JNK and p38 kinase.

VI. COUPLING OF Gα12/13 TO JNK

Activation of G proteins stimulates MAP kinases in a variety of cells (19,20). Whether MAP kinase pathways were connected with Gα12/13 was determined by using JNK1(−) cells or chemical inhibitors. iNOS protein was monitored in JNK1(−) cells transiently transfected with the plasmid expressing Gα12QL or Gα13QL. Expression of active mutant of Gα12/13 failed to induce iNOS in JNK1(−) cells. SB203580 slightly enhanced iNOS induction in cells expressing Gα12/13QL. PD98059 did not change the expression of iNOS by Gα12QL or Gα13QL.

Among the MAP kinases, the pathway of JNK was responsible for the induction of iNOS by thrombin, which was strongly supported by lack of induction of the enzyme in JNK1(−) cells. The pathway of p38 kinase oppositely regulated the induction of iNOS by thrombin. ERK1/2 activation was not responsible for the iNOS induction, as evidenced by the experiment using chemical inhibitor. Lack of iNOS induction by Gα12/13 in JNK1(–) cells supports the notion that JNK serves as the essential pathway downstream of Gα12/13 proteins. The induction of iNOS by thrombin was regulated by the opposed functions of JNK and p38 kinase downstream of Gα12/13.

VII. JNK-DEPENDENT DEGRADATION OF PHOSPHORYLATED I-κBα

To assess whether the pathway of JNK controlled NF-κB activation by thrombin, the nuclear extracts, prepared from cells treated with thrombin for 3 hr, were probed with the radiolabeled NF-κB consensus oligonucleotide. In JNK1(–) cells, thrombin did not increase NF-κB binding activity. This is consistent with the observation that the nuclear translocation of p65 was blocked by JNK1(–) stable transfection. Interestingly, I-κBα was not degraded by the presence of thrombin in JNK1 (–) cells. The level of phosphorylated I-κBα in JNK1(–) cells treated with thrombin was comparable to that of control. These data support the notion that the JNK pathway was responsible for the degradation of phosphorylated I-κBα (Fig. 3).

Gel shift and immunoblot analyses revealed that the pathway of JNK controlled NF-κB activation in response to thrombin. Increase in NF-κB DNA binding activity and nuclear translocation of p65 protein were both completely abolished in JNK1(–) cells. We found for the first time that thrombin failed to degrade I-κBα in JNK1(–) cells in spite of the increase in I-κBα phosphorylation, supporting the notion that the JNK pathway may be responsible for the degradation of phosphorylated I-κBα. The time course in phosphorylation and degradation of I-κBα by thrombin paralleled that in JNK activation. The ubiquitin–proteasome pathway controls the timed destruction of phosphorylated I-κBα for the activation of NF-κB (21,22). The accumulation of phosphorylated I-κBα and the failure of I-κBα degradation by thrombin in JNK1 (–) cells may result from the inhibition of the ubiquitin–proteasome pathway.

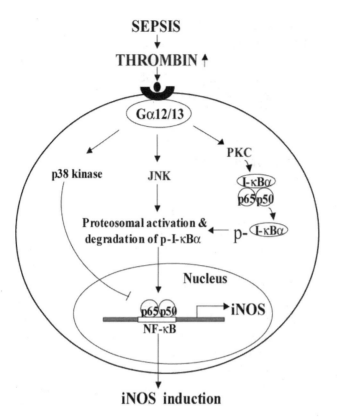

Figure 3 The proposed signaling pathways for the induction of iNOS by thrombin.

VIII. CONCLUSIONS

Thrombin plays an important role in vascular responsibility through the induction of iNOS, and thrombin activates the pathway coupled with Gα12/13 for iNOS induction. Activation of Gα12/13 then leads to PKC-dependent phosphorylation of I-κBα and JNK-mediated I-κBα degradation. The cellular signaling pathways, by which thrombin induce iNOS, may serve as important pharmacological targets for the prevention and treatment of vascular hyporeactivity in septic patients.

ACKNOWLEDGMENTS

This work was supported by National Research Laboratory Program (2001), KISTEP, The Ministry of Science and Technology, Republic of Korea.

REFERENCES

1. Wong JM, Billiar TR. Regulation and function of inducible nitric oxide synthase during sepsis and acute inflammation. Adv Pharmacol 1995; 34:155–170.

2. Szabo C, Thiemermann C. Invited opinion: role of nitric oxide in hemorrhagic, traumatic, and anaphylactic shock and thermal injury. Shock 1994; 2:145–155.

3. Esmon CT. Regulation of blood coagulation. Biochim Biophys Acta 2000; 1477:349–360.

4. Mavrommatis AC, Theodoridis T, Economou M, Kotanidou A, El Ali M, Christopoulou-Kokkinou V, Zakynthinos SG. Activation of the fibrinolytic system and utilization of the coagulation inhibitors in sepsis: comparison with severe sepsis and septic shock. Intensive Care Med 2001; 27:1853–1859.

5. Philippe J, Offner F, Declerck PJ, Leroux-Roels G, Vogelaers D, Baele G, Collen D. Fibrinolysis and coagulation in patients with infectious disease and sepsis. Thromb Haemost 1991; 65:291–295.

6. Hesselvik JF, Blomback M, Brodin B, Maller R. Coagulation, fibrinolysis, and kallikrein systems in sepsis: relation to outcome. Crit Care Med 1989; 17:724–733.

7. Nielsen JD. The effect of antithrombin on the systemic inflammatory response in disseminated intravascular coagulation. Blood Coagul Fibrinolysis 1998; (suppl 3):S11–S15.

8. Gohla A, Offermanns S, Wilkie TM, Schultz G. Differential involvement of $G\alpha12$ and $G\alpha13$ in receptor-mediated stress fiber formation. J Biol Chem 1999; 274:17901–17907.

9. Martin CB, Mahon GM, Klinger MB, Kay RJ, Symons M, Der CJ, Whitehead IP. The thrombin receptor, PAR-1, causes

transformation by activation of Rho-mediated signaling pathways. Oncogene. 2001; 20:1953–1963.

10. Nguyen QD, Faivre S, Bruyneel E, Rivat C, Seto M, Endo T, Mareel M, Emami S, Gespach C. RhoA- and RhoD-dependent regulatory switch of Gα subunit signaling by PAR-1 receptors in cellular invasion. FASEB J 2002; 16:565–576.

11. Meli R, Raso GM, Cicala C, Esposito E, Fiorino F, Cirino G. Thrombin and PAR-1 activating peptide increase iNOS expression in cytokine-stimulated C6 glioma cells. J Neurochem 2001; 79:556–563.

12. Maruyama I, Shigeta K, Miyahara H, Nakajima T, Shin H, Ide S, Kitajima I. Thrombin activates NF-κB through thrombin receptor and results in proliferation of vascular smooth muscle cells: role of thrombin in atherosclerosis and restenosis. Ann NY Acad Sci 1997; 811:429–436.

13. Bokarewa MI, Tarkowski A. Thrombin generation and mortality during *Staphylococcus aureus* sepsis. Microb Pathog 2001; 30:247–252.

14. Xie QW, Whisnant R, Nathan C. Promoter of the mouse gene encoding calcium-independent nitric oxide synthase confers inducibility by interferon-γ and bacterial lipopolysaccharide. J Exp Med 1993; 177:1779–1784.

15. Bogatkevich GS, Tourkina E, Silver RM, Ludwicka-Bradley A. Thrombin differentiates normal lung fibroblasts to a myofibroblast phenotype via the proteolytically activated receptor-1 and a protein kinase C-dependent pathway. J Biol Chem 2001; 276:45184–45192.

16. Offermanns S, Hu YH, Simon MI. Gα12 and Gα13 are phosphorylated during platelet activation. J Biol Chem 1996; 271:26044–26048.

17. Chen BC, Lin WW. PKC- and ERK-dependent activation of IκB kinase by lipopolysaccharide in macrophages: enhancement by P2Y receptor-mediated CaMK activation. Br J Pharmacol 2001; 134:1055–1065.

18. Kristof AS, Marks-Konczalik J, Moss J. Mitogen-activated protein kinases mediate activator protein-1-dependent human

inducible nitric-oxide synthase promoter activation. J Biol Chem 2001; 276:8445–8452.

19. Jho EH, Davis RJ, Malbon CC. c-Jun amino-terminal kinase is regulated by Gα12/Gα13 and obligate for differentiation of P19 embryonal carcinoma cells by retinoic acid. J Biol Chem 1997; 272:24468–24474.

20. Voyno-Yasenetskaya TA, Faure MP, Ahn NG, Bourne HR. Gα12 and Gα13 regulate extracellular signal-regulated kinase and c-Jun kinase pathways by different mechanisms in COS-7 cells. J Biol Chem 1996; 271:21081–21087.

21. Chen Z, Hagler J, Palombella VJ, Melandri F, Scherer D, Ballard D, Maniatis T. Signal-induced site-specific phosphorylation targets IκBα to the ubiquitin–proteasome pathway. Genes Dev 1995; 9:1586–1597.

22. Li CC, Dai RM, Longo DL. Inactivation of NF-κB inhibitor I-κBα: ubiquitin-dependent proteolysis and its degradation product. Biochem Biophys Res Commun 1995; 215:292–301.

9

Actions of 8-Hydroxyguanine in DNA

JIN WON HYUN

Department of Biochemistry, Cheju National University College of Medicine, Jeju, Jeju-do, Korea

MYUNG HEE CHUNG

Department of Pharmacology, Seoul National University College of Medicine, Chongno-gu, Seoul, Korea

I. INTRODUCTION

Mammals constantly form reactive oxygen species (ROS) by oxidative and reductive processes in mitochondria from oxygen (O_2) derived from respiration or by immune system exposed to foreign antigen, and externally by radiation or various chemicals. These ROS are; O_2^- (superoxide anion), $HO^\bullet$ (hydroxyl radical), $^1O_2$ (singlet oxygen), H_2O_2 (hydrogen peroxide) and $HOCl$ (hydrochlorous acid). However, antioxidant enzyme systems of body, which are superoxide dismutase, catalase, glutathione peroxidase, glutathione reductase, and various antioxidant chemicals such as vitamin C, vitamin E, uric acid, bilirubin constantly remove the ROS formed.

209

Therefore, a balance between the generation of ROS and its removal is made and then cellular functions are maintained by it. However, under conditions that produce much ROS or that suppress the antioxidant system, cells are attacked by ROS, a process referred to as "oxidative stress." In many clinical disorders, oxidative stress is increased, and a constant stream of reports concern the alleviation of these disorders by the administration of antioxidants to those suffering from disorders causing oxidative stress. For example, diabetic patients are subjected to much higher levels of oxidative stress than normal subjects, and some complications are relieved by antioxidant treatment (1). ROS are unstable and highly reactive species and damage cellular DNA, lipids, and proteins, thus impairing cellular functions. Therefore, ROS can cause various pathologic conditions such as, inflammation, aging, cancer, atherosclerosis, hypertension, and diabetes (2–8).

In particular, in the case of DNA damaged by ROS, the breakage of DNA strands or various modified bases may be generated. In this chapter, we explain the mechanism of cellular dysfunction by oxidative stress by focusing on 8-hydroxyguanine (8-oxoguanine; oh^{8}Gua), which is one of modified bases formed by oxidative stress.

II. 8-HYDROXYGUANINE (8-OXOGUANINE; oh^{8}Gua) AND ITS REPAIR ENZYMES

As described above, when DNA is damaged by oxidative stress, this damage may result in the modification of various bases. Of these, 8-hydroxyguanine (oh^{8}Gua) with a hydroxyl group at the eighth position of guanine is formed easily and abundantly by oxidative stress (9–11) and it can change to 8-oxoguanine with a keto group (Fig. 1). oh^{8}Gua is generally used as oxidative stress marker (12,13). During DNA synthesis, oh^{8}Gua can mismatch with adenine instead of cytosine, causing a GC → TA transversion, and finally mutagenesis or carcinogenesis. These relationships are demonstrated by in vitro data, reaction with DNA polymerase by using an

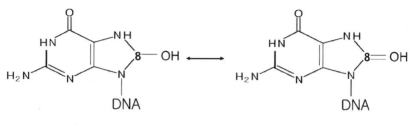

8-Hydroxyguanine   **8-Oxoguanine**

Figure 1 Structure of 8-hydroxyguanine and 8-oxoguanine.

oligonucleotide template containing oh[8]Gua, or by in vivo data, for example cells with a deficiency or supplemented with oh[8]Gua's repair enzyme show high or low mutation rates (14,15). However, repair enzymes can effectively remove oh[8]Gua in DNA and thus maintain genetic stability. These repair enzymes are three kinds (Fig. 2). First, 8-oxoguanine glycosylase 1 (OGG1) has dual functions, it acts by directly removing oh[8]Gua as a free base (glycosylase activity) and then cleaves 3′ and 5′ phosphodiester bonds to form an apurine/apyrimidine (AP) site (AP lyase activity) (Fig. 3) (16,17). Second, MYH (*Escherichia coli* MutY adenine-DNA glycosylase homolog) acts by removing adenine mismatching with oh[8]Gua (18). And third, MTH (*E. coli* MutT 8-oxo-dGTPase homolog) acts by hydrolyzing 8-oxo-dGTP to 8-oxo-dGMP, which prevents its incorporation into DNA (19). The fact that exist three specific enzymes to deal with one modified base suggest that oh[8]Gua in DNA is essentially detrimental to cellular functions.

III. MODEL FOR 8-HYDROXYGUANINE STUDIES

In addition to oh[8]Gua's oxidative stress marker or mutagenicity, some studies also have suggested that oh[8]Gua in DNA can affect cells. Many chemicals or radiations that produce ROS cause various cytotoxic effects in cells and increase the oh[8]Gua level in DNA (12,20). However, conclusive evidence of this mediation is unavailable because such ROS-producing

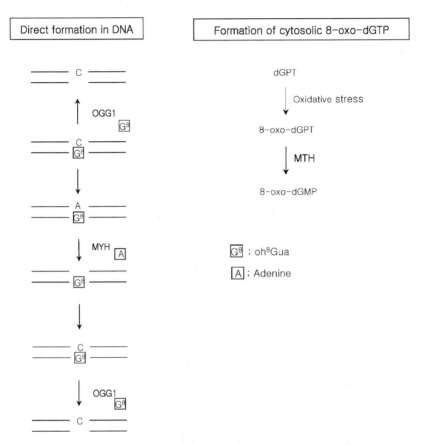

Figure 2   Modes of action of the 8-hydroxyguanine repair enzymes. OGG1 directly removes oh^{8}Gua from the oh^{8}Gua:C base pair in double stranded DNA induced by oxidative stress. MYH removes adenine mismatched with oh^{8}Gua, and MTH hydrolyzes 8-oxo-dGTP to 8-oxo-dGMP, which prevents its incorporation into DNA.

agents cause modifications in DNA other than the production of oh^{8}Gua. Therefore, to observe oh^{8}Gua's action on cells, a condition that can exclusively form oh^{8}Gua in cellular DNA should be established. Recently, we found that KG-1, an acute leukemia cell line, has no OGG1 activity (21). This cell line showed a point mutation in the OGG1 gene, namely, CGA → CAA (arginine → glutamine) at codon 229 of exon 4. Thus, the loss of the OGG1 enzyme activity in KG-1 may be

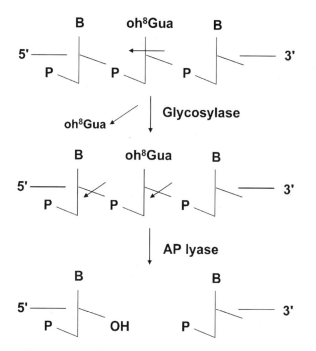

Figure 3 Actions of 8-oxoguanine glycosylase 1 (OGG1). OGG1 has dual functions, it removes oh^{8}Gua as a free base (glycosylase activity) and then cleaves the phosphodiester bonds 3′ and 5′ to the resulting apurine/apyrimidine (AP) site (AP lyase activity).

due to mutational impairment. In addition, oh^{8}Gua in cellular DNA increases in the presence of exogenous 8-hydroxydeoxy-guanosine (a nucleoside of oh^{8}Gua; oh^{8}dG) and causes cell death. We reported that senescence-accelerated mice (SAM) showed low OGG1 activity due to a point mutation of this gene and an increased amount of oh^{8}Gua in DNA (22). This means that impaired OGG1 activity may contribute to a high somatic mutation rate and accelerated senescence. OGG1 knock mice showed abolished oh^{8}Gua repair activity in a non-transcribed DNA sequence (23), and an elevated spontaneous mutation frequency in nonproliferating tissues vs. the wild type mice (17,24). This low OGG1 activity of KG-1 cells provides a useful in vitro system for studying OGG1, oxidative DNA damage, and oh^{8}Gua in DNA.

IV. 8-HYDROXYGUANINE IN DNA
AND APOPTOSIS

To investigate the effect of oh^8Gua on cells, we treated KG-1 cells with oh^8dG. We found that the concentration of oh^8Gua in DNA increased in a time- and dose-dependent pattern and that the viability of KG-1 cells was severely affected by oh^8dG (25). The nature of the killing action of oh^8dG on KG-1 cells was attributed to apoptosis, as demonstrated by increased sub-G_1 hypodiploid cells, DNA fragmentation, and apoptotic body formation, all of which are apoptosis indicators. The apoptosis induced by oh^8dG occurred through the downregulation of bcl-2, an antiapoptotic protein, and the activation of caspase 8 cleavage to Bid. This low level of bcl-2 and cleavage to Bid disturbed the mitochondrial transmembrane potential ($\Delta\varphi_m$), and downregulated level of cytochrome c in mitochondria and simultaneously increased its level in the cytosol, indicating the translocation of cytochrome c from mitochondria to the cytosol. This released cytochrome c activates caspase 9 and subsequently caspase 3, indicating that the apoptosis of KG-1 induced by oh^8dG occurs via a caspase dependent pathway (Fig. 4). However, this apoptosis appears not to be associated with Fas–Fas ligand, a death receptor pathway, because the expressions of these proteins are unchanged. Therefore, apoptosis of KG-1 by oh^8dG appears to take place via mitochondria mediated caspases activation not via a Fas associated death receptor pathway. In many cases, apoptosis is induced by ROS through oxidative stress. However, oh^8Gua in DNA itself do not cause oxidative stress in cells, because oh^8dG itself does not produce ROS. Therefore, apoptosis by oh^8dG is mediated by oh^8Gua itself rather than by indirect action through ROS. In addition to KG-1, another three susceptible cell lines to oh^8dG, i.e., CEM-CM3, Molt-4, and H9, also showed low OGG1 activities and an increased amount of oh^8Gua in their DNA. Therefore, oh^8Gua in DNA is toxic enough to kill cells and to exert a selective lethality in OGG1 deficient cells.

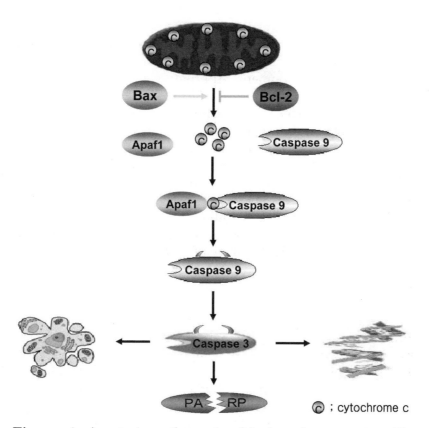

Figure 4 Apoptosis pathway by 8-hydroxydeoxyguanine. The apoptosis induced by oh^{8}dG occurred via the downregulation of bcl-2. This low level of bcl-2 disturbs the mitochondrial transmembrane potential ($\Delta\varphi_m$), leading to a decreased level of cytochrome c in mitochondria and a simultaneous increase in its level in the cytosol. The released cytochrome c activates caspase 9 and then caspase 3.

V. RADIATION, OGG1, AND 8-HYDROXYGUANINE

OGG1 activity was found to be enhanced by oxygen tension and redox-cycling compounds, for example, in rat kidneys treated with potassium bromate ($KBrO_3$) (26) or ferric nitrilotriacetate (27), in the ischemic-perfused rat heart (28), in mouse brain (29), and in cells exposed to ionizing radiation (30).

This adaptive response of oh^8Gua repair enzyme to oxidative stimuli in mammals indicates its importance in the maintenance of genetic integrity, which is constantly threatened in the aerobic environment. Alternatively and positively, the importance of oh^8Gua repair enzyme in cell survival under conditions of oxidative stress can be assessed by comparing the effects of such stress in cells or in animals deficient in this enzyme with that observed in wild type organisms. OGG1 was found to play an important role in protection against radiation induced cytotoxicities, during examinations of the functions of KG-1, a leukemia cell line without OGG1 activity. This conclusion was supported by the finding that after irradiation with γ-ray, KG-1 shows markedly inhibited cell growth, a dramatic arrest of the cell cycle at G_2/M and a high degree of apoptosis vs. cells with wild type OGG1 (31). Irradiated KG-1 showed a significant increase in oh^8Gua in DNA. However, regardless of the status of OGG1 activity, the degree of lipid peroxidation induced by irradiating the mutant and wild type cell lines was almost the same, indicating no difference in their antioxidant capacity. Thus, these findings indicate that OGG1 is important for radiation protection, as it repairs the DNA damage caused by ROS that has escaped the primary scavenging enzymes. This provides an insight into the role of OGG1 on cell survival under oxidative stress, and into the effect of oh^8Gua in DNA on cell function. Even irradiated KG-1 cells, which show marked increases of oh^8Gua in DNA, probably contain other lesions, since ionizing radiation can produce various lesions in DNA. However, lost oh^8Gua repair activity is the major difference observed between these mutant and wild type cell line, which suggests that OGG1 plays an important role in cell survival under conditions of oxidative stress, and that oh^8Gua in DNA is cytotoxic.

VI. CONCLUSION

It is known that oh^8Gua in DNA may be used as a marker of oxidative stress and that it may induce mutagenesis/carcinogenesis through GC $\rightarrow$ TA transversion. No other effects of oh^8Gua in DNA have been confirmed, because the exclusive production of oh^8Gua in DNA without contamination by other lesions remains impossible. However, it has been possible to elucidate the role of OGG1 upon oxidative DNA damage and the action of oh^8Gua in DNA using KG-1 cells, which show abrogated OGG1 activity resulting in increased oh^8Gua in DNA. KG-1 cells exogenously treated with oh^8dG show increase of oh^8Gua in DNA and apoptosis through a mitochondria associated caspases dependent pathway. This apoptosis induced by oh^8dG showed the selectivity to OGG1 deficient cells. Also, KG-1 cells show much higher sensitivity to γ-radiation than cells with normal OGG1 activity. On being exposed to γ-radiation, KG-1 shows higher oh^8Gua in DNA and undergoes apoptosis. This means that the impaired viability occurs via the elevation of oh^8Gua in DNA, and also strongly suggests that oh^8Gua in DNA is cytotoxic enough to kill cells. Thus, the cytotoxicity induced by either ROS producing agents or cell damage observed during pathological conditions accompanying high oxidative stress may contribute to this action of oh^8Gua in DNA. This toxicity of oh^8Gua may provide a mechanistic basis for understanding the toxicities of many ROS producing agents, and of the cell damage observed in many ROS-involved clinical conditions. Moreover, this action may provide a basic explanation for the beneficial effects of various antioxidants. Finally, we demonstrate that in addition to its mutagenic action, oh^8Gua in DNA disturbs cell viability.

REFERENCES

1. Ludvigsson J. Intervention at diagnosis of type I diabetes using either antioxidants or photopheresis. Diabetes Metab Rev 1993; 9(4):329–336.

2. Farinati F, Cardin R, Degan P, Rugge M, Mario FD, Bonvicini P, Naccarato R. Oxidative DNA damage accumulation in gastric carcinogenesis. Gut 1998; 42(3):351–356.

3. Cooke MS, Mistry N, Wood C, Herbert KE, Lunec J. Immunogenicity of DNA damaged by ROS implications for anti-DNA antibodies in Iupus. Free Rad Biol Med 1997; 22(1–2):151–159.

4. Darely-Usmer V, Halliwell B. Blood radicals: reactive nitrogen species, reactive oxygen species, transition metal

ions, and the vascular system. Pharm Res 1996; 13(5):
649–662.

5. Parthasarathy S, Steinberg D, Witztum JL. The role of oxidized LDL in the pathogenesis of atherosclerosis. Ann Rev
Med 1992; 43:219–225.

6. Laurindo FR, da Luz PL, Uint L, Rocha TF, Jaeger RG,
Lopes EA. Evidence for superoxide radical dependent coronary
artery vasospasm after angioplasty in intact dogs. Circulation
1991; 83(5):1705–1715.

7. Nakazono K, Watanabe N, Matsuno K, Sasaki J, Sato T,
Inoue M. Does superoxide underline the pathogenesis of
hypertension? Proc Natl Acad Sci USA 1991; 88(22):10045–10048.

8. Palinski W, Miller E, Witztum JL. Immunization of LDL recep-
tor-deficient rabbits with homologous MDA modified LDL
reduces atherogenesis. Proc Natl Acad Sci USA 1995; 92(3):
821–825.

9. Chung MH, Kasai H, Jones DS, Inoue H, Ishikawa H,
Ohtsuka E, Nishimura S. An endonuclease activity of *Escherichia coli* that specifically removes 8-hydroxyguanine residues
from DNA. Mutat Res 1991; 254:1–12.

10. Chung MH, Kim HS, Ohtsuka E, Kasai H, Yamamoto F,
Nishimura S. An endonuclease activity in human polymorphonuclear neutrophils that removes 8-hydroxyguanine residues
from DNA. Biochem Biophys Res Commun 1991; 178:1472–1478.

11. Yamamoto F, Kasai H, Bessho T, Chung MH, Inoue H,
Ohtsuka E, Hori T, Nishimura S. Ubiquitous presence in mammalian cells of enzymatic activity specifically cleaving 8-hydroxy-
guanine containing DNA. Jpn J Cancer Res 1992; 83:351–357.

12. Tsurudome Y, Hirano T, Yamato H, Tanaka I, Sagai M,
Hirano H, Nagata N, Itoh H, Kasai H. Changes in levels of
8-hydroxyguanine in DNA, its repair and OGG1 mRNA in
rat lungs after intratracheal administration of diesel exhaust
particles. Carcinogenesis 1999; 20(8):1573–1576.

13. Suzuki J, Inoue Y, Suzuki S. Changes in the urinary excretion
level of 8-hydroxyguanine by exposure to reactive oxygen-
generating substances. Free Radic Biol Med 1995; 18(3):431–436.

14. Moriya M. Single-strand shuttle phagemid for mutagenesis studies in mammalian cells: 8-oxoguanine induces targeted G.C-T.A transversion in simian kidney cells. Proc Natl Acad Sci USA 1993; 90:1122–1126.

15. Shibutani S, Takeshida M, Grollman AP. Insertion of specific bases during DNA synthesis past the oxidation-damaged base 8-oxoG. Nature 1991; 349:431–434.

16. Singh KK, Sigala B, Sikder HA, Schwimmwer C. Inactivation of *Saccaromyces cerevisiae* OGG1 DNA repair gene leads to an increased frequency of mitochondrial mutants. Nucleic Acid Res 2001; 29(6):1381–1388.

17. Klungland A, Rosewell I, Hollenbach S, Larsen E, Daly G, Epe B, Seeberg E, Lindahl T, Barnes DE. Accumulation of promutagenic DNA lesions in mice defective in removal of oxidative base damage. Proc Natl Acad Sci USA 1999; 96: 13300–13305.

18. Lee HM, Wang C, Hu Z, Greeley GH, Makalowski W, Hellmich HL, Englander EW. Hypoxia induces mitochondrial DNA damage and stimulates expression of a DNA repair enzyme, the *Escherichia coli* MutY DNA glycosylase homolog (MYH), in vivo, in the rat brain. Neurochemistry 2002; 80(5): 928–937.

19. Nakabeppu Y. Molecular genetics and structural biology of human MutT homolog, MTH1. Mutat Res 2001; 477(1–2): 59–70.

20. Ballmaier D, Epe B. Oxidative DNA damage induced by potassium bromate under cell free conditions and in mammalian cells. Carcinogenesis 1995; 16:335–342.

21. Hyun JW, Choi JY, Zeng HH, Lee YS, Kim HS, Yoon SH, Chung MH. Leukemic cell line, KG-1 has a functional loss of hOGG1 enzyme due to a point mutation and 8-hydroxydeoxyguanosine can kill KG-1. Oncogene 2000; 19:4476–4479.

22. Choi JY, Kim HS, Kang HK, Lee DW, Choi EM, Chung MH. Thermolabile 8-hydroxyguanine DNA glycosylase with low activity in senescence-accelerated mice due to a single base mutation. Free Radic Biol Med 1999; 27:848–854.

23. Le Page F, Klungland A, Barnes DE, Sarasin A, Boiteux S. Transcription coupled repair of 8-oxoguanine in murine cells: the ogg1 protein is required for repair in nontranscribed sequences but not in transcribed sequences. Proc Natl Acad Sci USA 2000; 97:8397–8402.

24. Minowa O, Arai T, Hirano M, Monden Y, Nakai S, Fukuda M, Itoh M, Takano H, Hippou Y, Aburatani H, Masumura K, Nohmi T, Nishimura S, Noda T. Mmh/ogg1 gene inactivation results in accumulation of 8-hydroxyguanine in mice. Proc Natl Acad Sci USA 2000; 97:4156–4161.

25. Hyun JW, Jung YC, Kim HS, Choi EY, Kim JE, Yoon BH, Yoon SH, Lee YS, Choi J, You HJ, Chung MH. 8-Hydroxydeoxyguanosine causes death of human leukemia cells deficient in 8-oxoguanine glycosylase 1 activity by inducing apoptosis. Mol Cancer Res 2003; 1(4):290–299.

26. Lee YS, Choi JY, Park MK, Choi EM, Kasai H, Chung MH. Induction of oh^{8}Gua glycosylase in rat kidneys by potassium bromate (KBrO$_3$), a renal oxidative carcinogen. Mutat Res 1996; 364(3):227–233.

27. Yamaguchi R, Hirano T, Asami S, Chung MH, Sugita A, Kasai H. Increased 8-hydroxyguanine levels in DNA and its repair activity in rat kidney after administration of a renal carcinogen, ferric nitrilotriacetate. Carcinogenesis 1996; 17(11):2419–2422.

28. You H, Kim G, Kim Y, Chun Y, Park J, Chung MH, Kim M. Increased 8-hydroxyguanine formation and endonuclease activity for its repair in ischemic-reperfused hearts of rats. J Mol Cell Cardiol 2000; 32(6):1053–1059.

29. Lin LH, Cao S, Yu L, Cui J, Hamilton WJ, Liu PK. Up-regulation of base excision repair activity for 8-hydroxy-2′-deoxyguanosine in the mouse brain after forebrain ischemia-reperfusion. J Neurochem 2000; 74(3):1098–1105.

30. Bases R, Franklin WA, Moy T, Mendez F. Enhanced excision repair activity in mammalian cells after ionizing radiation. Int J Radic Biol 1992; 62(4):427–441.

31. Hyun JW, Cheon GJ, Kim HS, Lee YS, Choi EY, Yoon BH, Kim JS, Chung MH. Radiation sensitivity depends on OGG1 activity status in human leukemia cell lines. Free Radic Biol Med 2002; 32:212–220.

10

Role of Oxidative Stress in Hypercholesterolemia-Induced Inflammation

KAREN Y. STOKES, ANITABEN TAILOR, and D. NEIL GRANGER

Department of Molecular and Cellular Physiology, Louisiana State University Health Sciences Center, Shreveport, Lousianna, U.S.A.

I. INTRODUCTION

It is widely recognized that a chronic inflammatory state ensues during hypercholesterolemia and atherosclerosis, which is characterized by enhanced oxidative stress, endothelial dysfunction, increased endothelial cell adhesion molecule (CAM) expression, and the activation and recruitment of circulating inflammatory blood cells (leukocytes and platelets). Cytokines such as tumor necrosis factor-α (TNF-α), interferon-γ (IFN-γ) and interleukin-12 (IL-12), released from both

221

infiltrating leukocytes and endothelial cells contribute to the development of the atherosclerotic lesion (1). While it is well established that large arteries assume an inflammatory phenotype following prolonged exposure to elevated cholesterol levels, it is becoming increasingly evident that acute exposure to hypercholesterolemia also induces a similar inflammatory phenotype in the microcirculation, long before the appearance of fatty streak lesions in larger arteries. In addition, it is becoming apparent that hypercholesterolemia also renders the microvasculature more vulnerable to the deleterious effects of inflammatory stimuli such as ischemia–reperfusion (I/R) (2).

The mechanism by which elevated cholesterol levels initiate the development of the atherosclerotic lesion has not been well elucidated, however, the oxidation of low-density lipoprotein (LDL) to form oxidized LDL (oxLDL) has been implicated. Several oxidizing species have been shown to be involved in the oxidation of LDL, including superoxide (O_2^-), hypochlorous acid (HOCl), and peroxynitrite ($ONOO^-$). Furthermore, the phenotypic changes that are observed in large arteries and the microcirculation during hypercholesterolemia can be mimicked by the administration of oxLDL, suggesting that this putative mediator of fatty streak formation may also participate in the pathogenesis of earlier microvascular dysfunction (3,4).

II. OXIDATIVE STRESS-GENERATING SYSTEMS IN HYPERCHOLESTEROLEMIA

Under normal physiological conditions, nitric oxide (NO) predominates over O_2^- in vascular endothelial cells. The excess NO regulates endothelial cell function, modulates CAM expression, inhibits leukocyte adhesion, and limits platelet activation (adhesion and aggregation). However, a large amount of evidence has amassed in the literature that implicates oxidative stress, a consequence of increased reactive oxygen species (ROS) generation and decreased NO bioavailability, in the chronic inflammatory responses observed in all segments of the microcirculation when blood cholesterol levels are elevated. Many of the enzymes involved in ROS production during hypercholesterolemia-induced oxidative stress have been identified, including NAD(P)H oxidase, xanthine oxidase (XO), myeloperoxidase, lipoxygenase, and endothelial nitric oxide synthase (eNOS). These pathways produce substantial quantities of ROS such as O_2^- and HOCl, overwhelming the capacity of endogenous antioxidants such as superoxide dismutase (SOD), catalase, and NO, thereby resulting in an oxidative stress. Evidence for the participation of the above mechanisms is derived from several animal models and human studies, using both direct measurements and pharmacological inhibitors. Indeed, the excess levels of oxidants may propagate the response to hypercholesterolemia by promoting the oxidative modification of LDL, which exacerbates many of the inflammatory consequences.

An increased production of O_2^- has been demonstrated in arteries of hypercholesterolemic animals (5). This can be attributed to hypercholesterolemia per se since it can be corrected by dietary means (6). The response is blunted when the vessels are denuded, implicating vascular endothelium as a major source of O_2^-. In order to elucidate the enzyme(s) responsible for the O_2^- production, several inhibitors have been tested. Treatment with either XO inhibitors (allopurinol or oxypurinol) (7) or heparin (which competes with XO for binding sites on the vascular endothelium), each diminish O_2^- production, implicating a role for XO. Studies using blockers of flavanoid-containing oxidases suggest that NAD(P)H oxidase may also be important in the O_2^- generation observed in hypercholesterolemic vessels of rabbits (8). Guzik et al. (5) demonstrated that hypercholesterolemia was independently associated with O_2^- generation from an NAD(P)H-containing enzyme in human saphenous veins, and that XO did not participate in this response. Furthermore, several studies using genetically engineered mice deficient in either p47[phox] or gp91[phox] (also known as NOX-2), subunits of NAD(P)H oxidase, have supported these findings.

During acute hypercholesterolemia, a p47[phox]-containing NAD(P)H oxidase enzyme has been implicated in the generation of O_2^- from both leukocytes and the vascular endothelium of postcapillary venules (9). Crossing atherosclerosis-prone ApoE-deficient mice with p47^{phox}-deficient mice led to reduced lesion development in the descending aorta but not at the origin of the aorta, suggesting that site-specific differences may exist for ROS production during hypercholesterolemia (10). No protection was seen in gp91$^{phox-/-}$ mice (11), although this may simply reflect the fact that many cells are capable of producing O_2^- via NAD(P)H oxidases containing other NOX isotypes. In humans, for instance, gp91^{phox} and NOX-4 are expressed in atherosclerotic lesions and are associated with macrophages and smooth muscle cells, respectively (12).

Nitric oxide presents a more complicated picture, such that under normal conditions, endothelial-derived NO exerts a homeostatic influence, generally viewed as anti-inflammatory. During inflammation, the bioavailability of NO is reduced, although this does not necessarily reflect decreased production. The imbalance between NO and O_2^- generation favors a proinflammatory environment (Fig. 1). The reduced bioavailability of NO may be explained by several mechanisms. Under normal physiological conditions, endothelium-derived NO counteracts the actions of O_2^-; however, during hypercholesterolemia, eNOS may become uncoupled. This could occur as a result of decreased availability of the substrate L-arginine, or diminished cofactors such as tetrahydrobiopterin (BH$_4$). During hypercholesterolemia, L-arginine levels remain within the normal range; however, L-arginine supplementation partially reverses the accompanying endothelial dysfunction (13). This suggests that the ability of endothelial cells to transport L-arginine to the cell interior may be impaired during hypercholesterolemia or that the endogenous inhibitor asymmetric dimethylarginine (ADMA) is competitively inhibiting the interaction between L-arginine and eNOS (14). The latter possibility is supported by the fact that ADMA levels are raised during hypercholesterolemia, perhaps as a result of diminished degradation by dimethylarginine dimethylaminohydrolase (DDAH), which is reduced in hypercholesterolemic rabbits.

Studies with BH$_4$-knockout mice have revealed a role for this cofactor in the pathogenesis of atherosclerosis, such that

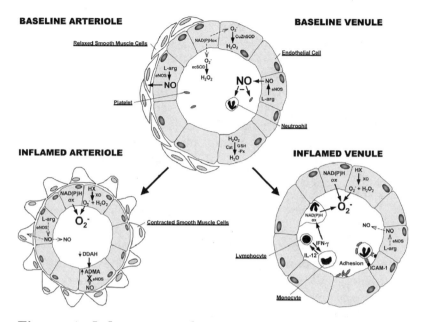

Figure 1 Inflammatory alterations in arterioles and venules elicited by hypercholesterolemia. Basal nitric oxide (NO) release maintains arteriolar smooth muscle tone and prevents cell–cell interactions in venules. During hypercholesterolemia, NO bioavailability is reduced and reactive oxygen species (ROS) generation is elevated, promoting smooth muscle contraction in arterioles and inducing a proinflammatory and prothrombogenic phenotype in the venular segment of the microcirculation. L-arginine (L-arg); endothelial nitric oxide synthase (eNOS); superoxide (O_2^-); catalase (cat); glutathione peroxidase (GSH); hydrogen peroxide (H_2O_2); superoxide dismutase (SOD); dimethylarginine dimethylaminohydrolase (DDAH); asymmetric dimethylarginine (ADMA); intercellular adhesion molecule-1 (ICAM-1); hypoxanthine (HX); xanthine oxidase (XO); interleukin-12 (IL-12); interferon-γ (IFN-γ).

these mice have accelerated lesion development. Supplementation of ApoE$^{-/-}$ mice with sepiapterin, a precursor of BH$_4$ (15), or infusion of BH$_4$ into hypercholesterolemic patients (16), blunts the endothelial cell dysfunction. It was also found that ONOO$^-$ and to a lesser degree O_2^- accelerated the decay of BH$_4$, suggesting that the interaction between O_2^- and NO serves to exacerbate the oxidative stress not only by its oxidizing action but also by promoting the uncoupling of eNOS (15). In a murine model of atherosclerosis, a protective role was found for eNOS, when eNOS/ApoE double-knockout mice demonstrated exaggerated atherosclerosis development (17). In contrast to eNOS, iNOS has been shown to exert a proinflammatory influence in many models of injury. Support for this comes from studies using mice that are genetically deficient in both iNOS and ApoE. These exhibit a reduction in fatty streak size when compared to $ApoE^{-/-}$ mice (18).

Despite the evidence implicating a role for oxidative stress in the pathogenesis of hypercholesterolemia-induced vascular responses, there are conflicting reports concerning the effects of hypercholesterolemia on endogenous antioxidant levels. Some studies have found antioxidants such as catalase, SOD, and glutathione peroxidase are decreased, while others have failed to find such a reduction. This may reflect differences between the cellular sources investigated, the time-point of assessment, or interspecies variability (19). Regardless of whether these enzymes are altered during hypercholesterolemia, it certainly appears that their antioxidant capacity is overwhelmed, thereby contributing to the oxidative stress found in hypercholesterolemia.

III. OXIDATIVE STRESS AND INFLAMMATORY RESPONSES DURING HYPERCHOLESTEROLEMIA

III.A. Arterioles

One of the earliest events known to occur in hypercholesterolemia is impairment of endothelium-dependent vasorelaxation, which is largely due to reduced elaboration of endothelium-derived NO from eNOS. Both arteries and arterioles, exhibit a profound attenuation in endothelium-dependent relaxation to vasodilator stimuli (such as acetylcholine and bradykinin) in both hypercholesterolemic humans and animals. Supportive evidence for this is derived from studies where either acute or chronic supplementation with L-arginine improves vascular responsiveness, suggesting a role for NO in mediating this response (20). Furthermore, relaxation to endothelium-independent NO donors (such as sodium nitroprusside) is unaffected, indicating the selective impairment of the L-arginine-NO pathway, rather than a generalized impairment of the responsiveness of vascular smooth muscle to NO. Similarly, the involvement of ROS in mediating the impaired endothelium-dependent vasorelaxation associated with hypercholesterolemia is blunted in $ApoE^{-/-}$ and $LDLr^{-/-}$ mice treated with ROS scavengers (SOD) or in human subjects treated with inhibitors of ROS-producing enzymes (oxypurinol) (21,22). Collectively, these data support the view that an accelerated production of O_2^- by arteriolar endothelial cells during hypercholesterolemia leads to the inactivation of endothelial cell-derived NO.

III.B. Venules

Hypercholesterolemia elevates the expression of P-selectin, intercellular adhesion molecule-1 (ICAM-1) and vascular cell adhesion molecule-1 (VCAM-1), resulting in increased leukocyte adhesion in venules of hypercholesterolemic rabbits (23). In vitro, human oxLDL promotes neutrophil adhesion to human umbilical vascular endothelial cell monolayers. Intra-arterial infusion of oxLDL (but not native LDL) also elicits the adhesion and emigration of leukocytes, mast cell degranulation, and albumin leakage in rat mesenteric venules. Infusion of mAbs directed against either CD11/CD18, ICAM-1, or P-selectin attenuates both the adhesion and albumin leakage responses to oxLDL (4), suggesting that oxLDL induces the loss of endothelial function in a leukocyte-dependent manner.

The importance of oxidant stress during hypercholesterolemia-induced venular responses has been supported by a number of studies implicating reduced bioavailability of NO as well as increased oxidant production. Leukocyte–endothelial interactions are increased in normocholesterolemic animals treated with the nonselective NOS inhibitor, L-NAME, whereas no enhancement of leukocyte adhesion was seen in hypercholesterolemic counterparts, suggesting that NO production was already compromised (23). This may be attributed to elevated circulating ADMA levels. It is noteworthy

that postcapillary venules exposed to an analog of ADMA
exhibit many of the endothelium-dependent inflammatory
responses that have been detected in venules of hypercholesterolemic animals, including leukocyte–endothelial cell
adhesion and endothelial barrier dysfunction (24). The
replenishment of NO in hypercholesterolemic rats abrogates
leukocyte–endothelial interactions in mesenteric venules
(25). In a model using oxLDL, the administration of NO
donors (sodium nitroprusside, spermine-NO, L-arginine), or
SOD, also decreased leukocyte recruitment and albumin leakage (3). A more recent study demonstrated that mice placed
on a high cholesterol diet for 2 weeks exhibited significant
increases in leukocyte adhesion and emigration in venules.
A role for O_2^- in mediating these responses was proposed
because mice that either overexpressed SOD or were genetically deficient in the p47^{phox} component of the NAD(P)H
oxidase exhibited a significant reduction of the hypercholesterolemia-induced leukocyte recruitment. Moreover, both
the endothelial cell- and leukocyte-associated NAD(P)H oxidase appeared to contribute to the increased ROS production
in this model as chimeric mice deficient in either vessel wall
or leukocyte p47^{phox} exhibited attenuated responses (9).

Although the oxidative stress associated with hyperchole-
terolemia elicits an inflammatory phenotype, the participation
of various circulating blood cell populations, in particular
T-lymphocytes, in the pathogenesis of hypercholesterolemia-
induced atherogenesis remains elusive. Some evidence reported in the literature implicates a role for lymphocytes in
the development of atherosclerotic lesions, whereas other studies have shown that lymphocytes do not play a role (26). Both
CD4$^+$ and CD8$^+$ T-lymphocytes are detected in the developing
atherosclerotic lesions of mice. Early findings using lympho-
cyte-deficient severe combined immunodeficient (SCID) or
recombinase activator gene-2-knockout (RAG2$^{-/-}$) mice crossed
with ApoE$^{-/-}$ mice failed to demonstrate an attenuation of
plaque formation. More recently, a 40% reduction in lesion size
was shown in RAG1$^{-/-}$ mice crossed with ApoE$^{-/-}$ mice (27).
Other studies have suggested that lymphocytes play an earlier
role in the initiation of atherosclerosis development (28). T-lym-

phocytes may also contribute to the hypercholesterolemic changes long before the development of the atherosclerotic lesions in larger blood vessels. Support for this contention is derived from a recent study, which demonstrated a role for T-lymphocytes ($CD4^+$ and $CD8^+$) in mediating hypercholesterolemia-induced blood leukocyte–endothelial cell adhesion and leukocyte emigration in hypercholesterolemic mice. Severe combined immunodeficient mice also exhibited attenuated leukocyte–endothelial interactions in hypercholesterolemic mice, which were restored when these mice received splenocytes from hypercholesterolemic wild-type mice (29).

Oxidative stress can also promote the adhesion of platelets to vascular endothelium, indicating that the arterial wall can assume both a proinflammatory and prothrombogenic phenotype when cholesterol levels are elevated. Both platelets and platelet–leukocyte aggregates have been identified in atherosclerotic lesions at all stages (30,31). Recent studies in $ApoE^{-/-}$ mice demonstrate that platelets are primed for adhesion to atherosclerotic lesion-prone sites and may participate in the initiation of atherosclerosis long before the appearance of fatty streaks. Furthermore, chronic blockade of platelet adhesion reduces both leukocyte accumulation and atherosclerotic lesion formation (31).

A number of studies from both animal models and human subjects have shown that platelets become hyperreactive to stimulating agents during hypercholesterolemia (32–34). Platelet activation in humans with elevated cholesterol levels is evident from the increased P-selectin expression on platelets (35,36), as well as elevated soluble P-selectin levels (37). While a large number of studies have demonstrated both platelets and leukocytes as major players in the inflammatory phenotype associated with hypercholesterolemia in larger vessels, information concerning a role for platelets in the microvasculature during hypercholesterolemia is limited. Vink et al. (38) demonstrated that oxLDL (but not native LDL) significantly increased platelet adhesion to the microvascular endothelium. A more recent study reported that both endothelial and platelet P-selectin were important in mediating platelet–endothelial adhesion in intestinal venules of mice on a cholesterol-enriched diet for 2 weeks (39). The adhesion of these activated platelets on the vessel wall may exacerbate the inflammatory phenotype of endothelial cells that is induced by hypercholesterolemia and thereby contribute to the development and progression of cardiovascular diseases, such as atherosclerosis and thrombosis.

In addition to promoting the proinflammatory and prothrombogenic phenotype in hypercholesterolemic venules via transcription-independent mechanisms (inactivation of NO), oxidant stress can also activate the redox-sensitive transcription factors, nuclear factor-kappa B (NF-κB) and activation protein-1 (AP-1), that govern the synthesis and expression of CAMs (ICAM-1, VCAM-1, P-selectin), and proinflammatory mediators (platelet-activating factor (PAF), leukotriene B$_4$ (LTB$_4$), TNF-α). Several studies have reported an inhibitory effect of NO on cytokine-induced NF-κB activation and subsequent CAM expression on endothelial cell monolayers (40,41). This action of NO appears to be mediated via the induction of inhibitor-κB (I-κB), the inhibitor of NF-κB (42). This inhibition of NF-κB by NO offers an explanation for the ability of NO donors to attenuate leukocyte adhesion by downregulating the expression of endothelial CAMs such as P-selectin and ICAM-1. The potential involvement of ROS in eliciting NF-κB activation and subsequent increased adhesion molecule expression in hypercholesterolemia is underscored by studies that have reported decreased NF-κB activation, hence reduced CAM expression, in response to antioxidants, oxygen radical scavengers, and iron chelators (43).

Numerous studies have also shown that oxidant stress induces the release of proinflammatory mediators, including PAF, LTB$_4$, and IL-8 from endothelial cells and leukocytes. Studies have revealed a role for monocyte chemotactic protein-1 (MCP-1), IL-6, IL-8, and TNF-α in hypercholesterolemic (ApoE$^{-/-}$) mice (44). More recent attention has focused on a role for T helper-1 (Th1) cell-derived cytokines such as IFN-γ and other cytokines such as IL-12 in regulating the inflammatory response during hypercholesterolemia. Both cytokines are detected in human atherosclerotic lesions (45,46) and the aortas of hypercholesterolemic ApoE$^{-/-}$ mice (47). Studies have demonstrated that IFN-γ promotes oxidative stress by activating enzymes such as NAD(P)H oxidase. Interferon-$\gamma^{-/-}$ mice are protected against atherosclerotic lesion development, whereas exogenous administration of this cytokine potentiates lesion development in hypercholesterolemic mice. More recently, the attenuated leukocyte–endothelium adhesion responses observed in postcapillary venules of hypercholesterolemic IFN-$\gamma^{-/-}$ mice were correlated with reduced oxidant stress (measured using an oxidant-sensitive fluorochrome). The inflammatory phenotype was restored when these mice were reconstituted with wild-type splenocytes (48). The production of IFN-γ is often associated with IL-12. In support of this, like IFN-γ, lack of IL-12 attenuates leukocyte recruitment and oxidative stress in postcapillary venules of hypercholesterolemic mice (49), suggesting a feedback loop may exist between these two cytokines.

There is a growing body of evidence in the literature that implicates a role for angiotensin II in the induction of the inflammatory phenotype associated with hypercholesterolemia. Physiologically, angiotensin II functions to regulate blood pressure and salt balance. Studies have shown that angiotensin II enhances oxidative stress as a consequence of increased ROS generation (50) and decreased NO. The oxidative stress is mediated via its interaction with the angiotensin II type 1 (AT1) receptor, with the subsequent activation of the NAD(P)H oxidase enzyme (51). Angiotensin II type 1 receptors are found on endothelial cells, granulocytes (neutrophils and monocytes), T-lymphocytes and platelets. During hypercholesterolemia, there is increased expression of AT1 receptors on endothelial cells, leukocytes, and platelets (52,53). Angiotensin II also stimulates NF-κB, and is likely to encode for the expression of CAMs, cytokines, and chemokines. Studies have shown that angiotensin II promotes leukocyte adhesion in a P-selectin and ROS-dependent manner (50,54). Furthermore, use of AT1 receptor antagonists improves hypercholesterolemia-associated endothelial dysfunction in humans (55) and rabbits (8,56) and attenuates hypercholesterolemia-induced NAD(P)H oxidase-dependent O_2^- production and the development of atherosclerotic lesions (8,56).

III.C. Stimulus-Induced Responses
in Hypercholesterolemia

Several studies have shown that the responses elicited by multiple proinflammatory mediators are greatly enhanced in hypercholesterolemic animals. Hypercholesterolemic rats and mice exhibit an exaggerated recruitment of adherent leukocytes to exogenously administered lipid mediators (PAF, LTB_4) and TNF-α (57), and in response to I/R (58–60). This exacerbation is accompanied by increased vascular leakage compared to normocholesterolemic animals. Furthermore, the enhanced leukocyte recruitment is significantly attenuated by a blocking mAb to ICAM-1 (57).

Although the mechanisms underlying the exaggerated responses to hypercholesterolemia remain poorly defined, there is evidence that implicates an enhanced oxidant stress in this response. Kurose et al. (59) demonstrated (using the oxidant-sensitive fluorochrome DHR-123) a more marked increase (12-fold) in oxidant production from venules after I/R in hypercholesterolemic mice compared to their normocholesterolemic (6-fold) counterparts. Furthermore, pretreatment of hypercholesterolemic animals with either oxypurinol or SOD largely prevented the enhanced oxidant stress and exaggerated leukocyte trafficking following I/R, suggesting that XO is the major source of ROS under these conditions. Since both I/R and hypercholesterolemia are also associated with a reduction in endothelial-derived NO, the enhanced oxidant stress observed in hypercholesterolemic animals exposed to I/R is likely a consequence of the combined effect of these two deleterious conditions. Oxidant stress is also often accompanied by enhanced PAF production, thus, this mechanism may account for the observation that PAF receptor antagonists exert a potent inhibitory effect on the inflammatory responses elicited by I/R in $LDLr^{-/-}$ mice (58). Furthermore, Liao et al. (61) demonstrated that intra-arterial infusion of oxLDL enhanced the I/R-induced adherence and emigration of leukocytes, without affecting albumin leakage and mast cell degranulation normally observed after I/R. This suggests that oxLDL may contribute to the exaggerated leukocyte recruitment response to I/R in hypercholesterolemic animals.

IV. LINKAGE BETWEEN MICROVASCULAR AND MACROVASCULAR RESPONSES TO HYPERCHOLESTEROLEMIA

The sections above provide an overview of the many inflammatory and thrombogenic changes that occur in the vasculature during hypercholesterolemia. These include alterations on both the arteriolar and venular sides of the microvasculature, unlike in large vessels where arteries (and not veins) appear to be the primary site of injury during hypercholesterolemia. In addition, the microcirculation responds to hypercholesterolemia within 1–2 weeks, whereas in atherosclerosis-prone areas of the aorta and other large arteries, changes are observed only after long-term hypercholesterolemia. Despite these differences, many similarities exist between the mechanisms involved in the microvascular and macrovascular responses to elevated cholesterol levels. These include the induction of CAMs, leukocyte, and platelet recruitment and the release of cytokines/chemokines. Of most relevance to the topic discussed here is the finding that some of the earliest detectable changes in both small and large vessels under hypercholesterolemic conditions are diminished NO bioavailability and the activation of several ROS-generating systems leading to oxidative stress. The demonstration that many of the microvascular alterations are similar to, but occur far in advance of the changes in the large vessels, leads to the possibility that the inflammatory and thrombogenic environment experienced during the initial phase of hypercholesterolemia may provide novel mechanistic insights into and may even play a role in the chronic modifications of the macrovasculature during the slow process of atherosclerotic lesion development. However, despite the mounting literature describing the pathophysiological alterations throughout the vasculature during hypercholesterolemia, a link between these two processes has not been investigated to date.

Regardless of the existence (and exact nature) of a link between the events that occur in the microvasculature and macrovasculature during hypercholesterolemia, it is likely that both acute and chronic elevations in blood cholesterol contribute to the cardiovascular consequences of this risk factor. Furthermore, several studies have demonstrated that the early microvascular changes exacerbate the vascular responses to other injuries such as ischemia by enhancing the vulnerability of the vessel wall and circulating cells to other stimuli. This has important implications in the pathogenesis of ischemic diseases in which a role for hypercholesterolemia has been suggested. Hence, therapeutic strategies that are directed against the early inflammatory alterations initiated in the microcirculation by hypercholesterolemia may be valuable in the quest to reduce the high mortality associated with ischemic tissue diseases. Agents that act to maintain the normal physiological balance between ROS and NO in the vascular wall may prove particularly useful in the prevention of ischemic diseases and in the reduction of other atherosclerosis-associated complications.

V. ANTIOXIDANT STRATEGIES IN HYPERCHOLESTEROLEMIA

By now, the reader will have gained an appreciation for the important role of oxidative stress in many of the responses observed in both acute and chronic hypercholesterolemic states. Thus, it follows that strategies targeting these oxidant-generating systems or their products could have immense therapeutic potential in the reduction of hypercholesterolemia-induced atherosclerosis. While animal studies have shown great promise, the use of antioxidant therapies, such as vitamins C and E, in humans has yielded disappointing results (62). Vitamins C and E have been shown to enhance antioxidant levels, prevent eNOS downregulation and improve endothelial function in several animal models (63). This may lead to reduced atherosclerotic plaque formation (64). While there have been relatively few positive findings or mechanistic insights with vitamin therapy in humans, the use of slow-release vitamins was recently found to delay the progression of atherosclerotic lesion development (65). However, high levels of certain vitamins may also alter the blood lipid profile in a manner that could outweigh any antioxidant benefit (66). Interestingly, evidence is emerging that drugs developed for other purposes, for example HMG-CoA reductases (statins), AT1 receptor antagonists, aspirin, and 17β-estradiol, possess properties that reduce oxidative stress independently or as a result of their targeted action. The most extensively investigated of these, i.e., statins and AT1 receptor antagonists, are discussed below.

Statins were first introduced as lipid-lowering drugs, by virtue of their blocking action on the conversion of HMG-CoA to mevalonoate, the substrate for cholesterol. However, recently, it has become increasingly apparent that statins exert many anti-inflammatory properties that extend beyond their cholesterol-lowering effect. For example, several statins have been shown to increase eNOS expression and subsequently endothelial NO production (67). Furthermore, there is evidence that statins may reduce oxidant production both by preventing enzyme activation, in particular NAD(P)H oxidase activation (68), and by increasing levels of antioxidants such as catalase (55). This would lead to a diminished oxidant stress, and possibly offer an explanation for the ability of statins to attenuate transcription factor activation, reduce CAM expression and cytokine release, and ultimately inhibit leukocyte and platelet recruitment in a cholesterol level-independent manner.

Interest in the angiotensin-converting enzyme inhibitors and AT1 receptor antagonists is increasing, as these agents appear to have therapeutic potential in areas other than hypertension. In particular hypercholesterolemia, which promotes AT1 receptor induction (52,53,69), may be a clinically viable target. In fact, AT1 receptor antagonists have been shown to attenuate atherosclerotic lesion development (56). Since angiotensin activates several subunits of

NAD(P)H oxidase via the AT1 receptor (51), and AT1 receptor antagonists can increase NO release from platelets and endothelial cells (70), this treatment may result in an overall reduction of oxidative stress during hypercholesterolemia. Therefore, the actions of this class of drugs may extend beyond their blood pressure-lowering effects. It is noteworthy that statins are capable of reducing the overexpression of AT1 receptors caused by angiotensin II in hypercholesterolemic humans (69).

VI. CONCLUSIONS

The literature strongly indicates that hypercholesterolemia induces an oxidative stress through the combined actions of increasing oxidant production and decreasing NO bioavailability. This provides a therapeutic window of opportunity for the treatment of cardiovascular diseases associated with hypercholesterolemia and other risk factors that are linked to oxidative stress such as smoking, hypertension, and diabetes. It is plausible that the use of existing pharmaceutical agents such as statins and AT1 receptor antagonists, or antioxidants in association with therapies that increase the availability of NO, may offer protection against the systemic inflammatory responses that accompany a variety of disease states. With the advent of improved targeting of drugs and genes to specific cell populations, there is also hope for the selective distribution of statin- and AT1 receptor antagonist-like drugs to the endothelial cells that mediate the vascular dysfunction that is associated with many cardiovascular diseases.

REFERENCES

1. Stokes KY, Cooper D, Tailor A, Granger DN. Hypercholesterolemia promotes inflammation and microvascular dysfunction: role of nitric oxide and superoxide. Free Radic Biol Med 2002; 33:1026–1036.

2. Granger DN. Ischemia–reperfusion: mechanisms of microvascular dysfunction and the influence of risk factors for cardiovascular disease. Microcirculation 1999; 6:167–178.

3. Liao L, Aw TY, Kvietys PR, Granger DN. Oxidized LDL-induced microvascular dysfunction. Dependence on oxidation procedure. Arterioscler Thromb Vasc Biol 1995; 15:2305–2311.

4. Liao L, Starzyk RM, Granger DN. Molecular determinants of oxidized low-density lipoprotein-induced leukocyte adhesion and microvascular dysfunction. Arterioscler Thromb Vasc Biol 1997; 17:437–444.

5. Guzik TJ, West NE, Black E, McDonald D, Ratnatunga C, Pillai R, Channon KM. Vascular superoxide production by NAD(P)H oxidase: association with endothelial dysfunction and clinical risk factors. Circ Res 2000; 86:E85–E90.

6. Ohara Y, Peterson TE, Sayegh HS, Subramanian RR, Wilcox JN, Harrison DG. Dietary correction of hypercholesterolemia in the rabbit normalizes endothelial superoxide anion production. Circulation 1995; 92:898–903.

7. White CR, Darley-Usmar V, Berrington WR, McAdams M, Gore JZ, Thompson JA, Parks DA, Tarpey MM, Freeman BA. Circulating plasma xanthine oxidase contributes to vascular dysfunction in hypercholesterolemic rabbits. Proc Natl Acad Sci USA 1996; 93:8745–8749.

8. Warnholtz A, Nickenig G, Schulz E, Macharzina R, Brasen JH, Skatchkov M, Heitzer T, Stasch JP, Griendling KK, Harrison DG, Bohm M, Meinertz T, Munzel T. Increased NADH-oxidase-mediated superoxide production in the early stages of atherosclerosis: evidence for involvement of the renin–angiotensin system. Circulation 1999; 99:2027–2033.

9. Stokes KY, Clanton EC, Russell JM, Ross CR, Granger DN. NAD(P)H oxidase-derived superoxide mediates hypercholesterolemia-induced leukocyte–endothelial cell adhesion. Circ Res 2001; 88:499–505.

10. Barry-Lane PA, Patterson C, van der Merwe M, Hu Z, Holland SM, Yeh ET, Runge MS. p47phox is required for atherosclerotic lesion progression in ApoE(−/−) mice. J Clin Invest 2001; 108:1513–1522.

11. Kirk EA, Dinauer MC, Rosen H, Chait A, Heinecke JW, LeBoeuf RC. Impaired superoxide production due to a deficiency in phagocyte NADPH oxidase fails to inhibit atherosclerosis in mice. Arterioscler Thromb Vasc Biol 2000; 20:1529–1535.

12. Sorescu D, Weiss D, Lassegue B, Clempus RE, Szocs K, Sorescu GP, Valppu L, Quinn MT, Lambeth JD, Vega JD, Taylor WR, Griendling KK. Superoxide production and expression of nox family proteins in human atherosclerosis. Circulation 2002; 105:1429–1435.

13. Kawano H, Motoyama T, Hirai N, Kugiyama K, Yasue H, Ogawa H. Endothelial dysfunction in hypercholesterolemia is improved by L-arginine administration: possible role of oxidative stress. Atherosclerosis 2002; 161:375–380.

14. Boger RH. The emerging role of asymmetric dimethylarginine as a novel cardiovascular risk factor. Cardiovasc Res 2003; 59:824–833.

15. Laursen JB, Somers M, Kurz S, McCann L, Warnholtz A, Freeman BA, Tarpey M, Fukai T, Harrison DG. Endothelial regulation of vasomotion in apoE-deficient mice: implications for interactions between peroxynitrite and tetrahydrobiopterin. Circulation 2001; 103:1282–1288.

16. Fukuda Y, Teragawa H, Matsuda K, Yamagata T, Matsuura H, Chayama K. Tetrahydrobiopterin restores endothelial function of coronary arteries in patients with hypercholesterolaemia. Heart 2002; 87:264–269.

17. Kuhlencordt PJ, Gyurko R, Han F, Scherrer-Crosbie M, Aretz TH, Hajjar R, Picard MH, Huang PL. Accelerated atherosclerosis, aortic aneurysm formation, and ischemic heart disease in apolipoprotein E/endothelial nitric oxide synthase double-knockout mice. Circulation 2001; 104:448–454.

18. Kuhlencordt PJ, Chen J, Han F, Astern J, Huang PL. Genetic deficiency of inducible nitric oxide synthase reduces atherosclerosis and lowers plasma lipid peroxides in apolipoprotein E-knockout mice. Circulation 2001; 103:3099–3104.

19. Mahfouz MM, Kummerow FA. Cholesterol-rich diets have different effects on lipid peroxidation, cholesterol oxides, and antioxidant enzymes in rats and rabbits. J Nutr Biochem 2000; 11:293–302.

20. Laroia ST, Ganti AK, Laroia AT, Tendulkar KK. Endothelium and the lipid metabolism: the current understanding. Int J Cardiol 2003; 88:1–9.

21. Harrison DG. Cellular and molecular mechanisms of endothelial cell dysfunction. J Clin Invest 1997; 100:2153–2157.

22. d'Uscio LV, Baker TA, Mantilla CB, Smith L, Weiler D, Sieck GC, Katusic ZS. Mechanism of endothelial dysfunction in apolipoprotein E-deficient mice. Arterioscler Thromb Vasc Biol 2001; 21:1017–1022.

23. Scalia R, Appel JZ III, Lefer AM. Leukocyte–endothelium interaction during the early stages of hypercholesterolemia in the rabbit: role of P-selectin, ICAM-1, and VCAM-1. Arterioscler Thromb Vasc Biol 1998; 18:1093–1100.

24. Kurose I, Wolf R, Grisham MB, Granger DN. Effects of an endogenous inhibitor of nitric oxide synthesis on postcapillary venules. Am J Physiol 1995; 268(6 Pt 2):H2224–H2231.

25. Gauthier TW, Scalia R, Murohara T, Guo JP, Lefer AM. Nitric oxide protects against leukocyte–endothelium interactions in the early stages of hypercholesterolemia. Arterioscler Thromb Vasc Biol 1995; 15:1652–1659.

26. Hansson GK. Immune mechanisms in atherosclerosis. Arterioscler Thromb Vasc Biol 2001; 21:1876–1890.

27. Dansky HM, Charlton SA, Harper MM, Smith JD. T and B lymphocytes play a minor role in atherosclerotic plaque formation in the apolipoprotein E-deficient mouse. Proc Natl Acad Sci USA 1997; 94:4642–4646.

28. Song L, Leung C, Schindler C. Lymphocytes are important in early atherosclerosis. J Clin Invest 2001; 108:251–259.

29. Stokes KY, Clanton EC, Bowles KS, Fuseler JW, Chervenak D, Chervenak R, Jennings SR, Granger DN. The role of T-lymphocytes in hypercholesterolemia-induced leukocyte–endothelial interactions. Microcirculation 2002; 9:407–417.

30. Ross R. Atherosclerosis is an inflammatory disease. Am Heart J 1999; 138(5 Pt 2):S419–S420.

31. Huo Y, Schober A, Forlow SB, Smith DF, Hyman MC, Jung S, Littman DR, Weber C, Ley K. Circulating activated platelets

exacerbate atherosclerosis in mice deficient in apolipoprotein E. Nat Med 2003; 9:61–67.

32. Davi G, Averna M, Catalano I, Barbagallo C, Ganci A, Notarbartolo A, Ciabattoni G, Patrono C. Increased thromboxane biosynthesis in type IIa hypercholesterolemia. Circulation 1992; 85:1792–1798.

33. Opper C, Clement C, Schwarz H, Krappe J, Steinmetz A, Schneider J, Wesemann W. Increased number of high sensitive platelets in hypercholesterolemia, cardiovascular diseases, and after incubation with cholesterol. Atherosclerosis 1995; 113: 211–217.

34. Tremoli E, Colli S, Maderna P, Baldassarre D, Di Minno G. Hypercholesterolemia and platelets. Semin Thromb Hemost 1993; 19:115–121.

35. Garlichs CD, John S, Schmeisser A, Eskafi S, Stumpf C, Karl M, Goppelt-Struebe M, Schmieder R, Daniel WG. Upregulation of CD40 and CD40 ligand (CD 154) in patients with moderate hypercholesterolemia. Circulation 2001; 104:2395–2400.

36. Ma LP, Nie DN, Hsu SX, Yin SM, Xu LZ, Nunes JV. Inhibition of platelet aggregation and expression of alpha granule membrane protein 140 and thromboxane B2 with pravastatin therapy for hypercholesterolemia. J Assoc Acad Minor Phys 2002; 13:23–26.

37. Davi G, Romano M, Mezzetti A, Procopio A, Iacobelli S, Antidormi T, Bucciarelli T, Alessandrini P, Cuccurullo F, Bittolo Bon G. Increased levels of soluble P-selectin in hypercholesterolemic patients. Circulation 1998; 97:953–957.

38. Vink H, Constantinescu AA, Spaan JA. Oxidized lipoproteins degrade the endothelial surface layer: implications for platelet–endothelial cell adhesion. Circulation 2000; 101: 1500–1502.

39. Tailor A. Hypercholesterolemia promotes P-selectin-dependent platelet–endothelial cell adhesion in postcapillary venules. Arterioscler Thromb Vasc Biol 2003; 23:675–680.

40. De Caterina R, Libby P, Peng HB, Thannickal VJ, Rajavashisth TB, Gimbrone MA Jr, Shin WS, Liao JK. Nitric oxide decreases cytokine-induced endothelial activation. Nitric oxide selectively

reduces endothelial expression of adhesion molecules and proinflammatory cytokines. J Clin Invest 1995; 96:60–68.

41. Khan BV, Harrison DG, Olbrych MT, Alexander RW, Medford RM. Nitric oxide regulates vascular cell adhesion molecule 1 gene expression and redox-sensitive transcriptional events in human vascular endothelial cells. Proc Natl Acad Sci USA 1996; 93:9114–9119.

42. Spiecker M, Peng HB, Liao JK. Inhibition of endothelial vascular cell adhesion molecule-1 expression by nitric oxide involves the induction and nuclear translocation of IkappaBalpha. J Biol Chem 1997; 272:30969–30974.

43. Weber C, Erl W, Pietsch A, Strobel M, Ziegler-Heitbrock HW, Weber PC. Antioxidants inhibit monocyte adhesion by suppressing nuclear factor-kappa B mobilization and induction of vascular cell adhesion molecule-1 in endothelial cells stimulated to generate radicals. Arterioscler Thromb 1994; 14: 1665–1673.

44. Zhou X, Hansson GK. Detection of B cells and proinflammatory cytokines in atherosclerotic plaques of hypercholesterolaemic apolipoprotein E knockout mice. Scand J Immunol 1999; 50: 25–30.

45. Frostegard J, Ulfgren AK, Nyberg P, Hedin U, Swedenborg J, Andersson U, Hansson GK. Cytokine expression in advanced human atherosclerotic plaques: dominance of pro-inflammatory (Thl) and macrophage-stimulating cytokines. Atherosclerosis 1999; 145:33–43.

46. Uyemura K, Demer LL, Castle SC, Jullien D, Berliner JA, Gately MK, Warrier RR, Pham N, Fogelman AM, Modlin RL. Cross-regulatory roles of interleukin (IL)-12 and IL-10 in atherosclerosis. J Clin Invest 1996; 97:2130–2138.

47. Gupta S, Pablo AM, Jiang X, Wang N, Tall AR, Schindler C. IFN-gamma potentiates atherosclerosis in ApoE knock-out mice. J Clin Invest 1997; 99:2752–2761.

48. Stokes KY, Clanton EC, Clements KP, Granger DN. Role of interferon-gamma in hypercholesterolemia-induced leukocyte–endothelial cell adhesion. Circulation 2003; 107:2140–2145.

49. Stokes KY, Clanton EC, Gehrig JL, Granger DN. Role of inter-leukin 12 in hypercholesterolemia-induced inflammation. Am J Physiol Heart Circ Physiol 2003; 285:H2623–H2629.

50. Alvarez A, Sanz MJ. Reactive oxygen species mediate angiotensin II-induced leukocyte–endothelial cell interactions in vivo. J Leukoc Biol 2001; 70:199–206.

51. Li JM, Shah AM. Mechanism of endothelial cell NADPH oxidase activation by angiotensin II. Role of the p47phox subunit. J Biol Chem 2003; 278:12094–12100.

52. Nickenig G, Bohm M. Regulation of the angiotensin AT1 receptor expression by hypercholesterolemia. Eur J Med Res 1997; 2:285–289.

53. Nickenig G, Jung O, Strehlow K, Zolk O, Linz W, Scholkens BA, Bohm M. Hypercholesterolemia is associated with enhanced angiotensin AT1-receptor expression. Am J Physiol 1997; 272(6 Pt 2):H2701–H2707.

54. Piqueras L, Kubes P, Alvarez A, O'Connor E, Issekutz AC, Esplugues JV, Sanz MJ. Angiotensin II induces leukocyte–endothelial cell interactions in vivo via AT(1) and AT(2) receptor-mediated P-selectin upregulation. Circulation 2000; 102:2118–2123.

55. Wassmann S, Laufs U, Muller K, Konkol C, Ahlbory K, Baumer AT, Linz W, Bohm M, Nickenig G. Cellular antioxidant effects of atorvastatin in vitro and in vivo. Arterioscler Thromb Vasc Biol 2002; 22:300–305.

56. Sun YP, Zhu BQ, Browne AE, Pulukurthy S, Chou TM, Sudhir K, Glantz SA, Deedwania PC, Chatterjee K, Parmley WW. Comparative effects of ACE inhibitors and an angiotensin receptor blocker on atherosclerosis and vascular function. J Cardiovasc Pharmacol Ther 2001; 6:175–181.

57. Henninger DD, Gerritsen ME, Granger DN. Low-density lipoprotein receptor knockout mice exhibit exaggerated microvascular responses to inflammatory stimuli. Circ Res 1997; 81:274–281.

58. Mori N, Horie Y, Gerritsen ME, Granger DN. Ischemia–reperfusion induced microvascular responses in LDL-receptor $-/-$ mice. Am J Physiol 1999; 276(5 Pt 2):H1647–H1654.

59. Kurose I, Wolf RE, Grisham MB, Granger DN. Hypercholesterolemia enhances oxidant production in mesenteric venules exposed to ischemia/reperfusion. Arterioscler Thromb Vasc Biol 1998; 18:1583–1588.

60. Cerwinka WH, Granger DN. Influence of hypercholesterolemia and hypertension on ischemia–reperfusion induced P-selectin expression. Atherosclerosis 2001; 154:337–344.

61. Liao L, Harris NR, Granger DN. Oxidized low-density lipoproteins and microvascular responses to ischemia–reperfusion. Am J Physiol 1996; 271(6 Pt 2):H2508–H2514.

62. Griendling KK, FitzGerald GA. Oxidative stress and cardiovascular injury. Part II: Animal and human studies. Circulation 2003; 108:2034–2040.

63. Rodriguez JA, Grau A, Eguinoa E, Nespereira B, Perez-Ilzarbe M, Arias R, Belzunce MS, Paramo JA, Martinez-Caro D. Dietary supplementation with vitamins C and E prevents downregulation of endothelial NOS expression in hypercholesterolemia in vivo and in vitro. Atherosclerosis 2002; 165:33–40.

64. Peluzio MC, Miguel E Jr, Drumond TC, Cesar GC, Santiago HC, Teixeira MM, Vieira EC, Arantes RM, Alvarez-Leite JI. Monocyte chemoattractant protein-1 involvement in the alpha-tocopherol-induced reduction of atherosclerotic lesions in apolipoprotein E knockout mice. Br J Nutr 2003; 90:3–11.

65. Salonen RM, Nyyssonen K, Kaikkonen J, Porkkala-Sarataho E, Voutilainen S, Rissanen TH, Tuomainen TP, Valkonen VP, Ristonmaa U, Lakka HM, Vanharanta M, Salonen JT, Poulsen HE. Antioxidant supplementation in atherosclerosis prevention. Six-year effect of combined vitamin C and E supplementation on atherosclerotic progression: the Antioxidant Supplementation in Atherosclerosis Prevention (ASAP) Study. Circulation 2003; 107:947–953.

66. Munday JS, Thompson KG, James KA, Manktelow BW. Dietary antioxidants do not reduce fatty streak formation in the C57BL/6 mouse atherosclerosis model. Arterioscler Thromb Vasc Biol 1998; 18:114–119.

67. Berkels R, Nouri SK, Taubert D, Bartels H, Roesen P, Roesen R, Klaus W. The HMG-CoA reductase inhibitor cerivastatin

enhances the nitric oxide bioavailability of the endothelium. J Cardiovasc Pharmacol 2003; 42:356–363.

68. Mitani H, Egashira K, Ohashi N, Yoshikawa M, Niwa S, Nonomura K, Nakashima A, Kimura M. Preservation of endothelial function by the HMG-CoA reductase inhibitor fluvastatin through its lipid-lowering independent antioxidant properties in atherosclerotic rabbits. Pharmacology 2003; 68: 121–130.

69. Nickenig G, Baumer AT, Temur Y, Kebben D, Jockenhovel F, Bohm M. Statin-sensitive dysregulated AT1 receptor function and density in hypercholesterolemic men. Circulation 1999; 100:2131–2134.

70. Kalinowski L, Matys T, Chabielska E, Buczko W, Malinski T. Angiotensin II ATI receptor antagonists inhibit platelet adhesion and aggregation by nitric oxide release. Hypertension 2002; 40:521–527.

11

Oxidative Stress in Rheumatoid Arthritis

TAKASHI OKAMOTO

Department of Molecular and Cellular Biology,
Nagoya City University, Nagoya, Japan

I. INTRODUCTION

Reactive oxygen species (ROS) are produced in the cells by various environmental stimuli such as infection of microbes (viruses, bacteria, etc.), ionizing and UV irradiation, and pollutants (i.e., oxidants), which are collectively called "oxidative stress." These environmental challenges elicit inflammatory and immune responses (1). Interestingly, these factors are also regarded as risk factors and disease-accelerating factors for autoimmune diseases including rheumatoid arthritis (RA) (Fig. 1). In addition, inflammatory responses in mammals are often associated with ROS production from neutrophils and macrophages. Therefore, this natural

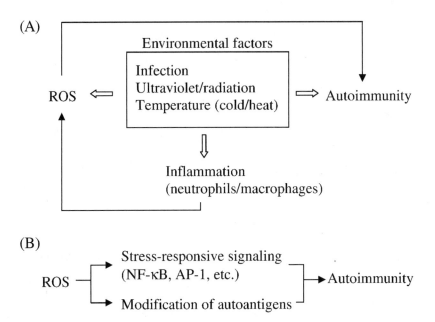

Figure 1 Involvement of ROS in autoimmunity and RA. (A) Reactive oxygen species production is either directly or indirectly induced by environmental factors during the inflammatory process. (B) Induction of autoimmune response by ROS through the stress-responsive signaling and oxidative modification of autoantigens.

defense system for the organism contains a potentially dangerous option conforming a positive feed-back loop which is considered responsible for further augmentation and expansion of the response if the feed-back regulatory system, antioxidant system, and negative-cytokine network do not work efficiently.

Rheumatoid arthritis is a common human autoimmune disease with a prevalence of about 1% (2). While there has been progress in defining its etiology and pathogenesis, these are still incompletely understood (3–5). Rheumatoid arthritis is characterized by a chronic inflammation of the synovial joints associated with proliferation of synovial cells and infiltration of activated immuno-inflammatory cells, including memory T cells, macrophages, and plasma cells (4–6), leading to progressive destruction of cartilage and bone. This process is considered to be mediated by a number of cytokines such as TNFα, IL1, IL-6, IL-8, IL-12, IL-16, IL-18, and IFNγ (reviewed in Refs. 2–5).

II. PATHOPHYSIOLOGY OF RA

Proposed causes for RA include: (i) genetic preposition; (ii) pathogenetic immuno-inflammatory responses triggered by environmental agents, particularly microbes; (iii) autoimmunity directed against components of synovium and cartilage; (iv) dysregulated production of cytokines; (v) recruitment of immuno-inflammatory cells through induction of inflammatory cell adhesion molecules (e.g., E-selectin, ICAM-1, and VCAM-1); and (vi) transformation of synovial cells into autonomously proliferating cells with tissue-infiltrating nature [often referred to as "transformed-like" phenotype (7)]. We have recently clarified the transformed-like nature of rheumatoid synoviocytes by performing gene expression profile analyses of synoviocytes and elucidated cellular genes specifically activated in RA synoviocytes (8,9). When compared with control synoviocytes obtained from healthy individuals (upon injury) or osteoarthritis (OA) patients, we found that both PDGF receptor α and SDF-1, a chemokine, genes are activated in RA synoviocytes without any external stimulus (8). Interestingly, from the gene knockout studies, it was shown that these factors are required for the development of limb joints. Moreover, when synoviocytes were stimulated with physiological concentration of TNFα, a principal proinflammatory cytokine, cell fate-determining factors, including Notch 1, Notch 4, and Jagged-2, a ligand for Notch proteins were activated only in RA synoviocytes (9). These findings indicate that RA synoviocytes may have reacquired the "revertant" phenotype mimicking the primordial synoviocytes that exhibit hyperproliferation and invasion, at least in a part. It is possible that this peculiar feature is caused by the long-term inflammatory stimulation, through which constitutive activation of particular signaling and transcription pathways lead to the change in "histone code" proposed by Allis (see Ref. 10 for review) and eventually change the epigenetic behavior of cells.

III. INVOLVEMENT OF NF-κB IN RA AS A PRIMARY PATHOGENIC DETERMINANT

Among the various signaling and transcription regulation pathways, nuclear factor-κB (NF-κB) and activator protein-1 (AP-1) are known to be the target of inflammatory responses. In fact, most of the factors involved in RA pathophysiology are under the control of these transcription factors (reviewed in Ref. 11). Particularly, various cytokines and cell adhesion molecules activated in the rheumatoid joints are under the transcriptional control of NF-κB. The self-perpetuating nature of rheumatoid inflammation is ascribable to TNFα and IL-1β, known to elicit the activation cascade for NF-κB and AP-1, as they constitute another positive feed-back loop in the logic of the inflammatory responses associated with RA (Fig. 2).

In addition, besides its action in upregulating inflammatory cytokines and cell adhesion molecules, NF-κB also

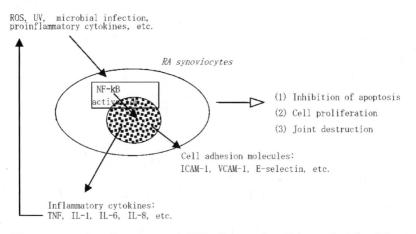

Figure 2 Involvement of NF-κB in the RA pathophysiology. Nuclear factor-κB induces gene expression of inflammatory mediators such as cytokines and cell adhesion molecules. As proinflammatory cytokines, TNF-α and IL-1β, stimulate the NF-κB activation cascade that induces expression of these cytokines, there will be a positive feed-back loop that perpetuates and expands the inflammatory responses even systemically. Nuclear factor-κB also stimulates synovial proliferation by inhibiting apoptosis. See also Fig. 9.

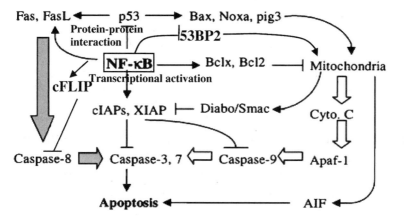

Figure 3 Antiapoptotic actions of NF-κB. Nuclear factor-κB inhibits apoptosis by: (i) transcriptional activation of antiapoptotic factors including cIAPs, XIAP, cFLIP, Bclx, and Bcl2, and (ii) direct inhibition of proapoptotic proteins such as p53 and 53BP2. Oxidation of NF-κB by gold cation. The active DNA binding form of NF-κB is hypothesized to contain zinc ions. When gold compound is added, Au(I) can take the electron from thiolate anions due to its higher oxidation potential compared with that of Zn^{2+}. Thus, Au(I) eventually oxidizes the thiolate anions of NF-κB into disulfide. The oxidation of NF-κB abolishes the DNA-binding activity.

induces gene expression of cell growth-promoting factors, such as cyclin D1 and c-Myc, and physiological inhibitors of apoptosis, such as cIAPs, Bcl-X_L, and cFLIP (12,13) (Fig. 3). Moreover, it is shown that NF-κB blocks apoptosis in the absence of de novo protein synthesis (14) through protein–protein interaction with p53 and proapoptotic protein 53BP2 (15,16).

These actions of NF-κB explain not only the inflammatory responses, but also the hyperproliferation of synovial tissues in RA, indicating that NF-κB acts as a major determinant for RA pathophysiology. Nuclear factor-κB induces TNFα and IL-1β gene expression, and both TNFα and IL-1β stimulate NF-κB signaling; a vicious cycle is formed to perpetuate and even expand the inflammatory responses (11). Thus, the intervention therapy against using anti-TNF antibody and IL-1β receptor antagonist has been developed (17,18). In addition, some of the drugs for RA have been shown to block

Table 1   List of RA Drugs that Inhibit NF-κB

Acetylsalicylic acid	Dexamethasone
Aurothioglucose	Ibuprofen
Aurothiomalate	Sodium salicylate
Auranofin	Sulfasalazine

NF-κB-activation cascade or its actions (Table 1) (19–22). However, there is no evidence to support the possibility that NF-κB or its signaling cascade is impaired in RA. Gene expression profile analysis using rheumatoid synoviocytes did not show any significant difference with regard to the responsiveness of NF-κB target genes (9). Thus, further studies are needed to elucidate a common mechanism by which NF-κB is activated in RA synovium.

IV. OXIDATIVE STRESS IN RA

It is well known that the synovial cavity of patients with RA is full of oxidative stress. First, Jayson and Dixon (23) found that the intra-articular pressure is much higher in RA joints as the result of a decreased compliance of the joint wall due to synovial membrane swelling and fibrosis of the capsule. Because of this elevated intra-articular pressure, the capillary flow rates of the inflamed joint tissues greatly fell, and reperfusion was delayed, thus associated with a decrease in the synovial O_2 tension (pO_2), an elevated pCO_2, an increase in the concentration of synovial fluid (SF) lactate, and a decrease in pH (24–26). Second, the synovial hypoxia was shown to cause accumulation of adenosine and its breakdown products including hypoxanthine and xanthine (27), which subsequently activates the xanthine oxidase system (28) leading to repeated episodes of oxidative injury in the rheumatoid joints. Third, ROS production was detected in the joints of RA patients including the direct measurement of superoxide anion by electron spin resonance (28), ROS-modified IgG (29), increase in lipid peroxidation product (26), depletion of ascorbate (29), and ROS-mediated fragmentation of glycosaminoglycans such as synovial hyaluronic acid (30).

As mentioned earlier, the degradation of hyaluronic acid is considered responsible for the decreased viscosity of joint fluid and the increase in intra-articular pressure, and the oxidatively damaged IgG accounts for the generation of reactive epitopes for the production of rheumatoid factor.

V. OXIDO-REDUCTION OF PROTEINS AS A SIGNAL: REDOX REGULATION

Reactive oxygen species are highly reactive with biological macromolecules to result in producing lipid peroxides (which are often radicals), inactivating proteins and mutating DNA (by producing 8-OH-dG or breaking nucleic acid chains). Therefore, cells must have acquired the multiplicated endogenous antioxidant system for the maintenance of a stable form of life under such harmful conditions. These defense mechanisms include reducing enzymes such as thioredoxin (Trx) and glutaredoxin (Grx) (31–33). Oxidized protein molecules by ROS are reversibly reduced by Trx or Grx. Importantly, this reversible oxidation and reduction involving Cys residues of a functional protein sometimes work as a regulatory modification that determines its biological/biochemical activities. This is analogous to the regulatory modification of proteins such as phosphorylation (involving Ser, Thr, and Tyr residues), acetylation and methylation (Lys), and ubiquitination (Lys) (Fig. 4). Thus, the term "redox regulation" has been proposed indicating the active role of oxido-reductive modifications of proteins in regulating their activities. In other words, oxidation and reduction of biomolecules can be regarded as "signals" through which the organism communicates with external environment. There are accumulating evidences indicating that such redox control system works for the maintenance of cellular homeostasis (11,33–35).

VI. SIGNALING CASCADE FOR NF-κB ACTIVATION

The members of the NF-κB family in mammalian cells include the proto-oncogene c-Rel, RelA (p65), RelB, NFkB1

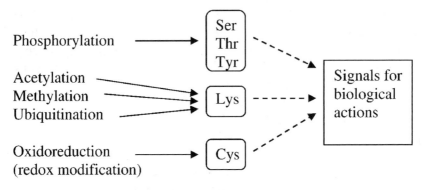

Figure 4 Biochemical modifications of proteins. Oxido-reduction (redox regulation) involves Cys residues. These amino acid modifications will generate biological signals.

(p50/105), and NFkB2 (p52/p100). These proteins share a conserved 300 amino acid region known as the Rel homology domain, which is responsible for DNA binding, dimerization, and nuclear translocation of NF-κB. In most cells, Rel family members form hetero- and homo-dimers with distinct specificities in various combinations (11,36–38). A common feature of the regulation of NF-κB family is their sequestration in the cytoplasm as inactive complexes with a class of inhibitory molecules known as IκBs (38,39). Upon stimulation of the cells by proinflammatory cytokines, IL-1β and TNF-α, IκBs are degraded and NF-κB is translocated to the nucleus and activates expression of target genes. In addition to these physiological stimuli, NF-κB activation cascade is also triggered by ionizing UV irradiation and oxidative reagents such as H_2O_2 (11,40–42). It is well established that two major signaling cascades are involved in NF-κB activation: kinase cascade and redox regulation.

VI.A. Kinase Cascades Involved in the NF-κB Activation Cascade

At least two distinct types of kinase pathways are known to be involved in NF-κB activation: IκB kinase and NF-κB kinase (Fig. 5). The IκB kinase complex, capable of specifically

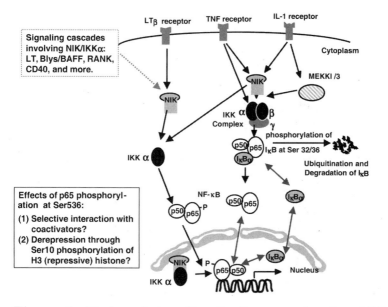

Figure 5 Nuclear factor-κB activation cascades. In addition to canonical pathway involving IκB phosphorylation and ubiquitination followed by its proteolytic degradation in 26S proteasome within the cytoplasm, there appears to be another cascade not involving IκB phosphorylation. Lymphotoxin β receptor signaling, CD40, RANK, and BlyS/BAFF stimulate the NIK–IKKα cascade that leads to p100/p52 processing and p65 phosphorylation at its C-terminal transactivation (Ser536). IKKα also phosphorylates histone H3 in the nucleus and de-represses the otherwise silent nucleosome, thus reactivating the dormant genes. The effect of p65 (Ser536) phosphorylation is considered to activate the transcriptional competence of NF-κB.

phosphorylating serines 32 and 36 of IκBα, was originally identified as ~700 kDa of high molecular complex (29,38,43). Subsequently, two catalytic subunits (IKKα and IKKβ) and a scaffold subunit of this complex (IKKγ/NEMO/IKKAP) were identified and cloned (37,44). The IκB kinase (IKK) complex, consisting of IKKα, β, and γ, can be activated by a variety of stimuli, including TNF-α, IL-1β, and LPS. Activation of the complex involves the phosphorylation of two serine residues located in the "activation loop" within the kinase domain of

IKKα and IKKβ. The IKK complex is stimulated by upstream kinases that belong to mitogen-activated protein kinase kinase kinases (MAP3Ks), such as MEKK1, MEKK2, MEKK3, and to NIK, capable of phosphorylating these serines in vitro, and activating NF-κB (62,65). Phosphorylation on specific serine residues of IκBs leads to ubiquitination of IκBs and subsequent degradation by proteasome complex.

There are accumulating evidences suggesting the involvement of additional kinases that phosphorylate the p65 (RelA) subunit of NF-κB and regulate its transcriptional competence (45–47). We recently found that IKKα is responsible for the p65 phosphorylation at Ser536 upon the lymphotoxin (LT)β receptor signaling mediated by NIK and induces NF-κB activation independently of the IκB phosphorylation and its degradation (48–50). Interestingly, this NIK–IKKα cascade is also involved in Blys/BAFF and RANK, and most likely CD40, signaling (50–52). Because Blys/BAFF and CD40 signaling cascades induce B-cell activation and RANK signaling is involved in osteoclast differentiation, the NIK–IKKα cascade is considered to play important roles in disease progression of RA. The TNF-α-dependent phosphorylation of serine 529 has also been demonstrated to increase the transcriptional activity of p65. For example, casein kinase II was implicated in the TNF-α-dependent phosphorylation of p65 on serine 529 (53). It was shown that serine residues 529 and 536 of p65 were required for the transcriptional activation of p65 by AKT and the IL-1β signaling (50,54). Although Ghosh and colleagues (47) have proposed a model in which the catalytic subunit of PKA (PKAc) is associated with the NF-κB/IκBα complex in the cytoplasm in an inactive form, and signal-induced degradation of IκBα allows PKAc to phosphorylate p65 on serine 276 for transcriptional activation, we and others (55–57) found that PKA activation did not stimulate NF-κB-dependent gene expression.

Inducible phosphorylation of p65 appears to function at many different levels, including conformational changes in the transcriptional activation domain and promoting association with coactivator proteins CBP/p300 (38). It is possible that the phosphorylation of p65 may lead to dissociation from corepressor proteins, such as histone deacetylases and Groucho proteins (TLE/AES), and selective interaction with FUS/TLS coactivator protein (58–60). Moreover, recent evidences have demonstrated that upon signaling IKKα translocates to the nucleus and phosphorylates Ser10 of the histone H3 component of nucleosome (61,62) (Fig. 5). Although the histone H3 with methylated Lys9 of H3 renders the local nucleosome to be "repressive," the adjacent Ser10-phosphorylation of H3 histone reverses this effect and de-represses the transcriptional activity of the genes located in the "de-repressed" nucleosome (10).

In addition, it was recently shown that ischemia/reperfusion injury and H_2O_2 induce Src family kinases that subsequently phosphorylate the Tyr42 of IκBα and induce NF-κB in the absence of ubiquitin-dependent degradation (63,64). It appears that Src family kinases act as redox sensors for NF-κB activation. Thus, IKK-independent pathway can function under specific redox-mediated stimuli to activate NF-κB. Although NF-κB activation may reduce tissue damage following the ischemia/reperfusion injury by blocking apoptosis, it may promote synovial cell proliferation in the affected joints of RA patients.

VI.B. Redox Regulation of NF-κB Activation

The induction of NF-κB following liver ischemia/reperfusion injury is regulated by acute redox-activated responses involving an NADPH oxidase Rac1 (65). Other evidences also indicate the involvement of Rac1 in the generation of ROS and activation of NF-κB (66,67). Divergent stimuli that activate NF-κB are considered to generate ROS on the basis of the facts that most such signaling cascades could be blocked by antioxidants, e.g., agents like *N*-acetyl-L-cysteine (NAC), PDTC, and α-lipoic acid were shown to block NF-κB activation in response to diverse stimuli (40,68–72). However, a recent study has clarified that NAC and PDTC block the NF-κB signaling by not necessarily blocking ROS but by lowering the affinity of TNF receptor to its ligand and inhibiting the ubiquitin ligase activity for IκB, respectively (73). No direct evidence of ROS production in response to various NF-κB stimulating agents, such as TNFα, IL-1β, and LPS, is yet to be obtained. Therefore, the general involvement of ROS in the NF-κB signaling is still elusive.

Intriguingly, there are accumulating evidences that support the positive effect of Trx in the NF-κB activation cascade (74–77). Thioredoxin is a cellular reducing catalyst and is known to participate in the redox regulation of cellular proteins by reducing the redox-active cysteins through reversible oxidation of the active center dithiol of Trx molecule to a disulfide. Interestingly, human Trx has been initially identified as a factor responsible for induction of the α subunit of the IL-2 receptor, which is known to be under the transcriptional control of NF-κB (78). We and others (74–76,79) have demonstrated in vitro that NF-κB cannot bind to the κB DNA sequence of the target genes until it is reduced.

Structural and biochemical approaches have provided evidences supporting the molecular model of the redox regulation of NF-κB by Trx (Fig. 6). Within the NF-κB DNA

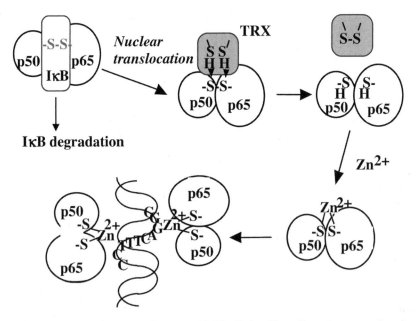

Figure 6 Redox regulation of NF-κB by Trx. See the text for the details.

recognition domain, there is a redox-sensitive Cys (74,75) in the loop of the β-barrel structure that makes a direct contact with the DNA (80,81). Qin et al. (82) has solved the 3D NMR structure of Trx molecule that is associated with the DNA-binding loop of p50 subunit of NF-κB and showed that a redox-active Cys located in the depth of the boot-shaped hollow on Trx surface is in the close proximity with the redox-sensitive Cys of the DNA-binding loop of p50 and likely to reduce the oxidized cysteine on p50 by donating protons in a structure-dependent fashion. However, the inter-molecular disulfide bridge bet-ween Trx and NF-κB must be transient because the binding of Trx to the NF-κB DNA-binding loop prevents the recognition of target DNA. On the basis of biochemical reactions, we have postulated that zinc ion replaces the inter-molecular disulfide bridge and dissociates NF-κB from Trx (11,20,41). In favor of this model, we have demonstrated with cultured rheumatoid synoviocytes that NF-κB and Trx concomitantly migrated to the nucleus during the early phase of the NF-κB activation process induced by TNFα (83). Thioredoxin was relocated in the cytoplasm after 30 min of stimulation, whereas the NF-κB was predominantly present at the nucleus for several hours. Thus, it is possible that NF-κB associates with Trx immediately after dissociation from IκB, translocates to the nucleus together, and dissociates from Trx through displacement of the inter-molecular disulfide by zinc ions.

VII. ROLES OF THIOREDOXIN IN RA PATHOPHYSIOLOGY

In order to examine the roles of Trx in RA, we have measured the Trx level in the joint fluid and explored its effects on the NF-κB activation cascade. Others have elucidated additional roles of Trx in the hyperproliferative nature of rheumatoid synoviocytes.

VII.A. Elevated Trx in the RA Joint Fluid

We found that the serum Trx level was elevated in patients with RA when compared with healthy individuals and

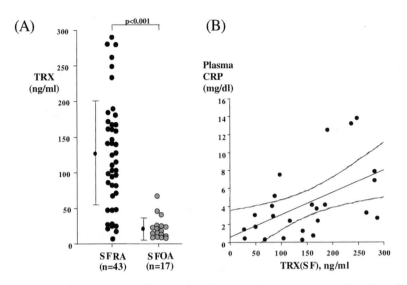

Figure 7 Elevated levels of Trx in RA patients. (A) The Trx concentration is elevated in the joint fluids of RA patients when compared with that of OA patients. (B) Positive correlation of Trx (SF) and serum CRP (its production is stimulated by IL-6 in the liver), indicating that Trx also plays a role in systemic inflammation.

patients with OA (Fig. 7) (77). Moreover, the Trx level in the SF was much greatly elevated in RA patients than in OA patients. In fact, Taniguchi et al. (84) identified a cis-regulatory element in response to oxidative stress within a promoter region of human Trx gene. Thus, the increase of Trx level in SF could be ascribed to the production of ROI by activated macrophages and by hypoxic-reperfusion injuries in the inflamed rheumatoid joints as discussed earlier (23–30). Moreover, multiple regression analysis revealed that the serum C-reactive protein (CRP) level, a clinical laboratory parameter of inflammation, was better correlated with the linear combination of TNF-α (SF) and Trx (SF) levels than TNF-α (SF) alone, which suggested that Trx might play a subsidiary role in the rheumatoid inflammation (Fig. 7) (77).

When the effect of Trx on the TNF-α-induced IL-6 and IL-8 production using rheumatoid synovial fibroblast cultures was examined, we found that the extents of IL-6 and IL-8

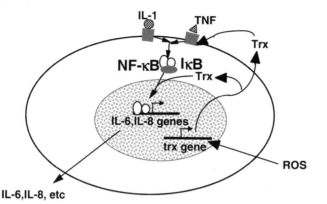

Figure 8 Effect of extracellular Trx on the NF-κB activation stimulated by proinflammatory cytokines. Reactive oxygen species stimulates Trx production in addition to NF-κB activation. See the text for the details.

production in response to TNF-α were greatly augmented by Trx when compared with TNF-α alone (Fig. 8). Furthermore, we found that Trx accelerated the nuclear translocation of NF-κB and facilitated the IκBα phosphorylation and subsequent degradation in response to TNF-α. The elevated Trx level indicates the persistent presence of oxidative stress in the joints of RA patients. Thus, these findings show that Trx has an active role in RA by augmenting the proinflammatory response of TNFα.

VII.B. Regulation of ASK1 Activity by Trx

More intriguingly, another piece of evidence linking the cellular redox status to specific signaling pathways came from the fact that the apoptosis signal regulated kinase 1 (ASK1) interacts with Trx. ASK1 is a member of the MAP3K family that activates downstream kinases including JNK and p38 MAPK. Screening for ASK1-associated proteins using the yeast two-hybrid system led to the identification of Trx as an ASK1-interacting molecule (85). When Trx binds to ASK1, the activity of ASK1 is inhibited (Fig. 9). The rise in ROS levels after TNF-α stimulation leads to activation of ASK1

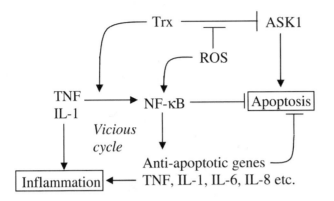

Figure 9 Thioredoxin (Trx) as a redox sensor and biological switch to control apoptosis. Trx inhibits the apoptosis-inducing ASK1 kinase. Reactive oxygen species ROS blocks the interaction between Trx and ASK1, thus stimulating apoptotic cascade. Nuclear factor-κB induces gene expression of various cytokines and antiapoptotic factors. In the presence of excessive ROS, both the anti- and proapoptotic processes are stimulated. Thus, the actual cell fate depends on the level of cellular antioxidant system. Signal transduction pathways for NF-κB activation. The first step involves kinase pathways such as IKKs and NF-κB kinases. The second step involves "redox regulation" by Trx. After the stimulation of the cells by TNF-α or IL-1β, for example, ROS are produced. ROS activate kinase cascades by direct or indirect mechanisms. Reactive oxygen species also induces production of Trx. Signals from TNF receptor or IL-1 receptor stimulate the downstream NF-κB activation pathway. Phosphorylation of NF-κB or IκBs will lead to dissociation of NF-κB from IκBs. The phosphorylated IκBs by IKK complex will be ubiquitinated and then degraded by proteasome. After liberated from IκBs, NF-κB must go through the Trx-mediated reduction of the "redox-sensitive" cysteine to recognize the target DNA sequence (κB site). The trans-activation potential of the NF-κB p65 subunit is modified by interaction between with coactivators or corepressors, including CBP/p300 or TLE/AES.

by dissociation of Trx from ASK1 (85,86). It appears that cellular reducing enzymes, such as Trx and Grx, could also function as redox sensor molecules. If the cellular antioxidant system is sufficient, Trx will be stably associated with ASK1 and act as an antiapoptotic factor, thus supporting the synoviocyte proliferation. However, when excessive ROS is present, due to I/R stress, for example, synoviocytes undergo apoptosis in the presence of severe inflammatory responses. In fact, apoptotic synoviocytes were often observed within the hyperproliferative synovial tissues.

VIII. CONCLUSION

Rheumatoid arthritis is a complexed process of chronic and progressive inflammation involving numerous transcription factors and signaling molecules. Although majority of pathologic changes of RA appeared to be limited in the joints, similar pathophysiological considerations are applicable to a wide variety of chronic and acute/severe inflammatory diseases including inflammatory bowel diseases, surgical inflammatory response syndrome, multiple sclerosis, and atherosclerosis. In terms of NF-κB involvement, the RA pathophysiology shares with HIV infection and cancer. Thus, findings and therapeutic measures discovered in RA are readily applicable to the understanding and the treatment of these fatal diseases. In other words, when disease processes are broken down into actions of each molecule that governs critical step of many events that build up the entire process, there will be no restriction in applying the concept obtained from the behavior of each molecule to other disease by crossing the border of different scientific disciplines.

ACKNOWLEDGMENTS

The author acknowledges the editors of this book for giving me this opportunity to write this review. I owe especially Dr. Lester Packer for his continuous encouragements and everlasting scientific stimulations. This work was supported by grants in aid from the Ministry of Health, Labor and Welfare, the Ministry of Education, Culture, Science and Technology, and from the Japan Health Sciences Foundation.

REFERENCES

1. Ermann J, Fathman CG. Autoimmune diseases: genes, bugs and failed regulation. Nat Immunol 2001; 2:759–761.

2. Lee DM, Weinblatt ME. Rheumatoid arthritis. Lancet 2001; 358: 903–911.

3. Feldmann M. Pathogenesis of arthritis: recent research progress. Nat Immunol 2001; 2:771–773.

4. Fox DA. Etiology and pathogenesis of rheumatoid arthritis. In: Koopman WJ, ed. Arthritis and Allied Conditions—A Textbook of Rheumatology. Baltimore: Williams & Wilkins, 1997: 1085–1101.

5. Firestein GS. Evolving concepts of rheumatoid arthritis. Nature 2003; 423:356–361.

6. Kinne RW, Brauer R, Stuhlmutter B, Palombo-Kinne E, Burmester GR. Macrophages in rheumatoid arthritis. Arthritis Res 2000; 3:189–202.

7. Fassbender HG, Simmling-Annefeld M. The potential aggressiveness of synovial tissue in rheumatoid arthritis. J Pathol 1983; 139:399–406.

8. Watanabe N, Ando K, Yoshida S, Inuzuka S, Kobayashi M, Matsui N, Okamoto T. Gene expression profile analysis of rheumatoid synovial fibroblast cultures revealing the overexpression of genes responsible for tumor-like growth of rheumatoid synovium. Biochem Biophys Res Commun 2002; 294: 1121–1129.

9. Ando K, Kanazawa S, Tetsuka T, Ohta S, Jiang X, Tada T, Kobayashi M, Matsui N, Okamoto T. Induction of Notch signaling by tumor necrosis factor in rheumatoid synovial fibroblasts. Oncogene 2003; 22:7796–7803.

10. Fischle W, Wang Y, Allis CD. Binary switches and modification cassettes in histone biology and beyond. Nature 2003; 425: 475–479.

11. Okamoto T, Sakurada S, Yang JP, Merin JP. Regulation of NF-κB and disease control: identification of a novel serine kinase and thioredoxin as effectors for signal transduction

pathway for NF-κB activation. Curr Top Cell Regul 1997; 35: 149–161.

12. Opferman JT, Korsmeyer SJ. Apoptosis in the development and maintenance of the immune system. Nat Immunol 2003; 4: 410–415.

13. Karin M, Lin A. NF-κB at the crossroads of life and death. Nat Immunol 2002; 3:221–227.

14. Kajino S, Suganuma M, Teranishi F, Takahashi N, Tetsuka T, Ohara H, Itoh M, Okamoto T. Evidence that de novo protein synthesis is dispensable for anti-apoptotic effects of NF-κB. Oncogene 2000; 19:2233–2239.

15. Yang JP, Hori M, Takahashi N, Kawabe T, Kato H, Okamoto T. NF-κB subunit p65 binds to 53BP2 and inhibits cell death induced by 53BP2. Oncogene 1999; 18:5177–5186.

16. Takahashi N, Kobayashi S, Jiang X, Kitagori K, Imai K, Hibi Y, Okamoto T. Expression of 53BP2 and ASPP2 proteins from TP53BP2 gene by alternative splicing. Biochem Biophys Res Commun 2004; 315:434–438.

17. Brennan FM, Chantry D, Jackson A, Maini R, Feldmann M. Inhibitory effect of TNF α antibodies on synovial cell interleukin-1 production in rheumatoid arthritis. Lancet 1989; 2: 244–247.

18. Bresnihan B, Alvaro-Gracia JM, Cobby M, Doherty M, Domljan Z, Emery P, Nuki G, Pavelka K, Rau R, Rozman B, Watt I, Williams B, Aitchison R, McCabe D, Musikic P. Treatment of rheumatoid arthritis with recombinant human interleukin-1 antagonist. Arthritis Rheum 1998; 41:2196–2204.

19. Yamamoto Y, Gaynor RB. Therapeutic potential of inhibition of the NF-κB pathway in the treatment of inflammation and cancer. J Clin Invest 2001; 107:135–142.

20. Yang JP, Merin JP, Nakano T, Kato T, Kitade Y, Okamoto T. Inhibition of the DNA-binding activity of NF-κB by gold compounds in vitro. FEBS Lett 1995; 361:89–96.

21. McKay LI, Cidlowski JA. Molecular control of immune/inflamma-inflammatory responses: interactions between nuclear factor-κB and steroid receptor-signaling pathways. Endocr Rev 1999; 20: 435–459.

22. Yoshida S, Kato T, Sakurada S, Kurono C, Yang JP, Matsui N, Soji T, Okamoto T. Inhibition of IL-6 and IL-8 induction from cultured rheumatoid synovial fibroblasts by treatment with aurothioglucose. Int Immunol 1999; 11:151–158.

23. Jayson MIV, Dixon ASJ. Intra-articular pressure in rheumatoid arthritis of the knee. Pressure changes during passive joint destruction. Ann Rheum Dis 1970; 29:261–265.

24. James MJ, Cleland LG, Rofe AM, Leslie AL. Intra-articular pressure and the relationship between synovial perfusion and metabolic demand. J Rheum 1990; 17:521–527.

25. Levick JR. Hypoxia and acidosis in chronic inflammatory arthritis, relation to vascular supply and dynamic effusion pressure. J Rheum 1990; 17:579–582.

26. Mapp PI, Grootveld MC, Blake DR. Hypoxia, oxidative stress and rheumatoid arthritis. Br Med Bull 1995; 51:419–436.

27. Herbert KE, Scott DL, Perrett D. Nucleosides and bases in synovial fluid from patients with rheumatoid arthritis and osteoarthritis. Clin Sci 1988; 74:97–99.

28. Allen RE, Blake DR, Nazhat NB, Jones P. Superoxide radical generation by inflamed human synovium after hypoxia. Lancet 1989; 2(8651):282–283.

29. Lunec J, Blake DR, Brailsford S, Bacon PA. Self perpetuating mechanisms of immunoglobulin G aggregation in rheumatoid inflammation. J Clin Invest 1985; 76:2984–2090.

30. Grootveld MC, Henderson EB, Farrell A, et al. Oxidative damage to hyaluronate and glucose in synovial fluid during exercise of the inflamed joint. Detection of low molecular mass metabolites by proton NMR spectroscopy. Biochem J 1991; 273:459–467.

31. Holmgren A. Thioredoxin. Annu Rev Biochem 1985; 54:237–271.

32. Holmgren A. Thioredoxin and glutaredoxin systems. J Biol Chem 1989; 264:13963–13966.

33. Holmgren A. Antioxidant function of thioredoxin and glutaredoxin systems. Antioxid Redox Signal 2000; 2:811–820.

34. Allen RG, Tresini M. Oxidative stress and gene regulation. Free Radic Biol Med 2000; 28:463–499.

35. Finkel T. Redox-dependent signal transduction. FEBS Lett 2000; 476:52–54.

36. Baldwin AS Jr. Series introduction: the transcription factor NF-κB and human disease. J Clin Invest 2001; 107:3–6.

37. Ghosh S, May MJ, Kopp EB. NF-κB and Rel proteins: evolutionarily conserved mediators of immune responses. Annu Rev Immunol 1998; 16:225–260.

38. Schmitz ML, Bacher S, Kracht M. IκB-independent control of NF-κB activity by modulatory phosphorylations. Trends Biochem Sci 2001; 26:186–190.

39. Karin M, Ben-Neriah Y. Phosphorylation meets ubiquitination: the control of NF-κB activity. Annu Rev Immunol 2000; 18:621–663.

40. Meyer M, Schreck R, Baeuerle PA. H_2O_2 and antioxidants have opposite effects on activation of NF-κB and AP-1 in intact cells: AP-1 as secondary antioxidant-responsive factor. EMBO J 1993; 12:2005–2015.

41. Okamoto T, Tetsuka T. Role of thioredoxin in the redox regulation of gene expression in inflammatory diseases. In: Winyard PG, Blake DR, Evans CH, eds. Free Radicals and Inflammation. Basel: Birkhuser Verlag, 2000:119–131.

42. Saliou C, Kitazawa M, Mclaughlin L, Yang JP, Lodge JK, Iwasaki K, Cillard J, Okamoto T, Packer L. Antioxidants modulate acute solar ultraviolet radiation-induced NF-κB activation in a human keratinocyte cell line. Free Radic Biol Med 1999; 26:174–183.

43. Chen ZJ, Parent L, Maniatis T. Site-specific phosphorylation of IκBα by a novel ubiquitination-dependent protein kinase activity. Cell 1996; 84:853–862.

44. Lee FS, Peters RT, Dang LC, Maniatis T. MEKK1 activates both IκB kinase α and IκB kinase β. Proc Natl Acad Sci USA 1998; 95:9319–9324.

45. Hayashi T, Sekine T, Okamoto T. Identification of a new serine kinase that activates NFκB by direct phosphorylation. J Biol Chem 1993; 268:26790–26795.

46. Naumann M, Scheidereit C. Activation of NF-κB in vivo is regulated by multiple phosphorylations. EMBO J 1994; 13:4597–4607.

47. Zhong H, Voll RE, Ghosh S. Phosphorylation of NF-κB p65 by PKA stimulates transcriptional activity by promoting a novel bivalent interaction with the coactivator CBP/p300. Mol Cell 1998; 1:661–671.

48. Jiang X, Takahashi N, Matsui N, Tetsuka T, Okamoto T. The NF-κB activation in the lymphotoxin beta receptor signaling depends on the phosphorylation of p65 at serine 536. J Biol Chem 2003; 278:919–926.

49. Jiang X, Takahashi N, Ando K, Otsuka T, Tetsuka T, Okamoto T. NF-κB p65 transacvation domain is involved in the NF-κB-inducing kinase (NIK) pathway. Biochem Biophys Res Commun 2003; 301:583–590.

50. Sakurai H, Chiba H, Miyoshi H, Sugita T, Toriumi W. IκB kinases phosphorylate NF-κB p65 subunit on serine 536 in the transactivation domain. J Biol Chem 1999; 274:30353–30356.

51. Cao Y, Bonizzi G, Seagroves TN, Greten FR, Johnson R, Schmidt EV, Karin M. IKKα provides an essential link between RANK signaling and cyclin D1 expression during mammary gland development. Cell 2001; 107:763–775.

52. Ghosh S, Karin M. Missing pieces in the NF-κB puzzle. Cell 2002; 109:S81–S96.

53. Wang D, Baldwin AS Jr. Activation of nuclear factor-κB-dependent transcription by tumor necrosis factor-α is mediated through phosphorylation of RelA/p65 on serine 529. J Biol Chem 1998; 273:29411–29416.

54. Madrid LV, Wang CY, Guttridge DC, Schottelius AJ, Baldwin AS Jr, Mayo MW. Akt suppresses apoptosis by stimulating the transactivation potential of the RelA/p65 subunit of NF-κB. Mol Cell Biol 2000; 20:1626–1638.

55. Neumann M, Grieshammer T, Chuvpilo S, Kneitz B, Lohoff M, Schimpl A, Franza BR Jr, Serfling E. RelA/p65 is a molecular target for the immunosuppressive action of protein kinase A. EMBO J 1995; 14:1991–2004.

56. Ollivier V, Parry GC, Cobb RR, de Prost D, Mackman N. Elevated cyclic AMP inhibits NF-κB-mediated transcription

in human monocytic cells and endothelial cells. J Biol Chem 1996; 271:20828–20835.

57. Takahashi N, Tetsuka T, Uranishi H, Okamoto T. Inhibition of NF-κB transcriptional activity by protein kinase A. Eur J Biochem 2002; 269:1–7.

58. Uranishi H, Tetsuka T, Yamashita M, Asamitsu K, Shimizu M, Itoh M, Okamoto T. Involvement of the pro-oncoprotein TLS (translocated in liposarcoma) in nuclear factor-κB p65-mediated transcription as a coactivator. J Biol Chem 2001; 276: 13395–13401.

59. Tetsuka T, Uranishi H, Imai H, Ono T, Sonta S, Takahashi N, Asamitsu K, Okamoto T. Inhibition of nuclear factor-κB-mediated transcription by association with the amino-terminal enhancer of split, a Groucho-related protein lacking WD40 repeats. J Biol Chem 2000; 275:4383–4390.

60. Ashburner BP, Westerheide SD, Baldwin AS Jr. The p65 (RelA) subunit of NF-κB interacts with the histone deacetylase (HDAC) corepressors HDAC1 and HDAC2 to negatively regulate gene expression. Mol Cell Biol 2001; 21:7065–7077.

61. Anest V, Hanson JL, Cogswell PC, Steinbrecher KA, Strahl BD, Baldwin AS. A nucleosomal function for IκB kinase-α in NF-κB-dependent gene expression. Nature 2003; 423:659–663.

62. Yamamoto Y, Verma UN, Prajapati S, Kwak YT, Gaynor RB. Histone H3 phosphorylation by IKK-a is critical for cytokine-induced gene expression. Nature 2003; 423:655–659.

63. Imbert V, et al. Tyrosine phosphorylation of IκBα activates NF-κB without proteolytic degradation of IκBα. Cell 1996; 86:787–798.

64. Livolsi A, Busuttil V, Imbert V, Abraham RT, Peyron JF. Tyrosine phosphorylation-dependent activation of NF-κB. Requirement for p56 LCK and ZAP-70 protein tyrosine kinase. Eur J Biochem 2001; 268:1508–1515.

65. Ozaki M, et al. Inhibition of the Rac1 GTPase protects against nonlethal ischemia/reperfusion-induced necrosis and apoptosis in vivo. FASEB J 2000; 14:418–429..

66. Sulciner DJ, Irani K, Yu ZX, Ferrans VJ, Goldschmidt-Clermont P, Finkel T. Rac1 regulates a cytokine-stimulated,

redox-dependent pathway necessary for NF-κB activation. Mol Cell Biol 1996; 16:7115–7121.

67. Sundaresan M, Yu ZX, Ferrans VJ, Sulciner DJ, Gutkind JS, Irani K, Goldschmidt-Clermont PJ, Finkel T. Regulation of reactive-oxygen-species generation in fibroblasts by Rac1. Biochem J 1996; 318(Pt 2):379–382.

68. Tozawa K, Sakurada S, Kohri K, Okamoto T. Effects of anti-nuclear factor κB reagents in blocking adhesion of human cancer cells to vascular endothelial cells. Cancer Res 1995; 55: 4162–4167.

69. Roederer M, Staal FJ, Raju PA, Ela SW, Herzenberg LA. Cytokine-stimulated human immunodeficiency virus replication is inhibited by N-acetyl-L-cysteine. Proc Natl Acad Sci USA 1990; 87:4884–4888.

70. Merin JP, Matsuyama M, Kira T, Baba M, Okamoto T. Alpha-lipoic acid blocks HIV-1 LTR-dependent expression of hygromycin resistance in THP-1 stable transformants. FEBS Lett 1996; 394:9–13.

71. Schmidt KN, Traenckner EB, Meier B, Baeuerle PA. Induction of oxidative stress by okadaic acid is required for activation of transcription factor NF-κB. J Biol Chem 1995; 270:27136–27142.

72. Suzuki YJ, Mizuno M, Packer L. Signal transduction for nuclear factor-κB activation. Proposed location of antioxidant-inhibitable step. J Immunol 1994; 153:5008–5015.

73. Hayakawa M, Miyashita H, Sakamoto I, Kitagawa M, Tanaka H, Yasuda H, Karin M, Kikugawa K. Evidence that reactive oxygen species do not mediate NF-κB activation. EMBO J 2003; 22: 3356–3366.

74. Okamoto T, Ogiwara H, Hayashi T, Mitsui A, Kawabe T, Yodoi J. Human thioredoxin/adult T cell leukemia-derived factor activates the enhancer binding protein of human immunodeficiency virus type 1 by thiol redox control mechanism. Int Immunol 1992; 4:811–819.

75. Hayashi T, Ueno Y, Okamoto T. Oxidoreductive regulation of nuclear factor κB. Involvement of a cellular reducing catalyst thioredoxin. J Biol Chem 1993; 268:11380–11388.

76. Matthews JR, Wakasugi N, Virelizier JL, Yodoi J, Hay RT. Thioredoxin regulates the DNA binding activity of NF-κB by reduction of a disulphide bond involving cysteine 62. Nucleic Acids Res 1992; 20:3821–3830.

77. Yoshida S, Katoh T, Tetsuka T, Uno K, Matsui N, Okamoto T. Involvement of thioredoxin in rheumatoid arthritis: its costimulatory roles in the TNF-α-induced production of IL-6 and IL-8 from cultured synovial fibroblasts. J Immunol 1999; 163:351–358.

78. Tagaya Y, Maeda Y, Mitsui A, Kondo N, Matsui H, Hamuro J, Brown N, Arai K, Yokota T, Wakasugi H, Yodoi J. ATL-derived factor (ADF), an IL-2 receptor/Tac inducer homologous to thioredoxin; possible involvement of dithiol-reduction in the IL-2 receptor induction. EMBO J 1989; 8:757–764.

79. Toledano MB, Leonard WJ. Modulation of transcription factor NF-κB binding activity by oxidation–reduction in vitro. Proc Natl Acad Sci USA 1991; 88:4328–4332.

80. Ghosh G, van Duyne G, Ghosh S, Sigler PB. Structure of NF-κB p50 homodimer bound to a κB site. Nature 1995; 373: 303–310.

81. Muller CW, Rey FA, Sodeoka M, Verdine GL, Harrison SC. Structure of the NF-κB p50 homodimer bound to DNA. Nature 1995; 373:311–317.

82. Qin J, Clore GM, Kennedy WM, Huth JR, Gronenborn AM. Solution structure of human thioredoxin in a mixed disulfide intermediate complex with its target peptide from the transcription factor NFκB. Structure 1995; 3:289–297.

83. Sakurada S, Kato T, Okamoto T. Induction of cytokines and ICAM-1 by proinflammatory cytokines in primary rheumatoid synovial fibroblasts and inhibition by N-acetyl-L-cysteine and aspirin. Int Immunol 1996; 8:1483–1493.

84. Taniguchi Y, Taniguchi-Ueda Y, Mori K, Yodoi J. A novel promoter sequence is involved in the oxidative stress-induced expression of the adult T-cell leukemia-derived factor (ADF)/human thioredoxin (Trx) gene. Nucleic Acids Res 1996; 24:2746–2752.

85. Saitoh M, Nishitoh H, Fujii M, Takeda K, Tobiume K, Sawada Y, Kawabata M, Miyazono K, Ichijo H. Mammalian thiore-

doxin is a direct inhibitor of apoptosis signal-regulating kinase (ASK) 1. EMBO J 1998; 17:2596–2606.

86. Gotoh Y, Cooper JA. Reactive oxygen species- and dimerization-induced activation of apoptosis signal-regulating kinase 1 in tumor necrosis factor-α signal transduction. J Biol Chem 1998; 273:17477–17482.

12

Roles of Carbon Monoxide and Nitric Oxide in Liver Damage

**HUN-TAEG CHUNG, HYUN-OCK PAE,
BYUNG-MIN CHOI, and SOO-CHEON CHAE**
Department of Microbiology and Immunology,
School of Medicine, Wonkwang University,
Iksan, South Korea

I. INTRODUCTION

Endogenous, as well as exogenous, nitric oxide (NO) and carbon monoxide (CO) can act as signaling molecules via cGMP-dependent and -independent manners. Both hepatic parenchymal cells and hepatic nonparenchymal cells can express iNOS, whereas eNOS is constitutively expressed in hepatic endothelial cells. Although iNOS is not thought to be expressed constitutively in healthy liver, it is readily upregulated in the liver under diseases and is considered to play an important role in the pathophysiological responses to various insulting damages of the liver. On the other hand, heme oxygenase-1 (HO-1) can be induced not only by heme-containing compounds, but also by non-heme-containing substances or environmental conditions such as NO, endotoxin, hypoxia, heat shock, and is thought to play a key role in maintaining homeostasis in the case of cellular injury.

In this chapter, the authors will describe the interaction of NO and CO in addition to their individual actions in the liver.

II. ROLES OF NO IN LIVER DAMAGE

II.A. General Aspects of NO in the Liver

Nitric oxide is synthesized endogenously from L-arginine via the action of NO synthase (NOS). The isoforms of NOS have been further subdivided and now fall into three basic categories: (i) endothelial NOS (eNOS), (ii) neuronal NOS (nNOS), and (iii) inducible NOS (iNOS). Unlike eNOS and nNOS, iNOS is not expressed constitutively, but rather is expressed in most cell types given the appropriate stimulatory conditions.

Cells in the liver can be divided into the hepatic parenchymal cells (hepatocytes) and the hepatic nonparenchymal cells, which are further subdivided into endothelial cells, smooth muscle cells, Kupffer cells, and hepatic stellate cells. Both hepatic parenchymal cells and hepatic nonparenchymal cells can express iNOS, whereas eNOS is constitutively expressed in hepatic endothelial cells. Although iNOS is not thought to be expressed constitutively in healthy liver, it is readily upregulated in the liver under a number of disease conditions, including ischemia–reperfusion injury, cirrhosis, hepatitis, and liver regeneration (1–4). The iNOS is also upregulated in vitro in hepatocytes and Kupper cells in response to endotoxin, proinflammatory cytokines, such as tumor necrosis factor-α (TNF-α), interleukin-1β (IL-1β), and interferon-γ, as well as their combinations (5). These stimuli often act synergistically to induce iNOS expression; however, IL-1β alone is an effective stimulator of iNOS in hepatocytes (6). The induction of iNOS gene by proinflammatory cytokines requires the activation of nuclear factor-κB (NF-κB) in human and rat hepatocytes (7). In addition, hepatic endothelial cells and stellate cells can also induce in vitro NO production through iNOS expression in response to proinflammatory cytokines (8). Therefore, in inflamed liver, hepatocytes are situated in an environment where NO is produced from surrounding cells, as well as themselves.

Nitric oxide is a highly labile molecule and, as such, is vulnerable to a lot of biological reaction once it is produced. NO, its oxidized form (NO^+), and its reduced form (NO^-) may all react with oxygen molecule and transition metals to form higher nitrogen oxides such as NO_2^- or with various metals to form nitrosyl-metal complexes. Nitric oxide activates the hem center of soluble guanylate cyclase (sGC), which leads to intracellular increases in cyclic guanosine monophosphate (cGMP). Nitric oxide can interact with thiol groups, including glutathione and cysteine (9). Nitrosylation of biological thiols can influence protein functions in important ways. Nitric oxide can also interact with the superoxide anion. Interaction with NO neutralizes superoxide, thereby decreasing oxidative stress. However, this reaction can also lead to formation of peroxynitrite ($ONOO^-$) that can cause nitration of protein tyrosines (nitrotyrosine), contributing to enzymatic and cellular dysfunction (10).

Because NO production is significantly elevated in the liver under disease conditions, the physiological and pathophysiological functions of NO in the liver have been widely studied. Both cytoprotective and cytotoxic effects of NO have been demonstrated in the liver (Fig. 1).

II.B. Roles of eNOS-Derived NO in the Liver

As a cGMP-dependent vasodilator, eNOS-derived NO has been shown to contribute to maintaining sinusoidal perfusion following liver injury during endotoxima. Treatment of rats with non-specific-NOS or selective-eNOS inhibitors results in rapid exacerbation of liver injury following endotoxin injection (11,12). This is associated with local failure of sinusoidal perfusion and the development of patchy necrosis. At the same time, specific iNOS inhibitors have little effect on liver perfusion (13). These data suggest that NO is necessary to maintain

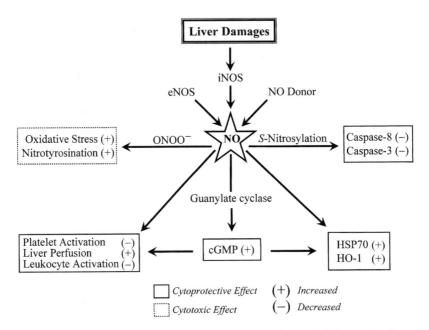

Figure 1 Cytoprotective and cytotoxic effects of NO in the liver. The eNOS-derived NO production is beneficial to the liver via its cGMP-dependent roles in maintaining sinusoidal perfusion and preventing platelet aggregation and accumulation of neutrophils. However, iNOS-derived NO production appears to have cytoprotective, as well as cytotoxic effects, depending on the types of liver damages. The cytotoxic effects of NO appear to include the redox state of liver. Both cGMP-independent inhibition of hepatic apoptosis and cGMP-dependent induction of protective gene may account for the cytoprotective effects of NO in the liver.

sinusoidal perfusion, but that the relatively small amount produced by eNOS is adequate. Nitric oxide produced by eNOS can also inhibit neutrophil adhesion to vascular endothelium and platelet aggregation (14), further contributing to the maintenance of sinusoidal perfusion.

II.C. Roles of iNOS-Derived NO in the Liver

Whereas eNOS-derived NO production is clearly beneficial to the liver for its roles in maintaining sinusoidal perfusion and preventing platelet aggregation and accumulation of neutrophils

(11–14), iNOS-derived NO production may have opposite effects, depending on the liver damages. The determinants for the effects of iNOS expression appear to include the redox status of liver and the simultaneous production of reactive oxygen species (ROS), indicating that NO, itself, may not be toxic to hepatocytes, but unexpected interactions between NO and toxic molecules, such as ROS, may result in massive necrosis of liver cells (15). The cytoprotective action of iNOS-derived NO in the liver may be due to the antiapoptotic properties of NO. Cultured hepatocytes undergo apoptosis by apoptosis-triggering agents, such as Fas and TNF-α, glucose deprivation, or prolonged cell culturing (16,17,32). It has been shown that in cultured hepatocytes, NO suppresses not only Fas- and TNF-α-mediated apoptosis, but also spontaneous apoptosis induced by glucose deprivation or prolonged culturing (16,17). These in vitro studies are further supported by in vivo results. A selective NO donor significantly reduced the hepatocyte apoptosis induced by TNF-α in rat (18). Furthermore, iNOS knockout animals exhibited a significant increase in apoptosis of hepatocytes after partial hepatomy (19), suggesting that iNOS-derived NO may be involved in preventing apoptosis during liver regeneration.

II.D. NO and Its Antiapoptotic Mechanisms in the Liver

The antiapoptotic mechanisms of NO actions in liver cells have been examined in some detail. It has been demonstrated that NO prevents partially, but not completely, TNF-α-induced apoptosis in hepatocytes via cGMP-dependent mechanism (20). As a cGMP-independent antiapoptotic mechanism, S-nitrosylation of liver-relevant thiols has been suggested by Kim et al. (16,21). S-nitrosylation of cysteine in caspase-8 and in caspase-9 inhibits these proteases and downstream apoptotic signaling in hepatocytes (21). Other possible antiapoptotic mechanisms of NO actions have been also examined. Nitric oxide potentially induces the expression of several cytoprotective genes inducing heat shock protein (HSP) 70 and, more recently, HO-1. HSP 70 inhibits oligomerization of Apaf-1 by associating with the caspase-recruitment domain of Apaf-1, resulting in the suppression of apoptosome formation (22). HSP 70 also involves the chaperone-mediated import of precursor proteins into the mitochondria, thus inhibiting cytochrome *c* release (22). Potential roles of NO-induced HO-1 in hepatocyte apoptosis are discussed later in this chapter.

III. ROLES OF CO IN LIVER DAMAGE

III.A. Heme Oxygenase

Heme oxygenases (HOs) are ubiquitous enzymes that catalyze the initial and rate-limiting steps in the oxidative degradation of heme into biliverdin, CO, and free iron (Fig. 2). Biliverdin is converted to bilirubin, a potent antioxidant, and CO functions

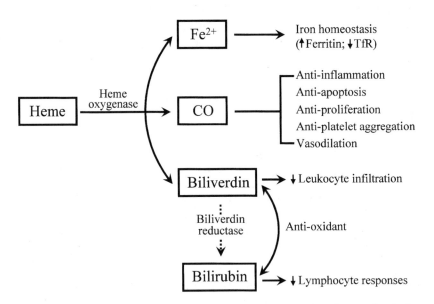

Figure 2 Mechanisms of cytoprotection rendered by heme degradation. Microsomal HO catalyzes the rate-limiting step in heme metabolism that produces CO, free iron (Fe^{2+}), and biliverdin. Carbon monoxide stimulates sGC of smooth muscle cells/platelets leading to vasodilatation or prevention of platelet aggregation and inhibits inflammation and apoptosis. Iron is directly sequestered by ferritin. Biliverdin gets converted into the antioxidant bilirubin by biliverdin reductase.

as a signaling molecule. Three HO isoforms have been identified: the inducible HO-1, also known as a HSP 32, the constitutive HO-2, and HO-3 (23,24). HO-3, which has been cloned recently, has a substantially lower catalytic activity than the isoforms 1 and 2. Although functions and regulation of HO-3 are incompletely understood, there is evidence to suggest a role in binding or transporting heme within the cell (25). HO-1 can be induced not only by heme-containing compounds, but also by non-heme-containing substances or environmental conditions, such as NO, endotoxin, hypoxia, ischemia, hyperthermia, and heat shock, and is thought to play a key role in maintaining antioxidant and oxidant homeostasis in the case of cellular oxidative injury (24,26). In contrast, HO-2 is constitutive; it is abundant in brain, testis, and unstimulated liver of rodents and humans (23). The differing regulation of the two isozymes is caused by their respective promoter regions. The HO-1 promoter contains heat shock elements, an activator protein-1 (AP-1), NF-κB, a cyclic nucleotide-response element, and metal regulatory elements (27). In contrast, the HO-2 promoter contains only a single glucocorticoid response element in the 5′-flanking region (28).

Although regulation of HO-1 gene expression in the intact liver is incompletely understood, the redox-sensitive transcription factor NF-κB and AP-1 might contribute to transcriptional activation of the HO-1 gene under appropriate conditions (29,30). A recent report indicates that cGMP, a second messenger upregulated by NO, can stimulate transcriptional regulation of HO-1 through activation of cyclic nucleotide-response element and AP-1 (31,32). In addition, HO-1 gene expression was inhibitable by dexamethasone(33). Similarly, data obtained by Oguro et al. (34) are consistent with a regulatory role of AP-1 binding for HO-1 induction in a model of glutathione depletion by phorone in the intact rat liver.

III.B. Biological Functions of Three Products of HO-1

The HO pathway has classically been considered as a sink to deal with excessive amounts of heme proteins accumulating on cellular damage. The products of this pathway received little attention because their biological functions were obscure.

The bile pigment biliverdin is produced as a consequence of the heme degradation by HO and is then reduced to bilirubin. Cytoprotection by bilirubin is not an intuitively intriguing concept to most hepatologists, and, as a sign in liver damages, removal of bile pigments is a potential beneficial factor of modern supportive treatment strategies in acute liver failure, such as albumin dialysis (35). On the other hand, cells primed by a wide variety of toxic compounds, such as heavy metals, develop tolerance to subsequent, otherwise lethal injury. Moreover, there is good evidence to suggest a significant role for bile pigments as a cellular antioxidant system (36).

The release of free iron during oxidative cleavage of the heme molecule is a particular matter of concern, because free iron catalyzes the formation of oxygen-free radicals in Fenton reaction. Therefore, the release of iron can mitigate any antioxidant actions of HO-1 gene expression and may explain the rather narrow threshold of overexpression of HO-1 conferring protection (37,38). Nevertheless, release of iron fosters gene expression of ferritin, a protein imparting additive cytoprotection against oxidative stress (39). The therapeutic uses of ferritin or bilirubin have also been shown to ameliorate hepatic injury in several models, including ischemia/reperfusion (I/R) (40,41).

The hazardous effects of CO have been known for many years but only recently have physiologic roles been ascribed to this gaseous molecule. Carbon monoxide may act as a regulatory molecule in different cellular and biological processes, similarly to NO (24). Both can activate sGC leading to smooth muscle relaxation, which contributes to endothelium vasodilatation, and may explain the CO-mediated cytoprotection mechanism against liver injury. Carbon monoxide allows the maintenance of microcirculatory blood flow by inhibiting the vasoconstriction seen otherwise during reperfusion (42). The activation of the cGMP pathway also provides CO with the ability to inhibit platelet aggregation, thus diminishing the microvascular thrombosis associated with I/R injury (43).

III.C. HO-1/CO and Hepatic Cytoprotection

As observed in many physiological, biological, and toxicological systems, overzealous production of effector molecules may be counterproductive for the maintenance of cellular and organismal homeostasis during both normal and pathophysiological conditions. HO-1 is upregulated by stress and, in general, is considered to be cytoprotective in the liver (44,45). Physiologically, HO-1 is involved in the modulation of hepatic sinusoidal tone and portal venous regulation (46,47). These effects are thought to be secondary to the generation of CO, activation of sGC, and subsequent production of cGMP, similar to that of NO. Whether the contractile properties of these cells actually underlie sinusoidal narrowing has not yet fully been documented ex vivo (46,48) or high-resolution analysis by intravital microscopy supports the concept (49). The importance of HO-1 in hepatobiliary vascular regulation is emphasized under pathologic conditions of portal hypertension in which expression of HO-1 is increased significantly (50). In these animal models of portal hypertension, enzymatic inhibition of HO-1 by tin-protoporphyrin exacerbates hypertension, whereas exogenous delivery of CO can decrease portal pressure.

Carbon monoxide allows the maintenance of microcirculatory blood flow by inhibiting the vasoconstriction seen otherwise during reperfusion (42). The activation of the cGMP pathway also provides CO with the ability to inhibit platelet aggregation, thus diminishing the microvascular thrombosis associated with I/R injury (43). Other CO cytoprotective mechanisms against hepatic I/R injury may include suppression of iNOS and downregulation of some procytokines via the mitogen-activated protein kinase (MAPK) pathways (51,52). Amersi et al. (53) have recently studied the effects and downstream mechanisms of CO on cold I/R injury in a clinically relevant isolated perfusion rat liver model. After 24 hr of cold storage, rat livers perfused ex vivo for 2 hr with blood supplemented with CO showed significantly decreased portal venous resistance, improved liver function and diminished histological features of hepatocyte injury. The CO-mediated cytoprotective effects were iNOS- and cGMP-independent, but p38 MAPK-dependent. Moreover, adjunctive use of zinc protoporphyrin, a HO inhibitor, has revealed that exogenous CO could fully substitute for endogenous HO-1 in preventing hepatic I/R insult. Hence, HO-1-mediated cytoprotection against hepatic I/R injury depends on the generation of exogenous CO.

In their disparate models of hepatocytes' death, we have recently shown that NO protect hepatocytes from cellular injury by glucose deprivation, presumably via upregulation of protective protein HO-1 (32). The protective effects of HO-1 are secondary to the generation of CO, which is one of the catalytic by-products in the breakdown of heme. The ERK MAPK signaling pathway represents the key downstream mechanism by which CO protects cells against glucose deprivation-induced cytotoxicity. In addition, recent data have also demonstrated that HO-1 induction protects mice from TNF-α-dependent liver injury and against TNF-α-independent liver injury caused by administration of anti-CD95 Abs and prolongs survival (54). Furthermore, the protective effect of HO-1 is also mediated through its reaction product, CO. Thus, induction of HO-1 or CO might be of a therapeutic modality for the prevention of fulminant hepatic failure in several pathologic conditions.

IV. INTERACTION BETWEEN CO AND NO IN THE LIVER

It is not surprising that conditions associated with increased production of ROS and NO favor the activation of HO-1/CO pathway in the liver, which is now regarded as an important cellular stratagem to overcome oxidative stress insults (26,27). The conception that HO-1 might function to counteract the potential toxic effects of ROS first emerged from the discovery that certain NO-releasing agents can induce HO-1 expression and HO activity in hepatocytes, resulting in cytoprotection against oxidative stress (16,55). Subsequent reports have confirmed these findings (56), and more recent works have established that CO production by HO-1 may account for the cytoprotective effects of NO on liver cells

(32). We had reported that deprivation of glucose markedly reduced the viability of hepatocytes via ROS-dependent pathway (32). Pretreatment with an NO donor protected hepatocytes from glucose deprivation-induced cytotoxicity; an inhibitor of HO was found to block the NO-induced cytoprotection. Nitric oxide increased the induction of HO-1 protein, as well as its activity, in hepatocytes. A cytoprotective effect comparable to SNP was observed when the cells were transfected with HO-1 gene or pre-incubated with a HO-1 inducer. Our additional experiments revealed the involvement of CO in the cytoprotective effect of NO/HO-1 in hepatocytes. Carbon monoxide mediated cytoprotective effect through suppression of ERK MAPK activation. Therefore, HO-1 might be an important cellular target of NO donor with clinical implications for the prevention of acute liver injury in several pathological conditions. Moreover, it has been proposed that CO-mediated protection operates by activating NF-κB, which in the presence of an inflammatory stimulus leads to the upregulation of iNOS with the consequent production of NO (57). In hepatocytes from iNOS knockout mice, CO does not protect TNF-α-induced apoptosis (57), suggesting that CO protects hepatocytes from TNF-α-induced cytotoxicity via iNOS/NO pathway. Interestingly, in hepatocytes from HO-1 knockout mice, both CO and NO are not protective from TNF-α-induced apoptosis (57), suggesting that HO-1 requires both CO- and NO-induced cytoprotection. On the basis of this finding, as well as our previous results (32), a putative CO/NO signaling in the liver is proposed in Fig. 3.

The role of NO in regulating liver injury through blood flow changes is more complicated by the recent reports that CO, which is produced by HO-1, also regulates hepatic perfusion (53). Carbon monoxide causes vasodilation by increasing cGMP in a manner similar to NO. In certain injuries, such as that accompanying hemorrhagic shock and resuscitation, inhibition of HO-1/CO impairs sinusoidal perfusion and exacerbates injury (58). It has also been reported that HO-1 induction is protective in sepsis (59). Thus, both CO and NO may help protect sinusoidal perfusion after liver injury, although their respective roles may vary with the type of injury.

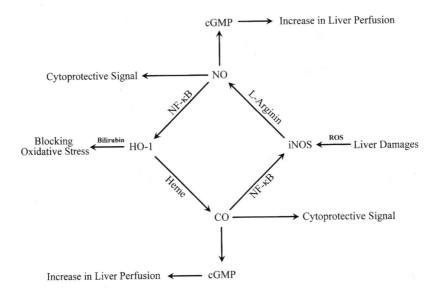

Figure 3 CO/NO signaling in the liver. The figure illustrates the concept of a cycle where CO and NO mediate their therapeutic effects through a series of steps, each of which is essential for hepatoprotection. At the initial step, iNOS/NO pathway is activated by liver damages, and, in sequence, HO-1/CO pathway is activated by NO. At the next step, iNOS/NO pathway is further activated by CO. At the final step, lack of a sign of liver damages may block iNOS/NO pathway.

V. CONCLUSION

In this chapter, we described the NO production in the liver under disease conditions, its cytoprotective and cytotoxic effects and the activation of HO-1/CO pathway in the conditions of increased ROS and NO productions, and cytoprotective effects of HO-1 products including CO and bilirubin. Finally, we discussed the sequential interactions of iNOS/NO and HO-1/CO pathways in the hepatoprotection for liver damages.

ACKNOWLEDGMENT

This work was supported by the Korea Research Foundation Grant (KRF-2004-005-200038).

REFERENCES

1. Teoh NC, Farrell GC. Hepatic ischemia reperfusion injury: pathogenic mechanisms and basis for hepatoprotection. J Gastroenterol Hepatol 2003; 18:891–902.

2. Wiest R, Groszmann RJ. The paradox of nitric oxide in cirrhosis and portal hypertension: too much, not enough. Hepatology 2002; 35:478–491.

3. Okamoto T, Yamamura K, Hino O. Expression of the inducible form of the nitric oxide synthase gene in the livers of mice with chronic hepatitis. Int J Mol Med 2000; 6:315–317.

4. Carnovale CE, Scapini C, Alvarez ML, Favre C, Monti J, Carrillo MC. Nitric oxide release and enhancement of lipid peroxidation in regenerating rat liver. J Hepatol 2000; 32: 798–804.

5. Geller DA, Nussler AK, Di Silvio M, Lowenstein CJ, Shapiro RA, Wang SC, Simmons RL, Billiar TR. Cytokines, endotoxin, and glucocorticoids regulate the expression of inducible nitric oxide synthase in hepatocytes. Proc Natl Acad Sci USA 1993; 90:522–526.

6. Schroeder RA, Gu JS, Kuo PC. Interleukin 1beta-stimulated production of nitric oxide in rat hepatocytes is mediated through endogenous synthesis of interferon gamma. Hepatology 1998; 27:711–719.

7. Hatano E, Bennett BL, Manning AM, Qian T, Lemasters JJ, Brenner DA. NF-κB stimulates inducible nitric oxide synthase to protect mouse hepatocytes from TNF-α- and Fas-mediated apoptosis. Gastroenterology 2001; 120(5):1251–1262.

8. Rockey DC, Chung JJ. Regulation of inducible nitric oxide synthase in hepatic sinusoidal endothelial cells. Am J Physiol 1996; 271:G260–G267.

9. Espey MG, Thomas DD, Miranda KM, Wink DA. Focusing of nitric oxide mediated nitrosation and oxidative nitrosylation as a consequence of reaction with superoxide. Proc Natl Acad Sci USA 2002; 99:11127–11132.

10. Jaeschke H, Gores GJ, Cederbaum AI, Hinson JA, Pessayre D, Lemasters JJ. Mechanisms of hepatotoxicity. Toxicol Sci 2002; 65:166–176.

11. Harbrecht BG, Wu B, Watkins SC, Marshall HP Jr, Peitzman AB, Billiar TR. Inhibition of nitric oxide synthase during hemorrhagic shock increases hepatic injury. Shock 1995; 4:332–337.

12. Harbrecht BG, Billiar TR, Stadler J, Demetris AJ, Ochoa J, Curran RD, Simmons RL. Inhibition of nitric oxide synthesis during endotoxemia promotes intrahepatic thrombosis and an oxygen radical-mediated hepatic injury. J Leukoc Biol 1992; 52:390–394.

13. Thiemermann C, Ruetten H, Wu CC, Vane JR. The multiple organ dysfunction syndrome caused by endotoxin in the rat: attenuation of liver dysfunction by inhibitors of nitric oxide synthase. Br J Pharmacol 1995; 116:2845–2851.

14. Clemens MG. Nitric oxide in liver injury. Hepatology 1999; 30:1–5.

15. Li J, Billiar TR. Nitric Oxide. IV. Determinants of nitric oxide protection and toxicity in liver. Am J Physiol 1999; 276: G1069–G1073.

16. Kim YM, Talanian RV, Billiar TR. Nitric oxide inhibits apoptosis by preventing increases in caspase-3-like activity via two distinct mechanisms. J Biol Chem 1997; 272: 31138–31148.

17. Pae HO, Kim HG, Paik YS, Paik SG, Kim YM, Oh GS, Chung HT. Nitric oxide protects murine embryonic liver cells (BNL CL.2) from cytotoxicity induced by glucose deprivation. Pharmacol Toxicol 2000; 86:140–144.

18. Saavedra JE, Billiar TR, Williams DL, Kim YM, Watkins SC, Keefer LK. Targeting nitric oxide (NO) delivery in vivo. Design of a liver-selective NO donor prodrug that blocks tumor necrosis factor-α-induced apoptosis and toxicity in the liver. J Med Chem 1997; 40:1947–1954.

19. Rai RM, Lee FY, Rosen A, Yang SQ, Lin HZ, Koteish A, Liew FY, Zaragoza C, Lowenstein C, Diehl AM. Impaired liver

regeneration in inducible nitric oxide synthase deficient mice. Proc Natl Acad Sci USA 1998; 95:13829–13834.

20. Wang Y, Vodovotz Y, Kim PK, Zamora R, Billiar TR. Mechanisms of hepatoprotection by nitric oxide. Ann NY Acad Sci 2002; 962:415–422.

21. Kim PK, Kwon YG, Chung HT, Kim YM. Regulation of caspases by nitric oxide. Ann NY Acad Sci 2002; 962:42–52.

22. Chung HT, Pae HO, Choi BM, Billiar TR, Kim YM. Nitric oxide as a bioregulator of apoptosis. Biochem Biophys Res Commun 2001; 282:1075–1079.

23. Maines MD. Heme oxygenase: function, multiplicity, regulatory mechanisms, and clinical applications. FASEB J 1988; 2: 2557–2568.

24. Maines MD. The heme oxygenase system: a regulator of second messenger gases. Annu Rev Pharmacol Toxicol 1997; 37:517–554.

25. McCoubrey WK Jr, Huang TJ, Maines MD. Isolation and characterization of a cDNA from the rat brain that encodes hemoprotein heme oxygenase-3. Eur J Biochem 1997; 24:725–732.

26. Choi AM, Alam J. Heme oxygenase-1: function, regulation, and implication of a novel stress-inducible protein in oxidant-induced lung injury. Am J Respir Cell Mol Biol 1996; 15(1):9–19.

27. Elbirt KK, Bonkovsky HL. Heme oxygenase: recent advances in understanding its regulation and role. Proc Assoc Am Physcians 1999; 111:438–447.

28. Raju VS, McCoubrey WK Jr, Maines MD. Regulation of heme oxygenase-2 by glucocorticoids in neonatal rat brain: characterization of a functional glucocorticoid response element. Biochim Biophys Acta 1997; 1351:89–104.

29. Lavrovsky Y, Song CS, Chatterjee B, Roy AK. Age-dependent increase of heme oxygenase-1 gene expression in the liver mediated by NFκB. Mech Ageing Dev 2000; 114:49–60.

30. Lu TH, Shan Y, Pepe J, Lambrecht RW, Bonkovsky HL. Upstream regulatory elements in chick heme oxygenase-1 promoter: a study in primary cultures of chick embryo liver cells. Mol Cell Biochem 2000; 209:17–27.

31. Immenschuh S, Hinke V, Ohlmann A, Gifhorn-Katz S, Katz N, Jungermann K, Kietzmann T. Transcriptional activation of the haem oxygenase-1 gene by cGMP via a cAMP response element/activator protein-1 element in primary cultures of rat hepatocytes. Biochem J 1998; 334:141–146.

32. Choi BM, Pae HO, Kim YM, Chung HT. Nitric oxide-mediated cytoprotection of hepatocytes from glucose deprivation-induced cytotoxicity: involvement of heme oxygenase-1. Hepatology 2003; 37:810–823.

33. Rensing H, Jaeschke H, Bauer I, Patau C, Datene V, Pannen BH, Bauer M. Differential activation pattern of redox-sensitive transcription factors and stress-inducible dilator systems heme oxygenase-1 and inducible nitric oxide synthase in hemorrhagic and endotoxic shock. Crit Care Med 2001; 29: 1962–1971.

34. Oguro T, Hayashi M, Numazawa S, Asakawa K, Yoshida T. Heme oxygenase-1 gene expression by a glutathione depletor, phorone, mediated through AP-1 activation in rats. Biochem Biophys Res Commun 1996; 221:259–265.

35. Mitzner SR, Stange J, Klammt S, Risler T, Erley CM, Bader BD, Berger ED, Lauchart W, Peszynski P, Freytag J, Hickstein H, Loock J, Lohr JM, Liebe S, Emmrich J, Korten G, Schmidt R. Improvement of hepatorenal syndrome with extracorporeal albumin dialysis MARS: results of a prospective, randomized, controlled clinical trial. Liver Transpl 2000; 6: 277–286.

36. Stocker R, Yamamoto Y, McDonagh AF, Glazer AN, Ames BN. Bilirubin is an antioxidant of possible physiological importance. Science 1987; 235:1043–1046.

37. Ryter SW, Tyrrell RM. The heme synthesis and degradation pathways: role in oxidant sensitivity. Heme oxygenase has both pro- and antioxidant properties. Free Radic Biol Med 2000; 28:289–309.

38. Suttner DM, Dennery PA. Reversal of HO-1 related cytoprotection with increased expression is due to reactive iron. FASEB J 1999; 13:1800–1809.

39. Regan RF, Kumar N, Gao F, Guo Y. Ferritin induction protects cortical astrocytes from heme-mediated oxidative injury. Neuroscience 2002; 113:985–994.

40. Tacchini L, Recalcati S, Bernelli-Zazzera A, Cairo G. Induction of ferritin synthesis in ischemic-reperfused rat liver: analysis of the molecular mechanisms. Gastroenterology 1997; 113: 946–953.

41. Yamaguchi T, Terakado M, Horio F, Aoki K, Tanaka M, Nakajima H. Role of bilirubin as an antioxidant in an ischemia–reperfusion of rat liver and induction of heme oxygenase. Biochem Biophys Res Commun 1996; 223: 129–135.

42. Furchgott RF, Jothianandan D. Endothelium-dependent and -independent vasodilation involving cyclic GMP: relaxation induced by nitric oxide, carbon monoxide and light. Blood Vessels 1991; 28:52–61.

43. Brune B, Ullrich V. Inhibition of platelet aggregation by carbon monoxide is mediated by activation of guanylate cyclase. Mol Pharmacol 1987; 32:497–504.

44. Otterbein LE, Choi AM. Heme oxygenase: colors of defense against cellular stress. Am J Physiol Lung Cell Mol Physiol 2000; 279:L1029–L1037.

45. Bauer I, Wanner GA, Rensing H, Alte C, Miescher EA, Wolf B, Pannen BH, Clemens MG, Bauer M. Expression pattern of heme oxygenase isoenzymes 1 and 2 in normal and stress-exposed rat liver. Hepatology 1998; 27:829–838.

46. Suematsu M, Goda N, Sano T, Kashiwagi S, Egawa T, Shinoda Y, Ishimura Y. Carbon monoxide: an endogenous modulator of sinusoidal tone in the perfused rat liver. J Clin Invest 1995; 96:2431–2437.

47. Wakabayashi Y, Takamiya R, Mizuki A, Kyokane T, Goda N, Yamaguchi T, Takeoka S, Tsuchida E, Suematsu M, Ishimura Y. Carbon monoxide overproduced by heme oxygenase-1 causes a reduction of vascular resistance in perfused rat liver. Am J Physiol 1999; 277:G1088–G1096.

48. Goda N, Suzuki K, Naito M, Takeoka S, Tsuchida E, Ishimura Y, Tamatani T, Suematsu M. Distribution of heme oxygenase isoforms in rat liver. Topographic basis for carbon monoxide-mediated microvascular relaxation. J Clin Invest 1998; 101: 604–612.

49. Bauer M, Zhang JX, Bauer I, Clemens MG. ET-1 induced alterations of hepatic microcirculation: sinusoidal and extrasinusoidal sites of action. Am J Physiol 1994; 267:G143–G149.

50. Makino N, Suematsu M, Sugiura Y, Morikawa H, Shiomi S, Goda N, Sano T, Nimura Y, Sugimachi K, Ishimura Y. Altered expression of heme oxygenase-1 in the livers of patients with portal hypertensive diseases. Hepatology 2001; 33:32–42.

51. Coito AJ, Buelow R, Shen XD, Amersi F, Moore C, Volk HD, Busuttil RW, Kupiec-Weglinski JW. Heme oxygenase-1 gene transfer inhibits inducible nitric oxide synthase expression and protects genetically fat Zucker rat livers from ischemia-reperfusion injury. Transplantation 2002; 74:96–102.

52. Otterbein LE, Bach FH, Alam J, Soares M, Tao LH, Wysk M, Davis RJ, Flavell RA, Choi AM. Carbon monoxide has anti-inflammatory effects involving the mitogen-activated protein kinase pathway. Nat Med 2000; 6:422–428.

53. Amersi F, Shen XD, Anselmo D, Melinek J, Iyer S, Southard DJ, Katori M, Volk HD, Busuttil RW, Buelow R, Kupiec-Weglinski JW. Ex vivo exposure to carbon monoxide prevents hepatic ischemia/reperfusion injury through p38 MAP kinase pathway. Hepatology 2002; 35:815–823.

54. Sass G, Soares MC, Yamashita K, Seyfried S, Zimmermann WH, Eschenhagen T, Kaczmarek E, Ritter T, Volk HD, Tiegs G. Heme oxygenase-1 and its reaction product, carbon monoxide, prevent inflammation-related apoptotic liver damage in mice. Hepatology 2003; 38:909–918.

55. Kim YM, Bergonia H, Lancaster JR Jr. Nitrogen oxide-induced autoprotection in isolated rat hepatocytes. FEBS Lett 1995; 374:228–232.

56. Hartsfield CL, Alam J, Cook JL, Choi AM. Regulation of heme oxygenase-1 gene expression in vascular smooth muscle cells by nitric oxide. Am J Physiol 1997; 273:L980–L988.

57. Zuckerbraun BS, Billiar TR, Otterbein SL, Kim PK, Liu F, Choi AM, Bach FH, Otterbein LE. Carbon monoxide protects against liver failure through nitric oxide-induced heme oxygenase 1. J Exp Med 2003; 198:1707–1716.

58. Billiar TR. The diverging roles of carbon monoxide and nitric oxide in resuscitated hemorrhagic shock. Crit Care Med 1999; 27:2842–2843.

59. Morse D, Choi AM. Heme oxygenase-1: the "emerging molecule" has arrived. Am J Respir Cell Mol Biol 2002; 27:8–16.

13

Role of Oxidative Stress and Inflammatory Response in Smokers and Chronic Obstructive Pulmonary Disease

IRFAN RAHMAN

Department of Environmental Medicine,
Division of Lung Biology and Disease,
University of Rochester Medical Center,
Rochester, New York, U.S.A.

ABSTRACT

Chronic inflammation and oxidative stress are important features in the pathogenesis of smoking-induced inflammatory lung diseases such as chronic obstructive pulmonary disease (COPD). The sources of the increased oxidative stress in patients with COPD derive from the increased burden of inhaled oxidants, and from the increased amounts of reactive oxygen species (ROS) generated by several inflammatory, immune, and various structural cells of the airways. The presence of oxidative stress has important consequences on several events of lung physiology and for the pathogenesis of COPD. These include increased sequestration of neutrophils in the pulmonary microvasculature, oxidative inactivation of antiproteases and surfactants, mucus hypersecretion, membrane lipid peroxidation, mitochondrial respiration, alveolar epithelial injury/permeability, breakdown/remodeling of extracellular matrix, and apoptosis. Increased levels of ROS produced in the airways are reflected by increased markers of oxidative stress in the airspaces, sputum, breath, lungs, and blood in patients with COPD. Reactive oxygen species may play a role in enhancing the inflammation through the activation of MAP kinases and redox sensitive transcription factors such as nuclear factor-κB and activator protein-1—either directly or via the formation of lipid peroxidation products, such as acrolein, 4-hydroxy-2-nonenal, and F_2-isoprostanes. Oxidative stress may alter the chromatin remodeling, leading to imbalance of gene expression of proinflammatory mediators and antioxidant enzymes in favor of inflammatory mediators in the lung. An effective wide spectrum antioxidant therapy that has good bioavailability and potency is urgently needed to control the localized oxidative and inflammatory processes that occur in the pathogenesis of COPD. In addition, development of such novel antioxidant compounds would be therapeutically useful in monitoring the oxidative and inflammatory biomarkers in the progression/severity of COPD.

I. INTRODUCTION

Reactive oxygen species (ROS) such as superoxide anion ($O_2^{\bullet-}$) and the hydroxyl radical ($^{\bullet}OH$) are highly unstable species with unpaired electrons capable of initiating oxidation. Biological systems are continuously exposed to oxidants, either generated endogenously by metabolic reactions (e.g., from mitochondrial electron transport during respiration or during activation of phagocytes) or exogenously such as through air pollutants or cigarette smoke. The lung exists in a high-oxygen environment, and, together with its large surface area and blood supply, is susceptible to injury mediated by ROS.

Production of ROS has been directly linked to oxidation of proteins, DNA, and lipids, which may cause direct lung injury or induce a variety of cellular responses, through the generation of secondary metabolic reactive species. Reactive oxygen species may alter remodeling of extracellular matrix (ECM) and blood vessels, stimulate mucus secretion, inactivate antiproteases, cause apoptosis, and regulate cell proliferation [1,2] (Fig. 1). Alveolar repair responses and immune modulation in the lung may also be influenced by ROS [1,2]. Furthermore, increased levels of ROS have been implicated in initiating inflammatory responses in the lungs through the activation of transcription factors such as nuclear factor-kappaB (NF-κB) and activator protein-1 (AP-1), signal transduction, and gene expression of proinflammatory mediators [3,4]. It is proposed that the ROS, produced by phagocytes that have been recruited to the sites of inflammation, are a major cause of cell and tissue damage associated with

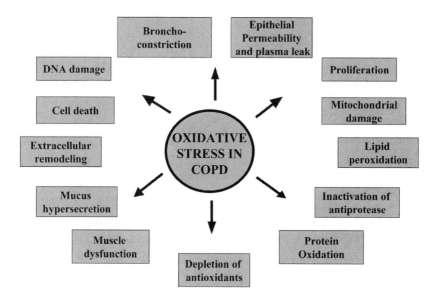

Figure 1 Reactive oxygen species-mediated cellular responses.

many chronic inflammatory lung diseases, including chronic obstructive pulmonary disease (COPD) (5–7).

This chapter reviews evidence for the role of ROS in the pathogenesis of COPD, discusses the cellular/molecular mechanisms and pathophysiological consequences of increased ROS release in COPD. The rationale for antioxidant therapeutic intervention in the light of oxidative stress will also be discussed.

II. CHRONIC OBSTRUCTIVE PULMONARY DISEASE

Chronic obstructive pulmonary disease is a gradual progressive condition characterized by airflow limitation, which is largely irreversible (8,9). Cigarette smoking is the major etiological factor in this condition. More than 90% of patients with COPD are smokers, but not all smokers develop COPD (10). However, 15–20% of cigarette smokers appear to be susceptible to its effects and show a rapid decline in forced expiratory volume in 1 sec (FEV_1) and develop the disease (10). An increased oxidant burden in smokers derives from the fact that cigarette smoke contains more than 10^{14} oxidants per puff, and many of these are relatively long lived such as tar-semiquinone, which can generate $^•OH$ and hydrogen peroxide (H_2O_2) by the Fenton reaction (11–13). Other factors, such as air pollutants, infections, and occupational dusts that may exacerbate COPD, also have the potential to produce oxidative stress (14,15) (Fig. 2).

III. CELL-DERIVED ROS

A common feature of COPD is the development of an inflammatory response, characterized by the activation of epithelial cells and resident macrophages, and the recruitment and activation of neutrophils, eosinophils, monocytes, and lymphocytes. Inflammatory cells once recruited in the airspace become activated and generate ROS in response to

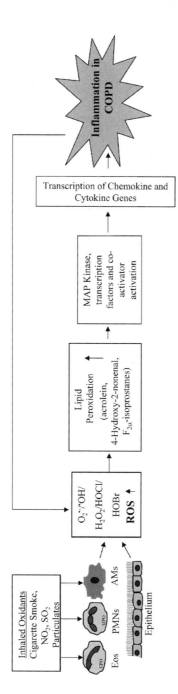

Figure 2 Mechanisms of ROS-mediated lung inflammation in COPD. Inflammatory response is mediated by oxidants either inhaled and/or released by the activated neutrophils, alveolar macrophages, eosinophils, and epithelial cells leading to production of ROS and membrane lipid peroxidation. Activation of transcription of the proinflammatory cytokine and chemokine genes, upregulation of adhesion molecules, and increased release of proinflammatory mediators that are involved in the inflammatory responses in patients with COPD.

inflammatory mediators. The activation of macrophages, neutrophils, and eosinophils generates $O_2^{\bullet-}$, which is rapidly converted to H_2O_2 under the influence of superoxide dismutase (SOD), and $^\bullet OH$ is formed nonenzymatically in the presence of Fe^{2+} as a secondary reaction. Reactive oxygen species and reactive nitrogen species (RNS) can also be generated intracellularly from several sources such as mitochondrial respiration, the NADPH oxidase system, and xanthine/xanthine oxidase (Fig. 3). However, the primary ROS-generating enzyme is NADPH oxidase, a complex enzyme system that is present in phagocytes and epithelial cells.

Activation of this enzyme involves a complex mechanism with the assembly of various cytosolic and membrane-associated subunits resulting in the one-electron reduction of oxygen to $O_2^{\bullet-}$ using NADPH as the electron donor. In addition to NADPH oxidase, the phagocytes employ other enzymes to produce ROS, which involve the activity of heme peroxidases, e.g., myeloperoxidase (MPO) or eosinophil peroxidase (EPO).

Input

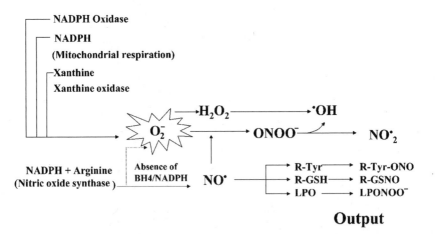

Output

Figure 3 Cellular generation of NO-derived ROS and RNS. Intracellular reactive oxygen species $O_2^{\bullet-}$, superoxide anion; NO, nitric oxide; H_2O_2, hydrogen peroxide; $^\bullet OH$, hydroxyl radical; NO_2, nitrogen dioxide; $ONOO^-$, peroxynitrite.

Activation of EPO results in the formation of the potent oxidant hypochlorous acid (HOCl) and hypobromous acid (HOBr) from H_2O_2 in the presence of chloride (Cl^-) and bromide (Br^-) ions, respectively (Fig. 4). It is believed that the oxidant burden produced by eosinophils is substantial because these cells possess several times greater capacity to generate $O_2^{\bullet-}$ and H_2O_2 than neutrophils, and the content of EPO in eosinophils is 3–10 times higher than the amount of MPO present in neutrophils (16–18).

The physiological consequences of EPO-dependent formation of brominated oxidants such as HOBr in vivo are unknown. Hypobromous acid reacts rapidly with a variety of nucleophilic targets, such as thiols, thiol ethers, amines, unsaturated groups, and aromatic compounds (19).

Several transition metal salts react with H_2O_2 to form $^{\bullet}OH$. Most attention in vivo for the generation of $^{\bullet}OH$ has focused on the role of iron (20). Iron is a critical element in

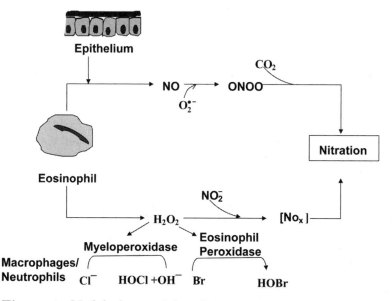

Figure 4 Model of potential pathways used by circulating cells for the generation of NO-derived reactive oxygen species and reactive halogen species.

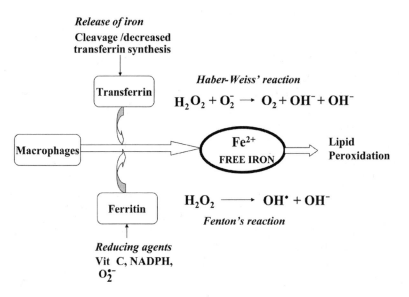

Figure 5 Free iron catalyzed membrane lipid peroxidation via Fenton and Haber–Weiss reactions.

many oxidative reactions. Free iron in the ferrous form catalyzes the Fenton reaction and the superoxide driven Haber–Weiss reaction, which generate the •OH, an ROS that damages tissues, particularly cell membranes by lipid peroxidation (Fig. 5).

IV. INHALED OXIDANTS AND CIGARETTE SMOKE

Cigarette smoking, inhalation of airborne pollutants, either oxidant gases [such as ozone, nitrogen dioxide (NO_2), and sulfur dioxide (SO_2)] or particulate air pollution, results in direct lung damage as well as in the activation of inflammatory responses in the lungs. Cigarette smoke is a complex mixture of over 4700 chemical compounds, including high concentrations of oxidants ($>10^{14}$ molecules/puff) (11) (Fig. 6). Short-lived oxidants such as $O_2^{\bullet-}$ and nitric oxide (NO) are predominantly found in the gas phase. Nitric oxide and $O_2^{\bullet-}$ immediately react to form

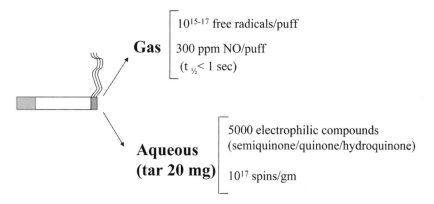

Figure 6 Cigarette smoke-derived gas and tar phase showing oxidative stress.

highly reactive peroxynitrite (ONOO⁻) molecule. The radicals in the tar phase of cigarette smoke are organic in nature, such as long-lived semiquinone radicals, which can react with $O_2^{\bullet-}$ to form $^{\bullet}OH$ and H_2O_2 (12). The aqueous phase of cigarette smoke condensate (CSC) may undergo redox recycling for a considerable period of time in the epithelial lining fluid (ELF) of smokers (13,21). The tar phase is also an effective metal chelator and can bind iron to produce tar-semiquinone + tar-Fe^{2+}, which can generate H_2O_2 continuously (13,21).

Quinone (Q), hydroquinone (QH_2), semiquinone ($QH^{\bullet}$) in the tar phase are present in equilibrium:

$$Q + QH_2 \rightarrow 2H^+ + Q^{\bullet-}$$

Aqueous extracts of cigarette tar contain the quinone radical ($Q^{\bullet-}$), which can reduce oxygen to form $O_2^{\bullet-}$ that may dismutate to form H_2O_2:

$$Q^{\bullet-} + O_2 \rightarrow Q + O_2^{\bullet-}$$

$$2O_2^{\bullet-} + 2H^+ \rightarrow O_2 + H_2O_2$$

Furthermore, since both cigarette tar and lung epithelial lining fluid contain metal ions, such as iron, Fenton chemistry will result in the production of the $^\bullet$OH, which is a highly reactive and potent ROS.

V. ALTERATIONS IN THE ALVEOLAR SPACE

The oxidant burden in the lungs is enhanced in smokers by the release of ROS from macrophages and neutrophils (5). Oxidants present in cigarette smoke can stimulate alveolar macrophages to produce ROS and to release a host of mediators, some of which attract neutrophils and other inflammatory cells into the lungs. Both neutrophils and macrophages, which are known to migrate in increased numbers into the lungs of cigarette smokers, compared with nonsmokers (5), can generate ROS via the NADPH oxidase system. Moreover, the lungs of smokers with airway obstruction have more neutrophils than smokers without airway obstruction (5). Circulating neutrophils from cigarette smokers and patients with exacerbations of COPD release more $O_2^{\bullet-}$ (22). Cigarette smoking is associated with increased content of MPO in neutrophils (23,24), which correlates with the degree of pulmonary dysfunction (23,24). Myeloperoxidase activity also has a negative correlation with FEV_1 in patients with COPD, suggesting that neutrophil MPO-mediated oxidative stress may play a role in the pathogenesis of COPD (25).

Alveolar macrophages obtained by bronchoalveolar lavage fluid (BALF) from the lungs of smokers are more activated compared with those obtained from nonsmokers (5). One manifestation of this is the release of increased amounts of ROS such as $O_2^{\bullet-}$ and H_2O_2 (5,26,27). Exposure to cigarette smoke in vitro has also been shown to increase the oxidative metabolism of alveolar macrophages (28). Subpopulations of alveolar macrophages with a higher granular density appear to be more prevalent in the lungs of smokers and are responsible for the increased $O_2^{\bullet-}$ production by the smoker's macrophages (28,29).

The generation of ROS in the epithelial lining fluid may be further enhanced by the presence of increased amounts of free iron in the airspaces in smokers (30,31). This is relevant to COPD since the intracellular iron content of alveolar macrophages is increased in cigarette smokers and is increased further in those who develop chronic bronchitis, compared with nonsmokers (32). In addition, macrophages obtained from smokers release more free iron in vitro than those from nonsmokers (33).

In some studies, both in stable (34) and mild exacerbations of bronchitis (35), eosinophils have been shown to be prominent in the airway walls. Bronchoalveolar lavage (BAL) from patients with COPD has also been shown to contain increased levels of eosinophilic cationic protein (24). Furthermore, peripheral blood eosinophilia is also considered to be a risk factor for the development of airway obstruction in patients with chronic bronchitis and is an adverse prognostic sign (36,37). However, despite the presence of increased number of eosinophils, EPO-mediated generation of specific 3-bromotyrosine has not been detected in COPD patients (19). This does not provide support for a role of brominating oxidants in eosinophil-mediated ROS damage in COPD.

VI. ALTERATIONS IN BLOOD

The neutrophil appears to be a critical cell in the pathogenesis of COPD (37). Previous epidemiological studies have shown a positive relationship between circulating neutrophil numbers and the decline in FEV_1 (38,39). Moreover, a relationship has also been shown between the change in peripheral blood neutrophil count and the change in airflow limitation over time (40). Similarly, a correlation between $O_2^{\bullet-}$ release by peripheral blood neutrophils and bronchial hyperreactivity in patients with COPD has been shown, suggesting a role for systemic ROS in the pathogenesis of the airway abnormalities in COPD (41). Another study has shown a relationship between peripheral blood neutrophil luminol enhanced chemiluminescence, as a measure of the release of ROS and measurements of airflow limitation in young cigarette smokers (42).

Various studies have demonstrated increased production of $O_2^{\bullet-}$ from peripheral blood neutrophils obtained from patients during acute exacerbations of COPD, which returned to normal when the patients were clinically stable (22,43,44). Other studies have shown that circulating neutrophils from patients with COPD show upregulation of their surface adhesion molecules, which may also be an oxidant-mediated effect (22,45). Activation may be even more pronounced in neutrophils, which are sequestered in the pulmonary microcirculation in smokers and in patients with COPD, since neutrophils that are sequestered in the pulmonary microcirculation in animal models of lung inflammation release more ROS than circulating neutrophils (46). Thus neutrophils, which are sequestered in the pulmonary microcirculation, may be a source of ROS, and may have a role in inducing endothelial adhesion molecule expression in COPD.

Superoxide anion and H_2O_2 can be generated by the xanthine/xanthine oxidase (XO) reaction. Xanthine oxidase activity has been shown to be increased in cell-free BALF and plasma from COPD patients, compared with normal subjects; and this has been associated with increased $O_2^{\bullet-}$ and lipid peroxide levels (47–49).

VII. REACTIVE OXYGEN AND REACTIVE NITROGEN SPECIES AS SURROGATE MARKERS IN PLASMA AND EXHALED BREATH CONDENSATE

There is a limitation in terms of the lack of specific biomarkers that correlate with disease severity or outcome. Biomarkers of COPD would be of great value for noninvasive investigations of the natural history and epidemiology of COPD, and for phenotyping in genetic studies. There are now a number of surrogate markers of oxidative stress in the lungs that have been measured in smokers and patients with COPD. Many of these markers have been measured in blood, urine, breath or breath condensate, or in induced or spontaneously produced sputum. Direct measurements of oxidative stress are difficult since ROS/free radicals are highly reactive and also short lived. An alternative has been to measure markers of the effects of radicals on lung biomolecules, such as lipids, proteins or DNA, or to measure the stress responses to an increased oxidant burden.

VII.A. (i) Exhaled ROS

Hydrogen peroxide, measured in exhaled breath, is a direct measurement of oxidant burden in the airspaces (Table 1). Smokers and patients with COPD have higher levels of exhaled H_2O_2 than nonsmokers (50–52), and levels are even

Table 1 ROS and Inflammatory Biomarkers in Smokers and COPD

Biochemical marker	COPD
Elevated breath hydrogen peroxide levels	50–52
Release of ROS from peripheral blood neutrophils, eosinophils, and macrophages	22,27,42–45
Increased release of ROS from alveolar macrophages, eosinophils, and neutrophils	29,41
Increased MPO and EPO levels	24,25,36
Increased BALF xanthine/xanthine oxidase activity	47–49
Elevated plasma and exhaled malondialdehyde, F_2-isoprostane, ethane, and pentane levels	74–78,249
Elevated plasma and exhaled lipid peroxide (TBARS) levels	26,31,78
Increased formation of 4-hydroxy-2-nonenal-protein adducts in lungs	70
Increased levels of IL-6 and leukotriene LTB_4	248,249
Elevated levels of plasma protein carbonyls	22
Increased exhaled carbon monoxide	59
Increased exhaled NO levels	53–56
Increased levels of 3-nitrotyrosine in plasma, BALF, and exhaled air	60,64

higher during exacerbations of COPD (52). The source of the increased H_2O_2 is unknown but may in part derive from the increased release of $O_2^{\cdot-}$ from alveolar macrophages in smokers (52). However, in one study smoking did not appear to influence the levels of exhaled H_2O_2 (50). The levels of exhaled H_2O_2 in this study correlated with the degree of airflow obstruction as measured by the FEV_1. However, the variability of the measurement of exhaled H_2O_2, along with the presence of other confounding factors, e.g., increased generation of ROS by cigarette smoke-mediated redox cycling, has led to concerns over its reproducibility as a marker for oxidative stress in smokers and in patients with COPD.

VII.B. (ii) Exhaled Reactive Nitrogen Species and Carbon Monoxide

Exhaled NO has been used as a marker of airway inflammation and indirectly as a measure of oxidative stress. There have been some reports of increased levels of NO in exhaled breath in patients with COPD, but not to the extent reported in asthmatics (53–55). Another study however failed to confirm this result (56). Smoking increases NO levels in breath (57) and the reaction of NO with $O_2^{\cdot-}$ (which forms $ONOO^-$) limits the usefulness of this marker in COPD, except perhaps to differentiate from asthma.

Carbon monoxide (CO) is another biomarker of oxidative stress that is generated by the induction of the stress responsive protein heme oxygenase-1 (HO-1) (58). Heme oxygenase-1 catalyzes the initial rate limiting step in the oxidative degradation of heme to bilirubin. Heme oxygenase (HO-1) catalyzes the breakdown of heme to biliverdin; this is then converted by biliverdin reductase to bilirubin, which has antioxidant properties. The reaction of HO-1 with heme releases iron and carbon monoxide, which can be measured in exhaled breath and has been shown to be elevated in patients with COPD (59).

Cigarette smoking increases the formation of RNS and results in nitration and oxidation of plasma proteins. The levels of nitrated proteins (fibrinogen, transferrin, plasminogen, and ceruloplasmin) were higher in smokers (60) compared to nonsmokers. Evidence of $NO/ONOO^-$ activity in plasma has been shown in cigarette smokers (60). In vitro exposure of gas-phase cigarette smoke caused increased lipid peroxidation and protein carbonyl formation in plasma (61). It is likely that alpha-, beta-unsaturated aldehydes (acrolein, acetaldehyde, and crotonaldehyde) that are abundantly present in cigarette smoke may react with protein-SH and $-NH_2$ groups leading to the formation of a protein-bound aldehyde functional group and is capable of converting tyrosine to 3-nitrotyrosine and dityrosine (62). Nitric oxide and ONOO-mediated formation of 3-nitrotyrosine in plasma and free catalytic iron (Fe^{2+}) levels in epithelial lining fluid are elevated in chronic smokers (60,63,64). Nitration of tyrosine residues on proteins in plasma leads to the production of 3-nitrotyrosine (60). The levels of nitrotyrosine and inducible nitric oxide synthase (iNOS) were higher in airway inflammatory cells obtained by induced sputum from patients with COPD, compared to those with asthma (60). The levels of nitrotyrosine were negatively correlated with the $FEV_1\%$. A recent study by Kanazawa et al. (65) has shown that increased levels of nitrogen oxides and reduced peroxynitrite inhibitory activity were present in induced sputum from patients with COPD. The increased level of reactive nitrogen species was inhibited following steroid therapy in patients with COPD, and the reduction in nitrotyrosine and iNOS immunoreactivity in sputum cells was correlated with the improvement in FEV_1 and airway responsiveness to histamine (66). These direct and indirect studies indicate that an increased RNS- and ROS-mediated protein nitration and lipid peroxidation, respectively, may play a role in the inflammatory response that occur in COPD.

VII.C. (iii) Lipid Peroxidation Products

Reactive oxygen species, such as $O_2^{\bullet-}$ and $^\bullet OH$, generated and released by activated immune and inflammatory cells are highly reactive, and when generated close to cell membranes oxidize membrane phospholipids (lipid peroxidation), a process that may continue as a chain reaction (Fig. 7). Thus, a single •OH can result in the formation of many molecules of lipid hydroperoxides in the cell membrane (67). The peroxidative breakdown of polyunsaturated fatty acids impairs membrane function, inactivates membrane-bound receptors and enzymes, and increases tissue permeability—processes that have been implicated in the pathogenesis of many forms of lung injury. There is increasing evidence that aldehydes, generated endogenously during the process of lipid peroxidation, are involved in many of the pathophysiological effects associated with oxidative stress in cells and tissues (67). Compared with free radicals, lipid peroxidation aldehydes are generally stable, can diffuse within, or even escape from the cell, and attack targets far from the site of the original free radical event. In addition to their cytotoxic properties, lipid peroxides are increasingly recognized as being important in signal transduction for a number of important events in the inflammatory response (68).

Many of the effects of ROS in airways may be mediated by the secondary release of inflammatory lipid mediators such as 4-hydroxy-2-nonenal (4-HNE). 4-Hydroxy-2-nonenal,

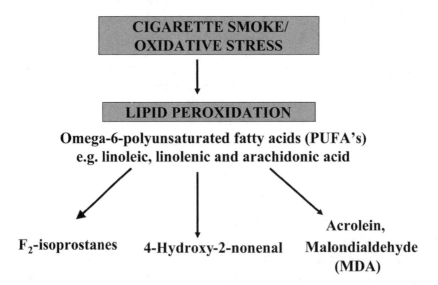

Figure 7 Membrane lipid peroxidation of polyunsaturated fatty acids leading to generation of various aldehydes.

a highly reactive diffusible end product of lipid peroxidation, is known to induce/regulate various cellular events, such as proliferation, apoptosis, and activation of signaling pathways (68,69). It has a high affinity toward cysteine, histidine, and lysine residues. It forms adducts with proteins, altering their function. Increased 4-HNE-modified protein levels are present in airway, alveolar epithelial cells and endothelial cells, and neutrophils in smokers with airway obstruction compared to subjects without airway obstruction (70) (Fig. 8a and b). This demonstrates not only the presence of 4-HNE, but also that 4-HNE modifies proteins in lung cells to a greater extent in patients with COPD. The increased level of 4-HNE-adducts in alveolar epithelium, airway endothelium, and neutrophils was inversely correlated with FEV_1, suggesting a role for 4-HNE in the pathogenesis of COPD (Fig. 8c).

Isoprostanes are products of nonenzymatic lipid peroxidation and have therefore been used as markers of oxidative stress (71). The isoprostanes (a member of F_2-isoprostane family) are ROS-catalyzed isomers of arachidonic acid and are stable lipid peroxidation products, which circulate in plasma and are excreted in the urine (72,73). The levels of lipid peroxides, such as 8-isoprostane and hydrocarbons, such as ethane and pentane are increased in exhaled air condensate in smokers and in patients with COPD (74–77). Furthermore, the increased levels of these markers of lipid peroxidation products have been correlated with airway obstruction (76). Urinary levels of isoprostane $F_2\alpha$-III have been shown to be elevated in patients with COPD compared with control subjects and are even more elevated during exacerbations of COPD(73). Similarly, Corradi et al. (78) have recently shown higher concentrations of malondialdehyde in the EBC of COPD. These studies indicate that there is increased lipid peroxidation in patients with COPD. However, it is not known whether the increased levels of lipid peroxidation products found in these diseases is the result of primary lung-associated processes (inflammatory responses), such as alveolar macrophage activation, neutrophil activity, or caused by the ongoing lipid peroxidative chain reaction in the alveoli, parenchyma, or airways, which are induced by inhaled oxidants/cigarette smoke.

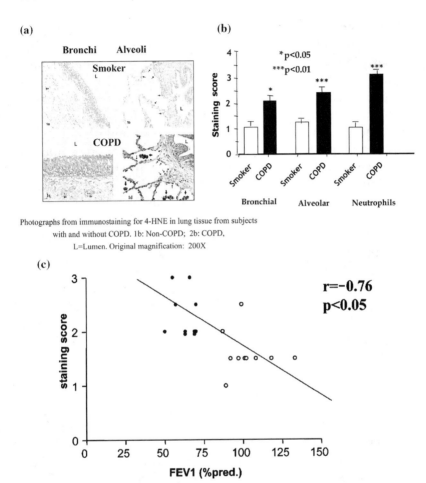

(a)

Bronchi   Alveoli

Smoker

COPD

Photographs from immunostaining for 4-HNE in lung tissue from subjects
with and without COPD. 1b: Non-COPD; 2b: COPD,
L=Lumen. Original magnification: 200X

(b)

Bronchial   Alveolar   Neutrophils

*p<0.05
***p<0.01

(c)

r=-0.76
p<0.05

Figure 8 (a) Photographs showing immunostaining for 4-HNE in lung tissue from subjects with and without COPD. 1A: Non-COPD, bronchial; 1B: non-COPD, alveolar; 2A: COPD, bronchial; 2B: COPD, alveolar. L, lumen. Original magnification: $200\times$. (b) Individual immunostaining scores for 4-HNE adducts in bronchial and alveolar lung tissue in epithelial cells, endothelial cells, and neutrophils. The mean is indicated, as well as significance levels (p) for differences between the indicated groups. (c) Correlations between the levels of 4-HNE adducts in alveolar epithelium with FEV_1 (percentage of prediction) levels in subjects with (closed circles) and without (open circles) airway obstruction. Correlation (r) and significance level (p value) are indicated.

Indirect and nonspecific measurements of lipid peroxidation products, such as thiobarbituric acid reactive substances (TBARS), have also been shown to be elevated in breath condensate and in lungs of patients with stable COPD (31,79,80). The levels of plasma lipid peroxides (TBARS) have been shown to be elevated in COPD, and negatively correlated with the FEV_1 (81). Oxidative stress, measured as lipid peroxidation products in plasma, has also been shown to correlate inversely with the percentage of predicted FEV_1 in a population study (82), suggesting that in patients with COPD, lipid peroxidation may play a role in the progression of the disease.

VIII. DEPLETION OF ANTIOXIDANTS

Smoking and exacerbations of COPD result in decreased antioxidant capacity in plasma (22,43), in association with depleted protein sulfydryls in the plasma (22,43). The decrease in antioxidant capacity in smokers occurred transiently during smoking and resolved rapidly after smoking cessation (Fig. 9). In exacerbations of COPD, however, the decrease in antioxidant capacity remained low for several days after the onset of the exacerbation, tending to return toward normal values at the time of recovery from the exacerbation (43). The depletion of antioxidant capacity could in part be explained by the increased release of ROS from peripheral blood neutrophils, as shown by a significant negative correlation between neutrophil superoxide anion release and plasma antioxidant capacity (22). Thus, there is clear evidence that oxidants in cigarette smoke markedly decrease plasma antioxidants (42).

Using an in vitro model, Eiserich et al. (62) showed that exposure of gas-phase cigarette smoke caused considerable depletion of antioxidants, including ascorbate, urate, ubiquinol-10, α-tocopherol, and a variety of carotenoids, including β-carotene that were associated with lipid peroxidation and protein carbonyl formation in plasma. They suggested that the alpha, beta-unsaturated aldehydes (acrolein and crotonaldehyde), abundantly present in cigarette smoke,

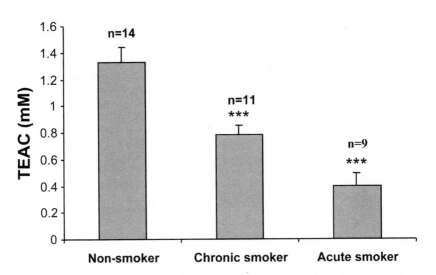

Figure 9 Plasma antioxidant capacity measured as trolox equivalent antioxidant capacity (TEAC) in healthy nonsmokers and healthy chronic smokers who have not smoked for 12 hr and smokers who smoked two cigarettes 1 hr prior to measurement (acute smoker). ***$p < 0.001$ compared with nonsmokers. Data are mean $\pm$ SEM.

may react with protein-SH and $-NH_2$ group leading to the formation of a protein-bound aldehyde functional group and is capable of converting tyrosine to 3-nitrotyrosine and dityrosine. Exposure of human erythrocytes to filtered cigarette smoke in vitro also depleted intracellular reduced glutathione (GSH) and protein thiols without any change in concentration of ascorbate (83).

Studies showing depletion of total antioxidant capacity in smokers are supported by earlier studies of measurements of the major plasma antioxidants in smokers (84–90). These studies show depletion of ascorbic acid, vitamin E, β-carotene, and selenium in the serum of chronic smokers (86–91). Moreover, decreased vitamin E and vitamin C levels were reported in leukocytes from smokers (92,93). However, circulating red blood cells from cigarette smokers contain increased levels of SOD and catalase, despite similar activity of glutathione peroxidase (GP), and are better able to protect endothelial cells from the effects of H_2O_2, when compared with cells from nonsmokers (94). Ascorbate appears to be a particularly important antioxidant in the plasma (Table 2). Cigarette smoke-induced lipid peroxidation of plasma in vitro is decreased by ascorbate (95).

Dietrich et al. (96) have recently shown that vitamin C or an antioxidant mixture containing vitamin C, α-lipoic acid, and vitamin E decrease plasma F_2-isoprostane levels in smokers with high body-mass index suggesting that smoking-mediated oxidative stress is involved in lipid peroxidation. Similarly, Zhang et al. (97) have reported that dietary supplementation of multiple antioxidants prevented lipid peroxidation and inflammatory responses (IL-6 release) induced by sidestream cigarette smoke in old mice. A recent study by Smith et al. (98) has shown that intratracheal instillation of a catalytic antioxidant, manganese (III) meso-tetrakis (N,N'-diethyl-1,3-imidazolium-2-yl) porphyrin (AEOL 10150), inhibited the cigarette smoke-induced inflammatory response (decreased number of neutrophils and macrophages) in rats after 2 days or 8 weeks (6 hr/day, 3 days/week) of exposure. This study suggested that cigarette smoking depletes a variety of multiple antioxidants that are needed to quench an array of free radicals present in cigarette smoke and inhibit inflammatory response induced by cigarette smoking.

Further support for the presence of systemic oxidative stress comes from the increased levels of hydrogen peroxide and lipid peroxidation products such as $F_2\alpha$-II isoprostane in

Table 2 Antioxidant Constituents of Plasma and Epithelial Lining Fluid

Antioxidant	Plasma (μM)	ELF (μM)
Ascorbic acid	40	100
Glutathione	1.5	100
Uric acid	300	90
Albumin-SH	500	70
Alpha-tocopherol	25	2.5
Beta-Carotene	0.4	–

plasma, breath condensate, and urine in patients with exacerbations of COPD (5,52,73,76,99). Nitric oxide- and peroxynitrite-mediated formation of 3-nitrotyrosine in plasma and free catalytic iron (Fe^{2+}) levels in ELF are elevated in chronic smokers (31,32,60). These direct and indirect studies indicate that an increased systemic and local pulmonary oxidant burden occurs in smokers and in patients with COPD (99).

IX. DEPLETION OF GLUTATHIONE

Several studies have suggested that GSH homeostasis may play a central role in the maintenance of the integrity of the lung airspace epithelial barrier. Decreasing the levels of GSH in epithelial cells leads to loss of barrier function and increased permeability (26,100).

There is limited information on the respiratory epithelial antioxidant defenses in smokers, and less for patients with COPD. Several studies have shown that GSH is elevated in BALF in chronic smokers (26,101,102). However, Harju et al. (103) have found that the γ-glutamylcysteine synthetase (γ-GCS) immunoreactivity was decreased (GSH levels were not measured) in the airways of smokers compared to non-smokers suggesting that cigarette smoke predisposes lung cells to ongoing oxidant stress. This suggests that the twofold increase in BALF GSH in chronic smokers may not be sufficient to deal with the excessive oxidant burden during smoking, when acute depletion of GSH may occur (104). The acute effects of CSC on GSH metabolism have been studied in a human alveolar epithelial cell line in vitro, and in rat lungs in vivo after intratracheal CSC instillation. Cigarette smoke condensate produced a dose- and time-dependent depletion of intracellular GSH, concomitant to the formation of GSH conjugates. Similar results were shown in animal lungs in vivo (100,104).

Thus, there appears to be no consistent change in antioxidant defenses in ELF in smokers. The apparent inconsistencies between these studies in the levels of the different antioxidants in ELF and alveolar macrophages may be due to differences in the smoking histories in chronic smokers, particularly the time of the last cigarette in relation to the sampling of BALF, which is rarely reported in these studies.

The activities of SOD and glutathione peroxidase (GP_x) have been shown to be higher in the lungs of rats exposed to cigarette smoke (105). McCusker and Hoidal (106) have also demonstrated enhanced antioxidant enzyme activities in alveolar macrophages in hamsters following cigarette smoke exposure, which resulted in reduced mortality when the animals were subsequently exposed to >95% oxygen. They speculated that alveolar macrophages undergo an adaptive response to chronic oxidant exposure that may ameliorate potential damage to lung cells from further oxidant stress. The mechanism(s) for the induction of antioxidants enzymes in erythrocytes (94), alveolar macrophages (106), and lungs (105) by cigarette smoke exposure is currently unknown. However, it is likely to be due to the induction of antioxidant genes (discussed later).

X. OTHER AUXILIARY ANTIOXIDANTS

Pacht et al. (107) demonstrated reduced levels of vitamin E in the BALF of smokers compared with nonsmokers, whereas Bui et al. (108) found a marginal increase in vitamin C in the BALF of smokers, compared to nonsmokers. Similarly, alveolar macrophages from smokers have both increased levels of ascorbic acid and augmented uptake of ascorbate (109). Increased activity of antioxidant enzymes (SOD and catalase) in alveolar macrophages from young smokers has also been reported (106). However, Kondo et al. (110) found that the increased superoxide generation by alveolar macrophages in elderly smokers was associated with decreased antioxidant enzyme activities when compared with nonsmokers. The activities of CuZnSOD, glutathione S-transferase, and GP are all decreased in alveolar macrophages from elderly smokers. However, this reduced activity was not associated with decreased gene expression, but was due to modification at the post-translational level (110).

XI. CONSEQUENCES OF
OXIDANT/ANTIOXIDANT IMBALANCE

XI.A. Protease/Antiprotease Imbalance

The development of a protease/antiprotease imbalance in the
lungs is a central hypothesis in the pathogenesis of emphysema
in smokers. This theory was developed from studies of early onset
emphysema in antitrypsin (α_1-AT) -deficient subjects. In the case
of smokers with normal levels of α_1-AT, the elastase burden may
be increased as a result of increased recruitment of leukocytes
to the lungs, and there may be a functional deficiency of α_1-AT,
due to the oxidative inactivation of α_1-AT in the lungs.

A large body of literature has been published in an
attempt to prove the protease/antiprotease theory of the
pathogenesis of emphysema. It is clear that an imbalance
between an increased elastase burden in the lungs and a functional "deficiency" of α_1-AT due to its inactivation by oxidants
is an oversimplification, since other proteinases and antipro-
teinases are also likely to have a role. Early studies showed
that the function of α_1-1-antitrypsin in the bronchoalveolar
lavage was decreased by around 40% in smokers, compared
with nonsmokers (111). This "functional α_1-1-AT deficiency"
is thought to be due to inactivation of the α_1-1-AT by oxidation
of the methionine residue at its active site (112,113) by
oxidants in cigarette smoke. Secretory leukoprotease
inhibitor (SLPI), another major inhibitor of neutrophil
elastase (NE), can also be inactivated by oxidants (114,115).

In vitro studies have also shown the loss of α_1-1-AT
inhibitory capacity when treated with oxidants (116), including
cigarette smoke (117). In addition, oxidation of the methionine
residue in α_1-1-AT was confirmed in the lungs of healthy
smokers (118). Furthermore, neutrophilic inactivation of α-1-
PI may be caused by hypochlorous acid or peroxynitrite (119).
It has been shown that stimulated alveolar type II epithelial
cells and alveolar macrophages (but not fibroblasts) from guinea
pigs inactivated α-1-PI in the presence of myeloperoxidase
(119,120). These studies supported the concept of inactivation
of α_1-1-AT by the oxidation of the active site of the protein. As
already discussed, macrophages from the lungs of smokers

release increased amounts of reactive oxygen species, which could also inactivate α_1-1-AT in vitro (112). However, most of the α_1-1-AT in the airspaces in cigarette smokers remains active and is therefore still capable of protecting against the increased protease burden. There are also conflicting data on whether functional activity of α_1-1-AT is altered in cigarette smokers (121), which may be due to technical differences between the studies that may have affected α_1-1-AT function. In addition, the original observation that oxidation of α_1-1-AT occurs in the bronchoalveolar lavage in smokers has not been confirmed (122).

The acute effects of cigarette smoking on the functional activity of α_1-1-AT in bronchoalveolar lavage fluid have also been studied and show transient, but nonsignificant fall, in the antiprotease activity of bronchoalveolar lavage fluid 1 hr after smoking (123). Thus, studies assessing the function of α_1-1-AT in either chronic or acute cigarette smoking have failed to produce a clear picture. A further hypothesis has been developed to explain these conflicting data, and has invoked a contributory role for other antiproteases, such as antileukoprotease, or by observing more subtle changes, e.g., a decrease in the association rate constant of α_1-1-AT for neutrophil elastase, which may contribute to elastin degradation (124). Thus, cigarette smoke-induced oxidative stress may reduce the antiprotease screen within the lung in favor of proteases and the development of emphysema.

XII. MUCUS HYPERSECRETION

Chronic bronchitis is associated with hyperplasia of both epithelial goblet cells and submucosal glands in the airways. Mucins, which are complex glycoproteins that provide the viscoelastic properties of mucus, are an essential protective mechanism in the upper airways. The regulation of mucins is altered in the lungs of COPD patients. The airways of smokers contain more goblet cells than do those of nonsmokers and goblet-cell activation results in mucus hypersecretion leading to airway plugging (125). Cigarette smoke can activate epidermal growth factor (EGF) receptors by tyrosine phosphorylation, resulting in the induction of mucin (MUCI-N5AC gene expression) synthesis in epithelial cells and in vivo in lungs (126). Cigarette smoke-mediated via MUC5AC gene expression was inhibited by selective EGF receptor tyrosine kinase inhibitors and antioxidants (127). Oxidant-generating systems, such as xanthine/xanthine oxidase, have been shown to cause the release of mucus from airway epithelial cells (128). It has also been shown that elastase released from neutrophils impairs mucociliary clearance and stimulates goblet-cell metaplasia and mucin production. Neutrophil elastase increases the expression of MUC5AC by enhancing mRNA stability (129). Understanding of the EGF receptor signaling pathway in cigarette smoke-mediated upregulation of mucin gene expression could lead to targeted inhibition of mucus hyperproduction in epithelial cells.

XIII. REMODELING OF EXTRACELLULAR MATRIX

Oxidative stress has been shown to be involved in the remodeling of extracellular matrix in lung injury (130). This was supported by two observations: (a) oxidant-induced lung injury was attenuated by the synthetic matrix metalloproteinase (MMP) inhibitor BB-3103 (131); and (b) depletion of intracellular GSH was associated with the activation of MMPs, thereby increasing degradation of the alveolar ECM in lungs (132). This breakdown of lung ECM (elastolytic activity) by MMP-9 and MMP-2 activation was blocked by increasing lung glutathione levels (133). Treatment of cigarette smoke to alveolar macrophages obtained from patients with COPD release increased amounts of MMP-9 compared to that of smokers suggesting the role of oxidative component of CS in increased elastolytic enzyme by alveolar macrophages of only patients with COPD (134). It has been shown that the oxidative stress caused by ozone and lipid peroxides induces matrix protein type I collagen and MMP-1 gene expression (135). Other forms of oxidative stress derived from *tert*-butyl hydroperoxide and iron can also modify collagen synthesis, by a mechanism presumably involving redox sensor/receptor (136).

XIV. APOPTOSIS

Apoptosis or programmed cell death of leukocytes is an important mechanism in the resolution of inflammation (137). However, structural lung cells may also undergo apoptosis. Reactive oxygen species are important molecules that have been implicated in mediating apoptotic processes. Hydrogen peroxide can induce apoptosis in airway epithelial cells (138). Recent evidence from both in vitro and in vivo studies in animals and in man has shown that apoptosis occurs in smoke-exposed macrophages and airway epithelial cells (139–141). The marked decrease in intracellular glutathione, which occurs upon exposure of cells to cigarette smoke exposure, may have a role in the regulation of apoptosis (142). Recently, a concept has been developed that pulmonary capillary endothelial cell apoptosis, induced by cigarette smoking, may be an early event in the process that leads to alveolar wall destruction and emphysema. Recent reports have shown both in vivo and in vitro that cigarette smoke exposure produces endothelial cell apoptosis (143,144) and that pulmonary vascular endothelial cell apoptosis is present in emphysematous lungs (145). This may be due to the formation of reactive carbonyls by cigarette smoke-mediated protein oxidation (146). Signaling pathways involving AP-1 and NF-κB and the downregulation of the vascular endothelial growth factor receptor KDR (VEGF-KDR) have been proposed as part of the mechanism (144,145). Recently, Tuder et al. (147) have shown that inhibition of KDR leads to increased oxidative stress which is mediated by reactive carbonyls and aldehydes leading to emphysema. Further studies are required to understand the mechanism of cigarette smoke-induced cell death using both in vitro and in vivo models.

XV. MUSCLE DYSFUNCTION

Dysfunction of the respiratory and of peripheral skeletal muscles is known to occur in patients with severe COPD. The underlying mechanisms of muscle dysfunction in COPD are not well understood. Skeletal muscles generate ROS at rest and ROS production increases during contractile activity. Oxidative stress occurs in skeletal muscle during skeletal muscle fatigue and sepsis-induced muscle dysfunction, accompanied by an increased load imposed on the diaphragm in patients with severe COPD (148). This may be due to hypoxia, impaired mitochondrial metabolism, and increased cytochrome C oxidase activity in skeletal muscle in patients with COPD (148–150). Engelen and coworkers have found reduced muscle glutamate (a precursor of glutathione) levels associated with increased muscle glycolytic metabolism in patients with severe COPD (151). Lowered levels of glutamate were associated with decreased GSH levels, suggesting that oxidant/antioxidant imbalance is involved in skeletal muscle dysfunction in patients with COPD. A causal relationship between abnormally low muscle redox potential at rest and the alterations of protein metabolism observed in patients with emphysema has been suggested. This is supported by Rabinovich, who showed decreased muscle redox capacity probably due to lower ability to synthesize GSH during endurance training in patients with COPD (148). Recently, Agusti et al. (152) have shown that apoptotic pathways may be involved in skeletal muscle atrophy in patients with COPD. However, it remains to be determined whether oxidative stress plays a central role in mediating muscle mass wasting/apoptosis, particularly in susceptible subsets of patients with COPD.

XVI. OXIDATIVE STRESS AND THE DEVELOPMENT OF AIRWAY OBSTRUCTION

The neutrophil appears to be a critical cell in the pathogenesis of COPD. Previous epidemiological studies have shown a relationship between circulating neutrophil numbers and the FEV_1 (38). Moreover, a relationship has also been shown between the change in peripheral blood neutrophil count and the change in airflow limitation over time (40). Another study has shown a relationship between peripheral blood neutrophil luminol enhanced chemiluminescence, as a measure of the release of reactive oxygen species and measurements of airflow limitation in young cigarette smokers (42). Even passive cigarette smoking has been associated with increased peripheral blood leukocyte counts and enhanced release of oxygen radicals (153). Oxidative stress, measured as lipid peroxidation products in plasma, has also been shown to correlate inversely with the percentage of predicted FEV_1 in a population study (154).

In the general population, there is a positive association between dietary intake of antioxidant vitamins and lung function. Epidemiological studies have demonstrated negative associations of dietary antioxidant intake with pulmonary function and with obstructive airway disease (155). Britton et al. (156), in a population of 2633 subjects, showed a positive association between dietary intake of the antioxidant vitamin E and lung function, supporting the hypothesis that this antioxidant may have a role in protecting against the development of COPD. Another study has suggested that antioxidant levels in the diet could be a possible explanation for differences in COPD mortality in different populations (157). Dietary polyunsaturated fatty acids may protect cigarette smokers against the development of COPD (158,159). These studies support the concept that dietary antioxidant supplementation may be a possible therapy to prevent the development of COPD (160). Such interventional studies have been difficult to carry out, but there is at least some evidence to suggest that antioxidant vitamin supplementation reduces oxidant stress in smokers, measured as a decrease in pentane levels in breath as an indication of lipid peroxides in the airways (75).

XVII. INFLAMMATION AND GENE EXPRESSION

XVII.A. NF-κB Activation

Many inflammatory mediator genes, such as those for the cytokines, IL-8, TNF-α, and nitric oxide are regulated by transcription factors such as NF-κB. Nuclear factor-κB is present in the cytosol in an inactive form linked to its inhibitory protein IκB. Many stimuli, including cytokines and oxidants, activate NF-κB, resulting in ubiquitination, cleaving of IκB from NF-κB, and the destruction of IκB in the proteasome (3) (Fig. 10). These critical events in the inflammatory response are redox sensitive.

Lipid peroxidation products, in particular aldehydes, derived from cigarette smoke have been shown to act as a signal for activation of transcription factors and gene expression, leading to inflammatory response (68,69). Cigarette smoke extract stimulates protein kinase C, possibly by the formation of aldehyde/lipid peroxidation products, in human bronchial epithelial cells (161). Di Stefano and colleagues have demonstrated increased expression of p65 protein of NF-κB in bronchial epithelium of smokers and patients with COPD (162). The increased expression of p65 in epithelial cells was correlated with the degree of airflow limitation in patients with COPD. Similarly, Caramori and coworkers have shown the p65 subunit of NF-κB was increased in sputum macrophages but not in sputum neutrophils during exacerbations of COPD suggesting that inflammatory response can be seen in diverse cell population (163). The activation of NF-κB in monocytes/macrophages can then trigger the release of proinflammatory mediators in lung fluid, which would then amplify the inflammatory cascade by activation of epithelial cells as well as recruitment of neutrophils in the airways (Fig. 11).

Mochida-Nishimura and coworkers have shown that cells obtained from BALF from smokers exhibited a 10-fold higher activation of NF-κB in response to lipopolysaccharide (LPS) compared to that of non-smokers (164). This may be due to the elevated release of inflammatory mediators that may activate NF-κB. However, the activation of MAP kinase—ERK, stress-activated protein kinase (SAPK), and p38 was differentially regulated. Activation of p38 was more rapid in the BAL cells from smokers compared to the activity of ERK and SAPK. They also suggested that the differences in activation of NF-κB and MAP kinases in BAL cells from smokers and nonsmokers may be related to the differences in their microenvironment, which is affected by chronic

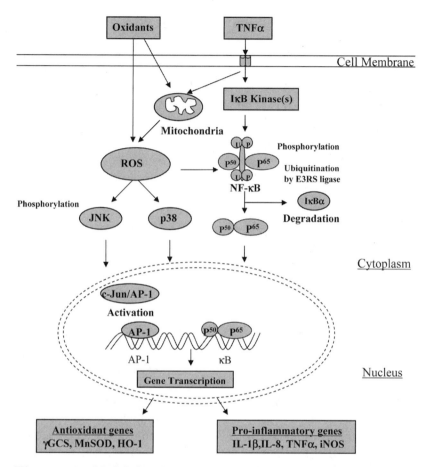

Figure 10 Model for the mechanism of NF-κB and AP-1 activation leading to the gene transcription in lung epithelial cells. Tumor necrosis factor-α/oxidants act on mitochondria to generate ROS that are involved in the activation of NF-κB and AP-1. Activation of NF-κB involves the phosphorylation, ubiquitination, and subsequent proteolytic degradation of the inhibitory protein IκB. FreeNF-κB then translocates into the nucleus and binds with its consensus sites. Similarly AP-1, either c-Jun/c-Jun (homodimer) or c-Fos/c-Jun (heterodimer), is activated by the phosphorylation of JNK pathway leading to the AP-1 activation and binds with its TRE consensus regions. Activation of NF-κB/AP-1 leads to the co-ordinate expression of antioxidant protective and proinflammatory genes.

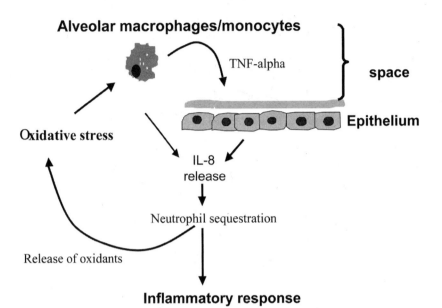

Figure 11 Oxidant-induced inflammation. Neutrophil recruitment to the lung is mediated by IL-8, a potent neutrophil chemotactic factor. Interleukin-8 may be secreted from immune and nonimmune cells in direct response to oxidant stimuli, or via a cytokine network. Through release of oxidants, neutrophils further the oxidative burden and consequent inflammatory response.

exposure to cigarette smoke. The activation of p38, therefore, may be responsible for the elevated levels of TNF-α and IL-8 seen in BALF and sputum of patients with COPD (165,166). Hellermann and colleagues have shown that exposure of normal bronchial epithelial cells and A549 cells to cigarette smoke condensate resulted in increased NF-κB activation and phosphorylation of ERKs 1 and 2 (167). This was associated with increased expression of sICAM-1, IL-1β, IL-8, and GM-CSF, suggesting that activation of NF-κB and MAP kinase pathway is involved in the proinflammatory effects of CSC in epithelial cells. Cigarette smoke condensate can also induce NF-κB in other cell lines, for example, in human histiocytic lymphoma U-937 cells, Jurkat T cells, and H1299 lung cells (168). The activation of NF-κB by CSC was correlated with time-dependent phosphorylation and degradation of

IκBα, and activation of IκBα kinase. In vitro experiments have demonstrated that CSC induces a distinct pattern of stress response in cultured epithelial cells, which may be related to the reported proinflammatory activities of CSC (via the formation of ROS/lipid peroxidation products) in vitro and in vivo (169). Cigarette smoke exposure led to a decreased DNA binding of NF-κB during the first 2 hr of exposure, followed by a more than twofold increase over control after 4–6 hr in Swiss 3T3 cells (169). This was not regulated by IκBα as evidenced by the lack of phosphorylation and degradation of IκB (170). It has been shown that thiol compounds such as thioredoxin can regulate NF-κB DNA binding in a redox-dependent manner (170). Therefore, the initial loss of DNA-binding activity may be due to the decreased level of reduced thioredoxin. However, within hours, the cells responded to this oxidative stress by induction of thioredoxin reductase mRNA, elevation of GSH levels, and restoration of NF-κB/thioredoxin complexes in nuclear extracts. This suggests that the binding of NF-κB in cigarette smoke-treated cells is subject to mainly a redox-controlled mechanism depending on the availability of reduced thioredoxin in the nucleus. Consequently, the result of this study would implicate a role for the redox status of thioredoxin or glutathione in the transcriptional regulation of gene expression by cigarette smoke. It is interesting to note that in vitro treatment with various constituents of cigarette smoke, such as nicotine, acrolein, hydroquinone, catechol, and 4-HNE, inhibits either basal or LPS-induced NF-κB activation and the expression of NF-κB-dependent genes, such as IL-1, IL-2, IF-γ, TNF-α, and IL-8 in the U937 cell line or in the peripheral blood-derived monocytes (141,171,172). It is postulated that the inhibitory effects of nicotine may contribute to cigarette smoke-induced immunosuppression. Similarly, Vayssier et al. have shown that cigarette smoke itself inhibits both spontaneous and LPS-induced NF-κB activation and cytokine release in human peripheral blood monocytes, which may be related to the altered cytokine profile observed in alveolar macrophages of smokers (173). Thus, it may be possible that cigarette smoke-mediated regulation of NF-κB is cell type specific.

XVII.A.A. Proinflammatory genes

Evidence from a large number of studies indicates that COPD
is associated with airway and airspace inflammation, as shown
in recent bronchial biopsy studies (35) and by the presence of
markers of inflammation, including IL-8 and TNF-α, which
are elevated in the sputum of patients with COPD (165). In vitro
studies show that treatment of macrophages, alveolar, and
bronchial epithelial cells with oxidants stimulates the release
of inflammatory mediators, such as IL-8, IL-1, and nitric oxide.
This is associated with increased expression of the mRNA for
the genes for these inflammatory mediators, and increased
nuclear binding and activation of NF-κB (174,175). Similarly,
several investigators have shown that cigarette smoke induces
IL-8 release from human bronchial and alveolar epithelial and
endothelial cells which may contribute to airway inflammation
in smokers (176–178). The increased IL-8 was significantly correlated with neutrophil counts in bronchial samples of BALF.
This cigarette smoke-induced IL-8 release may be dependent
on oxidative stress due to a Fenton reaction-mediated peroxidation reaction, since aldehydes such as acrolein and acetaldehyde
augmented IL-8 release (177,178). Saetta et al. (179) have
recently shown increased expression of the chemokine receptor
CXCR3 and its ligand CXCL10 in peripheral airways of smokers with COPD. This suggests that involvement of the
CXCR3/CXCL10 axis may be involved in T-lymphocyte
recruitment (increased levels of chemokines).

Cigarette smoke has been shown in vivo to be a cause of
increased adherence of leukocytes to vascular endothelium
(45). Shen et al. (180) have shown that cigarette smoke condensate induces the expression of a subset of cell adhesion molecules, such as intercellular adhesion molecule (ICAM-1),
endothelial leukocyte adhesion molecule 1 (ELAM-1), and
vascular cell adhesion molecule (VCAM-1) in human umbilical vascular endothelial cells associated with an increase in
the binding activity of NF-κB, suggesting the increased transendothelial migration of monocytes by cigarette smoking (180).
The release of proinflammatory mediators, such as IL-1β
and sICAM-1, was increased by cigarette smoke exposure

in bronchial epithelial cells cultured from biopsy materials obtained from patients with COPD compared to smokers (181). This suggests that patients with COPD have a greater susceptibility to the effects of cigarette smoke.

There may also be an interaction between oxidants and TNF that are both relevant mediators in COPD, producing synergistic activation of NF-κB, suggesting that anti-inflammatory and antioxidant therapies may be required together in order to prevent the activation of NF-κB in the lung inflammation that occurs in COPD (182).

XVIII. ANTIOXIDANT PROTECTIVE GENES

An important effect of oxidative stress in the lungs is the upregulation of protective antioxidant genes. The antioxidant glutathione (GSH) is concentrated in epithelial lining fluid compared with plasma (101) and has an important protective role, together with its redox enzymes, in the airspaces and intracellularly in epithelial cells. Human studies have shown elevated levels of glutathione in epithelial lining fluid in chronic cigarette smokers compared with nonsmokers (26,101). However, this increase is not present immediately after acute cigarette smoking (26).

The discrepancy between glutathione levels in epithelial lining fluid in chronic and acute cigarette smokers has been investigated in animal models and in vitro using cultured epithelial cells (100,104,182). Exposure of airspace epithelial cells to CSC in vitro produces an initial decrease in intracellular GSH with a rebound increase after 12 hr (183). This effect in vitro is mimicked by a similar change in glutathione in rat lungs in vivo following intratracheal instillation of cigarette smoke condensate (100). The initial fall in lung and intracellular glutathione after treatment with cigarette smoke condensate was associated with a decrease in the activity of glutamylcysteine ligase (GCL) formerly known as γ-GCS, the rate limiting enzyme for glutathione synthesis, with recovery of the activity by 24 hr (100,183). The increased levels of glutathione following cigarette smoke condensate exposure have been shown to be due to transcriptional

upregulation of the gene for GSH synthesis GCL by components within cigarette smoke (Fig. 12) (183,184). Thus, oxidative stress, including that produced by cigarette smoking,
causes upregulation of the gene involved in the synthesis of
glutathione as a protective mechanism against oxidative

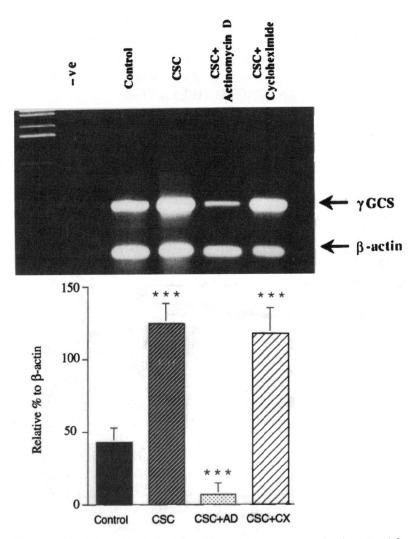

Figure 12 The effect of cigarette smoke exposure in A549 epithelial cells on γ-GCS mRNA by RT-PCR.

stress. These events are likely to account for the increased levels of glutathione seen in the epithelial lining fluid in chronic cigarette smokers (26,101). However, the injurious effects of cigarette smoke may occur repeatedly during and immediately after cigarette smoking when the lung is depleted of antioxidants, including glutathione (26).

The cytokine TNF, which is present as part of the airway inflammation in COPD (185), also decreases intracellular glutathione levels initially in epithelial cells by a mechanism involving the generation of intracellular oxidative stress, which is followed after 24 hr by a rebound increase in intracellular glutathione, as a result of AP-1 activation and an increased GCL mRNA expression (183,184). Corticosteroids have been used as anti-inflammatory agents in COPD, but there is still doubt over their effectiveness in reducing airway inflammation in COPD. Interestingly, dexamethasone also causes a decrease in intracellular glutathione in airspace epithelial cells, but no rebound increase compared with the effects of TNF (186). Moreover, the rebound increase in glutathione produced by TNF in epithelial cells is prevented by cotreatment with dexamethasone (186). These effects may have relevance for the treatment of COPD patients with corticosteroids.

Gilks et al. (187) have shown that rats exposed to whole cigarette smoke had increased expression of a number of antioxidant genes in the bronchial epithelial cells for up to 14 days (187). While, mRNA of manganese superoxide dismutase (MnSOD) and metallothionein (MT) was increased at 1–2 days and returned to normal by 7 days, mRNA for glutathione peroxidase did not increase until 7 days of exposure, suggesting the importance of the glutathione redox system as a mechanism for chronic protection against the effects of cigarette smoke (187).

The oncogene *cfos* belongs to a family of growth and differentiation-related immediate early genes, the expression of which generally represents the first measurable response to a variety of chemical and physical stimuli (173). Studies in various cell lines have shown enhanced gene expression of the *cfos* in response to cigarette smoke condensate (170,188). These effects of cigarette smoke condensate can be mimicked by peroxynitrite and smoke-related aldehydes

in concentrations that are present in cigarette smoke condensate (170). This effect can be enhanced by pretreatment of the
cells with buthionine sulfoximine to decrease intracellular
glutathione and can be prevented by treatment with the thiol
antioxidant N-acetylcysteine (170). These studies emphasize
the importance of intracellular levels of the antioxidant glutathione in regulating gene expression.

Ishii et al. (189) have shown that glutathione S-transferase
P1 (GSTP1) acts as a protective enzyme against cigarette smoke
in the airway cells. Similarly, Maestrelli et al. (190) have recently
shown that HO-1 is induced in alveolar spaces of smokers suggesting that oxidative stress due to cigarette smoke may increase
the gene expression of HO-1 leading to increased levels of exhaled
CO. Cigarette smoke also induces heat-shock protein 70 (HSP70)
in human monocytes and HO-1, which have been implicated in
the regulation of cell injury and cell death and, in particular, modulation of apoptosis in human endothelial cells and monocytes
(173,191). The induction of HSP70 may stabilize IκBα, possibly
through the prevention of IκB kinase activation (192).

Thus, oxidative stress, including that produced by cigarette
smoke, causes increased gene expression of both proinflammatory genes, by oxidant-mediated activation of NF-κB and also
activation of protective genes, such as γ-glutamylcysteine synthetase through other transcription factors (AP-1/ARE). A balance
may therefore exist between pro- and anti-inflammatory gene
expressions in response to cigarette smoke, which may be critical
to whether cell injury is induced by cigarette smoking (Fig. 13).
Such an imbalance of an array of redox-regulated antioxidant
versus proinflammatory genes might, therefore, be associated
with the susceptibility or tolerance to disease. Knowledge of the
molecular mechanisms that regulate these events may open
new therapeutic avenues in the treatment of COPD.

XIX. CHROMATIN REMODELING (HISTONE ACETYLATION AND DEACETYLATION) AND GLUCOCORTICOID INEFFICACY

Condensation of eukaryotic DNA in chromatin suppresses
gene activity through the coiling of DNA on the surface of

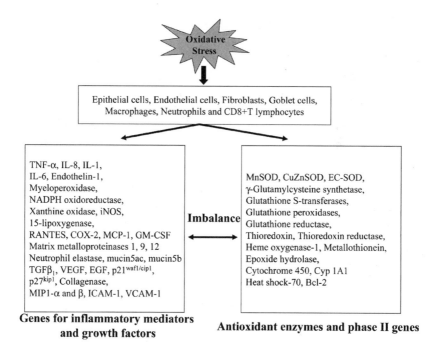

Figure 13 Various inflammatory and structural cells are activated by oxidative stress leading to transcription of various proinflammatory and antioxidant genes. It may be possible that induction of antioxidant enzymes may provide initial adaptive or protective responses against oxidative stress and inflammatory mediators. However, during sustained/chronic inflammation, the balance between genes for inflammatory mediators and antioxidant/phase II enzymes may be tipped in favor of proinflammatory mediators.

the nucleosome core and the folding of nucleosome assemblies, thus decreasing the accessibility to the transcriptional apparatus (193). Tightly bound DNA around a nucleosome core (histone residues H2A, H2B, H3, and H4) suppresses gene transcription by decreasing the accessibility of transcription factors, such as NF-κB and AP-1 to the transcriptional complex. Acetylation of lysine residues in the N-terminal tails of the core histone proteins results in uncoiling of the DNA, allowing increased accessibility for transcription factor binding (194) (Fig. 14) Acetylation of lysine (K) residues on histone

Model of histone acetylation and histone deacetylation

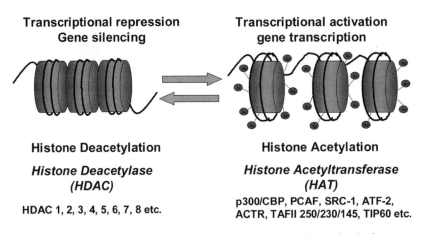

Transcriptional repression
Gene silencing

Transcriptional activation
gene transcription

Histone Deacetylation

Histone Acetylation

Histone Deacetylase
(HDAC)

HDAC 1, 2, 3, 4, 5, 6, 7, 8 etc.

Histone Acetyltransferase
(HAT)

p300/CBP, PCAF, SRC-1, ATF-2,
ACTR, TAFII 250/230/145, TIP60 etc.

Figure 14 Diagram illustrating the process by which histone acetylation causes chromatin remodeling. Acetylation causes the unwinding of DNA around the histone protein, making the chromatin less condensed, hence increasing the accessibility of transcription factors. This leads to an increase in gene transcription. Deacetylation results in the rewinding of the DNA around the histone proteins, decreasing gene transcription. There are eight different HDACs presently known, and many coactivators of the HAT complex that contain intrinsic HAT activity, such as SCR-1 and p300/CBP. (Adapted from Ref. 202.)

4 (K5, K8, K12, and K16) is thought to be directly related in the regulation of gene transcription (194,195) (Fig. 15). Histone acetylation is reversible and is regulated by a group of acetyltransferases (HATs), which promote acetylation, and deacetylases (HDACs) that promote deacetylation.

Oxidative stress and proinflammatory mediators have been suggested to influence histone acetylation and phosphorylation via a mechanism dependent on the activation of MAPK pathway (196–198). Recently, we and other investigators have shown that both H_2O_2 and TNF-α caused an increase in histone acetylation (HAT activity) leading to IL-8 expression in alveolar epithelial cells (199–204) (Fig. 16). Ito et al. (201) have shown a role for histone acetylation and deacetylation in IL-1β-induced TNF-α release in alveolar macrophages derived

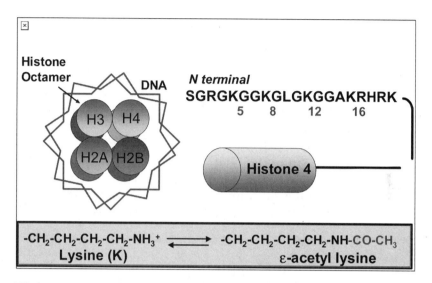

Figure 15 Diagram illustrating the lysine residues on the N-terminal of the histone 4 proteins that are targets for acetylation by HATs. (Adapted from Ref. 205.)

from cigarette smokers. They have also suggested that oxidants may play an important role in the modulation of HDAC and inflammatory cytokine gene transcription.

It has been suggested that oxidative stress may have a role in the poor efficacy of corticosteroids in COPD. It has

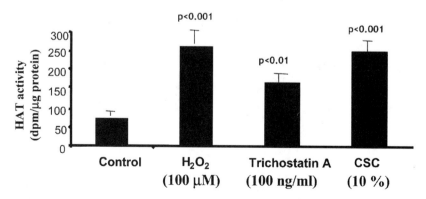

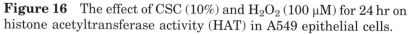

Figure 16 The effect of CSC (10%) and H_2O_2 (100 μM) for 24 hr on histone acetyltransferase activity (HAT) in A549 epithelial cells.

been shown that glucocorticoid suppression of inflammatory genes requires recruitment of HDAC-2 to the transcription activation complex by the glucocorticoid receptor (201,206). This results in deacetylation of histones and a decrease in inflammatory gene transcription. A reduced level of HDAC-2 was associated with increased proinflammatory response and reduced responsiveness to glucocorticoids in alveolar macrophages obtained from smokers (207). Cigarette smoke condensate also inhibited HDAC-2 levels in A549 cells (204) (Fig. 17). In a recent study, Culpitt et al. have shown that cigarette smoke solution-stimulated release of IL-8, and GM-CSF was not inhibited in alveolar macrophages obtained

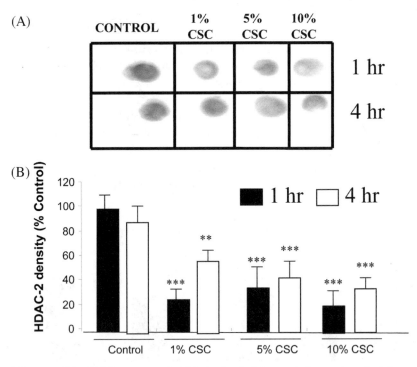

Figure 17 (A) Immuno-dot blot of soluble proteins extracted from A549 cells showing reduced levels of HDAC-2 proteins in response to CSC after 1 and 4 hr. (B) Histogram showing the relative density (percentage of control) of HDAC-2 in A549 cells after the above treatments. $^{**}p < 0.01$ and $^{***}p < 0.001$ compared to control values.

from patients with COPD compared to that of smokers (207). They suggested that the lack of efficacy of corticosteroids in COPD might be due to steroid insensitivity of macrophages in the respiratory tract. Thus, the cigarette smoke/oxidant-mediated reduction in HDAC-2 levels in alveolar epithelial cells and macrophages will not only increase inflammatory gene expression, but will also cause a decrease in glucocorticoid function in patients with COPD. This may be one of the potential reasons for the failure of glucocorticoids to function effectively in reducing inflammation in COPD (Fig. 18).

XX. INFLAMMATION AND ADENOVIRAL E1A

Adenoviral DNA can integrate itself into the human genome, and this can lead to the amplification of viral oncoproteins, such as adenoviral protein (E1A). The presence of the E1A has been suggested to be a possible factor in susceptibility to inflammation caused by cigarette smoke (208). The E1A gene has been found to be present at a higher frequency in the lungs of COPD patients than in similar smokers without COPD (209). A recent report by Retamales and colleagues suggests that the cigarette smoke-induced inflammatory process that underlies emphysematous destruction of the lung in COPD is amplified in smokers with advanced disease compared with those with similar smoking histories and preserved lung structure and function (210). This enhanced response to a similar degree of stimulation was associated with evidence of latent adenoviral infection of the alveolar surface epithelium (210). The presence of E1A also enhances the inflammatory response of cells to endotoxin and oxidative stress (211). The activation of the NF-κB and the expression of IL-8, TGF-β, and ICAM-1 were enhanced in E1A+ lung epithelial cells in response to oxidative stress, compared to E1A− cells (212). Similarly, Metcalf (213) found that stimulating E1A transfected cell lines (THP-1 and Jurkat) led to increased production of TNF-α, compared to cells transfected with a control plasmid. Thus, the presence of transactivating E1A primes the transcriptional machinery for oxidative stress signaling and, therefore, facilitates persistent amplification of

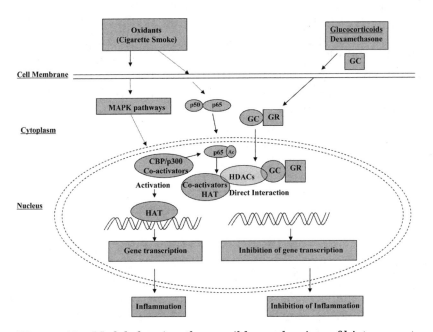

Figure 18 Model showing the possible mechanism of histone acetylation by oxidative stress and its repression by corticosteroids (GCs), leading to inhibition of gene transcription. Mitogen activated protein kinase signaling pathways may be activated by oxidative stress leading to histone acetylation. Direct interaction between coactivators (HAT), histone deacetylase and the glucocorticoid receptor (GR) may result in repression of the expression of proinflammatory genes. Human histone deacetylase forms a bridge with HAT to inhibit gene transcription. However, when the HDAC is inhibited by oxidants or the NF-κB subunit p65 is acetylated, steroids may not be able to recruit HDACs into the transcriptional complex to inhibit proinflammatory gene expression.

proinflammatory responses. This may be one of the susceptible factors in the pathogenesis of COPD.

XXI. OXIDATIVE STRESS AND SUSCEPTIBILITY TO COPD

Only 15–20% of cigarette smokers appear to be susceptible to the effects of cigarette smoke. These subjects show a rapid decline in FEV_1 and develop COPD (214). There has been considerable interest in identifying those who are susceptible and the mechanisms of that susceptibility (215–217), since this would provide an important insight into the pathogenesis of COPD as did the recognition of the association between α_1-antitrypsin and COPD.

Polymorphisms of various genes have been shown to be more prevalent in smokers who develop COPD (214). A number of these polymorphisms may have functional significance, such as the association between the TNF-α gene polymorphism (TNF2), which is associated with increased TNF levels in response to inflammation, and the development of chronic bronchitis (216). Relevant to the effects of cigarette smoke is a polymorphism in the gene for microsomal epoxide hydrolase, an enzyme involved in the metabolism of highly reactive epoxide intermediates that are present in cigarette smoke (218). The proportion of individuals with slow microsomal epoxide hydrolase activity (homozygotes) was significantly higher in patients with COPD and a subgroup of patients shown pathologically to have emphysema (COPD 22%; emphysema 19%) compared with control subjects (6%) (218). These data have, however, not been reproduced in other patient populations (219). Similarly, the polymorphism of another antioxidant gene glutathione *S*-transferase is associated with decline in lung function in smokers (220). Furthermore, the polymorphisms in the MMP1 or MMP12 genes, but not in MMP9 are either causative factors in smoking-related lung injury (221). It may be that a panel of "susceptibility" polymorphisms, of functional significance in enzymes involved in xenobiotic metabolism or antioxidant enzyme genes, may allow individuals to be identified as being susceptible to the effects of cigarette smoke.

XXII. ANTIOXIDANT THERAPEUTIC INTERVENTIONS

It is now evident that oxidant/antioxidant balance is altered in favor of oxidants in smokers, which play an important role in the pathogenesis of COPD. Therefore, it would be logical to propose the rationale of antioxidant therapy in ameliorating the increased inflammatory response in COPD. Furthermore, proof of concept of the role of oxidative stress in the pathogenesis of COPD will come from studies of effective antioxidant therapy. There are various options to enhance the lung antioxidant screen. One approach would be to use specific spin traps such as α-phenyl-*N*-*tert*-butyl nitrone to react directly with reactive oxygen and reactive nitrogen species at the site of inflammation (222). The therapeutic purposes of this drug are currently in clinical development. Inhibitors that have a double action, such as the inhibition of lipid peroxidation and quenching radicals, could be developed. Another approach could be the molecular manipulation of antioxidant genes, such as glutathione peroxidase or genes involved in the synthesis of glutathione, such as γ-GCS or by developing molecules with an activity similar to these antioxidant enzymes.

Recent animal studies have shown that recombinant SOD treatment can prevent the neutrophil influx to the airspaces and IL-8 release induced by cigarette smoking through a mechanism involving downregulation of NF-κB (223). This holds great promise if compounds can be developed with antioxidant enzyme properties, which may be able to act as novel anti-inflammatory drugs by regulating the molecular events in lung inflammation.

Another approach would simply be to administer antioxidant therapy. This has been attempted in cigarette smokers using various antioxidants such as vitamin C and vitamin E (224,225). The results have been rather disappointing, although as described above the antioxidant vitamin E has been shown to reduce oxidative stress in patients with COPD (226).

XXIII. THIOL COMPOUNDS

XXIII.A. *N*-acetyl-L-cysteine

N-acetyl-L-cysteine (NAC), a cysteine-donating reducing compound, acts as a cellular precursor of GSH and becomes deacetylated in the gut to cysteine following oral administration.

It may also reduce cystine to cysteine, which is an important mechanism for intracellular GSH elevation in vivo in lungs. It reduces disulfide bonds and also has the potential to interact directly with oxidants. N-acetyl-L-cysteine is also used as a mucolytic agent (to reduce mucus viscosity and to improve mucociliary clearance) (227).

Pharmacological approaches, particularly with thiol antioxidants, such as NAC have been used in an attempt to enhance lung GSH in patients with COPD with varying success (228–230). There have also been studies of patients with COPD where the administration of NAC has led to a conflicting result; the number of exacerbations of COPD having been modified (231,232). This probably arose as a result of differing dosage regimens and disease severity in these studies (230,233). A multicenter study using NAC by metered dose inhalers in patients with chronic cough failed to show a positive effect on well being, sensation of dyspnea, cough, or lung function (233). Van Schooten et al. (234) have reported that supplementation of oral dose of 600 mg twice daily for a period of 6 months in a randomized, double-blind, placebo-controlled, Phase II chemoprevention trial reduced various plasma and BALF oxidative biomarkers in smokers. Furthermore, the results of Phase III trial on the multicenter Bronchitis Randomized on NAC Cost-Utility Study (BRONCUS) will provide the effectiveness of NAC as an "antioxidant" and in altering the decline in FEV_1 exacerbation rate and quality of life in patients with moderate-to-severe COPD (235). In addition, the efficacy of oral NAC in COPD based on a quantitative systematic review and meta-analysis of published double-blind placebo-controlled clinical trials has been discussed (236,237). All of these studies have reported that treatment with NAC mucolytics when taken long term is associated with a reduction in risk of acute exacerbations, improves symptoms and days of illness in COPD.

XXIII.B. *N*-acystelyn

N-acystelyn (NAL), a lysine salt of N-acetyl-L-cysteine, is a mucolytic and antioxidant thiol compound. The advantage of NAL over NAC is that it has a neutral pH solution, whereas

NAC is acidic. *N*-acystelyn can be aerosolized into the lung without causing significant side effects (238). Gillissen et al. (239) compared the effect of NAL and NAC and found that both drugs enhance intracellular glutathione in alveolar epithelial cells and inhibited hydrogen peroxide and $O_2^{\bullet-}$ released from human blood-derived PMN from smokers with COPD. *N*-acystelyn also inhibited ROS generation induced by serum-opsonized zymosan by human polymorphonuclear neutrophils. This inhibitory response was comparable to the effects of NAC (238,239). Recently, Antonicelli et al. (174) have shown that NAL inhibited oxidant-mediated IL-8 release in A549 cells suggesting the anti-inflammatory effect of NAL. Therefore, NAL may represent an interesting alternative approach to augment the antioxidant screen in the lungs.

XXIII.C. *N*-isobutyrylcysteine

Because NAC becomes hydrolyzed in biological systems, the measured bioavailability of the drug is low. Thus, it was speculated that a drug might be synthesized that possessed greater bioavailability than NAC, and could be used as a more effective treatment for chronic bronchitis. *N*-isobutyrylcysteine (NIC) is an NAC-like compound that does not undergo effective first-pass hydrolysis, and therefore has a higher oral bioavailability than NAC. This oral bioavailability can be as high as 80%, dependent on food intake. However, when evaluated as a therapy for exasperations of chronic bronchitis, NIC performed no better than placebo drugs, and not as well as NAC. Recently, a study of *N*-isobutyrylcysteine, a derivative of *N*-acetylcysteine, also failed to reduce exacerbation rates in patients with COPD (240).

XXIII.D. Glutathione

Direct increase of lung cellular levels of GSH/antioxidant screen would be a logical approach in COPD. In fact, extracellular augmentation of GSH had been tried through intravenous administration of GSH, oral ingestion of GSH, and aerosol/inhalation of nebulized GSH in an attempt to reduce inflammation in various lung diseases (241–244). However, all these led to undesirable effects, suggesting that GSH aerosol therapy may not be an appropriate way of increasing GSH levels in lung ELF and cells in COPD. In all of these studies, the question was raised about the bioavailability of GSH, pH, and osmolality at the site of microenvironment and the resultant formation of toxic products (GSSG). It seems rational to suggest that neutralizing the pH, providing GSH in salt form, liposome-entrapped GSH delivery, and the maintenance of isotonicity would be useful in designing any GSH inhalation therapy in inflammatory lung diseases.

Increasing the activity of γ-GCS and glutathione synthetase by gene transfer techniques may increase cellular GSH levels. The induction of γ-GCS by molecular means increasing cellular GSH levels or that γ-GCS gene therapy also holds great promise in protection against chronic inflammation and oxidant-mediated injury in COPD.

XXIII.E. Glutathione Peroxidase Mimic

Ebselen is a seleno-organic compound, as it contains selenium, an important element in the glutathione peroxidase catalysis of the reaction between GSH and ROS (245). This increases the efficiency of GSH as an antioxidant, and can thus be used as a therapy against oxidative stress and inflammation (246,247). However, studies are needed to validate the bioavailability of these compounds in lung inflammation.

XXIII.F. Redox Sensor Molecules

There are other small redox molecules such as β-strand mimetic template MOL-294 and PNRI-299, which have been shown to inhibit NF-κB and AP-1-mediated transcription and blocks allergic airway inflammation in a mouse asthma model (250). The mechanism of inhibition is based on the reversible inhibition of redox sensor proteins (similar to redox effector factor-1). These redox compounds are novel and have been shown to reduce airway eosinophil infiltration, mucus hypersecretion, edema, and cytokine release in mouse asthma model. However, the use of these compounds against cigarette smoke-induced oxidative stress and the release of proinflammatory mediators have not been tested in vitro or in vivo.

XXIII.G. Polyphenols

Curcumin (diferuloyulmethane) is a naturally occurring flavonoid (polyphenol) present in the spice, turmeric, which has a long traditional use as a chemotherapeutic agent for many diseases. Curcumin is an active principal of the perennial herb *Curcuma longa* (commonly known as turmeric). Turmeric has a long traditional use in the Orient for many ailments, particularly as an anti-inflammatory agent. Recent studies have reported that curcumin inhibits NF-κB expression/activation, tivation, cyclo-oxygenase (COX-2), HO-1, 1L-8 release and neutrophil recruitment in the lungs (251). Curcumin has multiple properties to protect against cigarette smoke-mediated oxidative stress (251). It acts as an oxygen radical and a hydroxyl radical scavenger, which are formed by cigarette smoke, increases antioxidant glutathione levels by induction of γ-GCS, and plays the role of an anti-inflammatory agent through the inhibition of NF-κB and IL-8 release in lung cells (252). Resveratrol, a flavanoid found in red wine, is an effective inhibitor of inflammatory cytokine release from macrophages in COPD patients (253). This anti-inflammatory property of resveratrol may due to its ability to induce sirtuins, and HDAC activity   (254). The molecular mechanisms of anti-inflammatory properties of dietary polyphenols against cigarette smoke/oxidative stress have not yet been studied.

XXIV. CONCLUSION

There is now considerable evidence for the increased generation of ROS in COPD, which may be important in the pathogenesis of this condition. Reactive oxygen species may be critical in the inflammatory response to cigarette smoke/environmental oxidants, through the upregulation of redox-sensitive transcription factors and hence proinflammatory gene expression, but they are also involved in the protective mechanisms against the effects of cigarette smoke by the induction of antioxidant

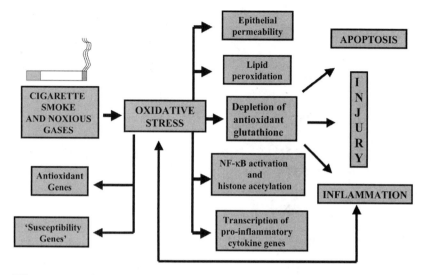

Figure 19 Summary diagram of cigarette smoke/oxidative stress-mediated lung injury and inflammation in smokers.

genes (Fig. 19). Further understanding of the effects of ROS in basic cellular functions such as amplification of proinflammatory and immunological responses, defective repair processes, signaling pathways, and apoptotic mechanisms will provide important information regarding basic pathological processes contributing to COPD. Further studies that investigate the mechanisms of the differential responses of immune-inflammatory and epithelial cells to inhaled oxidants or endogenous ROS are required in patients with COPD. More work in elucidating the mechanisms of ROS-mediated modulations of this process needs to be done before we can embark on the path of pharmacological therapy manipulating these processes in COPD. The role of oxidative stress in COPD will only be established when more potent antioxidants become available for clinical use.

REFERENCES

1. Gutteridge JM, Halliwell B. Free radicals and antioxidants in the year 2000. A historical look to the future. Ann NY Acad Sci 2000; 899:136–147.

2. Richer C, Cogvadze V, Laffranchi R, Schlapbach R, Schweizer M, Suter M. Oxidants in mitochondria: from physiology to diseases. Biochim Biophys Acta 1995; 127:67–74.

3. Rahman I, MacNee W. Role of transcription factors in inflammatory lung diseases. Thorax 1998; 53:601–612.

4. Guyton KZ, Liu Y, Gorospe M, et al. Activation of mitogen-activated protein kinase by H_2O_2. J Biol Chem 1996; 271: 4138–4142.

5. Rahman I, MacNee W. Role of oxidants/antioxidants in smoking-induced airways diseases. Free Radic Biol Med 1996; 21: 669–681.

6. Rahman I, MacNee W. Lung glutathione and oxidative stress: implications in cigarette smoke-induced airways disease. Am J Physiol 1999; 277:L1067–L1088.

7. Rahman I, MacNee W. Oxidative stress and regulation of glutathione synthesis in lung inflammation. Eur Respir J 2000; 16:534–554.

8. British Thoracic Society. British Thoracic Society Guidelines for the management of chronic obstructive pulmonary disease. Thorax 1997; 52:S1–S28.

9. American Thoracic Society. American Thoracic Society Standards for the diagnosis and care of patients with chronic obstructive pulmonary disease. Am J Respir Crit Care Med 1995; 152:S77–S120.

10. Snider G. Chronic obstructive pulmonary disease: risk factors, pathophysiology and pathogenesis. Ann Rev Med 1989; 40:411–429.

11. Church T, Pryor WA. Free radical chemistry of cigarette smoke and its toxicological implications. Environ Health Perspect 1985; 64:111–126.

12. Pryor WA, Stone K. Oxidants in cigarette smoke: radicals, hydrogen peroxides, peroxynitrate, and peroxynitrite. Ann NY Acad Sci 1993; 686:12–28.

13. Nakayama T, Church DF, Pryor WA. Quantitative analysis of the hydrogen peroxide formed in aqueous cigarette tar extracts. Free Radic Biol Med 1989; 7:9–15.

14. Repine JE, Bast A, Lankhorst I, the Oxidative Stress Study Group. Oxidative stress in chronic obstructive pulmonary disease. Am J Respir Crit Care Med 1997; 156:341–357.

15. Jorres RA, Magnussen H. Oxidative stress in COPD. Eur Respir Rev 1997; 7:131–135.

16. Walsh GM. Advances in the immunobiology of eosinophils and their role in disease. Crit Rev Clin Lab Sci 1999; 36:453–496.

17. Eiserich JP, Hristova M, Cross CE, Jones AD, Freeman BA, Halliwell B, van der Vliet A. Formation of nitric oxide-derived inflammatory oxidants by myeloperoxidase in neutrophils. Nature 1998; 391:393–397.

18. MacPherson JC, Comhair SAAA, Eruzurum SC, Klein DF, Lipscomb MF, Kavuru MS, Samoszuk MK, Hazen SL. Eosinophils are a major source of nitric oxide-derived oxidants in severe asthma: characterization of pathways available to eosinophils for generating reactive nitrogen species. J Immunol 2001; 166:5763–5772.

19. Wu W, Samoszuk MK, Comhair SAAA, Thomsassen MJ, Farver CF, Dweik RA, Kavuru MS, Erzurum SC, Hazen SL. Eosinophils generate brominating oxidants in allergen-induced asthma. J Clin Invest 2000; 105:1455–1463.

20. Halliwell B, Gutteridge JMC. Role of free radicals and catalytic metal ions in human disease: an overview. Methods Enzymol 1990; 186:1–85.

21. Zang LY, Stone K, Pryor WA. Detection of free radicals in aqueous extracts of cigarette tar by electron spin resonance. Free Radic Biol Med 1995; 19:161–167.

22. Rahman I, Morrison D, Donaldson K, MacNee W. Systemic oxidative stress in asthma, COPD, and smokers. Am J Respir Crit Care Med 1996; 154:1055–1060.

23. Aaron SD, Angel JB, Lunau M, Wright K, Fex C, Le Saux N, Dales RE. Granulocyte inflammatory markers and airway infection during acute exacerbation of chronic obstructive pulmonary disease. Am J Respir Crit Care Med 2001; 163:349–355.

24. Fiorini G, Crespi S, Rinaldi M, Oberti E, Vigorelli R, Palmieri G. Serum ECP and MPO are increased during exacerbations

of chronic bronchitis with airway obstruction. Biomed Pharmacother 2000; 54:274–278.

25. Gompertz S, Bayley DL, Hill SL, Stockley RA. Relationship between airway inflammation and the frequency of exacerbations in patients with smoking related COPD. Thorax 2001; 56:36–41.

26. Morrison D, Rahman I, Lannan S, MacNee W. Epithelial permeability, inflammation and oxidant stress in the airspaces of smokers. Am J Respir Crit Care Med 1999; 159:473–479.

27. Nakashima H, Ando M, Sugimoto M, Suga M, Soda K, Araki S. Receptor-mediated O_2^- release by alveolar macrophages and peripheral blood monocytes from smokers and nonsmokers. Am Rev Respir Dis 1987; 136:310–315.

28. Drath DB, Larnovsky ML, Huber GL. The effects of experimental exposure to tobacco smoke on the oxidative metabolism of alveolar macrophages. J Reticul Soc 1970; 25:597–604.

29. Schaberg T, Klein U, Rau M, Eller J, Lode H. Subpopulation of alveolar macrophages in smokers and nonsmokers: relation to the expression of CD11/CD18 molecules and superoxide anion production. Am J Respir Crit Care Med 1995; 151:1551–1558.

30. Mateos F, Brock JF, Perez-Arellano JL. Iron metabolism in the lower respiratory tract. Thorax 1998; 53:594–600.

31. Lapenna D, Gioia SD, Mezzetti A, Ciofani G, Consoli A, Marzio L, Cuccurullo F. Cigarette smoke, ferritin, and lipid peroxidation. Am J Respir Crit Care Med 1995; 151:431–435.

32. Thompson AB, Bohling T, Heires A, Linder J, Rennard SI. Lower respiratory tract iron burden is increased in association with cigarette smoking. J Lab Clin Med 1991; 117:494–499.

33. Wesselius LJ, Nelson ME, Skikne BS. Increased release of ferritin and iron by iron loaded alveolar macrophages in cigarette smokers. Am J Respir Crit Care Med 1994; 150:690–695.

34. Jeffery PK. Structural and inflammatory changes in COPD; a comparison with asthma. Thorax 1998; 53:129–136.

35. Saetta M. Airway inflammation in chronic obstructive pulmonary disease. Am J Respir Crit Care Med 1999; 160: S17–S20.

36. Lacoste JY, Bousquet J, Chanez P, Van Vgve T, Simony-Lafontaine J, Lequeu N, Vic P, Enander I, Godard P, Michel FB. Eosinophilic and neutrophilic inflammation in asthma, chronic bronchitis, and chronic obstructive pulmonary disease. J Allergy Clin Immunol 1993; 149:803–810.

37. Lebowitz MD, Postma DS. Adverse effects of smoking on the natural history of newly diagnosed chronic bronchitis. Chest 1995; 108:55–61.

38. Chan-Yeung M, Dybuncio A. Leucocyte count, smoking and lung function. Am J Med 1984; 76:31–37.

39. Van Antwerpen VL, Theron AJ, Richards GA, Steenkamp KJ, Van der Merwe CA, Van der Walt R, Anderson R. Vitamin E, pulmonary functions, and phagocyte-mediated oxidative stress in smokers and nonsmokers. Free Radic Biol Med 1995; 18:935–943.

40. Chan-Yeung M, Abboud R, Dybuncio A, Vedal S. Peripheral leucocyte count and longitudinal decline in lung function. Thorax 1988; 43:426–468.

41. Postma DS, Renkema TEJ, Noordhoek JA, Faber H, Sluiter HJ, Kauffman H. Association between nonspecific bronchial hyperreactivity and superoxide anion production by polymorphonuclear leukocytes in chronic airflow obstruction. Am Rev Respir Dis 1988; 137:57–61.

42. Richards GA, Theron AJ, van der Merwe CA, Anderson R. Spirometric abnormalities in young smokers correlate with increased chemiluminescence responses of activated blood phagocytes. Am Rev Respir Dis 1989; 139:181–187.

43. Rahman I, Skwarska E, MacNee W. Attenuation of oxidant/antioxidant imbalance during treatment of exacerbations of chronic obstructive pulmonary disease. Thorax 1997; 52: 565–568.

44. Muns G, Rubinstein I, Bergmann KC. Phagocytosis and oxidative bursts of blood phagocytes in chronic obstructive airway disease. Scand J Infect Dis 1995; 27:369–373.

45. Noguera A, Busquets X, Sauleda J, Villaverde JM, MacNee W, Agusti AG. Expression of adhesion molecules and G-proteins

in circulating neutrophils in COPD. Am J Respir Crit Care Med 1998; 158:1664–1668.

46. Brown DM, Drost E, Donaldson K, MacNee W. Deformability and CD11/CD18 expression of sequestered neutrophils in normal and inflamed lungs. Am J Respir Cell Mol Biol 1995; 13:531–539.

47. Pinamonti S, Muzzuli M, Chicca C, Papi A, Ravenna F, Fabri LM, Ciaccia A. Xanthine oxidase activity in bronchoalveolar lavage fluid from patients with chronic obstructive lung disease. Free Radic Biol Med 1996; 21:147–155.

48. Heunks LM, Vina J, van Herwaarden CL, Folgering HT, Gimeno A, Dekhuijzen PN. Xanthine oxidase is involved in exercise-induced oxidative stress in chronic obstructive pulmonary disease. Am J Physiol 1999; 277:R1697–R1704.

49. Pinamonti S, Leis M, Barbieri A, Leoni D, Muzzoli M, Sostero S, Chicca MC, Carrieri A, Ravenna F, Fabbri LM, Ciaccia A. Detection of xanthine oxidase activity products by EPR and HPLC in bronchoalveolar lavage fluid from patients with chronic obstructive pulmonary disease. Free Radic Biol Med 1998; 25:771–779.

50. Nowak D, Kasielski M, Pietras T, Bialasiewicz P, Antczak A. Cigarette smoking does not increase hydrogen peroxide levels in expired breath condensate of patients with stable COPD. Monaldi Arch Chest Dis 1998; 53:268–273.

51. Nowak D, Antczak A, Krol M, et al. Increased content of hydrogen peroxide in expired breath of cigarette smokers. Eur Respir J 1996; 9:652–657.

52. Dekhuijzen PNR, Aben KKH, Dekker I, et al. Increased exhalation of hydrogen peroxide in patients with stable and unstable chronic obstructive pulmonary disease. Am J Respir Crit Care Med 1996; 154:813–816.

53. Maziak W, Loukides S, Culpitt S, Sullivan P, Kharitonov SA, Barnes PJ. Exhaled nitric oxide in chronic obstructive pulmonary disease. Am J Respir Crit Care Med 1998; 157: 998–1002.

54. Corradi M, Majori M, Cacciani GC, Consigli GF, de'Munari E, Pesci A. Increased exhaled nitric oxide in patients with stable

chronic obstructive pulmonary disease. Thorax 1999; 54: 576–680.

55. Delen FM, Sippel JM, Osborne ML, Law S, Thukkani N, Holden WE. Increased exhaled nitric oxide in chronic bronchitis: comparison with asthma and COPD. Chest 2000; 117: 695–701.

56. Rutgers SR, van der Mark TW, Coers W, Moshage H, Timens W, Kauffman HF, Koeter GH, Postma DS. Markers of nitric oxide metabolism in sputum and exhaled air are not increased in chronic obstructive pulmonary disease. Thorax 1999; 54:576–680.

57. Robbins RA, Millatmal T, Lassi K, Rennard S, Daughton D. Smoking cessation is associated with an increase in exhaled nitric oxide. Chest 1997; 112:313–318.

58. Choi AM, Alam J. Heme oxygenase-1: function, regulation, and implication of a novel stress-inducible protein in oxidant-induced lung injury. Am J Respir Cell Mol Biol 1996; 15: 9–19.

59. Montuschi P, Kharitonov SA, Barnes PJ. Exhaled carbon monoxide and nitric oxide in COPD. Chest 2001; 120:496–501.

60. Petruzzelli S, Puntoni R, Mimotti P, Pulera N, Baliva F, Fornai E, Giuntini C. Plasma 3-nitrotyrosine in cigarette smokers. Am J Respir Crit Care Med 1997; 156:1902–1907.

61. Van der Vliet A, Smith D, O'Neill CA, Kaur H, Darley-Usmar V, Cross CE, Halliwell B. Interactions of peroxynitrite and human plasma and its constituents: oxidative damage and antioxidant depletion. Biochem J 1994; 303:295–301.

62. Eiserich JP, van der Vliet A, Handelman GJ, Halliwell B, Cross CE. Dietary antioxidants and cigarette smoke-induced biomolecular damage: a complex interaction. Am J Clin Nutr 1995; 62:1490S–1500S.

63. Ichinose M, Sugiura H, Yamagata S, Koarai A, Shirato K. Increase in reactive nitrogen species production in chronic obstructive pulmonary disease airways. Am J Respir Crit Care Med 2000; 162:701–706.

64. Pignatelli B, Li CG, Boffetta P, Chen Q, Ahrens W, Nyberg F, Mukeria A, Bruske-Hohlfeld I, Fortes C, Constantinescu V,

Ischiropoulos H, Ohshima H. Nitrated and oxidized plasma proteins in smokers and lung cancer patients. Cancer Res 2001; 61:778–784.

65. Kanazawa H, Shiraishi S, Hirata K, Yoshikawa J. Imbalance between levels of nitrogen oxides and peroxynitrite inhibitory activity in chronic obstructive pulmonary disease. Thorax 2003; 58:106–109.

66. Sugiura H, Ichinose M, Yamagata S, Koarai A, Shirato K, Hattori T. Correlation between change in pulmonary function and suppression of reactive nitrogen species production following steroid treatment in COPD. Thorax 2003; 58:299–305.

67. Gutteridge JMC. Lipid peroxidation and antioxidants as biomarkers of tissue damage. Clin Chem 1995; 41:1819–1828.

68. Uchida K, Shiraishi M, Naito Y, Torii N, Nakamura Y, Osawa T. Activation of stress signaling pathways by the end product of lipid peroxidation. J Biol Chem 1999; 274: 2234–2242.

69. Parola M, Bellomo G, Robino G, Barrera G, Dianzani MU. 4-Hydroxynonenal as a biological signal: molecular basis and pathophysiological implications. Antioxidant Redox Signal 1999; 1:255–284.

70. Rahman I, Van Schadewijk AA, Crowther A, Hiemstra PS, Stolk J, MacNee W, De Boer WI. 4-Hydroxy-2-nonenal, a specific lipid peroxidation product is elevated in lungs of patients with chronic obstructive pulmonary disease (COPD). Am J Respir Crit Care Med 2002; 166:490–495.

71. Morrow JD, Roberts LJ. The isoprostanes: unique bioactive products of lipid peroxidation. Prog Lipid Res 1997; 36:1–21.

72. Reilly M, Delanty N, Lawson JA, FitzGerald GA. Modulation of oxidant stress in vivo in chronic cigarette smokers. Circulation 1996; 94:19–25.

73. Pratico D, Basili S, Vieri M, Cordova C, Violi F, FitzGerald GA. Chronic obstructive pulmonary disease is associated with an increase in urinary levels of isoprostane $F_{2\alpha}$-III, an index of oxidant stress. Am J Respir Crit Care Med 1998; 158: 1709–1714.

74. Habib MP, Clements NC, Garewal HS. Cigarette smoking and ethane exhalation in humans. Am J Respir Crit Care Med 1995; 151:1368–1372.

75. Euler DE, Dave SJ, Guo H. Effect of cigarette smoking on pentane excretion in alveolar breath. Clin Chem 1996; 42:303–308.

76. Montuschi P, Collins JV, Ciabattoni G, Lazzeri N, Corradi M, Kharitonov SA, Barnes PJ. Exhaled 8-isoprostane as an in vivo biomarker of lung oxidative stress in patients with COPD and healthy smokers. Am J Respir Crit Care Med 2000; 162:1175–1177.

77. Paredi P, Kharitonov SA, Leak D, Ward S, Cramer D, Barnes PJ. Exhaled ethane, a marker of lipid peroxidation, is elevated in chronic obstructive pulmonary disease. Am J Respir Crit Care Med 2000; 162:369–673.

78. Corradi M, Rubinstein I, Andreoli RL, Manini P, Caglieri A, Poli D, Alinovi R, Mutti A. Aldehydes in exhaled breath condensate of patients with chronic obstructive pulmonary disease. Am J Respir Crit Care Med 2003; 167:1380–1386.

79. Nowak D, Kasielski M, Antczak A, Pietras T, Bialasiewicz P. Increased content of thiobarbituric acid reactive substances in hydrogen peroxide in the expired breath condensate of patients with stable chronic obstructive pulmonary disease: no significant effect of cigarette smoking. Respir Med 1999; 93:389–396.

80. Fahn H, Wang L, Kao S, Chang S, Huang M, Wei Y. Smoking-associated mitochondrial DNA mutation and lipid peroxidation in human lung tissue. Am J Respir Cell Mol Biol 1998; 19:901–909.

81. Tsukagoshi H, Shimizu Y, Iwamae S, Hisada T, Ishizuka T, Iizuka K, Dobashi K, More M. Evidence of oxidative stress in asthma and COPD: potential inhibitory effect of theophylline. Respir Med 2000; 94:584–588.

82. Britton JR, Pavord ID, Richards KA, Knox AJ, Wisniewski AF, Lewis SA, Tattersfield AE, Weiss ST. Dietary antioxidant vitamin intake and lung function in the general population. Am J Respir Crit Care Med 1995; 151:1383–1387.

83. Maranzana A, Mehlhorn RJ. Loss of glutathione, ascorbate recycling, and free radical scavenging in human erythrocytes exposed to filtered cigarette smoke. Arch Biochem Biophys 1998; 350:169–182.

84. Petruzzelli S, Hietanen E, Bartsch H, Camus AM, Mussi A, Angeletti CA, Saracci R, Giuntini C. Pulmonary lipid peroxidation in cigarette smokers and lung patients. Chest 1990; 98:930–935.

85. Bridges AB, Scott NA, Parry GJ, Belch JJF. Age, sex, cigarette smoking and indices of free radical activity in healthy humans. Eur J Med 1993; 2:205–208.

86. Duthie GG, Arthur JR, James WPT. Effects of smoking and vitamin E on blood antioxidant status. Am J Clin Nutr 1991; 53:1061S–1063S.

87. Mezzetti A, Lapenna D, Pierdomenico SD, Calafiore AM, Costantini F, Riario-Sforza G, Imbastaro T, Neri M, Cuccurullo F. Vitamins E, C and lipid peroxidation in plasma and arterial tissue of smokers and non-smokers. Atherosclerosis 1995; 112:91–99.

88. Antwerpen LV, Theron AJ, Myer MS, Richards GA, Wolmarans L, Booysen U. Cigarette smoke-mediated oxidant stress, phagocytes, vitamin C, vitamin E and tissue injury. Ann NY Acad Sci USA 1993; 686:53–65.

89. Pelletier O. Vitamin C status of cigarette smokers and nonsmokers. Am J Clin Nutr 1970; 23:520–528.

90. Chow CK, Thacker R, Bridges RB, Rehm SR, Humble J, Turbek J. Lower levels of vitamin C and carotenes in plasma of cigarette smokers. J Am Coll Nutr 1986; 5:305–312.

91. Bridges RB, Chow CK, Rehm SR. Micronutrients and immune functions in smokers. Ann N Y Acad Sci USA 1990; 587:218–231.

92. Theron AJ, Richards GA, Rensburg AJ, Van der Merwe CA, Anderson R. Investigation of the role of phagocytes and antioxidant nutrients in oxidant stress mediated by cigarette smoke. Int J Vitam Nutr Res 1990; 60:261–266.

93. Barton GM, Roath OS. Leukocytic ascorbic acid in abnormal leukocyte states. Int J Vitam Nutr Res 1976; 46:271–274.

94. Toth KM, Berger EM, Buhler CJ, Repine JE. Erythrocytes from cigarette smokers contain more glutathione and catalase and protect endothelial cells from hydrogen peroxide better than do erythrocytes from non-smokers. Am Rev Respir Dis 1986; 134:281–284.

95. Cross CE, O'Neill CA, Reznick AZ, Hu ML, Marcocci L, Packer L, Frei B. Cigarette smoke oxidation of human plasma constituents. Ann N Y Acad Sci USA 1993; 686:72–90.

96. Dietrich M, Block G, Hudes M, Morrow JD, Norkus EP, Traber MG, Cross CE, Packer L. Antioxidant supplementation decreases lipid peroxidation biomarker F(2)-isoprostanes in plasma of smokers. Cancer Epidemiol Biomarkers Prev 2002; 11:7–13.

97. Zhang J, Jiang S, Watson RR. Antioxidant supplementation prevents oxidation and inflammatory responses induced by sidestream cigarette smoke in old mice. Environ Health Perspect 2001; 109:1007–1009.

98. Smith KR, Uyeminami DL, Kodavanti UP, Crapo JD, Chang LY, Pinkerton KE. Inhibition of tobacco smoke-induced lung inflammation by a catalytic antioxidant. Free Radic Biol Med 2002; 15:1106–1114.

99. Rahman I, MacNee W. Oxidant/antioxidant imbalance in smokers and chronic obstructive pulmonary disease. Thorax 1996; 51:348–350.

100. Rahman I, Li XY, Donaldson K, Harrison DJ, MacNee W. Glutathione homeostasis in alveolar epithelial cells in vitro and lung in vivo under oxidative stress. Am J Physiol: Lung Cell Mol Biol 1995; 269:L285–L292.

101. Cantin AM, North SL, Hubbard RC, Crystal RG. Normal alveolar epithelial lung fluid contains high levels of glutathione. J Appl Physiol 1987; 63:152–157.

102. Linden M, Hakansson L, Ohlsson K, Sjodin K, Tegner H, Tunek A, Venge P. Glutathione in bronchoalveolar lavage fluid from smokers is related to humoral markers of inflammatory cell activity. Inflammation 1989; 13:651–658.

103. Harju T, Kaarteenaho-Wiik R, Soini Y, Sormunen R, Kunnula VL. Diminished immunoreactivity of γ-glutamylcys-

teine synthetase in the airways of smokers' lung. Am J Respir Crit Care Med 2002; 166:754–759.

104. Li XY, Donaldson K, Rahman I, MacNee W. An investigation of the role of glutathione in the increased epithelial permeability induced by cigarette smoke in vivo and in vitro. Am J Respir Crit Care Med 1994; 149:1518–1525.

105. York GK, Pierce TH, Schwartz LS, Cross CE. Stimulation by cigarette smoke of glutathione peroxidase system enzyme activities in rat lung. Arch Environ Health 1976; 31:286–290.

106. McCusker K, Hoidal J. Selective increase of antioxidant enzyme activity in the alveolar macrophages from cigarette smokers and smoke-exposed hamsters. Am Rev Respir Dis 1990; 141:678–682.

107. Pacht ER, Kaseki H, Mohammed JR, Cornwell DG, Davis WR. Deficiency of vitamin E in the alveolar fluid of cigarette smokers. Influence on alveolar macrophage cytotoxicity. J Clin Invest 1988; 77:789–796.

108. Bui MH, Sauty A, Collet F, Leuenberger P. Dietary vitamin C intake and concentrations in the body fluids and cells of male smokers and nonsmokers. J Nutr 1992; 122:312–336.

109. McGowan SE, Parenti CM, Hoidal JR, Niewoehner DW. Differences in ascorbic acid content and accumulation by alveolar macrophages from cigarette smokers and non-smokers. J Lab Clin Med 1984; 104:127–134.

110. Kondo T, Tagami S, Yoshioka A, Nishumura M, Kawakami Y. Current smoking of elderly men reduces antioxidants in alveolar macrophages. Am J Respir Crit Care Med 1994; 149: 178–182.

111. Gadek J, Fells GA, Crystal RG. Cigarette smoking induces functional antiprotease deficiency in the lower respiratory tract of humans. Science 1979; 206:1315–1316.

112. Carp H, Janoff A. Inactivation of bronchial mucous proteinase inhibitor by cigarette smoke and phagocyte-derived oxidants. Exp Lung Res 1980; 1:225–237.

113. Hubbard RC, Ogushi F, Fells GA, Cantin AM, Jallat S, Courtney M, Crystal RG. Oxidants spontaneously released by alveolar macrophages of cigarette smokers can inactivate

the active site of α-1-antitrypsin rendering it ineffective as an inhibitor of neutrophil elastase. J Clin Invest 1987; 80: 1289–1295.

114.   Kramps JA, Rudolphus A, Stolk J, Willems LNA, Dijkman JH. Role of antileukoprotease in the lung. Ann NY Acad Sci USA 1991; 624:97–108.

115.   Kramps JA, van Twisk C, Dijkman DH. Oxidative inactivation of antileukoprotease is triggered by polymorphonuclear leucocytes. Clin Sci 1988; 75:53–62.

116.   Johnson D, Travis J. The oxidative inactivation of human α1-proteinase inhibitor. Further evidence for methionine at the reactive center. J Biol Chem 1979; 254:4022–4026.

117.   Carp H, Janoff A. Possible mechanisms of emphysema in smokers: in vitro suppression of serum elastase-inhibitory capacity by fresh cigarette smoke and its prevention by antioxidants. Am Rev Respir Dis 1978; 118:617–621.

118.   Carp H, Miller F, Hoidal JR, Janoff A. Potential mechanisms of emphysema: α1-proteinase inhibitor recovered from lungs of cigarette smokers contains oxidised methionine and has decreased elastase inhibitory capacity. Proc Natl Acad Sci 1982; 79:2041–2045.

119.   Wallaert B, Gressier B, Marquette CH, Gosset P, Remy-Jardin M, Mizon J, Tonnel AB. Inactivation of alpha 1-proteinase inhibitor by alveolar inflammatory cells from smoking patients with or without emphysema. Am Rev Respir Dis 1993; 147:1537–1543.

120.   Wallaert B, Certs A, Gressier B, Gosset P, Voisen C. Oxidative inactivation of α-1-proteinase inhibitor by alveolar epithelial type II cells. J Appl Physiol 1993; 75:2376–2382.

121.   Stone P, Calore JD, McGowan SE, Bernardo J, Snider GL, Franzblau C. Functional alpha-1-protease inhibitor in the lower respiratory tract of smokers is not decreased. Science 1983; 221:1187–1189.

122.   Boudier C, Pelletier A, Pauli G, Bieth JG. The functional activity of alpha1-proteinase inhibitor in bronchoalveolar lavage fluids from healthy human smokers and non smokers. Clin Chim Acta 1983; 131:309–315.

123. Abboud RT, Fera T, Richter A, Tabona MZ, Johal S. Acute effect of smoking on the functional activity of alpha-1-protease inhibitor in bronchoalveolar lavage fluid. Am Rev Respir Dis 1985; 131:1187–1189.

124. Gadek JE, Hunninghake GW, Fells GA, Zimmerman RL, Keogh BA, Crystal RG. Evaluation of the protease-antiprotease theory of human destructive lung disease. Bull Eur Physiopathol Respir 1980; 16(suppl):27–40.

125. Nadel JA. Role of epidermal growth factor receptor activation in regulating mucin synthesis. Respir Res 2001; 2:85–89.

126. Takeyama K, Jung B, Shim JJ, Burgel PR, Dao-Pick T, Ueki IF, Protin U, Kroschel P, Nadel JA. Activation of epidermal growth factor receptors is responsible for mucin synthesis induced by cigarette smoke. Am J Physiol Lung Cell Mol Physiol 2001; 280:L165–L172.

127. Leikauf GD, Borchers MT, Prows DR, Simpson LG. Mucin apoprotein expression in COPD. Chest 2002; 121:166S–182S.

128. Adler KB, Holden-Stauffer WJ, Repine JE. Oxygen metabolites stimulate release of high-molecular-weight glycoconjugates by cell and organ cultures of rodent respiratory epithelium via an arachidonic acid-dependent mechanism. J Clin Invest 1990; 85: 75–85.

129. Fischer BM, Voynow JA. Neutrophil elastase induces MUC5AC gene expression in airway epithelium via a pathway involving reactive oxygen species. Am J Respir Cell Mol Biol 2002; 26:447–452.

130. Tyagi SC. Homocysteine redox receptor and regulation of extracellular matrix components in vascular cells. Am J Physiol 1998; 274:C396–C405.

131. Lois M, Brown LA, Moss IM, Roman J, Guidot DM. Ethanol ingestion increases activation of matrix metalloproteinases in rat lungs during acute endotoxemia. Am J Respir Crit Care Med 1999; 160:1354–1360.

132. Fod HD, Rollo EE, Brown P, Pakbaz H, Berisha H, Said SI, Zucker S. Attenuation of oxidant-induced lung injury by the synthetic matrix metalloproteinase inhibitor BB-3103. Ann NY Acad Sci 1999; 878:650–653.

133. Choi AM, Elbon CL, Bruce SA, Bassett DJ. Messenger RNA levels of lung extracellular matrix proteins during ozone exposure. Lung 1994; 172:15–30.

134. Russell REK, Culpitt SV, DeMatos C, Donnelly L, Smith M, Wiggins J, Barnes PJ. Release and activity of matrix metalloproteinase-9 and tissue inhibitor of metalloproteinase-2 by alveolar macrophages from patients with COPD. Am J Respir Cell Mol Biol 2002; 26:602–609.

135. Hagen K, Zhu C, Melefors O, Hultcrantz R. Susceptibility of cultured rat hepatocytes to oxidative stress by peroxides and iron. The extracellular matrix affects the toxicity of *tert*-butyl hydroperoxide. Int J Biochem Cell Biol 1999; 31:499–508.

136. Siwik DA, Pagano PJ, Colucci WS. Oxidative stress regulates collagen synthesis and matrix metalloproteinase activity in cardiac fibroblasts. Am J Physiol Cell Physiol 2001; 280:C53–C60.

137. Rossi AG, Haslett C. Inflammation, cell injury, and apoptosis. In: Said SI, ed. Lung Biology in Health and Disease. Proinflammatory and Antiinflammatory Peptides. New York: Marcel Dekker Inc., 1998:9–21.

138. Nakajima Y, Aoshiba K, Yasui S, Nagai A. H_2O_2 induces apoptosis in bovine tracheal epithelial cells in vitro. Life Sci 1999; 64:2489–2496.

139. Aoshiba K, Tamaoki J, Nagai A. Acute cigarette smoke exposure induces apoptosis of alveolar macrophages. Am J Physiol Lung Cell Mol Physiol 2001; 281:L1392–L1401.

140. Hoshino Y, Mio T, Nagai S, Miki H, Ito I, Izumi T. Cytotoxic effects of cigarette smoke extract on an alveolar type II cell-derived cell line. Am J Physiol Lung Cell Mol Physiol 2001; 281:L509–L516.

141. Vayssier M, Banzet N, Francois D, Bellmann K, Polla BS. Tobacco smoke induces both apoptosis and necrosis in mammalian cells: differential effects of HSP70. Am J Physiol 1998; 275:L771–L779.

142. Hall AG. Review: the role of glutathione in the regulation of apoptosis. Eur J Clin Invest 1999; 29:238–245.

143. Kasahara Y, Tuder RM, Cool CD, Lynch DA, Flores SC, Voelkel NF. Endothelial cell death and decreased expression

of vascular endothelial growth factor and vascular endothelial growth factor receptor 2 in emphysema. Am J Respir Crit Care Med 2001; 163:737–744.

144. Wang J, Wilcken DE, Wang XL. Cigarette smoke activates caspase-3 to induce apoptosis of human umbilical venous endothelial cells. Mol Genet Metab 2001; 72:82–88.

145. Kasahara Y, Tuder RM, Taraseviciene-Stewart L, Le Cras TD, Abman S, Hirth PK, Waltenberger J, Voelkel NF. Inhibition of VEGF receptors causes lung cell apoptosis and emphysema. J Clin Invest 2000; 106:1311–1319.

146. Mullick AE, McDonald JM, Melkonian G, Talbot P, Pinkerton KE, Rutledge JC. Reactive carbonyls from tobacco smoke increase arterial endothelial layer injury. Am J Physiol Heart Circ Physiol 2002; 283:H591–H597.

147. Tuder RM, Zhen L, Cho CY, Taraseviciene-Stewart L, Kasahara Y, Salvemini D, Voelkel NF, Flores SC. Oxidative stress and apoptosis interact and cause emphysema due to VEGF receptor blockade. Am J Respir Cell Mol Biol. 2003; 29:88–97.

148. Rabinovich RA, Ardite E, Trooster T, Carbo N, Alonso J, Gonzalex de Suso JM, Vilaro J, Barbera JA, Polo MF, Argiles JM, Fernandez-Checa JC, Roca J. Reduced muscle redox capacity after endurance training in patients with chronic obstructive pulmonary disease. Am J Respir Crit Care Med 2001; 164:1114–1118.

149. Heunks LM, Dekhuijzen PN. Respiratory muscle function and free radicals: from cell to COPD. Thorax 2000; 55: 704–716.

150. Sauleda J, Garcia-Palmer FJ, Gonzalez G, Palou A, Agusti AG. The activity of cytochrome oxidase is increased in circulating lymphocytes of patients with chronic obstructive pulmonary disease, asthma, and chronic arthritis. Am J Respir Crit Care Med 2000; 161:32–35.

151. Engelen MP, Schols AM, Does JD, Deutz NE, Wouters EF. Altered glutamate metabolism is associated with reduced muscle glutathione levels in patients with emphysema. Am J Respir Crit Care Med 2000; 161:98–103.

152. Agusti AG, Sauleda J, Miralles C, Gomez C, Togores B, Sala E, Batle S, Busquets X. Skeletal muscle apoptosis and weight loss in chronic obstructive pulmonary disease. Am J Respir Crit Care Med 2002; 166:485–489.

153. Anderson R, Theron AJ, Richards GA, Myer MS, van Rensburg AJ. Passive smoking by humans sensitizes circulating neutrophils. Am Rev Respir Dis 1991; 144:570–574.

154. Schunemann HJ, Muti P, Freudenheim JL, Armstrong D, Browne R, Klocke RA, Trevisan M. Oxidative stress and lung function. Am J Epidemiol 1997; 146:939–948.

155. Sargeant LA, Jaeckel A, Wareham NJ. Interaction of vitamin C with the relation between smoking and obstructive airways disease in EPIC Norfolk. European Prospective Investigation into Cancer and Nutrition. Eur Respir J 2000; 16:397–403.

156. Britton JR, Pavord ID, Richards KA, Knox AJ, Wisniewski AF, Lewis SA, Tattersfield AE, Weiss ST. Dietary antioxidant vitamin intake and lung function in the general population. Am J Respir Crit Care Med 1995; 151:1383–1387.

157. Grievink L, Smit HA, Ocke MC, van't Veer P, Kromhout D. Dietary intake of antioxidant (pro)-vitamins, respiratory symptoms and pulmonary function: the MORGEN Study. Thorax 1998; 53:166–171.

158. Shahar E, Folsom AR, Melnick SL, Tockman MS, Comstock GW, Gennaro V, Higgins MW, Sorlie PD, Ko WJ, Szklo M. Dietary n-3 polyunsaturated fatty acids and smoking-related chronic obstructive pulmonary disease. Atherosclerosis Risk in Communities Study Investigators. N Engl J Med 1994; 331:228–233.

159. Shahar E, Boland LL, Folsom AR, Tockman MS, McGovern PG, Eckfeldt JH. Docosahexaenoic acid and smoking related chronic obstructive pulmonary disease. Atherosclerosis Risk in Communities Study Investigators. Am J Respir Crit Care Med 1999; 159:1780–1785.

160. Sridhar MK, Galloway A, Lean MEJ, Banham SW. An outpatient nutritional supplementation programme in COPD patients. Eur Respir J 1994; 7:720–724.

161. Wyatt TA, Heires AJ, Sanderson SD, Floreani AA. Protein kinase C activation is required for cigarette smoke-enhanced

C5a-mediated release of interleukin-8 in human bronchial epithelial cells. Am J Respir Cell Mol Biol 1999; 21:283–288.

162. Di Stefano A, Caramore G, Oates T, Capelli A, Lusuardi M, Gnemmi I, Ioli F, Chung KF, Donner CF, Barnes PJ, Adcock IM. Increased expression of nuclear factor-κB in bronchial biopsies from smokers and patients with COPD. Eur Respir J 2002; 20:556–563.

163. Caramori G, Romagnoli M, Casolari P, Bellettato C, Casoni G, Boschetto P, Fan Chung K, Barnes PJ, Adcock IM, Ciaccia A, Fabbri LM, Papi A. Nuclear localization of p65 in sputum macrophages but not in sputum neutrophils during COPD exacerbations. Thorax 2003; 58:348–351.

164. Mochida-Nishimura K, Sureweicz K, Cross JV, Hejal R, Templeton D, Rich EA, Toossi Z. Differential activation of MAP kinase signaling pathways and nuclear factor-kappaB in bronchoalveolar cells of smokers and nonsmokers. Mol Med 2001; 7:177–185.

165. Keatings VM, Collins PD, Scott DM, Barnes PJ. Differences in interleukin-8 and tumor necrosis factor-alpha in induced sputum from patients with chronic obstructive pulmonary disease or asthma. Am J Respir Crit Care Med 1996; 153: 530–534.

166. Wesselius LJ, Nelson ME, Bailey K, O'Brien-Ladner AR. Rapid lung cytokine accumulation and neutrophil recruitment after lipopolysaccharide inhalation by cigarette smokers and nonsmokers. J Lab Clin Med 1997; 129:106–114.

167. Hellermann GR, Nagy SB, Kong X, Lockey RF, Mohapatra SS. Mechanism of cigarette smoke condensate-induced acute inflammatory response in human bronchial epithelial cells. Respir Res 2002; 3:22–30.

168. Anto RJ, Mukhopadhyay A, Shishodia S, Gairola CG, Aggarwal BB. Cigarette smoke condensate activates nuclear transcription factor κB through phosphorylation and degradation of IκBα: correlation with induction of cyclooxygenase-2. Carcinogenesis 2002; 23:1511–1518.

169. Gebel S, Muller T. The activity of NF-kappa B in Swiss 3T3 cells exposed to aqueous extracts of cigarette smoke is dependent on thioredoxin. Toxicol Sci 2001; 59:75–81.

170. Muller T, Gebel S. The cellular stress response induced by aqueous extracts of cigarette smoke is critically dependent on the intracellular glutathione concentration. Carcinogenesis 1998; 19:797–801.

171. Sugano N, Shimada K, Ito K, Murai S. Nicotine inhibits the production of inflammatory mediators in U937 cells through modulation of nuclear factor-kappaB activation. Biochem Biophys Res Commun 1998; 252:25–28.

172. Ouyang Y, Virasch N, Hao P, Aubrey MT, Mukerjee N, Bierer BE, Freed BM. Suppression of human IL-1beta, IL-2, IFN-gamma, and TNF-alpha production by cigarette smoke extracts. J Allergy Clin Immunol 2000; 16:280–287.

173. Vayssier M, Favatier F, Pinot F, Bachelet M, Poll BS. Tobacco smoke induces coordinate activation of HSF and inhibition of NF-κB in human monocytes: effects on TNFalpha release. Biochem Biophys Res Commun 1998; 252:249–256.

174. Antonicelli F, Parmentier M, Rahman I, Drost E, Hirani N, Donaldson K, MacNee W. *N*-acystelyn inhibits hydrogen peroxide mediated interleukin-8 expression in human alveolar epithelial cells. Free Radic Biol Med 2002; 32:492–502.

175. Parmentier M, Hirani N, Rahman I, Donaldson K, Antonicelli F. Regulation of LPS-mediated IL-1β by N-acetyl-L-cysteine in THP-1 cells. Eur Respir J 2000; 16:933–939.

176. Mio T, Romberger DJ, Thompson AB, Robbins RA, Heires A, Rennard SI. Cigarette smoke induces interleukin-8 release from human bronchial epithelial cells. Am J Respir Crit Care Med 1997; 155:1770–1776.

177. Wang H, Ye Y, Zh M, Cho C. Increased interleukin-8 expression by cigarette smoke extract in endothelial cells. Environ Toxicol Pharmacol 2000; 9:19–23.

178. Masubuchi T, Koyama S, Sato E, Takamizawa A, Kubo K, Sekiguchi M, Nagai S, Izumi T. Smoke extract stimulates lung epithelial cells to release neutrophil and monocyte chemotactic activity. Am J Pathol 1998; 153:1903–1912.

179. Saetta M, Mariani M, Panina-Bordignon P, Turato G, Buonsanti C, Baraldo S, Bellettato CM, Papi A, Corbetta L, Zuin R, Sinigaglia F, Fabbri LM. Increased expression of the

chemokine receptor CXCR3 and its ligand CXCL10 in peripheral airways of smokers with chronic obstructive pulmonary disease. Am J Respir Crit Care Med 2002; 165:1404–1409.

180. Shen Y, Rattan V, Sultana C, Kalra VK. Cigarette smoke condensate-induced adhesion molecule expression and transendothelial migration of monocytes. Am J Physiol Heart Circ Physiol 1996; 39:H1624–H1633.

181. Rusznak C, Mills PR, Devalia JL, Sapsford RJ, Davies RJ, Lozewicz S. Effect of cigarette smoke on the permeability and IL-1beta and sICAM-1 release from cultured human bronchial epithelial cells of never-smokers, smokers, and patients with chronic obstructive pulmonary disease. Am J Respir Cell Mol Biol 2000; 23:530–536.

182. Janssen-Heininger YMW, Macara I, Mossman BT. Cooperativity between oxidants and tumour necrosis factor in the activation of nuclear factor (NF)-κB. Requirement of Ras/ mitogen-activated protein kinases in the activation of NF-κB by oxidants. Am J Respir Cell Mol Biol 1999; 20:942–952.

183. Li XY, Rahman I, Donaldson K, MacNee W. Mechanisms of cigarette smoke induced increased airspace permeability. Thorax 1996; 51:465–471.

184. Rahman I, Smith CAD, Lawson M, Harrison DJ, MacNee W. Induction of gamma-glutamylcysteine synthetase by cigarette smoke condensate is associated with AP-1 in human alveolar epithelial cells. FEBS Lett 1996; 396:21–25.

185. Rahman I, Bel A, Mulier B, Lawson MF, Harrison DJ, MacNee W, Smith CAD. Transcriptional regulation of γ-glutamylcysteine synthetase-heavy subunit by oxidants inhuman alveolar epithelial cells. Biochem Biophys Res Commun 1996; 229:832–837.

186. Rahman I, Antonicelli F, MacNee W. Molecular mechanisms of the regulation of glutathione synthesis by tumour necrosis factor-α and dexamethasone in human alveolar epithelial cells. J Biol Chem 1999; 274:5088–5096.

187. Gilks CB, Price K, Wright JL, Churg A. Antioxidant gene expression in rat lung after exposure to cigarette smoke. Am J Pathol 1998; 152:269–278.

188. Muller T. Expression of *c-fos* in quiescent Swiss 3T3 exposed to aqueous cigarette smoke fractions. Cancer Res 1995; 55:1927–1932.

189. Ishii T, Matsuse T, Igarashi H, Masuda M, Teramoto S, Ouchi Y. Tobacco smoke reduces viability in human lung fibroblasts: protective effect of glutathione *S*-transferase P1. Am J Physiol Lung Cell Mol Physiol 2001; 280:L1189–L1195.

190. Maestrelli P, Messlemani AHE, Fina OD, Nowicki Y, Saetta M, Mapp C, Fabbri LM. Increased expression of heme oxygenase (HO)-1 in alveolar spaces and HO-2 in alveolar walls of smokers. Am J Respir Crit Care Med 2001; 164:1508–1513.

191. Favatier F, Polla BS. Tobacco-smoke-inducible human haem oxygenase-1 gene expression: role of distinct transcription factors and reactive oxygen intermediates. Biochem J 2001; 353:475–482.

192. Yoo CG, Lee S, Lee CT, Kim YW, Han SK, Shim YS. Anti-inflammatory effect of heat shock protein induction is related to stabilization of I kappa B alpha through preventing I kappa B kinase activation in respiratory epithelial cells. J Immunol 2000; 164:5416–5423.

193. Wu C. Chromatin remodeling and the control of gene expression. J Biol Chem 1997; 272:28171–28174.

194. Imhof A, Wolffe AP. Transcription: gene control by targeted histone acetylation. Curr Biol 1998; 8:R422–R424.

195. Bannister AJ, Miska EA. Regulation of gene expression by transcription factor acetylation. Cell Mol Life Sci 2000; 57:1184–1192.

196. Tikoo K, Lau SS, Monks TJ. Histone H3 phosphorylation is coupled to poly-(ADP-ribosylation) during reactive oxygen species-induces cell death in renal proximal tubular epithelial cells. Mol Pharmacol 2001; 60:394–402.

197. Miyata Y, Towatari M, Maeda T, Ozawa Y, Saito H. Histone acetylation induces by granulocyte colony-stimulating factor in a MAP kinase-dependent manner. Biochem Biophys Res Commun 2001; 283:655–660.

198. Bohm L, Schneeweiss FA, Sharan RN, Feinendegen LE. Influence of histone acetylation on the modification of cyto-

plasmic and nuclear proteins by ADP-ribosylation in response to free radicals. Biochim Biophys Acta 1997; 1334: 149–154.

199. Rahman I, Gilmour P, Jimenez LA, MacNee W. Oxidative stress and TNF-α induce histone acetylation and AP-1/NF-κB in alveolar epithelial cells: potential mechanism in inflammatory gene transcription. Mol Cell Biochem 2002; 234/235:239–248.

200. Berghe WV, Bosscher KD, Boone E, Plaisance S, Haegeman G. The nuclear factor-κB engages CBP/p300 and histone acetyltransferase activity for transcriptional activation of the interleukin-6 gene promoter. J Biol Chem 1999; 274:32091–32098.

201. Ito K, Lim G, Caramori G, Chung KF, Barnes PJ, Adcock IM. Cigarette smoking reduces histone deacetylase 2 expression, enhances cytokine expression, and inhibits glucocorticoid actions in alveolar macrophages. FASEB J 2001; 15:1110–1112.

202. Lakshminarayanan V, Drab-Weiss EA, Roebuck KA. H_2O_2 and TNF induce differential binding of the redox-responsive transcription factor AP-1 and NF-κB to the interleukin-8 promoter in endothelial and epithelial cells. J Biol Chem 1998; 273:32670–32678.

203. Tomita K, Barnes PJ, Adcock IM. The effect of oxidative stress on histone acetylation and IL-8 release. Biochem Biophys Res Commun 301; 2003:572–577

204. Moodie F, Marwick JA, Anderson C, Szulakowski P, Biswas S, Kilty I, Rahman I. Oxidative stress and cigarette smoke alter chromatin remodeling but differentially regulate NF-κB activation and pro-inflammatory cytokine release in alveolar epithelial cells. FASEB J 2004; 18:1897–1899.

205. Adcock IM, Caramori G. Cross-talk between pro-inflammatory transcription factors and glucocorticoids. Immunol Cell Biol 2001; 79:376–384.

206. Rahman I, Marwick JA, Kirkham PA. Redox modulation of histone acetylation and deacetylation in vitro and in vivo: modulation of NF-κB and pro-inflammatory genes. Biochem Pharmacol 2004; 68:1255–1267.

207. Culpitt SV, Rogers DF, Shah P, De Matos C, Russel RE, Donnelly LE, Barnes PJ. Impaired inhibition by dexametha-

sone of cytokine release by alveolar macrophages from COPD patients. Am J Respir Crit Care Med 2002; 167:24–31.

208. Hogg JC. Childhood viral infection and the pathogenesis of asthma and chronic obstructive lung disease. Am J Respir Crit Care Med 1999; 160:S26–S28.

209. Retamales I, Elliott WM, Meshi B, Coxson HO, Pare PD, Sciurba FC, Rogers RM, Hayashi S, Hogg JC. Implication of inflammation emphysema and its association with latent adenoviral infection. Am J Respir Crit Care Med 2001; 164: 469–473.

210. Keicho N, Higashimoto Y, Bondy GP, Elliott WM, Hogg JC, Hayashi S. Endotoxin-specific NF-kappaB activation in pulmonary epithelial cells harboring adenovirus E1A. Am J Physiol 1999; 277:L523–L532.

211. Higashimoto Y, Elliott WM, Behzad AR, Sedgwick EG, Takei T, Hogg JC, Hayashi S. Inflammatory mediator mRNA expression by adenovirus E1A-transfected bronchial epithelial cells. Am J Respir Crit Care Med 2002; 166:200–207.

212. Gilmour PS, Rahman I, Donaldson K, MacNee W. Environmental particle-mediated epithelial IL-8 release is regulated by histone acetylation. Am J Physiol: Lung Cell Mol Physiol 2003; 284:L533–L540.

213. Metcalf JP. Adenovirus E1A 13S gene product upregulates tumor necrosis factor gene. Am J Physiol 1996; 270: L535–L540.

214. Silverman EK, Speizer FE. Risk factors for the development of chronic obstructive pulmonary disease. Med Clin N Am 1996; 80:501–522.

215. Sandford AJ, Weir TD, Pare PD. Genetic risk factors for chronic obstructive pulmonary disease. Eur Resp J 1997; 10:1380–1391.

216. Barnes PJ. Genetics and pulmonary medicine. 9. Molecular genetics of chronic obstructive pulmonary disease. Thorax 1999; 54:245–252.

217. Huang S-L, Su C-H, Chang S-C. Tumour necrosis factor-α gene polymorphism in chronic bronchitis. Am J Respir Crit Care Med 1997; 156:1436–1439.

218. Smith CAD, Harrison DJ. Association between polymorphism in gene for microsomal epoxide hydrolase and susceptibility to emphysema. Lancet 1997; 350:630–633.

219. Yim JJ, Park GY, Lee CT, Kim YW, Han Sk, Shim YS, Yoo CG. Genetic susceptibility to chronic obstructive pulmonary disease in Koreans: combined analysis of polymorphic genotypes for microsomal epoxide hydrolase and glutathione *S*-transferase M1 and T1. Thorax 2000; 55:121–125.

220. He JQ, Juan J, Connett JE Anthonisen NR, Pare PD, Sandford AJ. Antioxidant gene polymorphisms and susceptibility to a rapid decline in lung function smokers. Am J Respir Crit Care Med 2002; 166:323–328.

221. Joos L, He JQ, Shepherdson MB, Connette JE, Nichonson R, Anothonisen P, Pare PD, Sandford AJ. The role of matrix metalloproteinase polymorphisms in the rate of decline in lung function. Hum Mol Genet 2002; 11:569–576.

222. Chabrier P-E, Auguet M, Spinnewyn B, Auvin S, Cornet S, Demerle-Pallardy C, Guilmard-Favre C, Marin J-G, Pignol B, Gillard-Roubert V, Roussillot-Charnet C, Schulz J, Voissat I, Bigg D, Moncada S. BN 80933, a dual inhibitor of neuronal nitric oxide synthase and lipid peroxidation: A promising neuroprotective strategy. Proc Natl Acad Sci USA 1999; 96: 10824–10829.

223. Nishikawa M, Kakemizu N, Ito T, Kudo M, Kaneko T, Suzuki M, Udaka N, Ikeda H, Okubo T. Superoxide mediates cigarette smoke-induced infiltration of neutrophils into the airways through nuclear factor-κB activation and IL-8 mRNA expression in guinea pigs in vivo. Am J Respir Cell Mol Biol 1999; 20:189–198.

224. Romieu I, Trenga C. Diet and obstructive lung diseases. Epidemiol Rev 2001; 23:268–287.

225. Lykkesfeldt J, Christen S, Wallock LM, Chang HH, Jacob RA, Ames BN. Ascorbate is depleted by smoking and repleted by moderate supplementation: a study in male smokers and non-smokers with matched dietary antioxidant intakes. Am J Clin Nutr 2000; 71:530–536.

226. Steinberg FM, Chait A. Antioxidant vitamin supplementation and lipid peroxidation in smokers. Am J Clin Nutr 1998; 68:319–327.

227. Olsson B, Johansson M, Gabrielson J, Bolme P. Pharmacokinetics of reduced and oxidised *N*-acetylcysteine. Eur J Clin Pharmacol 1988; 34:77–82.

228. Bridgeman MME, Marsden M, MacNee W, et al. Cysteine and glutathione concentrations in plasma and bronchoalveolar lavage fluid after treatment with *N*-acetylcysteine. Thorax 1991; 46:39–42.

229. Bridgemen MME, Marsden M, Selby C. Effect of *N*-acetyl cysteine on the concentrations of thiols in plasma bronchoalveolar lavage fluid lining tissue. Thorax 1994; 49:670–675.

230. Boman G, Backer U, Larsson S, et al. Oral acetylcysteine reduces exacerbation rate in chronic bronchitis. Eur J Respir Dis 1983; 64:405–415.

231. Rasmusse JB, Glennow C. Reduction in days of illness after long-term treatment with *N*-acetylcysteine controlled-release tablets in patients with chronic bronchitis. Eur J Respir Dis 1988; 1:351–355.

232. British Thoracic Society Research Committee. Oral *N*-acetylcysteine and exacerbation rates in patients with chronic bronchitis and severe airways obstruction. Thorax 1985; 40: 823–835.

233. Dueholm M, Nielson C, Thorshauge H, Evald T, Hansen NC, Madsen HD, Maltbeck N. *N*-acetylcysteine by metred dose inhaler in the treatment of chronic bronchitis: a multi-centre study. Respir Med 1992; 86:98–92.

234. Van Schooten FJ, Nia AB, De Flora S, D'Agostini F, Izzotti A, Camoirano A, Balm AJ, Dallinga JW, Bast A, Haenen GR, Van't Veer L, Baas P, Sakai H, Van Zandwijk N. Effects of oral administration of *N*-acetyl-L-cysteine: a multi-biomarker study in smokers. Cancer Epidemiol Biomarkers Prev 2002; 11: 167–175.

235. Decramer M, Dekhuijzen PN, Troosters T, van Herwaarden C, Rutten-van Molken M, van Schayck CP, Olivieri D, Lankhorst I, Ardia A. The Bronchitis Randomized on NAC Cost-Utility

Study (BRONCUS): hypothesis and design. BRONCUS-trial Committee. Eur Respir J 2001; 17:329–336.

236. Poole PJ, Black PN. Oral mucolytic drugs for exacerbations of chronic obstructive pulmonary disease: systematic review. Br Med J 2001; 322:1–16.

237. Stey C, Steurer J, Bachmann S, Medici TC, Tramer MR. The effect of oral *N*-acetylcysteine in chronic bronchitis: a quantitative systematic review. Eur Respir J 2000; 16:253–262.

238. Nagy AM, Vanderbist F, Parij N, et al. Effect of the mucoactive drug nacystelyn on the respiratory burst of human blood polymorphonuclear neutrophils. Pulm Pharmacol Ther 1997; 10:287–292.

239. Gillissen A, Jaworska M, Orth M, et al. Nacystelyn a novel lysine salt of *N*-acetylcysteine to augment cellular antioxidant defence in vitro. Respir Med 1997; 91:159–168.

240. Ekberg-Jansson A, Larson M, MacNee W, Tunek A, Wahlgren L, Wouters EFM, Larsson S for the *N*-isobutyrylcysteine Study Group. *N*-isobutyrylcysteine, a donor of systemic thiols, does not reduce the exacerbation rate in chronic bronchitis. Eur Respir J 1999; 13:829–834.

241. Marrades RM, Roca J, Barbera A, Jover L, MacNee W, Rodriguiez-Roisin R. Nebulised glutathione induces bronchoconstriction in patients with mild asthma. Am J Respir Crit Care Med 1997; 156:425–430.

242. Buhl R, Vogelmeier C, Critenden M, Hubbard RC, Hoyt RF, Wilson EM, Cantin AM, Crystal RG. Augmentation of glutathione in the fluid lining the epithelium of the lower respiratory tract by directly administering glutathione aerosol. Proc Natl Acad Sci USA 1990; 87:4063–4067.

243. Roum JH, Borok Z, McElvaney NG, Grimes GJ, Bokser AD, Buhl R, Crystal RG. Glutathione aerosol suppresses lung epithelial surface inflammatory cell-derived oxidants in cystic fibrosis. J Appl Physiol 1999; 87:438–443.

244. Borok Z, Buhl R, Grimes GJ, Bokser AD, Hubbard RC, Holryod KJ, Roum JH, Czerski DB, Cantin AM, Crystal RG. Effect of glutathione aerosol on oxidant–antioxidant imbalance in idiopathic pulmonary fibrosis. Lancet 1991; 338:215–216.

245. Wendel A, Fausel M, Safayhi H, Tiegs G, Otter R. A novel biologically active seleno-organic compound. II. Activity of PZ51 in relation to glutathione peroxidase. Biochem Pharmacol 1984; 33:3241–3245.

246. Haddad el-B, McCluskie K, Birrell MA, Dabrowski D, Pecoraro M, Underwood S, Chen B, De Sanctis GT, Webber SE, Foster ML, Belvisi MG. Differential effects of ebselen on neutrophil recruitment, chemokine, and inflammatory mediator expression in a rat model of lipopolysaccharide-induced pulmonary inflammation. J Immunol 2002; 169: 974–982.

247. Zhang M, Nomura A, Uchida Y, Iijima H, Sakamoto T, Iishii Y, Morishima Y, Mochizuki M, Masuyama K, Hirano K, Sekizawa K. Ebselen suppresses late airway responses and airway inflammation in guinea pigs. Free Radic Biol Med 2002; 32:454–464.

248. Carpagnano GE, Kharitonov SA, Foschino-Barbaro MP, Resta O, Gramiccioni E, Barnes PJ. Increased inflammatory markers in the exhaled breath condensate of cigarette smokers. Eur Respir J 2003; 21:589–593.

249. Biernacki WA, Kharitonov SA, Barnes PJ. Increased leukotriene B4 and 8-isoprostane in exhaled breath condensate of patients with exacerbations of COPD. Thorax 2003; 58: 294–298.

250. Henderson WR Jr, Chi EY, Teo JL, Nguyen C, Kahn M. A small molecule inhibitor of redox-regulated NF-kappa B and activator protein-1 transcription blocks allergic airway inflammation in a mouse asthma model. Immunology 2002; 169:5294–5299.

251. Shishodia S, Potdar P, Gairola CG, Aggarwal BB. Curcumin (diferuloylmethane) down-regulates cigarette smoke-induced NF-kappaB activation through inhibition of IkappaBalpha kinase in human lung epithelial cells: correlation with suppression of COX-2, MMP-9 and cyclin D1. Carcinogenesis 2003; 7:1269–1279.

252. Biswas SK, MacClure D, Jimenez LA, Megson IL, Rahman I. Curcumin induces glutathione biosynthesis and inhibits NF-κB activation and interleukin-8 release in alveolar epithelial

cells: mechanism of free radical scavenging activity. Antioxidant & Redox Signaling 2005; 7:32–41.

253. Culpitt SV, Rogers DF, Fenwick PS, Shah P, De Matos C, Russell RE, Barnes PJ, Donnelly LE. Inhibition by red wine extract, resveratrol, of cytokine release by alveolar macrophages in COPD. Thorax 2003; 58:942–946.

254. Howitz KT, Bitterman KJ, Cohen HY, Lamming DW, Lavu S, Wood JG, Zipkin RE, Chung P, Kisielewski A, Zhang LL, Scherer B, Sinclair DA. Small molecule activators of sirtuins extend *Saccharomyces cerevisiae* lifespan. Nature 2003; 425:191–196.

14

Association of Chronic Inflammation with Carcinogenesis: Implications of Anti-inflammatory Strategies for Cancer Prevention

MARIE YEO, SANG UK HAN, KI TAIK NAM, DAE YOUNG KIM, SUNG WON CHO, and KI-BAIK HAHM

Genomic Research Center for Gastroenterology,
Ajou University School of Medicine,
Suwon, Korea

ABSTRACT

Current evidences have expended the concept that chronic inflammation might play a crucial role in the development and progression of several gastrointestinal cancers. For instance, many cancers originated from gastrointestinal tissues are closely associated with chronic, persistent inflammation presenting as chronic *Helicobacter pylori*-infected gastritis, inflammatory bowel disease, Barrett's esophagus, and chronic viral hepatitis. Here, we discuss the molecular evidence that chronic inflammation is capable of inducing cancer and the pro-cancer microenvironment (PCM) favorable for survival of tumor cells and their growth. The explainable factors fostering the neoplastic process are including (1) the induction of neoplastic mutation by oxidative stress, (2) enhancement of cell proliferation and apoptosis inhibition, (3) the production of proteases and growth factors providing the environment for cell migration, and (4) the induction of angiogenesis. Thus, this evidence has helped to shed light on anti-inflammatory treatment or antioxidants as promising protective therapeutic approach for cancer prevention.

I. INTRODUCTION

Clinical and experimental data have suggested that chronic inflammation of the gastrointestinal tissues is associated with an increased risk of malignant transformation (1–6). Clinically, Barrett's esophagus resulting from chronic reflux esophagitis might progress to esophageal cancer and similar events were evident in case of longstanding ulcerative colitis and Crohn's disease regarding as very high risk factor for colon cancer. Also, *Helicobacter pylori* infection was defined as class I carcinogen, based on the finding that *H. pylori* definitely provoked chronic atrophic gastritis, intestinal metaplasia, and gastric cancer after longtime lapse of *H. pylori* infection in animal and even human. In addition, precancerous lesions surrounding tumor are usually associated with chronic inflammation at pathoclinical and molecular background. Early hypothesis to explain the role of chronic inflammation in carcinogenesis is focused on mutagenic action of free radicals released from inflammatory phagocytes (7,8). The free radicals induce the accumulation of aberrant mutation at oncogene or tumor suppressive genes resulting in a critical hit of malignant transformation. Moreover, in an aspect of tumor progression, chronic inflammation creates favorable conditions for tumor cell survival and growth by the lasting activation of proinflammatory factors and dysfunction of anti-inflammatory factors. In general, proinflammatory factors such as proinflammatory cytokines and nitric oxide (NO) potentiate the activation and proliferation of inflammatory cells and induce direct cytotoxic effects on target cells. The activated inflammatory response is downregulated by following activation of anti-inflammatory factors, resulting in repair or resolving of the inflammation. However, chronic, persistent inflammatory response provoke imbalance between proinflammatory and anti-inflammatory response, resulting in perturbation of inflammation accompanied with significant redox imbalance. It causes cells to resist against oxidative stress or apoptosis and to promote cell proliferation due to longstanding activation of proinflammatory mediators with the functional loss of anti-inflammatory factors. In this chapter, we discussed risk factors fostering the neoplastic process by chronic, longstanding inflammation and real evidences that modulating chronic inflammation could ameliorate carcinogenic process. These evidences suggest that appropriate treatment of inflammation should be further explored for the chemoprevention of cancer.

II. ASSOCIATION OF CHRONIC INFLAMMATION AND CANCER IN GASTROENTEROLOGY

Many malignancies arise in the areas of infection and inflammation (9–11). There is a growing body of evidences that chronic inflammation is strongly associated with incidence of cancer (Table 1). For example, colon cancer can arise from inflammatory bowel disease such as chronic ulcerative colitis and Crohn's disease persistent more than 10 years. Patients with chronic *H. pylori*-infected gastritis are often exposed to high risk of gastric adenocarcinoma. Barrett's esophagus can progress to esophageal cancer, and chronic hepatitis B virus-infection in liver predisposes to the development of hepatocellular cacinoma. Recently, it is clear that chronic infection with *H. pylori* leads to gastritis and the longstanding inflammation

Table 1 Association of Chronic Inflammation Associated With Carcinogenesis

Inflammatory condition (disease)	Oncogenic consequences (organ)	Aetiologic agent
Atrophic gastritis	Gastric cancer	Helicobacter pylori
Barrett's esophagus	Esophageal cancer	Gastric acid
Bronchitis	Lung cancer	Silica, smoking, asbestosis
Chronic hepatitis	Hepatic cancer	Hepatitis B/C virus
Eosinophlic cystitis	Bladder cancer	Urinary catheters
Leukoplakia	Skin cancer	Ultraviolet light
Pancreatitis	Panceratic cancer	Alcoholism
Pelvic inflammatory disease	Ovarian cancer	Human papillomavirus
Ulcerative colitis	Colon cancer	

evolves more severe diseases such as erosive gastritis, atrophic gastritis, intestinal metaplasia, adenocarcinoma, sequentially (Fig. 1A). Similar pathogenic phenomenon was observed in chronic hepatitis B virus-infected hepatitis, and it develops liver cirrhosis followed by hepatocellular cacinoma (Fig. 1B). These evidences all suggest that chronic inflammation not only directly induces cancer but also provides procancer microenvironment (PCM) favorable for survival and growth of tumor cells.

III. ROLE OF OXIDATIVE STRESS IN GI CARCINOGENESIS

Although the mechanisms of chronic inflammation leading to malignant transformation remain unresolved definitely, it has been known that chronic inflammation is associated with enhanced production of reactive oxygen metabolites and reactive nitrogen metabolites. Leukocytes and other phagocytic cells activated by inflammatory signals produce large amounts of reactive radical species. These reactive metabolites are an important part of adaptive host defense against various infections and inflammation, but the excess oxidative stress in

(A) **Steps for gastric carcinogenesis**

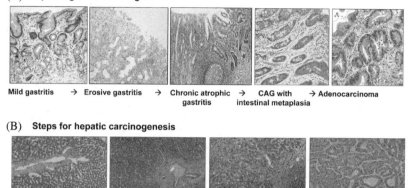

Mild gastritis → Erosive gastritis → Chronic atrophic → CAG with → Adenocarcinoma
gastritis intestinal metaplasia

(B) **Steps for hepatic carcinogenesis**

Normal liver → Chronic hepatitis → Liver cirrhosis → Hepatocellular carcinoma

Figure 1 Sequential promotion of neoplastic transformation by chronic inflammation in stomach (A) and liver (B). Inflammation was often induced by infection, especially *H. pylori* or hepatitis B virus for gastritis or hepatitis. Once the inflammation becomes chronic, more deteriorated conditions such as erosive gastritis, chronic atrophic gastritis, intestinal metaplasia, and adenoma can progress to gastric cancer. Longstanding inflammation induced by HBV also followed similar steps to hepatocellular carcinoma.

chronic inflammation makes cell to resist the cytotoxic effect of reactive species, in spite of cumulative DNA damage (Fig. 2). The cells undergo carcinogenic process via accumulation of cumulative DNA damage, carcinogenic compound formation, and transcriptional activation of genes which can promote cell proliferation. In chronic gastritis caused by infection by *H. pylori* infection, cytotoxin association gene A (*cag*A) product of the bacteria stimulates neutrophils to produce a lot of reactive oxygen species (ROS) (12,13). The excessive ROS induces oxidative stress to gastric mocosa, and damage cellular components including fatty acids, proteins, and DNA. In addition, NH_3 derived from *H. pylori* reacts with HOCl to yield NH_2Cl, which is more toxic to cells. These metabolites of oxygen and nitrogen oxidize intracellular components and directly mediate mutagenesis via DNA mutation and increase the expression of cell growth-related genes.

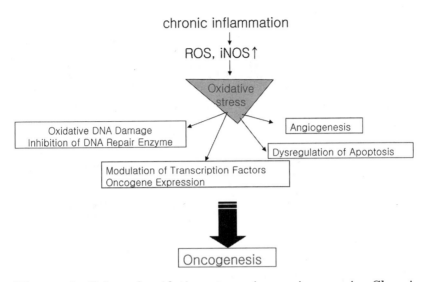

Figure 2 Roles of oxidative stress in carcinogenesis. Chronic inflammation provoked enhanced generation of ROS and increased expression of iNOS. The excessive oxidative stress can directly damage DNA, inhibit DNA repair, block apoptosis, and enhance angiogenesis, and provide the favorable conditions for carcinogenesis.

IV. PROCANCER MICROENVIRONMENT FORMATION BY CHRONIC, PERSISTENT INFLAMMATION

Inflammatory response is the event raised either to remove the cause of disturbance of body or to maintain and restore homoeostasis. Such an inflammatory response includes increases in neutrophils, the production of inflammatory mediators (cytokines, chemokines, reactive radical, etc.), phagocytosis of infected cells or damaged cells, and the regeneration of damaged tissues. The acute inflammatory response can be complete within a few days and homoeostasis is restored soon. For that, acute inflammation reaction results from activation of phagocyte-derived endogenous mediators, so-called cytokines such as tumor necrosis factor α (TNF-α), interleukin-1 (IL-1), and interferon-γ (IFN-γ). These cytokines directly stimulate the synthesis and release of second-phase cytokines or chemokine such as IL-6 or IL-8 from lymphocyte and stromal cells. Other products regulated by these cytokines are phosholipase A_2, cyclooxygenase 2 (COX-2), and inducible nitric oxide synthase (iNOS). These acute inflammatory responses are usually self-limiting and regulated in reciprocal manner, because the production of anti-inflammatory mediators (IL-1, IL-10, IL-13, TGF-β, etc.) immediately follows the proinflammatory cytokines. However, in the case of chronic inflammation, the response can persist due to lasting production of proinflammatory factors or a failure of mechanisms required for resolving the inflammatory response, and that seemed to contribute to construct procancer microenvironment and to develop cancer. As an example, chronic *H. pylori* infection provoked TNF-α, IL-1β, and IL-8 secretion from infected epithelial cells, resulting in NF-κB activation, which stimulates expression of iNOS, IL-8, and COX-2. Also *H. pylori* enhanced the expression of mRNA of chemokines like B cell attracting chemokine (BCA-1) and its receptor, BLR-1. These can contribute to the "mucosa associated lymphoid tissue originating tumor" (MALToma) in *H. pylori*-infected stomach (14,15) (Fig. 3), and which progress to the development of malignant lymphoma.

The inflammatory cytokines and chemokines directly or indirectly promote cell proliferation and growth coincident with the inhibition of apoptosis. The highlighting inflammatory molecules responsible for regulation of tumor cell proliferation are iNOS and COX-2 due to their over-expressions in many malignant tumors (16–21). TNF-α induces NO synthesis by nitric oxide synthase (NOS), and NO is a diffusing gas interacting with superoxide radicals as a result of which the peroxynitrite (ONOO$^-$) are generated. NO and ONOO$^-$ activate the inducible cyclooxygenase (COX-2) which is the key enzyme for the production of prostaglandins (PG) from arachidonic acid. COX-2 may protect cells against injury because it mediates stimulation of cellular proliferation and inhibition of apoptosis. However, sustained expression of the enzyme in chronic inflammation cause tremendous risk for neoplastic transformation (Fig. 4). Several groups reported that COX-2 over-express aberrantly in gastrointestinal carcinogenesis

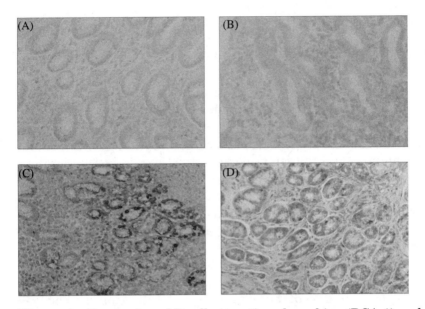

Figure 3 Expression of B cell attracting chemokine (BCA-1) and its receptor (BLR-1) in *H. pylori*-associated MALT lymphomas. Increased expressions of chemokines with their receptors might provide the molecular basis for *H. pylori*-associated MALTOMA development compared to non-MALTOMA patient with *H. pylori* (A and B), MALTOMA patient with *H. pylori* showed increased in situ expressions of BCA-1 (C) and its receptor (D) transcripts in the infiltrated lymphocytes and gastric glands.

such as familial adenomatous polyposis (FAP), ulcerative colitis-associated neoplasia, *H. pylori*-associated gastric cancer, and hepatocellular cacinoma. Based on these findings, the use of nonsteroidal anti-inflammatory drugs (NSAIDs) such as asprin, sulindac, and nimesulide, which are known to inhibit COX-2 enzymatic activity, reduces the relative risk of colorectal cancer by 40–50% (22,23). Crossing of COX-2 knockout mice with knockout of *APC* gene ameliorates the development of colonic polyposis (24). All these evidences suggest that COX-2 might play an essential role in gastrointestinal carcinogenesis.

Proinflammatory cytokines and chemokines released from diverse leukocytes enhance proliferation activity of cancer cell

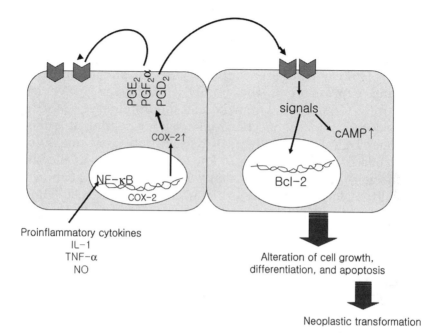

Figure 4 Associations of COX-2 with carcinogenesis. COX-2 is the enzyme for the high production of prostaglandins (PGE_2, PGF_2, and PGD_2) from arachidonic acid. The increased levels of COX-2 serve to lower the intracellular level of free arachidonic acid and therefore, prevent apoptosis. Prostanoids generated by COX-2 enable cell survival via autocrine or paracrine pathways. Especially, PGE_2 inhibits programmed cell death by inducing expression of Bcl-2 protooncogene and often elevate intracellular cyclic AMP concentration, which can suppress apoptosis.

as well as inflammatory cells via the secretion of growth factors. These cytokines and chemokines activate macrophages, which are main source of several growth factors (PDGF, bFGF, IGF, etc.) and cytokines, which can promote cell proliferation (25,26). Over-expression of these chemokine in a variety of tumor-derived cell lines enhances tumorigenicity in nude mice and also the blocking of these chemokines attenuate cell proliferation (27).

The several factors made by inflammatory cells or by stromal cells stimulated the recruitment, homing, and migration of inflammatory cells. Some tumor cells also directly use these factors to enhance tumor invasion and to facilitate tumor metastasis (28–31). In fact, cytokines like TNF-α and IL-1 mediate leukocyte adherence to the vascular endothelium by inducing expression of the family of adhesion molecules (selectin, VCAM-1) and matrix metalloproteinases (MMP). Evidence suggests the mechanisms used for homing of leukocytes may be appropriated for the dissemination of tumors via the bloodstream or lymphatics. For example, P- or E-selectin is adhesion receptor that normally recognize vascular mucin to facilitate leukocyte rolling along the blood vessels. Thus the molecule deficiency attenuates tumor growth and metastasis.

Alterations in the synthesis and dissolution of extracellular matrix (ECM) components are known to play a crucial role in tissue remodeling during inflammation and wound healing. The MMPs are a major family responsible for these remodeling of ECM. Matrix metalloproteinases are produced by inflammatory cells and by stromal cells responding to chemokines and cytokines produced by inflammatory cells. These inflammatory components are significant factors for invasion and dissemination of tumor cells. Recently, it is reported that the degree of expression of MMP in chronic inflammatory disease is considered as indicator of cancer and poor prognosis in colon and breast (32–35).

There are several lines of evidence that angiogenic actors derived from inflammatory reaction directly contribute for formation of procancer microenvironment favorable for tumor-associated angiogenesis and metastasis. Effector cells of inflammatory response like as mast cell released several angiogenic growth factors (VEGF, bFGF, PDGF, EGF, etc.) and some cytokines (IL-8, TNF-α) which act as angiogenic factor inducing expression of VEGF (36,37). Inflammatory mediators such as histamin, chymase, and NO regulate endothelial cell proliferation and degrade connective tissue matrix to provide space for neovascular sprouts (38,39).

The activated inflammatory response is also feedback-controlled by anti-inflammatory mediators such as tumor growth factor (TGF-β) and PGE2, which are released by phagocytosis of apoptotic cells from adjacent macrophage. These cytokines serve as an immunosuppressive cytokine, converting an active inflammatory response into normal physiologic condition by mechanism of repair and resolution. In contrast, these cytokines are often suppressed in chronic inflammation, and they cause the prolonged inflammation and enhance resistance to oxidative stress and apoptosis. For example, selective gastric expression of dominant negative form of TGF-β type II receptor exclusively the stomach under the promoter of pS2/TFF1 gene (pS2-dnRII) induced severe hyperplastic gastritis in chronic *H. pylori*-infected mice comparing to *H. pylori*-infected wild-type littermate mice (40). Furthermore, the *H. pylori*-infected pS2-dnRII transgenic mice developed gastric cancer, whereas *H. pylori*-infected mice expressing wild-type TGF-β type II receptor did not develop any malignant changes. These experiments definitely suggest that loss of TGF-β signaling, that is, loss of anti-inflammatory cytokine, contributes to form hyperplastic gastritis via loss of its anti-inflammatory effects. We also evaluated the role of TGF-β signaling in IBD mice model expressing the dominant negative form of TGF-β type II receptor selectively in the intestine under the control of ITF gene promoter (ITF dnRII) (Fig. 5) (41). The mice showed a significant increment of expression of several inflammatory cytokines (IL-1β, IL-2, INF-γ, TNF-α, and IL-10) and MMP in the colon. Moreover, loss of TGF-β signaling lead to appearance of multi foci with aberrant crypt foci (ACF) and polypoid colon adenoma, even in earlier time point after azoxymethane administration. This implied that TGF-β signaling is critical in the control of the development of cancer and maintenance between inflammatory stimuli and anti-inflammatory cytokine is essential for cancer prevention.

The signal imbalance between proinflammatory factors and anti-inflammatory factors provoke the disruption of self-limiting ability of inflammatory response. The dysregulation of inflammatory response contributes to the formation of procancer microenvironment presenting enhancing cell resistance to oxidative stress or apoptosis, survival of tumor cells, and their growth.

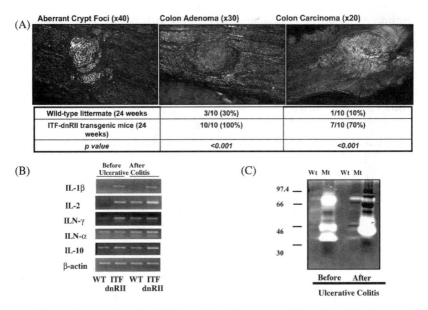

Figure 5 Loss of TGF-β signaling in the development of inflammation-associated carcinogenesis. TGF-β type II receptor transgenic mice were generated by selective expression of dominant negative form of TGF-β type II receptor under the promoter of ITF/TFF3 gene (ITF-dnRII). The loss of TGF-β signaling enhanced the development of multi foci showing ACF (aberrant crypt foci), polypoid colon adenoma, and colon carcinoma compared with wild-type littermates after azoxymethane administration (A):(24 weeks). The mice showed a significant increment of expression of several inflammatory cytokines (IL-1β, IL-2, INF-γ, TNF-α, and IL-10) (B) and MMP (C). All these finding suggest the increased inflammation conditions enhance the susceptibility to carcinogenesis.

V. ANTI-INFLAMMATORY TREATMENT AS CANCER PREVENTION STRATEGY

The best evidence for significance of inflammation during neoplastic progress comes from the study of attenuated cancer risk among long-term users of NSAIDs, cyclooxygenase inhibitor which blocks the conversion of arachdonic acid to prostaglandins. Much data indicate that use of these drugs could reduce colon cancer risk by 40–50%, and may be preventative especially for colon, skin, and esophageal cancer (42–44).

We evaluated the chemopreventive effects of NASID against *H. pylori*-associated gastric cancer of C57BL/6 mice (45). Gastric tumors were developed in 68% of mice administered with both MUN and *H. pylori*, whereas less than 10% developed gastric tumor treated with either MUN or *H. pylori* alone, suggesting *H. pylori* promoted carcinogen-induced gastric tumorigenesis in mice model. By the way, nimesulide, one of selective COX-2 inhibitor, significantly prevented the promotion of *H. pylori*-associated gastric tumorigenesis with prominent suppression of gastric carcinogenesis. Synergistic effects in inducing apoptosis and anti-proliferation were noted in HCC cell lines after the treatment COX-2 inhibitor. These finding indicated that COX-2 might be the potential target for the chemoprevention of gastric and liver cancer and the modulation of chronic inflammation also decreased the chance of carcinogenesis.

In same diseases model, the long-term administration of drug possessing anti-inflammatory and antioxidant action, rebamipide, significantly reduced mRNA expression of various cytokines (IL-1β, RANTES, TNFR p75, INF-γ, and TNF-α), NF-κB binding activity, and gastric mucosal expression of ICAM-1, HCAM, MMP (46–48). However, in contrast to NSAID, rebamipide could not decrease the incidence of *H. pylori*-associated gastric carcinogenesis. Although rebamipide cannot show the efficacy of reducing cancer development, they significantly reduced the development of precancerous lesions like chronic atrophic gastritis and intestinal metaplasia. Thus, long-term administration of anti-inflammatory drugs singly or combination with NSAID could be considered in the prevention of gastrointestinal cancer since they showed the definite molecular and biologic advantages. Conclusively, drugs like retinoic acid, interferon-γ, or tamoxifen were well documented as specific chemopreventive agent. Drugs like NSAID or antioxidants also could be considered as common chemopreventive agent because these drugs mechanistically modulated the inflammatory response responsible for carcinogenesis.

ACKNOWLEDGMENTS

This work was supported by a grant of the Korea Heath 21 R&D project, Ministry of Health and Welfare, Republic of Korea (1-PJ10-PG6-01GN14-0007). We would like to give sincere thanks to all researchers working at Genome Research Center for Gastroenterology, Ajou University Medical Center for their helps and assistances in many experiments.

REFERENCES

1. Kupper H, Adami HD, Trichopoulos D. Infections as a major preventable cause of human cancer. J Intern Med 2000; 248: 171–183.

2. Blaser MJ, Chyou HP, Nomura A. Age at establishment of *Helicobacter pylori* infection and gastric carcinoma, gastric ulcer, and duoden ulcer risk. Cancer Res 1995; 55:562–565.

3. Touati E, Michel V, Thiberge JM, Wusher N, Huerre M, Labigne A. Chronic *Helicobacter pylori* infections induce gastric mutations in mice. Gastroenterology 2003; 124:1408–1419.

4. Shacter E, Weitzman SA. Chronic inflammation and cancer. Oncology 2002; 16:217–226.

5. Gangarosa L, Halter S, Mertz H. Dysplastic gastroesophageal junction nodules—a precursor to junctional adenocarcinoma. Am J Gastroenterol 1999; 94:835–838.

6. Itzkowitz S. Colon carcinogenesis in inflammatory bowel diseases: applying molecular genetics to clinical practice. J Clin Gastroenterol 2003; 36:S70–S74.

7. Cerutti PA, Trump BF. Inflammation and oxidative stress in carcinogenesis. Cancer Cells 1991; 3:1–7.

8. Seril DN, Liao J, Yang GY, Yang CS. Oxidative stress and ulcerative colitis-associated carcinogenesis: studies humans and animal models. Carcinogenesis 2003; 24:353–362.

9. Koike K, Tsutsumi T, Fujie H, Shintani Y, Kyoji M. Molecular mechanism of viral hepatocarcinogenesis. Oncology 2002; 1: 29–37.

10. Ebert MP, Schandl L, Malfertheiner P. *Helicobacter pylori* infection and molecular changes in gastric carcinogenesis. J Gastroenterol 2002; 13:45–49.

11. Martinez-Maza O, Breen EC. B-cell activation and lymphoma in patients with HIV. Curr Opin Oncol 2002; 14:528–532.

12. Obst B, Wagner S, Sewing KF, Beli W. *Helicobacter pylori* causes DNA damage in gastric epithelial cells. Carcinogenesis 2000; 21:1111–1115.

13. Bagchi D, Bhattacharya G, Stohs GJ. Production of reactive oxygen species by gastric cells in association with *Helicobacter pylori*. Free Radic Res 1996; 24:439–450.

14. Enno A, O'Rourke JL, Howlett CR, Jack A, Dixon MF, Lee A. MALToma-like lesions in the murine gastric mucosa after long-term infection with *Helicobacter felis*. A mouse model of *Helicobacter* pylori-induced gastric lymphoma. Am J Pathol 1995; 147:217–222.

15. Biocchi M, De Vita S, Maestro D. Genetic abnormalities during transition from *Helicobacter pylori*-association gastritis to low-grade MALToma. Lancet 1995; 18:724.

16. Takahashi S, Fujita T, Yamamoto AA. Role of cyclooxygenase-2 in *Helicobacter pylori* induced gastritis in Mongolian gerbils. Am J Physiol Gastrointest Liver Physiol 2000; 279: G791–G798.

17. Chen CN, Sung CT, Lin MT, Lee PH, Chang KJ. Clinicopathologic association of cyclooxygenase 1 and cyclooxygenase 2 expression in gastric adenocarcinoma. Ann Surg 2001; 233: 183–188.

18. Romano M, Ricci V, Memoli A, Tuccillo C, Di Popolo A, Sommi P, Acquaviva AM, Del Vecchio Blanco C, Bruni CB, Zarrilli R. *Helicobacter pylori* up-regulates cyclooxygenase-2 mRNA expression and prostaglandin E2 synthesis in MKN 28 gastric mucosal cells in vitro. J Biol Chem 1998; 273:28560–28563.

19. Ajuebor MN, Singh A, Wallace JL. Cyclooxygenase-2-derived prostaglandin D2 is an early anti-inflammatory signal in experimental colitis. Am J Physiol Gastrointest Liver Physiol 2000; 279:238–244.

20. Bae SH, Jung ES, Park YM, Kim BS, Kim BK, Kim DG, Ryu WS. Expression of cyclooxygenase-2 (COX-2) in hepatocellular carcinoma and growth inhibition of hepatoma cell lines by a COX-2 inhibitor, NS-398. Clin Cancer Res 2001; 7: 1410–1418.

21. Kern MA, Schubert D, Sahi D, Schoneweiss MM, Moll I, Haugg AM, Dienes HP, Breuhahn K, Schirmacher P. Proapoptotic and antiproliferative potential of selective cyclooxygenase-2 inhibitors in human liver tumor cells. Hepatology 2002; 36:885–894.

22. Bresalier RS. Chemoprevention comes to clinical practice: COX-2 inhibition in familial adenomatous polyposis. Gastroenterology 2000; 119:1797–1798.

23. Thun MJ, Henley SJ, Patrono C. Nonsteroidal anti-inflammatory drugs as anticancer agents: mechanistic, pharmacologic, and clinical issues. J Natl Cancer Inst 2002; 94:252–266.

24. Oshima M, Dinchuk JE, Kargman SL, Oshima H, Hancock B, Kwong E, Trzaskos JM, Evans JF, Taketo MM. Suppression of intestinal polyposis in Apc delta 716 knockout mice by inhibitor of cyclooxygenase 2 (COX-2). Cell 1996; 29:803–809.

25. Brigati C, Noonan DM, Albini A, Benelli R. tumors and inflammatory infiltrates: friends or foes? Clin Exp Metastasis 2002; 19:247–258.

26. Owen JD, Strieter R, Burdick M, Haghnegahdar H, Nanney L, Shattuck-Brandt R, Richmond A. Enhanced tumor-forming capacity for immortalized melanocytes expressing melanoma growth stimulatory activity/growth-regulated sytokine beta and gamma proteins. Int J Cancer 1997; 73:94–103.

27. Nowicki A, Szenajch J, Ostrowska G, Wojtowicz A, Wojtowicz K, Kruszewski AA, Maruszynski M, Aukerman SL, Wiktor-Jedrzejczak W. Impaired tumor growth in colony-stimulating factor 1 (CSF-1)-deficient, macrophage deficient op/op mouse: evidence for a role of CSF-dependent macrophages in formation of tumor stroma. Int J Cancer 1996; 65:112–119.

28. Torisu H, Ono M, Kiryu H, Furue M, Ohmoto Y, Nakayama J, Nishioka Y, Sone S, Kuwano M. Macrophage infiltration correlates with tumor stage and angiogenesis in human malignant

melanoma: possible involvement of TNF-α and IL-1α. Int J Cancer 2000; 85:182–188.

29. Ono M, Torisu H, Fukushi J, Nishie A, Kuwano M. Biological implications of macrophage infiltrations in human tumor angiogenesis. Cancer Chemother Pharmacol 1999; 43:S69–S71.

30. Coussens LM, Raymond WW, Bergers G, Laig-Webster M, Behrendtsen O, Werb Z, Caughey GH, Hanahan D. Inflammatory mast cells up-regulate angiogenesis during squamous epithelial carcinogenesis. Genes Dev 1999; 13:1382–1397.

31. Coussens LM, Tinkle CL, Hanahan D, Werb Z. MMP-9 supplied by bone marrow-derived cells contributes to skin carcinogenesis. Cell 2000; 103:481–490.

32. Ougolkov AV, Yamashita K, Mai M, Minamoto T. Oncogenic beta-catenin and MMP-7 (matrilysin) cosegregate in late-stag clinical colon cancer. Gastroenterology 2002; 122:60–71.

33. Oberg A, Hoyhtya M, Tavelin B, Stenling R, Lindmark G. Limited value of preoperative serum analyses of matrix metalloproteinases (TIMP2, MMP-9) and tissue inhibitors of matrix metalloproteinases (TIMP-1, TIMP) in colorectal cancer. Anticancer Res 2000; 20:1085–1091.

34. Blot E, Chen W, Vasse M, Paysant J, Denoyelle C, Pille JY, Vincent L, Vannier JP, Soria J, Soria C. Cooperation between monocytes and breast cancer cells promotes factor involved in cancer aggressiveness. Br J Cancer 2003; 22:207–1212.

35. Baker EA, Stephenson TJ, Reed MW, Brown NJ. Expression of proteinases and inhibitors in human breast cancer progression and survival. Mol Pathol 2002; 55:300–304.

36. Sunderkotter C, Steinbrink K, Goebeler M, Bhardwaj R, Sorg C. Macrophages and angiogenesis. J Leukoc Biol 1994; 55:410–422.

37. Yoshida S, Ono M, Shono T, Izumi H, Ishibashi T, Suzuki H, Kuwano M. Involvement of interleukin-8, vascular endothelial growth factor, and basic fibroblast growth factor in tumor necrosis factor alpha-dependent angiogenesis. Mol Cell Biol 1997; 17:4015–4023.

38. Morbidelli L, Donnini S, Ziche M. Role of nitric oxide in the modulation of angiogenesis. Curr Pharm Des 2003; 9:521–530.

39. Ziche M, Morbidelli L. Nitric oxide and angiogenesis. J Neurooncol 2000; 50:139–148.

40. Hahm KB, Lee KM, Kim YB, Hong WS, Lee WH, Han SU, Kim MW, Ahn BO, Oh TY, Lee MH, Green J, Kim SJ. Conditional loss of TGF-beta signalling leads to increased susceptibility to gastrointestinal carcinogenesis in mice. Aliment Pharmacol Ther 2002; 16:115–127.

41. Hahm KB, Im YH, Parks TW, Park SH, Markowitz S, Jung HY, Green J, Kim SJ. Loss of transforming growth factor beta signalling in the intestine contribute tissue injury in inflammatory bowel disease. Gut 2001; 49:190–198.

42. Husain SS, Szabo IL, Yamawski AS. NSAID inhibition of GI cancer growth: clinical implications and molecular mechanisms of action. Am J Gastroenterol 2002; 97:542–553.

43. Dannenberg AJ, Altorki NK, Boyle JO, Dang C, Howe LR, Weksler BB, Subbaramaiah K. Cyclo-oxygenase 2: a pharmacological target for the prevention of cancer. Lancet Oncol 2001; 2:544–551.

44. Hahm KB, Song YJ, Oh TY, Lee JS, Surh YJ, Kim YB, Yoo BM, Kim JH, Han SU, Nahm KT, Kim MW, Kim DY, Cho SW. Chemoprevention of *Helicobacter pylori*-associated gastric carcinogenesis in mouse model: is it possible? J Biochem Mol Biol 2003; 36:82–94

45. Nam KT, Hahm KB, Oh SY, Yeo M, Han SU, Ahn B, Kim YB, Kang JS, Jang DD, Yang KH, Kim DY. The selective cyciooxygenase-2 inhibitor nimesulide prevents Helicobacter pylori-associated gastric cancer development in a mouse model. Clin Cancer Res 2004; 10:8105–8113.

46. Choi KW, Lee YC, Chung IS, Lee JJ, Chung MH, Kim NY, Kim SW, Kim JG, Roe IH, Lee SW, Jung HY, Choi MG, Hahm KB, Hong WS, Kim JH. Effect of rebamipide in treatment of *Helicobacter pylori*-associated duodenal ulcer: attenuation of cytokine expression and nitrosative damage. Dig Dis Sci 2002; 47:283–291.

47. Lee JS, Oh TY, Ahn BO, Cho H, Kim WB, Kim YB, Surh YJ, Kim HJ, Hahm KB. Involvement of oxidative stress in experimentally induced reflux esophageal Barrett's esophagus: clue

for the chemoprevention of esphageal carcinogenesis antioxidants. Mutat Res 2001; 480:189–200.

48. Oh TY, Lee JS, Ahn BO, Cho H, Kim WB, Kim YB, Surh YJ, Cho SW, Hahm KB. Oxidative damages are critical in pathogenesis of reflux esophagitis: implication of antioxidants in its treatment. Free Radic Biol Med 2001; 30:905–915.

15

Molecular Inflammation as an Underlying Mechanism of Aging: The Anti-inflammatory Action of Calorie Restriction

HAE YOUNG CHUNG and KYUNG JIN JUNG

College of Pharmacy, Aging Tissue Bank, Longevity Life Science and Technology Institutes, Pusan National University, Pusan, South Korea

BYUNG PAL YU

Department of Physiology, The University of Texas Health Science Center at San Antonio, San Antonio, Texas, U.S.A.

I. INTRODUCTION

A general characteristic of the aging process is a progressive, physiopathological deterioration of an organism with time. One probable cause for the biological phenomenon has been formulated as the oxidative stress of aging. According to this hypothesis, accumulated oxidatively altered cellular constituents, a disturbed redox balance, and weakened antioxidative

389

defense systems arise over time cause functional deficits, i.e., aging and age-related degenerative diseases (1,2).

Among the several other hypotheses of aging, the oxidative stress hypothesis currently offers the most plausible molecular and mechanistic elucidations of the time-dependent biological dysfunctions and the accompanying pathological processes that ensue with age. Evidence accumulated from aging research casts strong support for oxidative stress as the fundamental force influencing the global phenomenon of aging (2–4).

Although it seems paradoxical, the organism utilizes potentially harmful free radicals and oxidants for its own benefit as evolutionarily adapted trait. For instance, an organism's activation of various immune cells depends on the utilization of various reactive species (RS), including reactive oxygen species (ROS), and reactive nitrogen species (RNS). Reactive oxygen species, such as $^{\bullet}O_2^-$, $^{\bullet}OH$, and H_2O_2, and RNS, such as NO and $ONOO^-$, are essential for normal biological processes similar to the way vascular tone depends on the presence of NO. One common age-related dysfunction is the organism's inability to regulate these RS, as expressed as the inflammatory process.

The strict and tight control of these RS is a mandate for the proper maintenance of redox balance. Overproduced and/or unregulated ROS/RNS are a major causative factor in tissue injury (4,5). In aged organisms, uncontrolled oxidative stress activates redox-sensitive transcription factors involved in the inflammation process, thereby leading to chronically proinflammatory state of aging (6). In this chapter, we chose to use the term, "molecular inflammation" to emphasize proinflammatory molecular alterations occurring under normal conditions prior to a fully expressed inflammatory disease state.

To date, caloric restriction (CR) is the only paradigm shown to affect both median and maximum life spans of aging animals (2). It has been claimed that CR's life-prolonging effect is its efficient antioxidative property, with boosted defenses to resist to diseases and stress (7). The cellular and molecular mechanisms underpinning this resistive action against age-related oxidative stress has been reviewed (8,9). Unlike other interventions, CR exhibits an exceptionally broad protective action against oxidative threats, while maintaining counter-acting antioxidative defense systems to sustain a well-balanced redox state during aging (2). As described later, the basic mechanisms underlying this CR's antiaging action provides the basis for the anti-inflammatory action of CR (2,6).

Recently, based on what was learned from CR's potent anti-inflammatory action at the gene level, we proposed the molecular inflammatory process (see Fig. 1) as cross-talk that links normal aging with age-associated disease processes (3,4). In this chapter, we review the molecular processes implicated in the age-associated proinflammatory status that

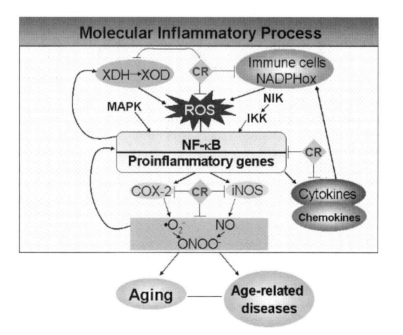

Figure 1 Schematic presentation of molecular inflammation hypothesis. XDH, xanthine dehydrogenase; XOD, xanthine oxidase; NADPHox, NADPH oxidase; ROS, reactive oxygen species; COX-2, cyclo-oxygenase-2; iNOS, inducible NO synthase; ONOO⁻, peroxynitrite; CR, calorie restriction.

bridges between normal biological aging and pathological processes. We begin with the involvement of the most common ROS/RNS in the molecular inflammatory and aging processes, the role of redox-responsive NF-κB, and the regulation of proinflammatory genes through cyclo-oxygenase (COX-2), inducible NO synthase (iNOS), and cytokines. Then, we describe how the modulation of these activities can be a major factor in mediating the anti-inflammatory action of CR.

II. BIOLOGICAL GENERATION SITES OF REACTIVE SPECIES

Although mitochondria are recognized as a major source of RS, they are generated for a variety of cellular sources, including oxidative enzymes, microsomes, peroxisomes, lysosomes, and plasma membranes (1,2). Most biological systems are vulnerable to oxidation due to the high oxidation potential of RS (2,10). Additional RS sources are lipo-oxygenase (LO), cyclo-oxygenase (COX), and NADPH oxidase in the plasma membrane; ubiquinone, NADH dehydrogenases, cytochrome P_{450}, cytochrome b5, and microsomal electron transport; flavoproteins and oxidases in peroxisome; and xanthine oxidase (XOD) in cytosol (5). However, quantitatively speaking, mitochondria and microsomes are two major sites for the bulk of intracellular ROS production (11,12).

One other important source of ROS production that was less appreciated until recently is the biosynthetic pathway of prostaglandins (PGs). They are arachidonic acid-derived lipid metabolites with potent proinflammatory and potentially pathogenic activities. Certain PG metabolites are important mediators of inflammation in their own right. However, as describe below, the ROS production from PG metabolism exacerbates inflammatory conditions and affects tissue damage (13,14).

Cyclo-oxygenase, a key enzyme in the PG synthetic pathway, converts arachidonic acid to prostaglandin H_2 (PGH$_2$) (15). During the conversion of PGG$_2$ to PGH$_2$ by COX, ROS are generated (16). The ROS production from the PG synthesis pathway can be significant for the overall ROS pool under both normal and pathological conditions, particularly during aging. For this regard, the activation of COX and its essential role are highlighted in the molecular inflammation hypothesis of aging (17). Although the increase of ROS reactions and their involvements in cell and tissue damage have received much attention lately, their involvement in the age-related inflammation has attracted less attention. In below, we describe COX-2, as a key enzyme that plays a central role in modulating the age-related proinflammatory process (Fig. 1).

III. INFLAMMATION CAUSED BY OXIDATIVE STRESS

III.A. The Roles of PGs and COX in the Aging Process

The exact status of COX in PG synthesis during aging has not been fully explored, although PG production in aged animals have been reported by several labs (18). Data clearly indicate the increased production of proinflammatory PGs, while showing cytoprotective PGs decreased with advancing age (19). For instance, macrophages from old mice are reported to produce more proinflammatory PGE_2 than young mice, which provides some basis for the age-related increased proinflammatory process (20).

The effect of age on PG synthetic activity was reported from Ridker's lab (21), showing that PG-derived ROS generation gradually increased with age in the heart, lung, kidney, and brain. In our lab, the extent of COX-derived ROS generated from PG during aging was assessed (22,23) utilizing the COX-specific inhibitor, indomethacin in rats. Results showed a steady increase of PG-dependent ROS generation with age, and at 24 months, a significant increase (187% higher than the 6-month-old rats) was noted (Fig. 2). To examine the involvement of COX, changes in COX activity during aging were monitored. Data showed COX activity increased with age: at 24 months of age, the activity was 87% higher than

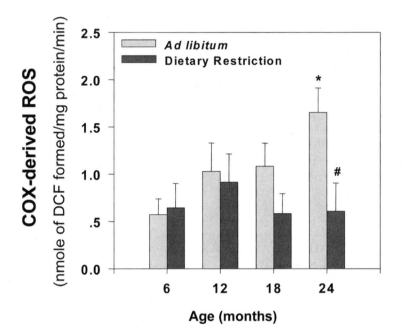

Figure 2 COX-derived ROS generation by aging and CR effect.

animals at 6 months of age. Moreover, the mRNA expression of COX-2 was shown to increase with age (17), setting a conducive condition for the age-associated inflammatory activation.

To examine the effect of oxidative stress on COX-2, Mn-superoxide dismutase (Mn-SOD) knockout mice were tested (22). Results showed that the heart postmitochondrial and mitochondrial fractions from these mice, whose antioxidative defenses are compromised, had ROS levels significantly higher than the wild-type mice. Furthermore, cardiac and renal mRNA levels of COX-2 in these knockout mice were markedly higher than the wild-type mice (22). All these results are consistent with the finding that ROS levels are closely linked with COX-2 activity during aging (23).

As substantiated by Feng et al. (20), oxidative stress is an important, specific inducer of COX-2 gene expression. These investigators found hydrogen peroxide and superoxide readily induce COX-2 expression, while the addition of radical scavengers suppresses IL-1β-, TNFα-, and LPS-induced COX-2 expression. These findings and others strongly support the notion that oxidatively activated COX-2 in aged animals is likely responsible for increased sensitivity to the proinflammatory insults (17).

III.B. Changes in XHD/XOD Status with Aging

Although much attention was given to XOD as a major source of ROS under ischemia/reperfusion conditions, XOD in relation to the aging process has received little attention. An earlier report on XOD in aging showed an age-related increase of XOD activity in rat liver (24). More recently, Janssen et al. (25) reported XOD and xanthine dehydrogenase (XDH) activity in myocardial tissue of Wistar rats and found a slight decrease in XDH activity during aging, and marked increase in XOD activity at 18 months compared to young rats. Thus, it would be interesting to know how the ratio of these two inter-convertible enzyme XOD/XDH alters during aging.

Evidence showed that the ratio of XOD/XDH changes with age in favor of XOD during aging as shown by the XOD activity that was significant at 24 months in kidney isolated from Fischer 344 rats (24,26). The biological consequence of increased XDH conversion to XOD signifies that the contribution of the XOD system to the overall oxidative stress status increases during aging (26). This increase was estimated to be approximately 26% of the total ROS generated by various sources in these rats. Thus, the age-related shift in XOD/XDH could be an important contributing factor to over-all age-related oxidative stress.

III.C. Status of NO and ONOO⁻ in Inflammatory Process

NO is synthesized enzymatically from L-arginine by NOS. Both constitutive and inducible isoforms have been characterized and differentiated by their dependence on Ca^{2+}/calmodulin. The expression of iNOS occurs in many cell types including macrophage, epithelial, and chondrocyte, in response to inflammatory and immunological stimuli, such as cytokines (27,28). Excessive production of NO by iNOS cause the pathogenesis and destruction of various tissues due to chronic inflammatory processes including those attributed to vascular diseases, diabetes, arthritis, dementia, transplant rejection, and irritable bowel syndrome (28,29).

The stimulatory ability of inflammation to induce iNOS to generate large amounts of NO, strongly implicates this enzyme as a major player in chronic inflammatory diseases. Cernadas et al. (30) reported aged vessel walls in rats with enhanced expression of eNOS and iNOS. Our laboratory, while investigating age-related changes in iNOS, found that iNOS gene expression increased in the kidney of lupus nephritis-prone mice during aging (31), and in a more recent work with Fischer 344 rats, increased iNOS gene expression in kidney with age was observed (4).

From the standpoint of the inflammation process, NO can be regarded as a proinflammatory factor, because of its enhanced activity during inflammatory conditions that caused increased vascular permeability. In addition, NO is a potent inducer of COX and the production of IL-1β, TNFα, and NADPH oxidase activity. In the presence of $^{\bullet}O_2^-$, NO combines with $^{\bullet}O_2^-$ to form peroxynitrite (ONOO$^-$), which is a nonradical, strong oxidant (32).

Recent evidence makes clear that NO and peroxynitrite bring about covalent modifications of cellular macromolecules, leading to various abnormal physiological and pathological consequences. For instance, modifications of proteins containing iron and thiols are the major targets of RNS. Research showed that iron (Fe^{2+} or Fe^{3+}) in hemes or iron–sulfur clusters can easily be targeted to become S-nitro modifications (33,34), and modifications of thiols at active sites can inhibit many enzymatic function (35–37). Furthermore, covalent modifications by nitration process of DNA, amino acids, and protein tyrosine residues are well documented to cause irreversible functional alterations, as seen in the altered signal transduction pathway (38).

Much of the involvement of NO in the inflammatory responses relates to its ability to form ONOO$^-$. During inflammatory responses, a series of events leads to the activation of residential macrophages and the recruitment of leukocytes from the circulation system to the injured sites with increased ONOO⁻. ONOO⁻ inhibits mitochondrial respiratory chain enzymes, oxidizes various proteins, and possibly triggers DNA strand breakage. Nitrated proteins by ONOO⁻ are detected widely in chronic inflammation, atherosclerotic lesions, coronary arteries, ischemia-reperfusion, shock, and cancer (39,40).

IV. STATUS OF NF-κB IN INFLAMMATION AND AGING

NF-κB was first described as a B-cell-specific factor that bound to a short DNA sequence motif located in the immunoglobulin k light chain enhancer, but it is now known that NF-κB is expressed in all cell types and plays a central role in gene transcription (41–43). Activation of NF-κ B and its dependent genes have been associated with various pathological processes including inflammatory conditions, radiation damage, atherosclerosis, cancer, and aging (41–43).

Because of its sensitivity to oxidative status, regulation of NF-κB is greatly influenced by the intracellular redox status. Under normal, unstimulated condition, NF-κB is held in the cytoplasm in an inactive state by the inhibitory subunit, IκB. With oxidative stimuli, the phosphorylation of IκB and the subsequent degradation of inhibitory IκB allow the translocation of NF-κB to the nucleus, and permit the binding of NF-κB to regulatory elements in DNA enhancers and promoters subsequent to gene expression (44). The high sensitivity to oxidative stress makes NF-κB as the key transcription factor in the regulation of inflammation processes during aging (45).

IV.A. Role of NF-κB in Molecular Inflammation

As mentioned earlier, NF-κB plays a key role in the expression of many genes that are important to the inflammatory response (43,46,47). Under challenged conditions, proinflammatory genes are activated and their encoded proteins are expressed. These newly synthesized proteins participate in organism's defense. Among them, cellular signaling proteins, such as cytokines, growth factors, or chemokines, are essential members of the defense systems. NF-κB is also known to regulate the transcription of tumor necrosis factor (TNFα and TNFβ), interleukins (IL-1β, IL-2, and IL-6), chemokines (IL-8 and RANTES), adhesion molecules (ICAM-1, VCAM, E-selectin), and enzymes like iNOS and COX-2 (48,49).

In general, the NF-κB activation in response to proinflammatory signals is short-lived, and the reaction stops quickly. However, when the signal persists, proinflammatory condition as in aging (see below for details) would occur. Some of the NF-κB-induced proteins are know to act as potent NF-κB activators that generate an auto-activating loop (50,51). In this case, more inflammatory mediators are synthesized and long-term inflammatory conditions would prevail.

IV.B. Altered NF-κB During Aging

Several lines of evidence show that increased levels of oxidative stress are the most important force accelerating the aging process (52–54). Many studies have explored whether aging also affects the regulation of the oxidative stress-responsive transcription factor NF-κB. Consensual findings show enhanced levels of NF-κB activity during aging in old rodents, as found in heart, liver, kidney, and brain tissues, and high NF-κB binding activity compared to the young rodents (55). Korhonen et al. also reported a significant upregulation of NF-κB in rat brain (53).

The upregulation of NF-κB activity accompanied by increased ROS production during aging was recently reported (23), which was effectively suppressed by CR via its antioxidative action (56). These results provided a strong experimental evidence supporting the notion that increased NF-κB activity in aged animals is likely due to a shift in intracellular redox balance during aging. As shown in Fig. 3, increased oxidative damage and the proinflammatory status are two most important occurrences observed during aging (3,4).

Much of the available evidence shows that NF-κB activators induce IκB phosphorylation and degradation by the activation of IκB kinase, triggering the activation of NF-κB.

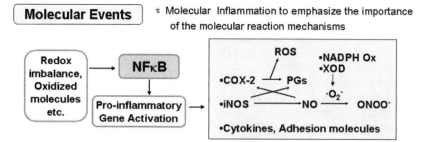

Figure 3 Molecular changes of inflammation.

Evidence was reported that increased NF-κB binding activity during aging is elicited through the phosphorylation of IκB kinase, causing a degradation of IκBα and IκBβ. Data on CR, in contrast, showed that CR inhibited IκB kinase activation, down-regulating NF-κB activation as evidenced by increased bound IκBα, and IκBβ proteins in cytoplasm (57) (Fig. 4). These observations led to conclusions that the antioxidative action of CR is likely a major force responsible for the maintenance of properly functioning NF-κB/IKK/NIK and mitogen-activated protein kinases (MAPKs), which might be associated with CR's life-prolonging action (57,58).

V. REGULATION OF NF-κB ACTIVATION BY IKK/NIK AND MAPK

V.A. IKK/NIK

Recent efforts investigating the regulation of NF-κB have focused on the phosphorylation of IκBs. So far, two closely related kinases, IκB kinase α (IKKα) and IKKβ, have been identified as a multi-protein 700 kDa IKK complex (59–63). Molecular probing indicates that the predominant form of IKK is an IKKα/IKKβ heterodimer through their COOH-terminal leucine zipper motifs, and that this dimer is associated with the regulatory subunit, IKKγ (64). In response to stimuli, upstream kinases are activated and recruited to the complex via IKKγ, resulting in the phosphorylation of IKKβ, i.e., activation of IKK. The activated IKK complexes

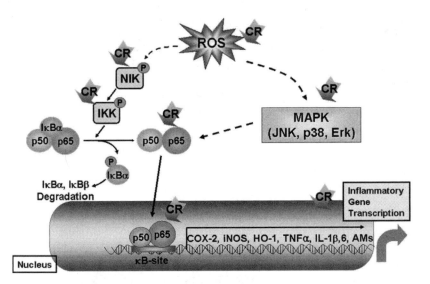

Figure 4   Modulation of NF-κB activation by age and CR. ROS, reactive oxygen species; NF-κB, nuclear factor kappaB; AP-1, activator protein-1; IKK, IkB kinase; MAPK, mitogen-activated protein kinase; Erk, extracellular signal-related kinase; JNK, c-Jun N-terminal kinases; CR, calorie restriction.

phosphorylate IκB subunits of NF-κB/IκB to trigger the degradation of IκB and the activation of NF-κB (65,66). A previous report revealed that the upregulation of NF-κB activity is elicited through the enhanced degradation of IκBα and IκBβ induced from phosphorylation of IKK during aging (57). Furthermore, data clearly showed CR's ability to suppress age-dependent increases of NF-κB via the upregulation of cytoplasmic IκBα and IκBβ by inhibiting IKK activity.

Because the phosphorylation of IκB is an essential step in the NF-κB activation process, the age-related status of IKK and its modulation by CR were examined (57). Data show that IKK activity increased with age and this increase was suppressed by CR (Fig. 4), which was the first report on IKK modulation by age and CR. The IKK complex is an important convergence site of signals elicited by various stimuli to activate NF-κB, showing the major role IKKβ plays in responding to proinflammatory stimuli (65,67).

Previous studies indicate a link between the exquisitely sensitive nature of the NF-κB pathway and the pro-oxidant status (68,69). The identification of IKK as an upstream redox sensor of NF-κB activation was reported by Chen et al. (68). These investigators demonstrated that IKKβ activity is significantly elevated in macrophages exposed to the pro-oxidant, vanadate. Korn et al. (69) also demonstrated that H_2O_2 is capable of directly inactivating IKK because of the redox sensitivity of the complex.

Studies exploring the regulation of NF-κB found that the nuclear DNA-binding activity of the transcription factor is strongly enhanced during aging in heart, liver, and brain tissues (55,53). Thus, upregulated NF-κB activity seen in aged animals is significant because NF-κB is a critical transcription factor participating in the pathogenesis of many age-related disorders (70,71). Further exploration of the underpinning mechanisms of NF-κB in age-related dysfunctions and the identification of specific NF-κB modulators could provide further molecular insights into the altered cellular signal pathways in aged organisms and in age-related diseases.

A knockout mice model provided useful information to gain molecular insights into the role of NF-κB in a normal physiological condition. Knockout models have been created for RelA, IκBα, p50, c-Rel, and RelB (72–74). Evidence showed that deficiencies of p50, c-Rel, or RelB resulted in developmentally normal mice and that a RelA deficiency resulted in embryonic lethality from liver apoptosis (72). Further, IκBα-deficient mice were apparently normal at birth, but during the postnatal period, their growth ceased and they died at 7–10 day of age (72,74). These studies suggest that NF-κB is an indispensable factor for normal physiological function, yet its unregulated chronic activation can be devastating.

V.B. MAPKs

All eukaryotic cells possess multiple MAPK pathways, for each of stimuli, thereby allowing the cell to respond to multiple, divergent input in a well-coordinated manner. Recent evidence led to conclusions that free radicals are closely implicated in MAPK activation (75–77). Mitogen-activated protein kinases are serine/threonine-specific, proline-directed protein kinases, enzymes that are demonstrated to play a central role in transducing extracellular signals to the nucleus (78). As part of the kinase cascade, MAPKs serve as information relays that connect extracellular stimuli with specific transcription factors, thereby allowing extracellular signals to regulate specific gene expression intracellularly.

Mitogen-activated protein kinases have three subtypes: extracellular signal regulated kinase (ERK), c-Jun N-terminal kinase (JNK), and p38 MAPK. Current evidence indicates that ERK is predominantly stimulated by mitogens and that JNK and p38 MAPK are a part of the stress response pathways activated by cellular stress induced by agents such as heat, UV and ionizing radiation, and inflammatory cytokines that mediate the inhibition of cell proliferation or cell death (79,80). Fas/APO-1/CD95 receptor (FasR) activates all major signaling pathways that belong to the family of the MAPK pathway, by either caspase-independent or -dependent mechanisms. Furthermore, phosphorylation-based signaling serves as a potent modifier of FasR responses (81).

The effects of aging on cellular signaling pathways and protein kinases, as well as their patho-physiological implications in age-related diseases have been investigated (76,77, 82). Several investigations showed that impaired ERK activation in response to stimulation by epidermal growth factor in hepatocytes from aged rats (83) and in T lymphocytes obtained from elderly subjects (84), indicating that this signaling kinase system is compromised during aging. In addition, reduced ERK and p38 MAPK activities have been reported in brains of aged animals (85). However, alterations in renal ERK, JNK, and p38 MAPK by aging and by the antiaging action of CR have not previously been reported (6).

Age-related activation of MAPK with a corresponding increase in ROS was consistently reduced by CR at all ages studied (58). Because of the redox sensitivity of transcription factors, increased ROS during aging activates MAPKs, and the activation process is blocked by the antioxidative action of CR (58).

The importance of the involvement of MAPK family members in the regulation of various cellular responses as seen in inflammation is increasingly appreciated in recent years. For instance, it is now known that TNFα can induce rapid and transient activation of ERK, JNK, and p38 MAPK (86,87). Further, ERK, JNK, and p38 MAPK pathways are reported to control the transcription and synthesis of TNFα (88) (Fig. 3). In the latter case, the situation becomes more complicated because MAPKs regulate multiple transcription factors including NF-κB and AP-1.

NF-κB activation also may cause coordinated expression of the genes for several cytokines, chemokines, enzymes, and adhesion molecules that are essential to the inflammatory response (89). In addition, AP-1 activation may have relevance to tissue repair or disease development such as tumor promotion (90). Thus, it is expected that a slight change in the level in MAPK activity could produce widespread changes in the cellular responses as seen in persistent MAPK activation, leading to chronic inflammatory conditions (91).

A recent study showed that AP-1 binding activity markedly increased with age (92). This finding indicated that age-induced AP-1 activation is closely related to MAPK activation and aged rats showed persistently increased MAPK activity. Thus, aged animals become more susceptible to various age-related diseases because of their proinflammatory state, as hypothesized in a recent publication (6).

It is now well documented that MAPKs' activity and the regulation of their activation by the antioxidative action of CR that are under the influence of increased oxidative stress during aging can be modified. Currently, available data indicate not only a correlation between increased age-related MAPK activities (ERK, JNK, and p38 MAPK) and the nuclear translocation process, but also with their modulation by the anti-inflammatory action of CR. At present, the precise mechanism underlying the activation of MAPKs remains to be elucidated in relation to cellular redox imbalance during aging.

However, contradicting data were also reported on the changes in MAPK activity during aging (85,93–95). For instance, ERK and p38 MAPK activities were reported to be impaired in aged rat brain (85). Other studies found a reduced MAPK pathway in isolated T cell and splenic lymphocyte (93,94). It was found (95) that JNK1/2 is increased in old, compared with young, human skin in vivo, which is consistent with data on age-related increase of MAPK activity. Thus, it appears that the aging effect on MAPK activity may be cell- and tissue-specific.

Available evidence also strongly implicate the involvement of aberrant MAPK activation in various age-associated disorders (72,76,83,96–99). For instance, Xhu et al. reported that JNK is activated and redistributed in the degenerating neurons of Alzheimer's disease (AD), suggesting that JNK dysregulation, probably resulting from oxidative stress, plays an important role in the increased phosphorylation of cytoskeletal proteins found in AD (100). Gupta et al. reported that increased ROS levels contribute to MAPK activity in a malignantly progressed mouse keratinocyte cell line (101). Peters et al. suggested the involvement of MAPK activation in ROS-induced endothelial dysfunction and decreased contractile function (102).

In summary, it should be noted that the chronic activation of MAPKs, as seen in chronic inflammation of old animals, likely leads to age-dependent functional alterations (17,58). Accumulating data showed the age-related activation of ERK, JNK, and p38 MAPK in aged tissues under increased oxidative stress, and demonstrated the ability of CR to prevent these activation processes (58). The significance of normal and selective modulation of MAPKs in response to cellular stress by CR may well be related to the organism's defense mechanism of the organism for survival (6).

VI. BASES FOR MOLECULAR INFLAMMATION IN AGING

Evidence showing CR's anti-inflammatory action on the suppression of COX-derived ROS generation in the presence of indomethacin is of importance (17,23). Data showed that CR suppresses COX-derived ROS generation from 26.36% in AL to 16.55% in the CR group, and concomitantly blunted COX activity and the production of TXA_2 and PGI_2. These findings provide additional evidence to support what is known about CR's anti-inflammatory and antioxidative actions at molecular levels.

The beneficial effects of CR can be further extended to its ability to regulate gene expression as evidenced by the modulation of COX-2 mRNA and protein levels by manipulating NF-κB. Other proinflammatory proteins, such as, IL-1β, IL-6, TNFα, and iNOS, are all similarly modulated by CR (4), and are summarized in Table 1.

There is additional involvement of the ERK, JNK, and p38 MAPK pathways that control essential NF-κB dependent transcription during the inflammatory response. Recently, the aging process was shown to strongly enhanced all three ERK, JNK, and p38 of the MAPK activities studied which was parallel to increased ROS status. In contrast, CR markedly suppressed the age-related activation of MAPKs (58). Based on these and other available data, a molecular inflammation hypothesis of aging was proposed (see schematic presentation in Fig. 1).

VII. SUPPORTING EVIDENCE FOR THE MOLECULAR INFLAMMATION HYPOTHESIS

VII.A. Anti-Inflammatory Action of CR

Caloric restriction remains the only known robust means of increasing maximum life span and delaying age-related physiological changes, and the onset of a wide range of age-related diseases in rodents (4,8,9). Characteristics of CR's mode of action on the retardation of aging are well documented (2). The enhanced metabolic efficiency that CR produces and its ability to increase stress resistance over the lifespan have been hypothesized to be at the core of CR's antiaging effects (8,9). Two recent publications on CR based on genomic profiling support this view (102,104). Accumulating evidence

Table 1 Changes of Proinflammatory Factors During Aging

		Inflammatory process	Aging process	CR
Redox state	Reactive oxygen species	↑	↑	⊢
	Reactive nitrogen species	↑	↑	⊢
	Catalase, superoxide dismutase	↓	↓	⊢
	GSH peroxidase, GSH/GSSG	↓	↓	⊢
Proinflammatory enzymes	Inducible NO synthase	↑	↑	⊢
	Heme oxygenase-1	↑	↑	⊢
	Cyclo-oxygenase-2	↑	↑	⊢
	Conversion of xanthine dehydrogease to xanthine oxidase	↑	↑	⊢
Proinflammatory cytokines	IL-1β	↑	↑	⊢
	IL-6	↑	↑	⊢
	TNF-α	↑	↑	⊢
NF-κB activation	NF-κB DNA binding activity	↑	↑	⊢
	NIK/IKK activation	↑	↑	⊢
	Phosphorylation of IκBκ	↑	↑	⊢
	Degradation of IκBα and IκBβ in cytoplasm	↑	↑	⊢
	Nuclear translocation of p65 and p50	↑	↑	⊢
	NF-κB-dependent gene expression	↑	↑	⊢
	Active MAPKs (ERK, JNK, p38 MAPK)	↑	↑	⊢
Adhesion molecules	E-selectin	↑	↑	⊢
	P-selectin	↑	↑	⊢
	VCAM-1	↑	↑	⊢
	ICAM-1	↑	↑	⊢
Hypoxic markers	HIF-1α	↑	↑	⊢
	VEGF	↑	↑	⊢
	EPO	↑	↑	⊢

↑, Increased; ↓, decreased; ⊢, blunted.

indicates that CR protects homeostatic integrity by prioritizing energy allocation to increase the resistance capacity against both intrinsic and extrinsic stresses (9).

Among its diverse effects, the anti-inflammatory action of CR provides interesting molecular insights (2,3) (Figs. 1 and 3). Recent reports indicate that the brains of CR animals are protected from the oxidative stress of ischemia/reperfusion (I/R) (105) as well as neurotoxic damage (106). The resolution of foot-pad edema after carrageenan injection (a common model for of inflammatory study) is much more rapid for CR mice than for ad libitum-fed mice (107). Kari et al. (108) reported that CR mitigates ozone-induced lung inflammation in rats, in part, by increasing pulmonary storage of ascorbate. More recently, Chandrasekar et al. (109) reported that a moderately calorie restricted diet attenuates oxidative stress and damage during cardiac I/R injury as evidenced by more rapid recovery of GSH levels, attenuated expression of proinflammatory cytokines, and a rapid response by the free-radical scavenging system.

Chung et al. (4) found that CR suppresses the production of the proinflammatory cytokines, TNFα, IL-1β, and IL-6. In another study using alveolar macrophage, CR enhanced phagocytic activity and improved resistance to a challenge of Gram-positive bacteria, while suppressing the production of inflammatory cytokines, such as TNFα and IL-6 (110). Spaulding et al. (111) reported that the levels of the LPS-induced release of IL-1β, IL-6, and TNFα were significantly lower in the CR mice compared with controls. Consistent with decreased inflammatory response gene expression from CR, Cherry et al. (112) reported a delayed onset and diminished severity of autoimmune, inflammatory disease in mice.

As mentioned earlier, the excessive and/or unregulated production of NO by iNOS is known to cause a widespread tissue injury including inflammation. Macrophages, well-known sources of NO, generate large amounts of NO upon exposure to cytokines or LPS (113,114). However, the effective suppression of NO production in alveolar macrophage by CR is significant for both baseline and LPS-stimulated levels, clearly exhibiting CR's anti-inflammatory action (110).

A recent study of Cao et al. (103) using genomic profiling
to study short- and long-term CR effects on aging mice liver
provides molecular information on CR's anti-inflammatory
action. In their study, both long-term (until 27 months after
weaning) and short-term [4 weeks restriction] suppressed
the age-related increase of inflammatory response genes.
Similar findings on CR were reported earlier by the suppression of age-related induction of stress response genes, as determined by the genomic profiling of muscle (104). Further
support for genomic changes by CR in aged animals was documented in a recent publication on neocortex and cerebellum
(115). Thus, resistance to stress and inflammatory responses
seems to be a part of major strategies supporting the antiaging
effect of CR.

VII.B. Regulation of Proinflammatory Transcription Factors by CR

Characteristics of CR's action for the aging retardation are its
diverse antioxidative effects (2). Our laboratory was among
the first to document evidence on CR's attenuation of the age-
related activation of the redox-sensitive transcription factors,
NF-κB, AP-1, and hypoxia-inducible factor-1 (HIF-1). A preliminary study revealed that the age-related up-regulation of
NF-κB, AP-1, and HIF-1 activities can be blunted by CR's
antioxidative action (116). Because phosphorylation of IκB
is an essential step in the NF-κB activation process, we investigated the status of the key IKK, during aging and its modulation by CR were investigated, and found evidence
showing that IKK increased with age and CR that suppressed
this increase. Due to the sensitive nature of the NF-κB pathway to oxidants, the activation of IKK may well be related closely to the age-related increase of RS, although no data
detailing on molecular regulation are available at this time.

Another redox-sensitive transcription factor, AP-1, acted
similarly to NF-κB, i.e., increased with age, but was blunted
by CR. The regulation of AP-1 activity occurs at two separate,
but interrelated levels, namely transcriptional and post-
translational regulation (117), and most AP-1 proteins are

generally regulated by the control of gene transcription levels (117). The c-Jun and c-Fos genes are expressed in many cell types at low levels, but their expression is rapidly elevated when responding to various stimuli (118). Phosphorylation of c-Jun at Ser-73 and Ser-63, and c-Fos at The-232, located within its transactivation domain, potentiates their transcriptional activities (119).

It has been suggested that the age-related increases of nuclear c-Jun and c-Fox may be major contributing factors to increased AP-1 binding and phosphorlylation of c-Jun, which in turn leads to the further increase of AP-1 transcriptional activity. Caloric restriction's action was to prevent AP-1 activation in the nucleus of aged rat kidney by both suppressing the nuclear levels of c-Jun and c-Fos, and inhibiting the phorphorylation of the c-Jun protein (116).

Hypoxia-inducible factor-1, another redox-related transcription factor, known to increase during aging, is suppressed by CR. Interestingly, the HIF-1 protein increases without changes in its mRNA amount, indicating that HIF-1 protein stabilization is regulated likely by a post-translational modification process (116). The signaling pathway regulating HIF-1 activity contains several intermediates, such as PI_3K/Akt and Erk, which are known to be involved in the activation of HIF-1 during hypoxia. In addition, these factors also have been implicated in the regulation of HIF-1, even in normoxic cells (120,121).

Current evidence suggests that age-related activation of the Raf-1/MEK1/Erk pathway, but not the PI_3K/Akt pathway, might contribute to age-related HIF-1 activation. Furthermore, studies also suggest that increased nitric oxide in aged animals could enhance HIF-1 stabilization via nitrosylation (122).

If aging is the result of a long-term cellular oxidative imbalance as proposed in oxidative stress hypothesis of aging (1), then it is likely that the antioxidative action of CR is the best antiaging intervention. Several studies have reached the conclusion that the increased activities of NF-κB, AP-1, and HIF-1 in aged rats were due to a shift in intracellular redox balance during aging (57,92,116), and that the inappropriately activated these transcription factors during aging can set the stage for various age-related pathogenic incidences. This is why CR can be an effective intervention of many age-related diseases by controlling the activity of NF-κB, AP-1, and HIF-1.

VIII. CONCLUSION

Existing evidences are strongly supportive of molecular inflammation as a major biological alteration underpinning the aging process and age-related diseases. The key players in the inflammatory reaction are the age-related upregulation of NF-κB, IL-1β, IL-6, TNFα, COX-2, and iNOS. Furthermore, NF-κB occupied a central position in the inflammatory process that is activated by phosphorylation by IKK/NIK and MAPKs, all of which are modulated by CR. It is hoped that the molecular inflammation hypothesis proposal will provide further insights into a better understanding of the aging process and age-related diseases. The current view regarding major chronic diseases such as atherosclerosis, arthritis, cancer, dementia, and vascular diseases as "inflammatory diseases" is in line with what was proposal for the molecular inflammation hypothesis (6).

REFERENCES

1. Yu BP, Yang R. Critical evaluation of the free radical theory of aging. A proposal for the oxidative stress hypothesis. Ann NY Acad Sci 1996; 786:1–11.

2. Yu BP. Aging and oxidative stress: modulation by dietary restriction. Free Radic Biol Med 1996; 21:651–668.

3. Chung HY, Kim HJ, Jung KJ, Yoon JS, Yoo MA, Kim KW, Yu BP. The inflammatory process in aging. Rev Clin Gerontol 2000; 10:207–222.

4. Chung HY, Kim HJ, Kim JW, Yu BP. The inflammation hypothesis of aging: molecular modulation by calorie restriction. Ann NY Acad Sci 2001; 928:327–335.

5. Bodamyali T, Stevens CR, Blake DR, Winyard PG. Reactive oxygen/nitrogen species and acute inflammation: a physiological process. In: Winyard PG, Blake DR, Evans CH, eds. Free Radicals and Inflammation. Basel: Birkhäuser Verlag, 2000:11–16.

6. Chung HY, Kim HJ, Kim KW, Choi JS, Yu BP. Molecular inflammation hypothesis of aging based on the anti-aging mechanism of calorie restriction. Microsc Res Tech 2002; 59: 264–272.

7. Frame LT, Hart RW, Leakey JE. Caloric restriction as a mechanism mediating resistance to environmental disease. Environ Health Perspect 1998; 106:313–324.

8. Yu BP, Chung HY. Oxidative stress and vascular aging. Diabetes Res Clin Pract 2001; 54:S73–S80.

9. Yu BP, Chung HY. Stress resistance of calorie restriction for longevity. Ann NY Acad Sci 2001; 928:39–47.

10. Halliwell B, Gutteridge JMC. Free Radicals in Biology and Medicine. 2nd ed. London: Oxford University Press, 1999.

11. Kehrer JP, Lund LG. Cellular reducing equivalents and oxidative stress. Free Radic Biol Med 1994; 17:65–75.

12. Cadenas E, Davies KJ. Mitochondrial free radical generation, oxidative stress, and aging. Free Radic Biol Med 2000; 29: 222–230.

13. Kim JW, Baek BS, Kim YK, Herlihy JT, Ikeno Y, Yu BP, Chung HY. Gene expression of cyclooxygenases in the aging heart. J Gerontol 2001; 56:B350–B355.

14. Baek BS, Kwon HJ, Lee KH, Yoo MA, Kim KW, Ikeno Y, Yu BP, Chung HY. Regional difference of ROS generation, lipid peroxidation, and antioxidant enzyme activity in rat brain and their dietary modulation. Arch Pharm Res 1999; 22:361–366.

15. Ikai K, Ujihara M, Urade Y. Changes of the activities of enzymes involved in prostaglandin synthesis in rat skin during development and aging. Arch Dermatol Res 1989; 281: 433–436.

16. Bazan NG. Metabolism of arachidonic acid in the retina and retinal pigment epithelium: biological effects of oxygenated

metabolites of arachidonic acid. Prog Clin Biol Res 1989; 312: 15–37.

17. Chung HY, Kim HJ, Shim KH, Kim KW. Dietary modulation of prostanoid synthesis in the aging process: role of cyclooxygenase-2. Mech Ageing Dev 1999; 111:97–106.

18. Adelizzi RA. COX-1 and COX-2 in health and disease. J Am Osteopath Assoc 1999; 99:S7–S12.

19. Choi JH, Yu BP. Dietary restriction as a modulator of age-related changes in rat kidney prostaglandin production. J Nutr Health Aging 1998; 2:167–171.

20. Feng L, Xia Y, Garcia GE, Hwang D, Wilson CB. Involvement of reactive oxygen intermediates in cyclooxygenase-2 expression induced by interleukin-1 tumor necrosis factor-alpha, and lipopolysaccharide. J Clin Invest 1995; 95:1669–1675.

21. Ridker PM. Inflammation, infection, and cardiovascular risk: how good is the clinical evidence? Circulation 1998; 97: 1671–1674.

22. Kim JW. Attenuation of free radical generation related to cardiac arachidonate cascade by dietary restriction during aging. MS. Thesis, Pusan National University, Busan, Korea, 1998.

23. Kim HJ, Kim KW, Yu BP, Chung HY. The effect of age on cyclooxygenases-2 gene expression: NF-κB activation and IκBα degradation. Free Radic Biol Med 2000; 28:683–692.

24. Chung HY, Yu BP. Change of the rat xanthine oxidase/ dehydrogenase and uric acid formation by aging and dietary restriction in the liver. J Am Aging Assoc 2000; 24:161–169.

25. Janssen M, Tavenier M, Koster JF, de Jong JW. In vitro and ex vivo xanthine oxidoreductase activity in rat and guinea-pig hearts using hypoxanthine or xanthine as substrate. Biochim Biophys Acta 1993; 1156:307–312.

26. Chung HY, Song SH, Kim HJ, Ikeno Y, Yu BP. Modulation of renal xanthine oxidoreductase in aging: gene expression and reactive oxygen species generation. J Nutr Health Aging 1999; 3:19–23.

27. Vernet D, Bonavera JJ, Swerdloff RS, Gonzalez-Cadavid NF, Wang C. Spontaneous expression of inducible nitric oxide synthase in the hypothalamus and other brain regions of aging rats. Endocrinology 1998; 139:3254–3261.

28. Clancy RM, Abramson SB. Nitric oxide: novel mediator of inflammation. Proc Soc Exp Biol Med 1995; 210:93–101.

29. McInnes IB, Leung BP, Field M, Wei XQ, Huang FP, Sturrock RD, Kinninmonth A, Weidner J, Mumford R, Liew FY. Production of nitric oxide in the synovial membrane of rheumatoid and osteoarthritis patients. J Exp Med 1996; 184: 1519–1524.

30. Cernadas MR, Sanchez de Miguel L, Garcia-Duran M, Gonzalez-Fernandez F, Millas I, Monton M, Rodrigo J, Rico L, Fernandez P, de Frutos T, Rodriguez-Feo JA, Guerra J, Caramelo C, Casado S, Lopez-Farre A. Expression of constitutive and inducible nitric oxide synthases in the vascular wall of young and aging rats. Circ Res 1998; 83:279–286.

31. Kim YJ, Yokozawa T, Chung HY. Effects of energy restriction and fish oil supplementation on renal guanidino levels and antioxidant defenses in aged lupusprone B/W mice. Br J Nutr 2005 (in press).

32. Stampler JS, Single D, Loscalzo J. Biochemistry of nitric oxide and its redox-activated forms. Science 1992; 258:1898–1902.

33. Ignarro LJ. Signal transduction mechanism involving nitric oxide. Biochem Pharmacol 1991; 41:485–490.

34. Bredt DS, Synder SH. Nitric oxide: a physiologic messenger molecule. Annu Rev Biochem 1994; 63:175–195.

35. Stampler JS. Redox signaling: nitrosylation and related target interactions of nitric oxide. Cell 1994; 78:931–936.

36. Lander HM, Milbank AJ, Tauras JM, Hajjar DP, Hempstead BL, Schwartz GD, Kraemer RT, Mirza UA, Chait BT, Burk SC, Quilliam LA. Redox regulation of cell signalling. Nature 1996; 381:380–381.

37. Xu L, Eu JP, Meissner G, Stampler JS. Activation of cardiac calcium release channel (ryanodine receptor) by poly-S-nitrosylation. Science 1998; 279:234–237.

38. Deora AA, Lander HM. Regulation of signal transduction and gene expression by reactive nitrogen species. In: Sen CK, Sies H, Baeuerle PA, eds. Antioxidant and Redox Regulation of Genes. San Diego: Academic Press, 2000:147–178.

39. Weinberg JB. Nitric oxide as an inflammatory mediator in autoimmune MRL-1pr/1pr mice. Environ Health Perspect 1998; 106:1131–1137.

40. Smith MA, Richey Harris PL, Sayre LM, Beckman JS, Perry G. Widespread peroxynitrite-mediated damage in Alzheimer's disease. J Neurosci 1997; 17:2653–2657.

41. Baeuerle PA, Baltimore D. NF-κB: ten years after. Cell 1996; 87:13–20.

42. Baeuerle PA, Henkel T. Function and activation of NF-κB in the immune system. Annu Rev Immunol 1994; 12:141–179.

43. Baldwin AS Jr. The NF-κB and IκB proteins: new discoveries and insights. Annu Rev Immunol 1996; 14:649–683.

44. Bowie A, O'Neill LA. Oxidative stress and nuclear factor-kappaB activation: a reassessment of the evidence in the light of recent discoveries. Biochem Pharmacol 2000; 59:13–23.

45. Kwon HJ, Sung BK, Kim JW, Lee JH, Kim ND, Yoo MA, Kang HS, Baek HS, Bae SJ, Choi JS, Takahashi R, Goto S, Chung HY. The effect of lipopolysaccharide on enhanced inflammatory process with age: modulation of NF-κB. J Am Aging Assoc 2001; 24:161–169.

46. Baeuerle PA, Baichwal VR. NF-κB as a frequent target for immunosuppressive and anti-inflammatory molecules. Adv Immunol 1997; 65:111–137.

47. Manning AM, Bell FP, Rosenbloom CL, Chosay JG, Simmons CA, Northrup JL, Shebuski RJ, Dunn CJ, Anderson DC. NF-κB is activated during acute inflammation in vivo in association with elevated endothelial cell adhesion molecule gene expression and leukocyte recruitment. J Inflamm 1995; 45: 283–296.

48. Brand K, Page S, Rogler G, Bartsch A, Brandl R, Knuechel R, Page M, Kaltschmidt C, Baeuerle PA, Neumeier D. Activated transcription factor nuclear factor-kappa B is present in the atherosclerotic lesion. J Clin Invest 1996; 97:1715–1722.

49. Bohrer H, Qiu F, Zimmermann T, Zhang Y, Jllmer T, Mannel D, Bottiger BW, Stern DM, Waldherr R, Saeger HD, Ziegler R, Bierhaus A, Martin E, Nawroth PP. Role of NFκB in the mortality of sepsis. J Clin Invest 1997; 100:972–985.

50. Handel ML, McMorrow LB, Gravallese EM. Nuclear factor-kappa B in rheumatoid synovium. Localization of p50 and p65. Arthritis Rheum 1995; 38:1762–1770.

51. Fisher GJ, Datta SC, Talwar HS, Wang ZQ, Varani J, Kang S, Voorhees JJ. Molecular basis of sun-induced premature skin ageing and retinoid antagonism. Nature 1996; 379:335–339.

52. Helenius M, Hanninen M, Lehtinen SK, Salminen A. Aging induced up-regulation of nuclear binding activities of oxidative stress responsive NF-κB transcription factor in mouse cardiac muscle. J Mol Cell Cardiol 1996; 28:487–498.

53. Korhonen P, Helenius M, Salminen A. Age-related changes in the regulation of NF-κB in rat brain. Neurosci Lett 1997; 225:61–64.

54. Menzel EJ, Sobal G, Staudinger A. The role of oxidative stress in the long-term glycation of LDL. Biofactors 1997; 6: 111–124.

55. Helenius M, Hanninen M, Lehtinen SK, Salminen A. Changes associated with aging and replicative senescence in the regulation of transcription factor nuclear factor-kappa B. Biochem J 1996; 318:603–608.

56. Kim HJ, Jung KJ, Yu BP, Cho CG, Choi JS, Chung HY. Modulation of redox-sensitive transcription factors by calorie restriction during aging. Mech Ageing Dev 2002; 123: 1589–1595.

57. Kim HJ, Yu BP, Chung HY. Molecular exploration of age related NF-κB/IKK downregulation by calorie restriction in rat kidney. Free Radic Biol Med 2002; 10:991–1005.

58. Kim HJ, Jung KJ, Yu BP, Cho CG, Chung HY. Influence of aging and calorie restriction on MAPKs activity in rat kidney. Exp Gerontol 2002; 37:1041–1053.

59. DiDonato JA, Hayakawa M, Rothwarf DM, Zandi E, Karin M. A cytokine responsive IκB kinase that activates the transcription factor NF-κB. Nature 1997; 388:548–554.

60. Mercurio F, Zhu H, Murray BW, Shevchenko A, Bennett BL, Li J, Young DB, Barbosa M, Mann M, Manning A, Rao A. IKK-1 and IKK-2: cytokine-activated IκB kinases essential for NF-κB activation. Science 1997; 278:860–866.

61. Zandi E, Rothwarf DM, Delhase M, Hayakawa M, Karin M. The IκB kinase complex (IKK) contains two kinase subunits, IKKα and IKKβ, necessary for IκB phosphorylation and NF-κB activation. Cell 1997; 91:243–252.

62. Woronicz JD, Gao X, Cao Z, Rothe M, Goeddel DV. IκB kinase-β: NF-κB activation and complex formation with IκB kinase-α and NIK. Science 1997; 278:866–869.

63. Regnier CH, Song HY, Gao X, Goeddel DV, Cao Z, Rothe M. Identification and characterization of an IκB kinase. Cell 1997; 90:373–383.

64. Rothwarf DM, Zandi E, Natoli G, Karin M. IKKgamma is an essential regulatory subunit of the IκB kinase complex. Nature 1998; 395:297–301.

65. Delhase M, Hayakawa M, Chen Y, Karin M. Positive and negative regulation of IκB kinase activity through IKKβ subunit phosphorylation. Science 1999; 284:309–313.

66. Karin M. The beginning of the end: IκB kinase (IKK) and NF-κB activation. J Biol Chem 1999; 274:27339–27342.

67. Marshall HE, Merchant K, Stamler JS. Nitrosation and oxidation in the regulation of gene expression. FASEB J 2000; 14:1889–1900.

68. Chen F, Demers LM, Vallyathan V, Ding M, Lu Y, Castranova V, Shi X. Vanadate induction of NF-κB involves IκB kinase β and SAPK/ERK kinase 1 in macrophages. J Biol Chem 1999; 274:20307–20312.

69. Korn SH, Wouters EF, Vos N, Janssen-Heininger YM. Cytokine-induced activation of nuclear factor-κB is inhibited by hydrogen peroxide through oxidative inactivation of IκB kinase. J Biol Chem 2001; 276:35693–35700.

70. Lum H, Roebuck KA. Oxidant stress and endothelial cell dysfunction. Am J Physiol Cell Physiol 2001; 280:719–741.

71. Fukagawa NK. Aging: is oxidative stress a marker or is it causal? Proc Soc Exp Biol Med 1999; 222:293–298.

72. Beg AA, Sha WC, Bronson RT, Ghosh S, Baltimore D. Embryonic lethality and liver degeneration in mice lacking the RelA component of NF-κB. Nature 1995; 376:167–170.

73. Beg AA, Sha WC, Bronson RT, Baltimore D. Constitutive NF-κB activation, enhanced granulopoiesis, and neonatal lethality in IκBα-deficient mice. Genes Dev 1995; 9:2736–2746.

74. Klement JF, Ricc NR, Car BD, Abbondanzo SJ, Powers GD, Bhatt H, Chen CH, Rosen CA, Stewart CL. IκBα deficiency results in a sustained NF-κB response and severe widespread dermatitis in mice. Mol Cell Biol 1996; 16:2341–2349.

75. Irani K, Xia Y, Zweier JL, Sollott SJ, Der CJ, Fearson ER, Sundaresan M, Finkel T, Goldschmidt-Clermont PJ. Mitogenic signaling mediated by oxidants in ras-transformed fibroblasts. Science 1997; 275:1649–1652.

76. Ambrecht JJ, Nemani RK, Wongsurawat N. Protein phosphorylation changes with age-related diseases. J Am Geriatr Soc 1993; 41:873–879.

77. Quadri RA, Plastre O, Phelouzat MA, Arbogast A, Proust JJ. Age-related tyrosine-specific protein phosphorylation defect in human T lymphocytes activated through CD3, CD4, CD8, or IL-2 receptor. Mech Ageing Dev 1996; 88:125–138.

78. Su B, Karin M. Mitogen-activated protein kinase cascades and regulation of gene expression. Curr Opin Immunol 1996; 8:402–411.

79. Xia Z, Dickens M, Raingeaud J, Davis RJ, Greenberg ME. Opposing effects of ERK and JNK-p38 MAP kinases on apoptosis. Science 1995; 270:1326–1331.

80. Wang X, Martindale JL, Liu Y, Holbrook NJ. The cellular response to oxidative stress: Influences of mitogen activated protein kinase signaling pathways on cell survival. Biochem J 1998; 333:291–300.

81. Holmstrom TH, Eriksson JE. Phosphorylation-based signaling in Fas receptor-mediated apoptosis. Crit Rev Immunol 2000; 20:121–152.

82. Pisano M, Wang HY, Friedman E. Protein kinase activity changes in the aging brain. Biomed Env Sci 1991; 4:173–181.

83. Liu Y, Guyton KZ, Gorospe M, Xu Q, Kokkonen GC, Mock YD, Roth GS, Holbrook NJ. Age-related decline in mitogen-activated protein kinase activity in epidermal growth factor-stimulated rat hepatocytes. J Biol Chem 1996; 271: 3604–3607.

84. Whisler RL, Newhous YG, Bagenstose SE. Age-related reductions in the activation of mitogen-activated protein kinase p44mapk/ERK and p42mapk/ERK2 in human T cells stimulated via ligation of the T cell receptor complex. Cell Immunol 1996; 168:201–210.

85. Zhen X, Uryu K, Cai G, Johnson GP, Friedman E. Age-associated impairment in brain MAPK signal pathways and the effect of calorie restriction in Fischer 344 rats. J Gerontol 1999; 54:539–548.

86. Wang Z, Brecher P. Salicylate inhibition of extracellular signal-regulated kinases and inducible nitric oxide synthase. Hypertension 1999; 34:1259–1264.

87. Wang Z, Brecher P. Salicylate inhibits phosphorylation of the nonreceptor tyrosine kinases, proline-rich tyrosine kinase 2 and c-Src. Hypertension 2001; 37:148–153.

88. Hoffmeyer A, Grosse-Wilde A, Flory E, Neufeld B, Kunz M, Rapp UR, Ludwig S. Different mitogen-activated protein kinase signaling pathways cooperate to regulate tumor necrosis factor alpha gene expression in T lymphocytes. J Biol Chem 1999; 274:4319–4327.

89. Barnes PJ, Karin M. Nuclear factor-kappaB: a pivotal transcription factor in chronic inflammatory disease. N Engl J Med 1997; 336:1066–1071.

90. Dong Z, Huang C, Brown RE, Ma WY. Inhibition of activator protein 1 activity and neoplastic transformation by aspirin. J Biol Chem 1997; 272:9962–9970.

91. Underwood DC, Osborn RR, Kotzer CJ, Adams JL, Lee JC, Webb EF, Carpenter DC, Bochnowicz S, Thomas HC, Hay DW, Griswold DE. SB 239063, a potent p38 MAP kinase inhibitor, reduces inflammatory cytokine production, airways

eosinophil infiltration, and persistence. J Pharmacol Exp Ther 2000; 293:281–288.

92. Kim HJ, Jung KJ, Seo, AY, Choi JS, Yu BP, Chung HY. Calorie restriction modulates redox-sensitive AP-1 during the aging process. J Am Aging Assoc 2002; 25:123–130.

93. Pahlavani MA, Vargas DM. Influence of aging and caloric restriction on activation of Ras/MAPK, calcineurin, and CaMKIV activities in rat T cells. Proc Soc Exp Biol Med 2000; 223:163–169.

94. Li M, Walter R, Torres C, Sierra F. Impaired signal transduction in mitogen activated rat splenic lymphocytes during aging. Mech Ageing Dev 2000; 113:85–99.

95. Chung JH, Kang S, Varani J, Lin J, Fisher GJ, Voorhees JJ. Decreased extracellular-signal-regulated kinase and increased stress-activated MAP kinase activities in aged human skin in vivo. J Invest Dermatol 2000; 114:177–182.

96. Friedman E, Wang HY. Effect of age on brain cortical protein kianse C and its mediation of 5-hydroxytryptamine release. J Neurochem 1989; 52:187–192.

97. Battaini F, Lucchi L, Bergamaschi S, Ladisa V, Trabucchi M, Govoni S. Intracellular signaling in the aging brain. The role of protein kinase Cα and its calcium-dependent isoforms. Ann N Y Acad Sci 1994; 719:271–284.

98. Wang HY, Pisano MR, Friedman E. Attenuated protein kinase C activity and translocation in Alzheimer's disease brain. Neurobiol Aging 1994; 15:293–298.

99. Jindal HK, Ai ZW, Gascard P, Horton C, Cohen CM. Specific loss of protein kinase activities in senescent erythrocytes. Blood 1996; 88:1479–1487.

100. Xhu X, Raina AK, Rottkamp CA, Aliev G, Perry G, Boux H, Smith MA. Activation and redistribution of c-Jun N-terminal kinase/stress activated protein kinase in degenerating neurons in Alzheimer's disease. J Neurochem 2001; 76:435–441.

101. Gupta A, Rosenberonger SF, Bowden GT. Increased ROS levels contribute to elevated transcription factor and MAP kinase activities in malignantly progressed mouse keratinocyte cell lines. Carcinogenesis 1999; 20:2063–2073.

102. Peters SL, Mathy MJ, Pfaffendorf M, van Zwieten PA. Reactive oxygen species-induced aortic vasoconstriction and deterioration of functional integrity. Naunyn Schmiedebergs Arch Pharmacol 2000; 361:127–133.

103. Cao SX, Dhahbi JM, Mote PL, Spindler SR. Genomic profiling of short- and long-term caloric restriction effects in the liver of aging mice. Proc Natl Acad Sci USA 2001; 98:10630–10635.

104. Lee CK, Klopp RG, Weindruch R, Prolla TA. Gene expression profile of aging and its retardation by caloric restriction. Science 1999; 285:1390–1393.

105. Yu ZF, Mattson MP. Dietary restriction and 2-deoxyglucose administration reduces focal ischemic brain damage and improves behavioral outcome: evidence for a preconditioning mechanism. J Neurosci Res 1999; 57:830–839.

106. Bruce-Keller AJ, Umberger G, McFall R, Mattson MP. Food restriction reduces brain damage and improves behavioral outcome following excitotoxic and metabolic insults. Ann Neurol 1999; 45:8–15.

107. Klebanov S, Diais S, Stavinoha WB, Suh Y, Nelson JF. Hyperadrenocorticism, attenuated inflammation, and the life-prolonging action of food restriction in mice. J Gerontol A Biol Sci Med Sci 1995; 50:79–82.

108. Kari F, Hatch G, Slade R, Crissman K, Simeonova PP, Luster M. Dietary restriction mitigates ozone-induced lung inflammation in rats: a role for endogenous antioxidants. Am J Respir Cell Mol Biol 1997; 17:740–747.

109. Chandrasekar B, Nelson JF, Colston JT, Freeman GL. Calorie restriction attenuates inflammatory responses to myocardial ischemia-reperfusion injury. Am J Physiol Heart Circ Physiol 2001; 280:2094–2102.

110. Dong W, Selgrade MK, Gilmour IM, Lange RW, Park P, Luster MI, Kari FW. Altered alveolar macrophage function in calorie restricted rats. Am J Respir Cell Mol Biol 1998; 19:462–469.

111. Spaulding CC, Walford RL, Effros RB. Calorie restriction inhibits the age-related dysregulation of the cytokines TNF-α and IL-6 in C3B10RF1 mice. Mech Ageing Dev 1997; 93:87–94.

112. Cherry, Engelman RW, Wang BY, Kinjoh K, El-Badri NS, Good RA. Calorie restriction delays the crescentic glomerulonephritis of SCG/Kj mice. Proc Soc Exp Biol Med 1998; 218:218–222.

113. Drapier JC, Hibbs JB Jr. Differentiation of murine macrophages to express nonspecific cytotoxicity for tumor cells results in L-arginine-dependent inhibition of mitochondrial iron-sulfur enzymes in the macrophage effector cells. J Immunol 1988; 140:2829–2838.

114. Stuehr DJ, Marletta MA. Synthesis of nitrite and nitrate in murine macrophage cell lines. Cancer Res 1987; 47: 5590–5594.

115. Jiang CH, Tsien JZ, Schultz PG, Hu Y. The effects of aging on gene expression in the hypothalamus and cortex of mice. Proc Natl Acad Sci USA 2001; 98:1930–1934.

116. Kim HJ. Regulation of redox-sensitive transcription factor by Age and Calorie restriction. Ph.D. Dissertation, Pusan National University, Busan, Korea, 2002.

117. Karin M. The regulation of AP-1 activity by mitogenactivated protein kinases. J Biol Chem 1995; 270:16483–16486.

118. Karin M, Liu ZG, Zandi E. AP-1 function and regulation. Curr Opin Cell Biol 1997; 9:240–246.

119. Smeal T, Hibi M, Karin M. Altering the specificity of signal transduction cascades: positive regulation of c-Jun transcriptional activity by protein kinase A. EMBO J 1994; 13:6006–6010.

120. Richard DE, Berra E, Gothi E, Roux D, Pouysségur J. p42/p44 Mitogen-activated protein kinases phosphorylate hypoxia-inducible factor 1a (HIF-1a) and enhance the transcriptional activity of HIF-1. J Biol Chem 1999; 274:32631–32637.

121. Hirota K, Semenza GL. Rac1 activity is required for the activation of hypoxia-inducible factor 1. J Biol Chem 2001; 276: 21166–21172.

122. Palmer LA, Gaston B, Johns RA. Normoxic stabilization of hypoxia-inducible factor-1 expression and activity: redox-dependent effect of nitrogen oxides. Mol Pharmacol 2000; 58:1197–1203.

16

Oxidative Stress in Mitochondria: The Involvement in Neurodegenerative Diseases

MAKOTO NAOI

Department of Neurosciences, Gifu International Institute of Biotechnology, Kakamigahara, Gifu, Japan

WAKAKO MARUYAMA and MASAYO SHAMOTO-NAGAI

Laboratory of Biochemistry and Metabolism, Department of Basic Gerontology, National Institute for Longevity Sciences, Obu Aichi, Japan

YOJI KATO

School of Humanity for Enviromental Policy and Technology, Himeji Institute of Technology, Himeji, Hyogo, Japan

MASASHI TANAKA

Department of Gene Therapy, Gifu International Institute of Biotechology, Kakamigahara, Gifu, Japan

ABSTRACT

In mitochondria, oxidative phosphorylation and enzymatic oxidation of biogenic amines by monoamine oxidase produces reactive oxygen and nitrogen species, which may account for neuronal cell death in neurodegenerative disorders, including Parkinson's and Alzheimer's disease. In these disorders, inclusion body composed of oxidation-modified proteins and lipids is detected specifically for distinct diseases, such as the Lewy body for Parkinson's disease. The relationship between mitochondrial dysfunction, increased oxidative stress, accumulation of oxidation-modified protein, and final cell death of definite neurons in the brain remains to be clarified. In this paper, we review our recent results on interaction among these factors in neurons, using a cellular model of apoptosis induced by peroxynitrite-generating N-Morpholino sydnonimine (SIN-1) and an inhibitor of complex I, rotenone in human dopaminergic SH-SY5Y cells. In control cells, 3-nitrotyrosine-containing protein produced by peroxynitrite was detected, suggesting that neurons exist in a state of constant oxidative stress. N-Morpholino sydnonimine induced apoptosis and reduction in ATP level, which is, increased further by an inhibitor of proteasome, carbobenzoxy-L-isoleucyl-γ-t-butyl-glutamyl-L-alanyl-L-leucinal (PSI). The subunits of mitochondrial complex I were found to contain 3-nitrotyrosine, suggesting that peroxynitrite prefers these enzymes. In addition, rotenone induced mitochondrial dysfunction, and accumulation and aggregation of protein modified with acrolein, an aldehyde product of lipid peroxidation. Rotenone treatment reduced the enzymatic activity of the proteasome system, a major organelle in the degradation of oxidation-modified protein, and it was due to the oxidative modification of 20S β subunit of the proteasome. These results are discussed in relation to the interaction between mitochondrial dysfunction, oxidative stress, and proteasome inactivation, resulting in neuronal cell death in neurodegenerative disorders, such as Parkinson's and Alzheimer's disease.

I. OXIDATIVE STRESS AND MODIFIED PROTEIN AS THE MARKER

Oxidative stress has been proposed to induce neuronal death in aging and age-associated disorders (1,2), and mitochondria are a major source of reactive oxygen and nitrogen species (ROS–RNS). The superoxide anion radical generated by oxidative phosphorylation in the mitochondria is one of the most potent ROS and reacts with nitric oxide (NO) to form peroxynitrite (ONOO$^-$), whereas oxidation of biogenic amines by monoamine oxidase in mitochondrial outer membrane produces hydrogen peroxide. Mitochondria are now considered to play a pivotal role in apoptosis (3), which emerges as a common death type of neurons in neurodegenerative disorders, including Parkinson's (PD) and Alzheimer's diseases (AD) (4,5). The role of mitochondria in the process of apoptotic commitment is recognized. In mitochondria, impairment of energy charge and redox, permeability transition (PT), disruption of membrane potential, $\Delta\Psi$m, and release of cytochrome c are observed prior to the fragmentation of nuclear DNA, a hallmark of apoptotic morphological features.

Neurodegenerative disorders are characterized by a decline of specified neurons associated with protein deposits typical for each disease. In PD, dopamine neurons in the substantia nigra degenerate progressively with the formation of the Lewy bodies (LB). The pathogenesis of PD remains unknown, and the gene responsible for the sporadic cases has not been identified. PD is considered to represent the final outcome of various genetic and environmental interactions. The vulnerability of dopamine neurons is a consequence of the increased generation of ROS and RNS, reduced antioxidant capacity, high content of iron and dopamine, and possible defect in mitochondrial function. ROS and RNS generated in mitochondria modify bioactive molecules, such as lipids, proteins, DNA, and carbohydrates, either directly or indirectly with peroxidation products of lipids or carbohydrates. Several kinds of modified bioactive molecules have been proposed as markers of oxidative modification by ROS and RNS, as summarized in Table 1. Hydroxyl radicals

Table 1 Oxidative Modification of Protein

Direct modification	Secondary modification
Polymerization (cross reaction)	Modification by lipid peroxidation
Aggregation	Aldehydes
Fragmentation	4-Hydroxynonenal
	Acrolein
Inactivation or activation of enzymes	Malondialdehyde
	Hydroperoxide
Modification of amino acids	Carbonyl production
3-Nitrotyrosine	
Dityrosine	Modification by glycosylation
	Aldehydes
Carbonyl production	Carbonyls

modify tyrosine, phenylalanine, tryptophan, histidine, methionine, and cysteine residues as preferred targets. Under anaerobic conditions, the hydroxyl radicals promote protein–protein crosslinking through –S–S– and –tyrosyl-tyrosyl–(dityrosine) bonding, and under aerobic conditions, peroxyradicals induce fragmentation of the polypeptide chain. In addition, proline, arginine, and lysine are particularly sensitive to metal-catalyzed oxidation and are converted to carbonyl derivatives.

Oxidative modification produces aggregated and cross-linked proteins, which are resistant to proteolytic degeneration and are difficult to be removed from the cells. Accumulation of the modified proteins may impact on a variety of cellular pathways by changing the enzymatic, regulatory, and transporting potencies of cellular specific protein, in addition to taking up space in limited cellular volume. The level of the oxidized protein may reflect the balance between the generation of ROS–RNS and degradation of modified protein, in which the ubiquitin–proteasome system plays a key role (6).

One of the most active RNS is peroxynitrite ($ONOO^-$) (7), which is unstable, but its protonated peroxynitrous acid (ONOOH) is extremely reactive (8), which generates hydroxyl radical by homolytic cleavage (9). The main targets of nitration are sulfhydryl and hydroxyl residues in cysteine, methionine, phenylalanine, and tyrosine. It inactivates the membrane function and key enzymes (see reviews 7, 10). As shown in Fig. 1, 3-nitrotyrosine (3-NT) is synthesized by the nitration of tyrosine residues in protein and a marker for the oxidative stress induced by peroxynitrite in vivo (11). 3-Nitrotyrosine containing protein (3-NT protein) was detected in atherosclerosis (12) and neurodegenerative disorders, such as amyotrophic lateral sclerosis (ALS) (7), AD, (13,14), and PD (15). Figure 2 shows the immuno-histochemical detection of 3-NT protein in pyramidal hippocampal neurons, using anti-3-NT protein antibody (16). Another oxidation product of tyrosine is dityrosine, which is produced from free and protein-bound tyrosine in the presence of hydrogen peroxide

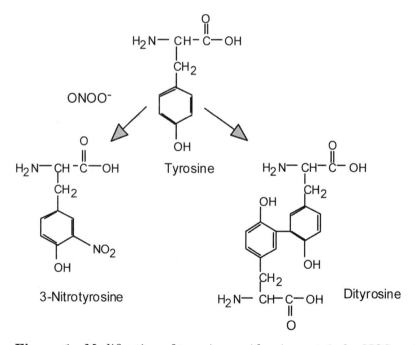

Figure 1 Modification of tyrosine residue in protein by NOS and RNS. Peroxynitrite modifies tyrosine residues to 3-nitrotyrosine in protein, and hydrogen peroxide and peroxidase, and irradiation produces tyrosyl radicals, yielding dityrosine with tyrosine residues in protein or free tyrosine.

(A)   (B)

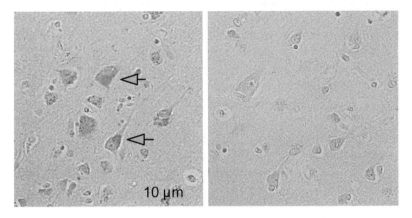

Figure 2 Immuno-histochemical detection of 3-NT-containing protein in lipofuscin in pyramidal hippocampal neurons. The brain was obtained from a 72-year-old male patient without neurological or psychiatric disorders. The tissue samples were incubated with (A) anti-3-NT protein antibody (diluted 1:200 with bovine serum albumin) or (B) bovine serum albumin alone, followed by the treatment of peroxidase-labeled antirabbit IgG. Lipofuscin stained was observed as brown granule as indicated by arrows.

and myeloperoxidase (17) (Fig. 1), and is detected in atherosclerotic plaques (18) and lipofuscin pigments in the aged human brain (19).

On the other hand, lipid peroxidation generates various reactive aldehydes, including 4-hydroxynonenal (4-HNE) and acrolein as shown in Fig. 3 (20). 4-Hydroxynonenal reacts with sulfhydryl and amino groups and leads to inactivation of DNA polymerases, dehydrogenases, and various transporters, and also to cell cycle arrest and apoptosis. Proteins modified with 4-HNE and malondialdehyde were detected in nigro-striatal dopamine neurons in PD (21), neurofibral tangles in AD (22,23), and the spinal cord of ALS patients (24). Acrolein, $CH_2=CH–CHO$, is ubiquitously generated in the biological system and is the most reactive α,β-unsaturated aldehyde product of lipid peroxidation. It is incorporated into proteins easily and accumulates as protein adducts

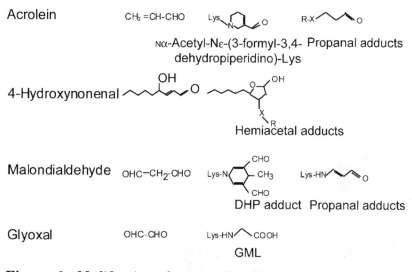

Figure 3 Modification of protein by aldehyde products of lipid peroxidation. Aldehyde products of lipid and carbohydrate peroxidation, acrolein, 4-hydroxynonenal, malondialdehyde, and glycol, modify lysine residues in proteins. DHP, dihydropyridine, GML, glyoxal modified lysine.

after reacting with lysine and histidine residues by forming Michaelis-type acrolein–amino acid complexes (25). Acrolein-modified protein was detected in oxidized low-density lipoproteins (26) and the brain of patients with AD (27). Figure 4 shows the histochemical staining of the substantia nigra in a parkinsonian brain with an antibody against 4-HNE-modified protein. Dopamine neurons containing neuromelanin from parkinsonian patients were stained more markedly than those in normal control and nondopaminergic cells. These results indicate that the oxidative stress increases markedly in nigro-striatal dopamine neurons of a parkinsonian brain.

II. MITOCHONDRIAL COMPLEX I SUBUNITS ARE NITRATED BY ONOO⁻

In the brain, NO has been considered to be produced in microglia and astrocytes and transported to neurons, where it

Control   Parkinson's disease

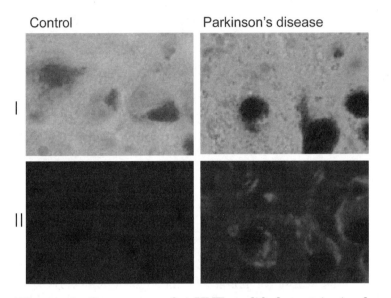

Figure 4 Occurrence of 4-HNE-modified protein in dopamine neurons in substantia nigra of the brain from the patient with Parkinson's disease. Only in the substantia nigra of the brain from the parkinsonian patient, but not from control, dopamine neurons containing neuromelanin are positively stained with antibody against 4-HNE modified protein. Cells other than dopamine neurons are not stained with the antibody.

reacts with the superoxide yielding $ONOO^-$. However, SH-SY5Y cells produce NO and $ONOO^-$ in situ, as confirmed by the use of $2',7'$-dichlorodihydrofluorescein diacetate (H_2DCFDA) (28) and inhibitors of nitric oxide synthase (NOS); H_2DCFDA is cleaved into $2',7'$-dichlorofluorescein by hydroxyl radical and $ONOO^-$; and NOS inhibitors, N^5-(1-iminoethyl) -L-ornithine (L-NIO) and N^5-nitro-L-arginine methyl ester (L-NAME), reduced DCF to about a half.

Using an antibody against the 3-NT protein (16), nitrated proteins were detected in human dopaminergic SH-SY5Y cells. The lysate from these cells was subjected to western blot analysis as shown in Fig. 5A, and the molecular weight of major 3-NT proteins was estimated to be 33, 21, 15, and 11 kDa. The nitrated protein bands were detected even in control under physiological conditions, suggesting that the cells were under

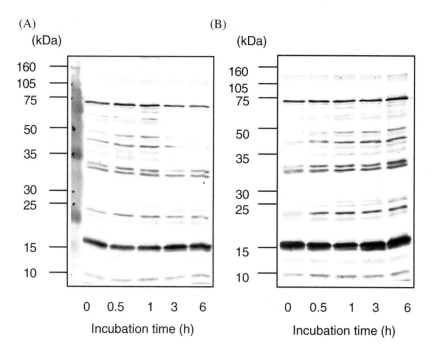

Figure 5 Formation of 3-NT protein in SH-SY5Y cells and the effects of peroxynitrite-generating SIN-1 and a proteasome inhibitor, PSI. (A) Cells were treated with 250 μM of SIN-1, then applied to the immunoblotting with antibody against 3-NT proteins. After the treatment with SIN-1, the intensity of 3-NT proteins increased, but the number did not change markedly. At the left lane in A, the molecular markers are shown. (B) The cells were treated with SIN-1 (250 μM) in the presence a proteasome inhibitor, PSI (10 μM). 3-Nitrotyrosine protein increased in amount according to the incubation time.

constant oxidative stress. However, the cells are intact in growth and proliferation, suggesting the functioning of an active mechanism to eliminate modified protein from the cells. The mitochondrial fraction of the SH-SY5Y cells was the one most intensively stained with the anti-3-NT antibody, and was subjected to western blot analysis using antibodies against 3-NT protein, and against mitochondria complex I, II, III, and IV (29). Some of the nitrated proteins were identified to be

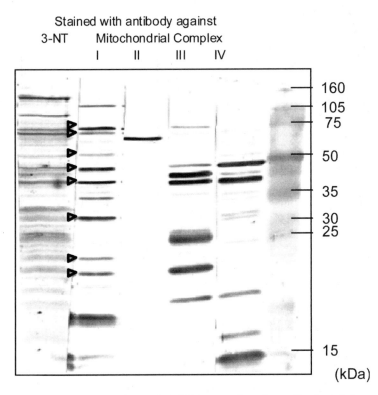

Figure 6 Detection of 3-NT proteins in mitochondria complex I–IV subunits. The cells were treated with 250 µM SIN-1 for 1 hr, and P2 fraction was isolated. The samples were subjected to SDS-PAGE and detected by the immunoblot assay using the antibody against 3-NT , or complex I (I), complex II (II), complex III (III), and complex IV (IV), respectively. Molecular markers are shown in the right column. Arrows show the protein bands positively stained with anti-3-NT and anticomplex I antibody.

the subunits of complex I, as assigned based on the staining with anticomplex I antibody (Fig. 6). These results clearly show that the preferential nitration of complex I subunits may contribute to mitochondrial dysfunction observed in the nigro-striatum of parkinsonian brain (30,31).

Treatment of ONOO⁻-generating *N*-morpholino sydnonimine (SIN-1) induced apoptosis in the SH-SY5Y cells (32–35).

A step-wise activation of apoptotic cascade was observed: decline in ΔΨm, activation of caspase 3, and phosphorylation of p38 mitogen-activated phosphokinase (MAP) (33). Nitric oxide and ONOO⁻ were reported to induce apoptosis by nitration of tyrosine residues to release cytochrome c (36), or of cytochrome c itself (37). In addition, mitochondrial ATP synthesis was inhibited markedly by ONOO⁻, as shown in Fig. 7, which may be due to its reversible binding to cytochrome oxidase or inactivation of complex I and II, and ATPase (38–41).

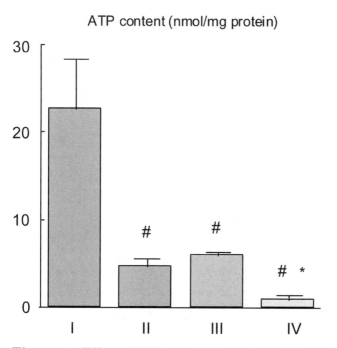

Figure 7 Effect of SIN-1 and PSI on the ATP levels in SH-SY5Y cells. The cells were incubated without or with 250 μM SIN-1 in the absence and presence of 10 μM PSI for 2 hr. The ATP content was measured by the luminofluorometric method. The column and bar represent the mean and SD of triplicate measurements of three independent experiments. I, control cells; II, cells treated with PSI alone; III, SIN-1 alone; and IV, SIN-1 and PSI. # Difference from control (I) is statistically significant ($p < 0.05$). * Difference from SIN-1 alone (III) is statistically significant ($p < 0.05$).

As shown in Fig. 5A, SIN-1 treatment increased 3-NT protein only in distinct protein bands, suggesting the preference of some proteins to ONOO$^-$ modification. In addition, increase in the amount 3-NT protein was not so significant.

III. PROTEASOME: ITS ROLE IN ACCUMULATION OF OXIDIZED PROTEINS

Insoluble intracellular protein aggregates, such as Lewy body in PD and senile plaques composed of β-amyloid in AD, are hallmarks of neurodegeneration. Although it remains unclear whether protein aggregates cause directly neuronal cell death or are the results of deteriorated cellular homeostasis in dying neurons, the analysis of the constituents of inclusion bodies may suggest the molecular mechanism leading their formation. Protein aggregation is a manifestation of disturbed cellular protein-folding homeostasis maintained by the ubiquitin–proteasome system. In LB, ubiquitin, and proteasome, subunits are major components (42,43), in addition to α-synuclein, Parkin, and ubiquitin C-terminal hydrolase-L1 (UCH-1) (44–47). They were modified with ROS–RNS as nitrated synuclein (48) and dityrosine (49).

To clarify the interactions among oxidative stress, dysfunction of the proteasome system, and formation of the inclusion body, the effects of a proteasome inhibitor, carbobenzoxy-L-isoleucyl-γ-t-butyl-L-alanyl-L-leucinal (PSI) were examined on the deposition of modified proteins and the cell vulnerability (34,50). PSI increased the amount of 3-NT proteins in the SH-SY5Y cells, but the number of 3-NT protein bands was almost the same as in the control (Fig. 5B). The functional deterioration of mitochondria was also enhanced by PSI, as shown by the severe reduction of ATP synthesis (Fig. 7). In addition, the number of apoptotic cells increased significantly by PSI, whereas that of necrotic cells remained almost the same (Fig. 8A). At the same time, the acrolein-modified protein increased significantly in subcellular fractions of the SH-SY5Y cells after being treated with PSI (Fig. 8B). These results clearly demonstrate that the inhibition of proteasome activity

(A)

(B)

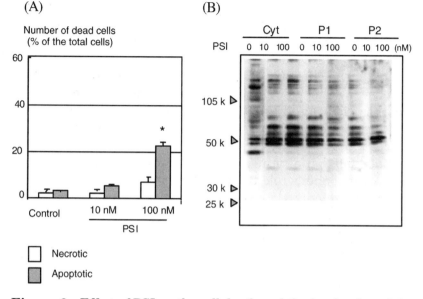

Figure 8 Effect of PSI on the cell death and the levels of acrolein-modified protein. (A) Apoptotic and necrotic cell death after treatment with 10 and 100 nM PSI was quantified by staining the cells with 10 nM Hoechst 33342. The column and the bar represent the mean and SD of the number of dead cells, expressed as the percentage of the total. The open and the filled columns represent necrotic and apoptotic dead cells, respectively. The difference from the control cells is significant ($p < 0.05$) by ANOVA. (B) The effects of PSI on the level of acrolein-modified proteins were examined by immunoblot with antiacrolein antibody. The cytoplasmic (Cyt) P1 and P2 fractions of the cells treated with 10 and 100 nM of PSI for 24 hr were prepared by centrifugation at 800 g and 13,000 g, respectively.

may play a key role in the accumulation of oxidation-modified protein and neuronal death.

IV. MITOCHONDRIAL DYSFUNCTION INHIBITS THE PROTEASOME SYSTEM BY OXIDATIVE MODIFICATION

Systemic reduction in complex I of the mitochondrial electron transfer chains was reported in the nigro-striatum of patients with PD (30,31,51). Recently, an inhibitor of complex I, rotenone, was reported to induce parkinsonism in rodents, and fibrillar cytoplasmic inclusions containing ubiquitin and α-synuclein were detected after the continuous administration (52). The effects of mitochondrial dysfunction on the proteasome system were studied by the use of rotenone at a concentration to induce apoptosis in the SH-SY5Y cells (50). ATP levels reduced markedly, and apoptosis was induced in the cells after 4–5 days treatment with rotenone. The oxidative modification of proteins was followed by the use of antibody against the acrolein-modified protein (19). The levels of the acrolein-modified protein increased markedly by rotenone treatment (Fig. 9).

The ubiquitin–proteasome system is a major site to remove damaged or mutated proteins and also regulatory proteins controlling cell cycle and signal transduction. In the nigro-striatum of the parkinsonian brain, the decreased activity of the proteasome was reported, suggesting its involvement in the pathological features (53). The oxidized protein is preferentially degraded in vitro by 20S proteasome in an ATP-independent way. 20S proteasome is composed of four rings to make a cylindrical structure, and each ring is made of seven α and β 20S subunits (54). Noncatalytic α-heptameric rings form each of the two outer rings and catalytic β-heptamers form the two inner rings. Binding of a regulatory subunit named 19S complex (ATPase, PA700) to both ends of the 20S cylinder produces 26S proteasome with a higher catalytic activity than the 20S proteasome. The 26S proteasome can degrade polyubiquitinated proteins and ornithine decarboxylase in an ATP-dependent process.

In the SH-SY5Y cells, rotenone treatment increased ROS–RNS levels only transitionally and slightly, whereas the oxidized protein levels increased progressively, suggesting that degradation of modified proteins may be mainly involved inthe deposition. The enzymatic activity of proteasome measured with an artificial fluorescent substrate, carbobenzoxy-L-leucyl-L-leucyl-L-glutamic acid α-(4-methyl-coumary1-7-amide) (Z-Leu-Leu-Glu-MCA), reduced in a time- and dose-dependent way, and virtually was not detected after 4 days treatment with

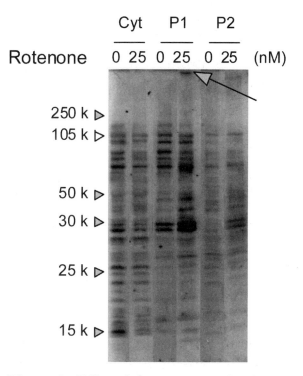

Figure 9 Effect of rotenone on oxidized protein level in SH-SY5Y cells. The cells were treated without or with 25 nM of rotenone for 5 days. The cells were fractionated into the cytoplasmic (Cyt), P1 and P2 fractions, and each fraction was subjected to immunoblotting, using antibody against acrolein-modified proteins. The aggregated proteins were observed in the stacking gel of the P1 fraction (arrow).

rotenone, as shown in Table 2. The reduction in the enzymatic activity was neither due to ATP depletion nor to the reduction of the protein level, indicating that the homospecific activity of proteasome decreased by the treatment. Indeed, the 20S β subunit protein was modified by acrolein, as shown by immunoprecipitation with anti-20S β subunit and by western blot analysis with the antiacrolein antibody. In this immunoprecipitant coprecipitated with 20S β subunit, oxidized proteins immunoreactive to antiacrolein antibody were detected in the cells treated with rotenone, suggesting that the acrolein-conjugated protein

Table 2 Effects of Rotenone Treatment on the Enzymatic Activity of Proteasome in SH-SY5Y Cells

Rotenone (25 nM)	ATP (2 mM)	Proteasome activity in SH-SY5Y cells treated with rotenone (pmol/min/μg protein)	
		After 6 hr	After 96 hr
Control	−	0.89 ± 0.11*	5.19 ± 0.47
	+	1.76 ± 0.02	8.21 ± 0.11
Sample	−	0.97 ± 0.02	0.96 ± 0.04
	+	2.25 ± 0.02	1.80 0.13

Proteasome activity in the cytoplasmic fraction of the cells treated with 25 nM rotenone was measured fluorometrically in the absence or presence of exogenously added ATP (2 mM).
*Mean ± SD of four experiments.

may be a substrate of 20S proteasome. However, further studies are required to clarify whether the acrolein-modified protein can directly inhibit proteasome activity, as in the case of the 4-HNE proteins (55). The increase in acrolein adducts after rotenone treatment may be comparable with that observed in the presence of PSI, as described earlier.

V. CONCLUSION

The results of this study suggest that mitochondrial dysfunction may induce degeneration of dopamine neurons through modification and inactivation of the proteasome system and subsequent aggregation of the oxidized proteins. The precise mechanism behind the cell death induced by oxidative stress requires further investigation, but it may be quite relevant to consider that the inactivation of proteasome may play a critical role in the activation of apoptotic signal. In neurodegenerative disorders, the environmental and genetic factors may induce a malignant cycle between mitochondrial dysfunction, increased oxidative stress, and reduced activity of proteasome in neurons, leading to the typical pathological characteristics, cell death and formation of inclusion body. Our recent studies on the ubiquitin–proteasome system in the substantia nigra of human brains present further support to our hypothesis. In the parkinsonian brain, the 20S β subunit precipitated with the specified antibody was modified with acrolein, whereas that from control was not modified. Future studies may give a new insight into the mechanism of neuronal cell death in age and age-related disorders, and may give us clues for a new therapeutic strategy to protect neurons from apoptosis.

ACKNOWLEDGMENT

This work was supported by a Grant-in-Aid for Scientific Research on Priority Areas (C) from the Japan Society for the Promotion of Science (W.M.), a Grant for Clinical Research for Evidence Based Medicine for W.M. and M.N. from the Ministry of Health, Labor and Welfare, Japan.

REFERENCES

1. Halliwell B. Reactive oxygen species and the central nervous system. J Neurochem 1992; 59:1609–1623.

2. Stadtman ER. Protein oxidation and aging. Science 1992; 257:1220–1224.

3. Kroemer G, Dallaporta B, Resch-Rigon M. The mitochondrial death/life regulator in apoptosis and necrosis. Annu Rev Physiol 1998; 60:619–642.

4. Thompson CG. Apoptosis in the pathogenesis and treatment of disease. Science 1995; 267:1456–1462.

5. Cotman CW, Su JH. Mechanisms of neuronal death in Alzheimer's disease. Brain Pathol 1996; 6:493–506.

6. Kaytor MD, Warren ST. Aberrant deposition and neurological disease. J Biol Chem 1999; 274:37507–37510.

7. Beckman JS. Oxidative damage and tyrosine nitration from peroxynitrite. Chem Res Toxicol 1996; 9:836–844.

8. Pryor WA, Squandrito GL. The chemistry of peroxynitrite: a product from the reaction of nitric oxide with superoxide. Am J Physiol 1995; 268:L699–L722.

9. Koppenol WH, Moreno JJ, Pryor WA, Ischiropoulous H, Beckman JS. Peroxynitrite, a cloaked oxidant formed by nitric oxide and superoxide. Chem Res Toxicol 1992; 5:834–842.

10. Kröncke KD, Fehsel K, Kolb-Bachofen V. Nitric oxide: cytotoxicity versus cytoprotection—how, why, when, and where? Nitric Oxide 1997; 1:107–120.

11. Halliwell B. What nitrates tyrosine? Is nitrotyrosine specific as a biomarker of peroxynitrite formation in vivo? FEBS Lett 1997; 411:157–160.

12. White R, Brock T, Cjang L, Crapo J, Briscoe P, Ku D, Bradley W, Gianturco S, Gore J, Freeman BA, Tarpey MM. Superoxide and peroxynitrite in atherosclerosis. Proc Natl Acad Sci USA 1994; 91:1044–1048.

13. Koppal T, Draake J, Yatin S, Jordan B, Varadarajan S, Bettenhausen L, Butterfield A. Peroxynitrite-induced alternation in synaptosomal membrane proteins: insight into oxidative stress in Alzheimer's disease. J Neurochem 1999; 72:310–317.

14. Smith M, Richey A, Harris PL, Sayre LM, Beckman JS, Perry G. Widespread peroxynitrite-mediated damage in Alzheimer's disease. J Neurosci 1997; 17:2653–2657.

15. Good PF, Hsu A, Werner P, Perl DP, Olanow CW. Protein nitration in Parkinson's disease. J Neuropathol Exp Neurol 1998; 57:338–342.

16. Kato K, Ogino Y, Aoki T, Uchida K, Kawakishi S, Osawa T. Phenolic anti-oxidants prevent peroxynitrite-derived collagen modification in vitro. J Agric Food Chem 1997; 45:3004–3009.

17. Heinecke JW, Li W, Daehnke HL, Goldstein JA. Dityrosine, a specific marker of oxidant, is synthesized by the myeloperoxidase–hydrogen peroxide system of human neutrophils and macrophages. J Biol Chem 1993; 268:4069–4077.

18. Leeuwenburgh C, Rasmussen JE, Hsu FF, Mueller DM, Pannathur S, Heinecke JW. Mass spectrometric quantitation of markers for protein oxidation by tyrosyl radical, copper, and hydroxyl radical in low density lipoprotein isolated from

human atherosclerotic plaques. J Biol Chem 1997; 272: 3520–3526.

19. Kato Y, Maruyama W, Naoi M, Hashizume Y, Osawa T. Immunohistochemical detection of dityrosine in lipofuscin pigments in the aged human brain. FEBS Lett 1998; 439:231–234.

20. Esterbauer H, Schauer RJ, Zollner H. Chemistry and biochemistry of 4-hydroxynonenal, malonaldehyde and related aldehydes. Free Radic Biol Med 1991; 11:81–128.

21. Yoritaka A, Hattori N, Uchida K, Tanaka M, Stadtman ER, Mizuno Y. Immunohistochemical detection of 4-hydroxynonenal protein adducts in Parkinson disease. Proc Natl Acad Sci USA 1996; 93:2696–2701.

22. Sayre LM, Zelasko DA, Harris PLR, Perry G, Salomon RG, Smith MA. 4-Hydroxynonenal-derived advanced lipid peroxidation and products are increased in Alzheimer's disease. J Neurochem 1997; 68:2092–2097.

23. Yan SD, Chen X, Schmidt AM, Brett J, Godman G, Zou YS, Scott CW, Caputo C, Frappier T, Smith MA, Perry G, Yen SH, Stern G. Glycated tau protein in Alzheimer's disease: a mechanism for induction of oxidative stress. Proc Natl Acad Sci USA 1994; 91:7787–7791.

24. Pedersen WA, Fu W, Keller JN, Markesbery WR, Appel S, Smith PG, Kasarkis E, Mattson MP. Protein modification by the lipid peroxidation product 4-hydroxynonenal in the spinal cords of amyotrophic lateral sclerosis patients. Ann Neurol 1998; 44:819–824.

25. Gan JC, Cardasan A, Ansari GA. In vitro modification of serum albumin by acrolein. Chemosphere 1999; 23:939–947.

26. Uchida K, Kanematsu M, Motimistu Y, Osawa T, Noguchi N, Niki F. Acrolein a product of lipid peroxidation: formation of free acrolein and its conjugate with lysine residues in oxidized low density lipoproteins. J Biol Chem 1998; 273:16058–16066.

27. Calingoson NY, Uchida K, Gibson GE. Protein-bound acrolein: a novel marker of oxidative stress in Alzheimer's disease. J Neurochem 1999; 72:751–756.

28. Crow JP. Dichlorodihydrofluorescein and dihydrorhodamine 123 are sensitive indicators of peroxynitrite in vitro: implications for

intracellular measurement of reactive nitrogen and oxygen species. Nitric Oxide 1997; 1:145–157.

29. Tanaka M, Miyabayashi S, Nishikimi M, Suzuki H, Shimomura Y, Ito K, Narisawa K, Tada K, Ozawa K. Extensive defects of mitochondrial electron-transfer chain in muscular cytochrome *c* oxidase deficiency. Pediatr Res 1988; 24:447–454.

30. Schapira AHV, Cooper JM, Dexter D, Jenner P, Clark JB, Marsden CD. Mitochondrial complex I deficiency in Parkinson's disease. Lancet 1989; 1:1269.

31. Mizuno Y, Ohta S, Tanaka M, Takemiya S, Suzuki K, Sato T, Oya H, Ozawa T, Kagawa Y. Deficiencies in complex I subunits of the respiratory chain in Parkinson's disease. Biochem Biophys Res Commun 1989; 163:1450–1455.

32. Maruyama W, Takahashi T, Naoi M. (−)Deprenyl protects human dopaminergic neuroblastoma SH-SY5Y cells from apoptosis induced by peroxynitrite nitric oxide. J Neurochem 1998; 70:2510–2515.

33. Oh-hashi K, Maruyama W, Yi H, Takahashi T, Naoi M, Isobe M. Mitogen-activated protein kinase pathway mediates peroxynitrite-induced apoptosis in human dopaminergic neuroblastoma SH-SY5Y cells. Biochem Biophys Res Commun 1999; 263:504–509.

34. Maruyama W, Kato Y, Yamamoto T, Oh-hashi K, Hashizume Y, Naoi M. Peroxynitrite induced neuronal cell death in aging and age-associated disorders. J Am Aging Assoc 2001; 24:11–18.

35. Yamamoto T, Maruyama W, Kato Y, Yi H, Shamoto-Nagai M, Tanaka M, Sato Y, Naoi M. Selective nitration of mitochondrial complex I by peroxynitrite: involvement in mitochondrial dysfunction and cell death of dopaminergic SH-SY5Y cells. J Neural Transm 2002; 109:1–13.

36. Hortelano S, Alvarez AM, Bosca L. Nitric oxide induces tyrosine nitration and release of cytochrome c preceding an increase of mitochondrial transmembrane potential in macrophases. FASEB J 1999; 13:2311–2317.

37. Cassina AM, Hondara R, Souza JM, Thomson I, Castro L, Ischiropoulos H, Freeman BA, Radi R. Cytochrome *c* nitration by peroxynitrite. J Biol Chem 2000; 275:21409–21415.

38. Radi R, Rodriguez M, Castro L, Telleri R. Inhibition of mito-chondrial electron transport by peroxynitrite. Arch Biochem Biophys 1994; 308:89–95.

39. Brorson JB, Schumacker PT, Zhang H. Nitric oxide acutely inhi-bits neuronal energy production. J Neurosci 1999; 19:147–158.

40. Bolanos JP, Almeida A, Stewart V, Peuchen S, Land JM, Clark JB, Heales SJR. Nitric oxide-mediated mitochondrial damage in the brain: mechanisms and implications for neuro-degenerative disease. J Neurochem 1997; 68:2227–2240.

41. Beltran B, Quinetro M, Garcia-Zaragoza E, O'Connor E, Esplugues JV, Moncada S. Inhibition of mitochondrial respira-tion by endogenous nitric oxide: a critical step in Fas signaling. Proc Natl Acad Sci USA 2002; 99:8892–8897.

42. Iwatsubo T, Yamaguchi H, Fujikura M, Yokosawa H, Ihara Y, Trojanowski JQ, Lee VM. Purification and characterization of Lewy bodies from the brains of patients with diffuse Lewy body disease. Am J Pathol 1996; 148:1517–1529.

43. Ii K, Ito H, Tanaka K, Hirano A. Immunocytochemical co-localization of the proteasome in ubiquitinated structures in neurodegenerative disease and the elderly. J Neuropathol Exp Neurol 1997; 56:125–131.

44. Spillantini MG, Schmidt ML, Lee VM-Y, Trojanowski JQ, Jakes R, Goedert M. α-Synuclein in Lewy bodies. Nature 1997; 388:837–840.

45. Choi P, Osrerova-Golts N, Sparkman D, Cochran E, Lee JM, Wolozin B. Parkin is metabolized by the ubiquitin/proteasome roteasome system. Neuroreport 2000; 11:2635–2638.

46. Shimura H, Schlossmacher MG, Hattori N, Frosch MP, Trockenbacher A, Schneider R, Mizuno Y, Kosik KS, Selkoe DJ. Ubiquitination of a new form of alpha-synuclein by Parkin from human brain: implication for Parkinson's disease. Science 2001; 293:263–269.

47. Tanaka Y, Engelender S, Igarashi S, Rao PK, Wanner T, Tanzi RE, Sawa A, Dawson VL, Dawson TM, Ross CA. Inducible expression of mutant alpha-synuclein decreases proteasome activity and increases sensitivity to mitochondria-dependent apoptosis. Hum Mol Genet 2001; 10:919–926.

48. Trojanowski JQ, Goedert M, Iwatsubo T, Lee VM-Y. Fatal attractions: abnormal protein aggregation and neuron death in Parkinson's disease and Lewy body dementia. Cell Death Differ 1998; 8:832–837.

49. Souza JM, Giasson BI, Chen Q, Lee M-Y. Dityrosine cross-linkage promotes formation of stable α-synuclein polymers. J Biol Chem 2000; 275:18344–18349.

50. Shamoto-Nagai M, Maruyama W, Kato Y, Isobe K, Tanaka M, Naoi M, Osawa T. An inhibitor of mitochondrial complex I, rotenone, inactivates proteasome by oxidative modification and induces aggregation of oxidized proteins in SH-SY5Y cells. J Neurosci Res 2003; 74:589–597.

51. Hattori N, Tanaka M, Ozawa T, Mizuno Y. Immunohistochemical studies on complex I, II, III, and IV of mitochondria in Parkinson's disease. Ann Neurol 1991; 30:563–571.

52. Betarbert R, Sherer TB, MacKenzie G, Garcia-Osuna M, Panov AV, Greenamyre JT. Chronic systemic pesticide exposure reproduces features of Parkinson's disease. Nat Neurosci 2000; 3:1301–1306.

53. McNaught KSP, Jenner P. Proteasomal function is impaired in substantia nigra in Parkinson's disease. Neurosci Lett 2001; 297:191–194.

54. Hough R, Pratt G, Rechsteiner M. Purification of two high molecular weight proteasomes from rabbit reticulocyte lysate. J Biol Chem 1987; 262:8303–8313.

55. Okada K, Wangpoengtrakui C, Osawa T, Toyokuni S, Tanak K, Ucida K. 4-Hydroxy-2-nonenal-mediated impairment of intracellular proteolysis during oxidative stress. J Biol Chem 1999; 274:23787–23793.

17

Implications of Inflammatory Stress in Alzheimer's Disease

JUNG-HEE JANG and YOUNG-JOON SURH

Laboratory of Biochemistry and Molecular
Toxicology, College of Pharmacy, Seoul National
University, Seoul 151-742, South Korea

I. INTRODUCTION

Alzheimer's disease (AD) is a neurodegenerative disorder characterized by progressive degeneration and loss of neurons in the brain, which correlates with the appearance of neurofibrillary tangles and senile plaques, the two neuropathological hallmarks of AD. Beta-amyloid peptide (Aβ) is the major component of senile plaques and considered to have a causal role in development and progress of AD. This hydrophobic polypeptide consisting of 39- to 43- amino acid residues is proteolytically produced by β- and γ-secretases from a single transmembrane polypeptide collectively referred to as amyloid precursor protein (APP). Experimental data from both in vitro and in vivo studies indicate that different molecular forms of Aβ affect a wide array of neuronal and glial functions, thereby leading to neuronal cell death. Several lines of evidence suggest that enhanced oxidative stress and dysregulation of the normal inflammatory processes are associated with the pathogenesis and/or progression of AD. Elevated expression of immune response-related molecules and their receptors has been observed throughout the AD brain, and recent research has suggested that such brain-derived immune factors disrupt normal neurophysiology and contribute to cognitive and behavioral dysfunction. This chapter focuses on the putative molecular mechanisms of neuroinflammatory processes implicated in AD. The role of inducible nitric oxide synthase (iNOS) and cyclooxygenase-2 (COX-2) signaling pathways in mediating neuroinflammation and their regulation by upstream transcription factors and mitogen-activated protein kinases (MAPK) are also discussed.

II. ELEVATED INFLAMMATORY RESPONSES IN AD

II.A. Cytokines, Chemokines, Acute Phase Proteins, and Complements

In AD, several components of inflammation-associated elements are activated in a tissue-restricted manner. Extensive activation of astrocytes and microglia, the two primary mediators of neuroinflammation, has been observed in the CNS of AD patients. Microglia and astrocytes produce a variety of cytokines and chemokines, including interleukin-1β (IL-1β), IL-6, and tumor necrosis factor-α (TNF-α), known to be associated with immunologic and inflammatory reactions and found to be upregulated in AD-affected brains (1). Furthermore, IL-1β and TNF-α activate microglia and astrocytes and induce a number of acute phase proteins such as α_1-antichymotrypsin, which is found in amyloid plaques and may play a role in processing APP (2). Besides a large number of inflammatory mediators, the levels of components of the complement cascades are elevated in the AD brain and colocalized with both plaques and tangles. The synthesis of the complement component C1q is markedly enhanced in AD (3,4). Binding of C1q to Aβ may facilitate Aβ aggregation. Furthermore, in cultured microglia and astrocytes, Aβ and its subfragments induced or enhanced the production of TNF-α, IL-1, and complement C3 (5,6). In astrocytes, Aβ potentiated IL-6 and TNF-α secretion by lipopolysaccharide (LPS) (7).

II.B. Cyclooxygenase and Prostaglandins

Cyclooxygenase (COX) is a key enzyme that catalyzes the rate-limiting step in prostaglandin (PG) biosynthesis, i.e., the conversion of arachidonic acid via oxygenation to PG endoperoxide intermediates, such as PGG_2 and PGH_2. Although both isoforms (COX-1 and COX-2) are involved in the formation of PG endoperoxides, they are likely to have fundamentally different biological roles. COX-1 is a housekeeping enzyme, which is constitutively expressed in most mammalian tissues and is thought to be involved in maintaining physiological functions. In contrast, COX-2 is barely detectable under normal physiological conditions, but can be induced rapidly and transiently by proinflammatory mediators and mitogenic stimuli. In AD, the expression of COX-2 is upregulated (8,9). In cultured SH-SY5Y neuroblastoma, synthetic Aβ induced COX-2 expression and subsequent PGE_2 production (9). An expanding body of evidence suggests that PGs are involved in pathogenesis of AD. Elevated production of PGs, especially PGE_2, has been found in the brains of patients with AD (10). In addition, the level of phospholipase A_2 ($cPLA_2$), an enzyme which catalyzes the hydrolysis of membrane phospholipids to free arachidonic acid, has been shown to increase in AD brains, specifically in the astrocytes in regions that contain Aβ deposits (11).

II.C. Nitric Oxide and Peroxynitrite

Another enzyme that plays a pivotal role in mediating inflammation is nitric oxide synthase (NOS). Three forms of NOS are recognized. These include endothelial (eNOS), neuronal (nNOS), and inducible (iNOS) enzymes, with distinct cellular distribution patterns and physiological functions. nNOS is expressed in neurons of diverse locations with different densities and eNOS has been identified not only in endothelial cells but also in some neurons. iNOS is mainly found in astrocytes and microglia and catalyzes the oxidative deamination of L-arginine to produce nitric oxide (NO), a potent proinflammatory mediator. NO is a free radical that is particularly adapted to serve as an intra- and intercellular messenger in a variety of biological systems. NO exerts a number of activities including endothelium-dependent relaxation, neurotransmission, and cell-mediated immune response. In addition, NO can affect cellular functions through posttranslational modifications of proteins directly (nitrosylation and nitration) or indirectly (methylation and ribosylation) (12). Under certain conditions, NO reacts rapidly with superoxide anion to produce extremely powerful oxidant, peroxynitrite ($ONOO^-$), which probably accounts for much of NO neurotoxicity. Peroxynitrite is capable of reacting with the majority of the cellular components, including thiol, thiol ethers, iron sulfur centers, and zinc fingers (13). In addition, it combines with free and protein-bound tyrosine to form a stable product, nitrotyrosine (13). Though reactive per se, peroxynitrite can also form another extremely reactive oxygen species, hydroxyl radical (14). Several studies have demonstrated that Aβ stimulates microglial and astrocytic iNOS induction and subsequent NO production (6,15,16). Aβ has been shown to exert synergistic action with glutamate to induce neuronal damage via an NO-dependent pathway (17). Aβ-induced TNF-α production has been reported as a crucial event in the NO production in the microglia and murine-derived monocytes/macrophage J774 cell line (16,18). Peroxynitrite production in AD is evidenced by increased levels of nitrotyrosine residues in neurofibrillary tangles of AD patients (19). Hensley et al. (20) quantified the nitrotyrosyl content in the hippocampus, neurocortical regions, and ventricular cerebrospinal fluid by HPLC with an electrochemical array detection system. Representative

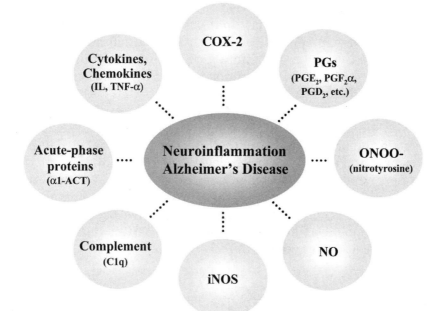

Figure 1 Representative proinflammatory markers of neuroinflammatory processes in AD.

proinflammatory markers reflecting the neuroinflammatory process in AD are summarized in Fig. 1.

III. ANTI-INFLAMMATORY THERAPY FOR AD

Interest in treatment of AD with nonsteroidal anti-inflammatory drugs (NSAIDs) has dramatically increased in the past few years as several epidemiologic studies have suggested beneficial effects associated with chronic NSAID use. NSAIDs are believed to act by inhibiting COX and subsequent PGE_2 production. Among twin and sibling pairs, prior use of NSAIDs has been associated with delayed onset of AD (21). A number of cross-sectional and longitudinal epidemiologic surveys also indicate that repeated use of NSAIDs may retard cognitive decline associated with AD in an elderly population (22). In a small controlled trial, indomethacin or diclofenac delayed cognitive deterioration (23,24). There are, however, side effects of NSAIDs that limit their use to some patients. Most common among these side effects are irritation and damage to the gastro-intestinal mucosa. Recent findings suggest that selective COX-2 inhibitors, such as celecoxib, may have an advantage over nonselective NSAIDs as potential therapeutic agents for AD. It has been demonstrated that COX-2 inhibitors protect neuronal cells form amyloid toxicity in vitro and promote neuronal cell survival in animal models of ischemic and excitotoxic neurodegeneration (25,26). Recently, attention has been focused on dietary constituents with antioxidative and anti-inflammatory properties. For instance, garlic extracts, green tea, and ginseng have been reported to have beneficial effects in improving certain age-related neurological dysfunctions (27).

IV. ROLES OF COX-2 AND NITROSATIVE STRESS IN NEURONAL CELL DEATH

Recent research into mechanisms of neuronal cell death in AD has been mainly directed towards understanding of how apoptosis and necrosis contribute to neurodegeneration. Apoptosis refers to a form of programmed cell death displaying characteristic morphologic features and executed by a series of cascades of cellular events. Neuropathologic studies show an increased rate of apoptotic neurons in postmortem samples of AD patients as compared with controls. Proteins associated with apoptosis, such as caspase, p53, Bax, c-Jun, and c-Jun NH_2-terminal kinase (JNK), have been found to be elevated in the AD (28).

IV.A. COX-2 and Apoptosis

There are multiple lines of compelling evidence supporting that COX-2 plays a role in the apoptosis. Observations in genetically engineered cells and transgenic animals verify a role of COX-2 in neuronal cell death. Aβ induced COX-2 expression in SH-SY5Y cells, and transient overexpression of COX-2 in these cells potentiated Aβ-mediated redox impairment (9). Neurons derived from transgenic mice with neuronal COX-2 overexpression were more vulnerable to $A\beta$-mediated neurotoxicity (29). Overexpression of human COX-2 in neurons of APPswe/PS1-A246E/hCOX-2 triple transgenics induced S795 Rb phosphorylation and caspase-3 activation (30). Pasinetti and Aisen (9) have demonstrated that upregulation of COX-2 expression overlapped the cellular morphological features of apoptosis in frontal cortex of AD brain. In a neuroectodermal P19 cell line, the induction of COX-2 preceded apoptosis in response to serum deprivation (31). Conversely, the inhibition of COX-2 expression and PGE_2 synthesis by NSAIDs or selective COX-2 inhibitors led to cell survival (25,26,32). PGE_2, a major eicosanoid involved in many inflammatory conditions, induced DNA fragmentation through activation of caspase-3 (33). PGE_2 could exacerbate neurotoxicity by augmenting astrocytic glutamate release (34). $PGF_2\alpha$ stimulated generation of hydroxyl radical, increased the lipid peroxidation, and eventually induced apoptosis (35). It has been shown that PGD_2 readily undergoes dehydration in vivo and in vitro to yield biologically active PGs of the J series (cyPG), such as PGJ_2, Δ^{12}-PGJ_2, $\Delta^{12,14}$-PGJ_2, and 15-deoxy-$\Delta^{12,14}$-PGJ_2 (15d-PGJ_2). 15d-PGJ_2 caused apoptotic cell death in SH-SY5Y cells through mitochondrial damage, intracellular accumulation of reactive oxygen species (ROS), and depletion of reduced glutathione (36). Moreover, 15d-PGJ_2 induced phosphorylation/accumulation of p53 and activation of $p21^{Waf1}$ and $p27^{Kip1}$ (37,38). In contrast, several studies in the literature indicate that COX-2 expression plays a role in preventing apoptosis in a number of cancer cells and macrophages. For example, human hepatocellular carcinoma cells overexpressing COX-2 have enhanced phosphorylation of Akt/protein kinase B and are resistant to apoptosis (39). Furthermore, the inhibition of COX-2 by SC236, a selective COX-2 inhibitor, has been shown to cause apoptosis in gastric cancer cells (40). Thus, the association between COX-2 expression and cell death/survival seems to be cell type specific. It may also depend on a variety of PGE_2-induced cellular responses mediated through EP receptor subtypes/isoforms and the kinetics and duration of PGE_2 activation. Whereas higher concentrations of PGE$_2$ aggravated neurotoxicity induced by several toxic stimuli, submicromolar concentrations of PGE$_2$ have been reported to be protective through heme oxygenase-1 (HO-1) induction (41). Similarly, 15d-PGJ$_2$ elicits protective effects by upregulating HO-1 expression and glutathione synthesis (42).

IV.B. Nitrosative Stress and Apoptosis

Increased NO production can lead to cellular toxicity and death via a variety of mechanisms. NO induced apoptosis in several cell types including macrophage, astrocytes, microglia, and tumor cells through p53 accumulation, p21 upregulation, cytochrome c release, and caspase activation (43). NO can cause cell death by apoptosis. Under certain conditions, however, NO can induce poly(ADP-ribose)polymerase cleavage, thereby causing DNA damage and neuronal NADPH/ATP depletion which eventually leads to necrotic cell death. A high concentration of sodium nitroprusside (SNP) caused caspase-independent cell death, probably due to primary or secondary necrosis (44). Some of NO-mediated cytotoxic effects may be direct while others may arise from the reaction of NO with superoxide to form peroxynitrite. Peroxynitrite reacts in complex ways with different biomolecules, inducing lipid peroxidation, DNA strand scission, oxidation of cystein, lysine, methionine, and histidine residues, and nitration of heterocyclic compounds like tryptophan and guanine or tyrosine. These alterations will contribute to neuronal cell death (45). Moreover, NO can inhibit mitochondrial respiration at the respiratory complex IV, whereas peroxynitrite can inhibit mitochondrial aconitase and complex I, II, and V and open the mitochondrial permeability pore (46). To clarify the causal link among protein nitration, neuronal degeneration, and DNA damage in AD, Su et al. (47) compared fragmented DNA detected by the deoxynucleotidyl transferase histochemical technique and tyrosine-nitrated proteins and found that the majority of TdT-labeled nuclei coincided with upregulation of nitrotyrosine expression. Aβ stimulation of microglia and monocytes resulted in TNF-α-dependent expression of iNOS, nitrotyrosine formation and neuronal apoptosis in THP-1 cells and microglia (6,16). Aβ-induced NO release in primary rat cortical mixed cultures led to apoptosis which was inhibited by iNOS inhibitors (e.g., L-NIL and 1400W), NO scavengers (e.g., carboxy-PTIO), and peroxynitrite decomposition catalysts (FeTMPyP and FeTPPS) (48). However, depending upon intracellular conditions, NO can either be neurotoxic or neurotrophic. In certain redox states, NO can decrease Ca^{2+} influx and counteract N-methyl-D-aspartate (NMDA) receptor-mediated glutamate cytotoxicity (49). Long-lasting production of NO acts as a proapoptotic modulator, whereas low or physiological concentrations of NO prevent cells form apoptosis induced by trophic factor withdrawal, Fas, TNF-α, and LPS (43,50). The antiapoptotic action of NO is attributable to expression of protective genes including those encoding heat shock proteins as well as *bcl-2* and direct inhibition of the caspase family proteases through S-nitrosylation of the cysteine thiol (43,50). Possible proinflammatory/ oxidative stress-mediated molecular mechanisms involved in Aβ-induced apoptosis are illustrated in Fig. 2.

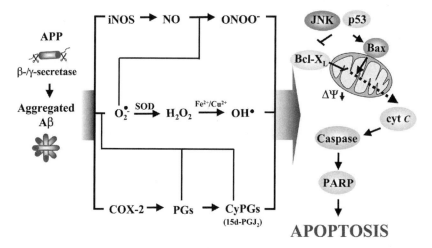

Figure 2 Possible inflammatory/oxidative stress-mediated molecular mechanisms involved in Aβ-induced apoptosis.

V. CROSS-TALK BETWEEN iNOS AND COX PATHWAYS IN INFLAMMATION

Under certain physiological conditions, NO and PGs appear to work cooperatively and synergistically (Fig. 3). In addition, coinduction or coregulation of COX-2 and iNOS has been demonstrated in a number of cell culture studies and animal models.

V.A. Modulation of the COX Pathway by Nitrosative Stress

There has been an increasing body of evidence supporting the roles of endogenous or exogenous NO in PG biosynthesis. One of the most plausible mechanisms by which NO activates COX-2 is direct stimulation of the enzyme (51). NO has affinity for iron or iron-containing enzymes, which contributes to the vast majority of pharmacologic effects it exerts, and it is noticeable that the active site of COX-2 contains an iron

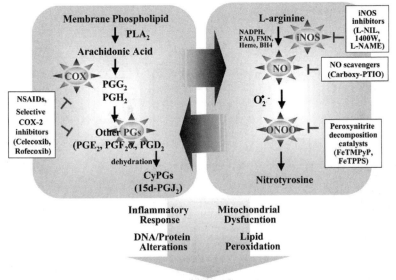

Figure 3 Cross-talk between COX-2 and iNOS signaling leading to neuronal dysfunctions and cell death.

heme center. It has been reported that NO donors, such as, SNP, S-nitroso-penicillamine (SNAP), LPS/INF-γ/NMDA, and S-nitrosoguanosine (GSNO), stimulate the production of PGE_2 via the induction of COX-2 in neuronal or cancer cell lines, chondrocytes, and macrophage, which was inhibited by NO scavengers or iNOS inhibitors (52–54). However, NO is reported to exert dual effects on COX-2 activation, depending on the reactivity of NO and peroxynitrite. Low concentrations of NO or peroxynitrite stimulate synthesis of PG via S-nitrosylation of COX-2 thiols, whereas high concentrations of peroxynitrite inhibit COX-2 via nitration of critical tyrosyl residues (55). Therefore, the amount of NO generated in different biological system seems to be one of the critical factors affecting the action of NO on PG generation.

V.B. Modulation of NO by the COX Pathway

The cross-talk between NOS and COX pathways is evident in the action of NO on PG biosynthesis and vice versa. While NO can act as a regulator of COX activity and/or expression, little is known about the involvement of COX products, such as PGE_2, in the modulation of the NOS pathway. PGE_2 reduced apoptotic thresholds of the chondrocytes against NO-induced cytotoxicity, including direct DNA damage, the generation of peroxynitrite, and the inactivation of antioxidant enzymes (56). However, the role of COX enzymes or their products in the regulation of the NOS pathway is still controversial. PGE_2, via cAMP formation subsequent to activation of G protein-coupled prostanoid receptors, has both stimulatory and inhibitory effects on iNOS expression and activity depending on the in vitro concentration and the cell type. Limited levels of cAMP stimulate iNOS expression via protein kinase A phosphorylation of transcription factors, yet high levels of cAMP inhibit release of cytokines that ordinarily promote iNOS expression (57). In addition, 15d-PGJ_2 inhibited LPS- or IL-1β-induced expression of iNOS and COX-2 by interfering with nuclear factor κB (NF-κB) and activator protein-1 (AP-1) transcriptional activity in activated microglia and chondrocytes (58,59).

VI. TRANSCRIPTION FACTORS AND UPSTREAM INTRACELLULAR SIGNALING MOLECULES IN Aβ-INDUCED INFLAMMATION

VI.A. Transcription Factors

Several consensus sequences, including binding sites for NF-κB, CCAAT/enhancer binding protein (C/EBP), cyclic AMP response element (CRE), AP-1, activator protein 2 (AP-2), hypoxia inducible factor (HIF), Erg1, signal transduced and activator of transcription (STAT), and specificity protein-1 (Sp-1) are found in 5′ region of the *Cox-2* gene and regulate COX-2 expression in response to a variety of stimuli in different cell types (60,61). The 5′-flanking region of iNOS has been cloned and sequenced, and putative binding sites for NF-κB, Sp-1, AP-1, AP-2, CRE, C/EBP, interferon γ response element (γ-IRE), interferon α-stimulated element (ISRE), tumor necrosis factor response element (TNR-RE), HIF, γ-activated site (GAS) and octamer-like sequence (Oct-1) have been identified in the promoter of *iNos* (62,63).

NF-κB is one of the most ubiquitous eukaryotic transcription factors that regulate expression of genes involved in controlling cellular proliferation/growth and inflammatory responses. The functionally active NF-κB exists mainly as a heterodimer of Rel family proteins, which is normally sequestered in the cytoplasm as an inactive complex with inhibitory protein, IκB. Activation of an IκB-specific protein kinase pathway by diverse stimuli leads to phosphorylation and degradation of IκB. Removal of IκB allows NF-κB to translocate into the nucleus, interact with the DNA-binding motif, and regulate the activation or expression of a variety of target genes. Highly close relationship between NF-κB activation and the COX-2 mRNA expression was observed in sporadic AD superior temporal lobe cortex (64). Conversely, systematic deletion of NF-κB DNA-binding sites in the COX-2 promoter construct attenuated transcription of COX-2 induced by mediators of inflammation in a variety of neuronal cell types (64). Aβ treatment of primary mouse microglia and the human monocytic cell line induced a time-dependent activation of NF-κB, with subsequent increases in IL-1β, TNFα, and NO production (16). iNOS immunoreactivity has been observed in the region near Aβ neuritic plaques, and iNOS expression can be stimulated in cultured astrocytes or microglia by Aβ through an NF-κB-dependent mechanism (15).

AP-1 is another transcription factor that regulates expression of genes involved in cellular adaptation, differentiation, proliferation, inflammation, and oxidative response. AP-1 consists of either homo- or heterodimers between members of the Jun (c-Jun, JunB, and JunD) and Fos (c-Fos, FosB, Fra-1, and Fra-2) families, Jun dimerization partners (JDP1 and JDP2) and the closely related activating factors (ATF2, LRF1/ATF3 and B-ATF) which interact via a leucine-zipper domain. Similar to NF-κB, AP-1 activates transcription of genes involved in inflammatory disorders. The 5'-flanking region of Rhesus monkey APP gene has been cloned and sequenced, and putative binding sites for AP-1, AP-2, and NF-κB have been identified (65). NF-κB and AP-1 binding sites were also identified in the IL-1/TNF-α-responsive enhancer of the acute phase protein, α_1-antichymotrypsin gene (2). Furthermore, the neuroprotective activities of anti-inflammatory phytochemicals have been ascribed to their inhibitory effects on COX-2 and iNOS expression, which are considered to be mediated by downregulation of NF-κB or AP-1.

The C/EBP family of transcription factors comprises pleiotropic proteins involved in tissue-specific metabolic gene transcription, signal transduction activated by several cytokines, and cell differentiation (66). The C/EBP family includes C/EBPα, C/EBPβ (also known as NF-IL6, IL-6-DBP, LAP, AGP/EBP, or CRP2), C/EBPδ (also called CRP3, CELF, or NF-IL6β), C/EBPγ (Ig/EBP), C/EBPε (CRP1), and C/EBPζ (CHOP, Gadd153) which form homo- or heterodimers and bind to similar cis-regulatory elements with various affinities. C/EBPα, -β, and -δ are involved in terminal differentiation of a variety of cells, and C/EBPβ and -δ activate various genes involved in acute phase, inflammatory, and immune responses. Especially, C/EBPβ regulates the expression of several genes that are important for inflammation and immunity, including IL-1β, IL-6, TNFα, iNOS, and COX-2. In the brain tissue, C/EBP has been reported to be induced by hypoxic–ischemic brain injury and proinflammatory cytokines, such as IL-1β and TNF-α (67,68).

Another major transcription factor targeted by inflammatory signaling cascades is CREB. CREB is a member of the CREB/activating transcription factor 1/CRE modulator (CREM) family of transcription factor. CREB mediates cAMP-, growth factor-, and calcium-dependent gene expression through the CRE. The biological effects of CREB are mediated by phosphorylation at serine 133, which results in cellular gene expression critical for proliferation, differentiation, and inflammation. C-Terminal fragment of APP, which is another constituent of senile plaque and an abnormal product of APP metabolism, induced CREB activation and TNF-α expression (69). Conversely, a cell-permeable cAMP analog dibutyryl cAMP suppressed CREB DNA-binding activity, interaction with NF-κB, and subsequent TNF-α expression. The transcription factors that play a role in regulating Aβ-mediated inflammatory cascades are schematically represented in Fig. 4.

VI.B. Upstream Regulators of Transcription Factors: Mitogen-Activated Protein Kinases

Numerous intracellular signal-transduction pathways converge with the activation of the transcription factors, which act independently or coordinately to regulate target gene expression. The most extensively investigated intracellular signaling cascades involved in proinflammatory responses is probably the one mediated by mitogen-activated protein kinases (MAPKs). Three distinct groups of well-characterized major MAPK subfamily members include extracellular signal-regulated protein kinase (ERK), JNK, and p38 MAPK which are serine/threonine protein kinases. The activated form of each of these signaling enzymes then phosphorylates and activates other kinases or transcription factors, thereby altering the expression of the target genes.

ERK has been proposed to act by simulating NF-κB, AP-1, C/EBP, and CREB. The ERK signaling pathway has also

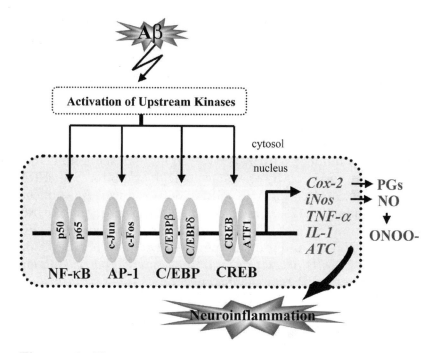

Figure 4 The representative transcription factors that may be involved in intracellular signaling cascades leading to neuroinflammation.

been linked to NF-κB activation. Thus, the ERK regulated kinase, 90-κDa ribosomal S kinase (p90^{RSK}) can phosphorylate IκB (70). The ERK pathway may also be responsible for AP-1 activation. The phosphorylation of Elk-1 by ERK resulted in a potentiated interaction with serum response element (SRE) and enhanced binding to SRE on the promoter region of *c-fos* for elevated expression (71,72). Moreover, the ERK pathway is also involved in the phosphorylation of CREB and C/EBP (73,74). The p38 MAPK can phosphorylate and activate Elk-1, resulting in enhanced SRE-dependent *c-fos* expression (71,72). In addition, p38 MAPK is involved in ATF2 phosphorylation through p38 MAPK-activated protein kinase-2 (MAPKAPK-2) activation (75). Furthermore, p38 MAPK mediates CREB phosphorylation through mitogen- and stress-activated protein kinases (MSK-1) or MAPKAPK-2 (75). JNK, on the other hand, phosphorylates the stimulatory site of c-Jun and ATF2 (71,72). ATF-2 can form a heterodimer with c-Jun and induce activation of AP-1. JNK is also capable of activating TCF/Elk-2, suggesting it may be involved in *c-fos* induction under certain circumstances.

Microglial cells stimulated with Aβ showed rapid activation of MAPKs followed by increased IL-1β levels, which was reduced by MAPKs inhibitors (76). In a primary rat microglia culture, Aβ induced MAPK activation and subsequent TNF-α production, and both pharmacological inhibitors of ERK (PD098059) and p38 MAPK (SB203580) reduced Aβ-induced TNF-α production (77). The exposure of rat primary astrocytes to C-terminal fragment of APP resulted in activation of MAPK as well as NF-κB, and pretreatment of PD098059 or SB203580 decreased NF-κB activation and subsequent NO production (78). Mutant APP transgenic mice have higher activity of

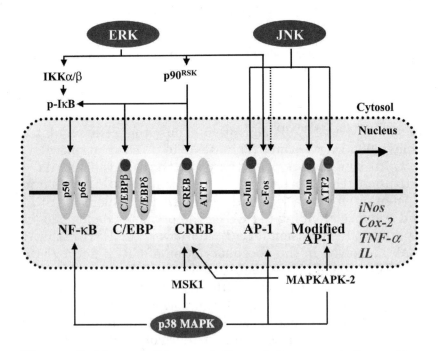

Figure 5 The possible regulation of the neuroinflammation-related transcription factors by MAPKs.

p38, AP-1, and iNOS, which were suppressed by SD-282, a selective inhibitor of p38 MAPK (79). Injection of preaggregated Aβ into the nuclear basalis of the rat caused a strong inflammatory reaction characterized by elevated IL-1β production and increased COX-2 and iNOS expression through p38 MAPK (80). Freshly solubilized Aβ increased COX-2 expression and PGE$_2$ production, which were prevented by inhibition of p38 MAPK with SB202190 (81). The possible involvement of upstream MAPKs in Aβ-mediated inflammatory stress is schematically proposed in Fig. 5.

VII. CONCLUSION AND FUTURE DIRECTIONS

Some PGs generated by COX-2 and NO or peroxynitrite produced by iNOS may play a pivotal role in the neuroinflammation of AD. However, the complete molecular events involved in these processes remain to be elucidated. A wide array of upstream kinases and transcription factors are involved in COX-2 and iNOS expression. Understanding of cellular and molecular regulatory mechanisms of these proinflammatory enzymes may provide better therapeutic opportunities for the management of AD. Continued attempts to identify the novel molecular targets for the neuroinflammation and to clarify their cross-talk with upstream and downstream signaling molecules will pave the way to exploiting the preventive or therapeutic strategies for fighting inflammatory neurodegenerative disorders.

ACKNOWLEDGMENT

This work was supported by the grant from Korea Research Foundation (ES0022).

REFERENCES

1. Reale M, Iarlori C, Gambi F, Feliciani C, Salone A, Toma L, DeLuca G, Salvatore M, Conti P, Gambi D. Treatment with an acetylcholinesterase inhibitor in Alzheimer patients

modulates the expression and production of the pro-inflammatory and anti-inflammatory cytokines. J Neuroimmunol 2004;
148:162–171.

2.   Kordula T, Bugno M, Rydel RE, Travis J. Mechanism of
     interleukin-1-β and tumor necrosis factor alpha-dependent
     regulation of the alpha 1-antichymotrypsin gene in human
     astrocytes. J Neurosci 2000; 20:7510–7516.

3.   McGeer PL, Rogers J, McGeer EG. Neuroimmune mechanisms
     in Alzheimer disease pathogenesis. Alzheimer Dis Assoc Disord
     1994; 8:149–158.

4.   Velazquez P, Cribbs DH, Poulos TL, Tenner AJ. Aspartate
     residue 7 in amyloid beta-protein is critical for classical
     complement pathway activation: implications for Alzheimer's
     disease pathogenesis. Nat Med 1997; 3:77–79.

5.   Haga S, Ikeda K, Sato M, Ishii T. Synthetic Alzheimer amyloid
     beta/A4 peptides enhance production of complement C3 com-
     ponent by cultured microglial cells. Brain Res 1993; 601:88–94.

6.   Akama KT, Van Eldik LJ. Beta-amyloid stimulation of induci-
     ble nitric-oxide synthase in astrocytes is interleukin-1 beta-
     and tumor necrosis factor-alpha (TNFalpha)-dependent, and
     involves a TNF alpha receptor-associated factor- and NF
     kappaB-inducing kinase-dependent signaling mechanism.
     J Biol Chem 2000; 275:7918–7924.

7.   Forloni G, Mangiarotti F, Angeretti N, Lucca E, De Simoni MG.
     Beta-amyloid fragment potentiates IL-6 and TNF-alpha secre-
     tion by LPS in astrocytes but not in microglia. Cytokine 1997; 9:
     759–762.

8.   Hoozemans JJ, Rozemuller AJ, Janssen I, De Groot CJ,
     Veerhuis R, Eikelenboom P. Cyclooxygenase expression in
     microglia and neurons in Alzheimer's disease and control
     brain. Acta Neuropathol 2001; 101:2–8.

9.   Pasinetti GM, Aisen PS. Cyclooxygenase-2 expression is
     increased in frontal cortex of Alzheimer's disease brain.
     Neuroscience 1998; 87:319–324.

10. Montine TJ, Sidell KR, Crews BC, Markesbery WR, Marnett LJ,
     Roberts LJ, Morrow JD. Elevated CSF prostaglandin E_2 levels
     in patients with probable AD. Neurology 1999; 53:1495–1498.

11. Stephenson DT, Lemere CA, Selkoe DJ, Clemens JA. Cytosolic phospholipase A_2 ($cPLA_2$) immunoreactivity is elevated in Alzheimer's disease brain. Neurobiol Dis 1996; 3:51–63.

12. Ischiropoulos H. Biological selectivity and functional aspects of protein tyrosine nitration. Biochem Biophys Res Commun 2003; 305:776–783.

13. Beckman JS. Oxidative damage and tyrosine nitration from peroxynitrite. Chem Res Toxicol 1996; 9:836–844.

14. Beckman JS, Beckman TW, Chen J, Marshall PA, Freeman BA. Apparent hydroxyl radical production by peroxynitrite: implications for endothelial injury from nitric oxide and superoxide. Proc Natl Acad Sci USA 1990; 87:1620–1624.

15. Akama KT, Albanese C, Pestell RG, Van Eldik LJ. Amyloid beta-peptide stimulates nitric oxide production in astrocytes through an NF kappaB-dependent mechanism. Proc Natl Acad Sci USA 1998; 95:5795–5800.

16. Combs CK, Karlo JC, Kao SC, Landreth GE. Beta-amyloid stimulation of microglia and monocytes results in TNF alpha-dependent expression of inducible nitric oxide synthase and neuronal apoptosis. J Neurosci 2001; 21:1179–1188.

17. Yang SN, Hsieh WY, Liu DD, Tsai LM, Tung CS, Wu JN. The involvement of nitric oxide in synergistic neuronal damage induced by beta-amyloid peptide and glutamate in primary rat cortical neurons. Chin J Physiol 1998; 41:175–179.

18. Shalit F, Sredni B, Rosenblatt-Bin H, Kazimirsky G, Brodie C, Huberman M. Beta-amyloid peptide induces tumor necrosis factor-alpha and nitric oxide production in murine macrophage cultures. Neuroreport 1997; 8:3577–3580.

19. Smith MA, Richey Harris PL, Sayre LM, Beckman JS, Perry G. Widespread peroxynitrite-mediated damage in Alzheimer's disease. J Neurosci 1997; 17:2653–2657.

20. Hensley K, Maidt ML, Yu Z, Sang H, Markesbery WR, Floyd RA. Electrochemical analysis of protein nitrotyrosine and dityrosine in the Alzheimer brain indicates region-specific accumulation. J Neurosci 1998; 18:8126–8132.

21. Breitner JC, Gau BA, Welsh KA, Plassman BL, McDonald WM, Helms MJ, Anthony JC. Inverse association of anti-

inflammatory treatments and Alzheimer's disease: initial results of a co-twin control study. Neurology 1994; 44:227–232.

22. Delanty N, Vaughan C. Risk of Alzheimer's disease and duration of NSAID use. Neurology 1998; 51:652.

23. Rich JB, Rasmusson DX, Folstein MF, Carson KA, Kawas C, Brandt J. Nonsteroidal anti-inflammatory drugs in Alzheimer's disease. Neurology 1995; 45:51–55.

24. Hull M, Lieb K, Fiebich BL. Pathways of inflammatory activation in Alzheimer's disease: potential targets for disease modifying drugs. Curr Med Chem 2002; 9:83–88.

25. Giovannini MG, Scali C, Prosperi C, Bellucci A, Vannucchi MG, Rosi S, Pepeu G, Casamenti F. Beta-amyloid-induced inflammation and cholinergic hypofunction in the rat brain in vivo: involvement of the p38MAPK pathway. Neurobiol Dis 2002; 11: 257–274.

26. Govoni S, Masoero E, Favalli L, Rozza A, Scelsi R, Viappiani S, Buccellati C, Sala A, Folco G. The cyclooxygenase-2 inhibitor SC58236 is neuroprotective in an in vivo model of focal ischemia in the rat. Neurosci Lett 2001; 303:91–94.

27. Youdim KA, Joseph JA. A possible emerging role of phytochemicals in improving age-related neurological dysfunctions: a multiplicity of effects. Free Radic Biol Med 2001; 30:583–594.

28. Tamagno E, Parola M, Guglielmotto M, Santoro G, Bardini P, Marra L, Tabaton M, Danni O. Multiple signaling events in amyloid beta-induced, oxidative stress-dependent neuronal apoptosis. Free Radic Biol Med 2003; 35:45–58.

29. Ho L, Pieroni C, Winger D, Purohit DP, Aisen PS, Pasinetti GM. Regional distribution of cyclooxygenase-2 in the hippocampal formation in Alzheimer's disease. J Neurosci Res 1999; 57: 295–303.

30. Xiang Z, Ho L, Valdellon J, Borchelt D, Kelley K, Spielman L, Aisen PS, Pasinetti GM. Cyclooxygenase (COX)-2 and cell cycle activity in a transgenic mouse model of Alzheimer's disease neuropathology. Neurobiol Aging 2000; 23:327–334.

31. Ho L, Osaka H, Aisen PS, Pasinetti GM. Induction of cyclooxygenase (COX)-2 but not COX-1 gene expression in apoptotic cell death. J Neuroimmunol 1998; 89:142–149.

32. Bate C, Veerhuis R, Eikelenboom P, Williams A. Neurones treated with cyclo-oxygenase-1 inhibitors are resistant to amyloid-beta 1–42. Neuroreport 2003; 14:2099–2103.

33. Takadera T, Yumoto H, Tozuka Y, Ohyashiki T. Prostaglandin E_2 induces caspase-dependent apoptosis in rat cortical cells. Neurosci Lett 2002; 317:61–64.

34. Bezzi P, Carmignoto G, Pasti L, Vesce S, Rossi D, Rizzini BL, Pozzan T, Volterra A. Prostaglandins stimulate calcium-dependent glutamate release in astrocytes. Nature 1998; 391: 281–285.

35. Liu D, Li L, Augustus L. Prostaglandin release by spinal cord injury mediates production of hydroxyl radical, malondialdehyde and cell death: a site of the neuroprotective action of methylprednisolone. J Neurochem 2001; 77:1036–1047.

36. Kondo M, Oya-Ito T, Kumagai T, Osawa T, Uchida K. Cyclopentenone prostaglandins as potential inducers of intracellular oxidative stress. J Biol Chem 2001; 276:12076–12083.

37. Han C, Demetris AJ, Michalopoulos GK, Zhan Q, Shelhamer JH, Wu T. PPARγ ligands inhibit cholangiocarcinoma cell growth through p53-dependent GADD45 and p21 pathway. Hepatology 2003; 38:167–177.

38. Hashimoto K, Ethridge RT, Evers BM. Peroxisome proliferator-activated receptor gamma ligand inhibits cell growth and invasion of human pancreatic cancer cells. Int J Gastrointest Cancer 2002; 32:7–22.

39. Leng J, Han C, Demetris AJ, Michalopoulos GK, Wu T. Cyclooxygenase-2 promotes hepatocellular carcinoma cell growth through Akt activation: evidence for Akt inhibition in celecoxib-induced apoptosis. Hepatology 2003; 38:756–768.

40. Jiang XH, Lam SK, Lin MC, Jiang SH, Kung HF, Slosberg ED, Soh JW, Weinstein IB, Wong BC. Novel target for induction of apoptosis by cyclo-oxygenase-2 inhibitor SC-236 through a protein kinase C-β_1-dependent pathway. Oncogene 2002; 21: 6113–6122.

41. Chen YC, Shen SC, Lee WR, Lin HY, Ko CH, Lee TJ. Nitric oxide and prostaglandin E_2 participate in lipopolysaccharide/interferon-gamma-induced heme oxygenase 1 and prevent

RAW264.7 macrophages from UV-irradiation-induced cell death. Cell Biochem 2002; 86:331–339.

42. Alvarez-Maqueda M, El Bekay R, Alba G, Monteseirin J, Chacon P, Vega A, Martin-Nieto J, Bedoya FJ, Pintado E, Sobrino F. 15-Deoxy-delta 12,14-prostaglandin J_2 induces heme oxygenase-1 gene expression in a reactive oxygen species-dependent manner in human lymphocytes. J Biol Chem 2004; 279:21929–21937.

43. Chung HT, Pae HO, Choi BM, Billiar TR, Kim YM. Nitric oxide as a bioregulator of apoptosis. Biochem Biophys Res Commun 2001; 282:1075–1079.

44. Kuhn K, Lotz M. Mechanisms of sodium nitroprusside-induced death in human chondrocytes. Rheumatol Int 2003; 23: 241–247.

45. Demiryurek AT, Cakici I, Kanzik I. Peroxynitrite: a putative cytotoxin. Pharmacol Toxicol 1998; 82:113–117.

46. Brown GC, Borutaite V. Nitric oxide, cytochrome c and mitochondria. Biochem Soc Symp 1999; 66:17–25.

47. Su JH, Deng G, Cotman CW. Neuronal DNA damage precedes tangle formation and is associated with up-regulation of nitrotyrosine in Alzheimer's disease brain. Brain Res 1997; 774: 193–199.

48. Law A, Gauthier S, Quirion R. Neuroprotective and neurorescuing effects of isoform-specific nitric oxide synthase inhibitors, nitric oxide scavenger, and antioxidant against beta-amyloid toxicity. Br J Pharmacol 2001; 133:1114–1124.

49. Lei SZ, Pan ZH, Aggarwal SK, Chen HS, Hartman J, Sucher NJ, Lipton SA. Effect of nitric oxide production on the redox modulatory site of the NMDA receptor–channel complex. Neuron 1992; 8:1087–1099.

50. Boyd CS, Cadenas E. Nitric oxide and cell signaling pathways in mitochondrial-dependent apoptosis. Biol Chem 2002; 383: 411–423.

51. Salvemini D, Misko TP, Masferrer JL, Seibert K, Currie MG, Needleman P. Nitric oxide activates cyclooxygenase enzymes. Proc Natl Acad Sci USA 1993; 90:7240–7244.

52. Lim SY, Jang JH, Surh YJ. Induction of cyclooxygenase-2 and peroxisome proliferator-activated receptor-γ during nitric oxide-induced apoptotic PC12 cell death. Ann N Y Acad Sci 2003; 1010:648–658.

53. von Knethen A, Callsen D, Brune B. NF-κB and AP-1 activation by nitric oxide attenuated apoptotic cell death in RAW 264.7 macrophages. Mol Biol Cell 1999; 10:361–372.

54. Park SW, Lee SG, Song SH, Heo DS, Park BJ, Lee DW, Kim KH, Sung MW. The effect of nitric oxide on cyclooxygenase-2 (COX-2) overexpression in head and neck cancer cell lines. Int J Cancer 2003; 107:729–738.

55. Boulos C, Jiang H, Balazy M. Diffusion of peroxynitrite into the human platelet inhibits cyclooxygenase via nitration of tyrosine residues. J Pharmacol Exp Ther 2000; 293:222–229.

56. Notoya K, Jovanovic DV, Reboul P, Martel-Pelletier J, Mineau F, Pelletier JP. The induction of cell death in human osteoarthritis chondrocytes by nitric oxide is related to the production of prostaglandin E_2 via the induction of cyclooxygenase-2. J Immunol 2000; 165:3402–3410.

57. Galea E, Feinstein DL. Regulation of the expression of the inflammatory nitric oxide synthase (NOS2) by cyclic AMP. FASEB J 1999; 13:2125–2137.

58. Petrova TV, Akama KT, Van Eldik LJ. Cyclopentenone prostaglandins suppress activation of microglia: down-regulation of inducible nitric-oxide synthase by 15-deoxy-$\Delta^{12,14}$-prostaglandin J_2. Proc Natl Acad Sci USA 1999; 96:4668–4673.

59. Boyault S, Simonin MA, Bianchi A, Compe E, Liagre B, Mainard D, Becuwe P, Dauca M, Netter P, Terlain B, Bordji K. 15-Deoxy-$\Delta^{12,14}$-prostaglandin J_2, but not troglitazone, modulates IL-1β effects in human chondrocytes by inhibiting NF-κB and AP-1 activation pathways. FEBS Lett 2001; 501:24–30.

60. Kosaka T, Miyata A, Ihara H, Hara S, Sugimoto T, Takeda O, Takahashi E, Tanabe T. Characterization of the human gene (PTGS2) encoding prostaglandin-endoperoxide synthase 2. Eur J Biochem 1994; 221:889–897.

61. Tazawa R, Xu XM, Wu KK, Wang LH. Characterization of the genomic structure, chromosomal location and promoter of

human prostaglandin H synthase-2 gene. Biochem Biophys Res Commun 1994; 203:190–199.

62. Eberhardt W, Kunz D, Hummel R, Pfeilschifter J. Molecular cloning of the rat inducible nitric oxide synthase gene promoter. Biochem Biophys Res Commun 1996; 223:752–756.

63. Keinanen R, Vartiainen N, Koistinaho J. Molecular cloning and characterization of the rat inducible nitric oxide synthase (iNOS) gene. Gene 1999; 234:297–305.

64. Lukiw WJ, Bazan NG. Strong nuclear factor-kappaB-DNA binding parallels cyclooxygenase-2 gene transcription in aging and in sporadic Alzheimer's disease superior temporal lobe neocortex. J Neurosci Res 1998; 53:583–592.

65. Song W, Lahiri DK. Molecular cloning of the promoter of the gene encoding the Rhesus monkey beta-amyloid precursor protein: structural characterization and a comparative study with other species. Gene 1998; 217:151–164.

66. Ramji DP, Foka P. CCAAT/enhancer-binding proteins: structure, function and regulation. Biochem J 2002; 365(Pt 3): 561–575.

67. Cardinaux JR, Allaman I, Magistretti PJ. Pro-inflammatory cytokines induce the transcription factors C/EBPbeta and C/EBPdelta in astrocytes. Glia 2000; 29:91–97.

68. Walton M, Saura J, Young D, MacGibbon G, Hansen W, Lawlor P, Sirimanne E, Gluckman P, Dragunow M. CCAAT-enhancer binding protein alpha is expressed in activated microglial cells after brain injury. Brain Res Mol Brain Res 1998; 61:11–22.

69. Chong YH, Shin YJ, Suh YH. Cyclic AMP inhibition of tumor necrosis factor alpha production induced by amyloidogenic C-terminal peptide of Alzheimer's amyloid precursor protein in macrophages: involvement of multiple intracellular pathways and cyclic AMP response element binding protein. Mol Pharmacol 2003; 63:690–698.

70. Schouten GJ, Vertegaal AC, Whiteside ST, Israel A, Toebes M, Dorsman JC, van der Eb AJ, Zantema A. IkappaB alpha is a

target for the mitogen-activated 90 kDa ribosomal S6 kinase. EMBO J 1997; 16:3133–3144.

71. Karin M. The regulation of AP-1 activity by mitogen-activated protein kinases. J Biol Chem 1995; 270:16483–16486.

72. Whitmarsh AJ, Davis RJ. Transcription factor AP-1 regulation by mitogen-activated protein kinase signal transduction pathways. J Mol Med 1996; 74:589–607.

73. Nakajima T, Kinoshita S, Sasagawa T, Sasaki K, Naruto M, Kishimoto T, Akira S. Phosphorylation at threonine-235 by a ras-dependent mitogen-activated protein kinase cascade is essential for transcription factor NF-IL6. Proc Natl Acad Sci USA 1993; 90:2207–2211.

74. Mora-Garcia P, Cheng J, Crans-Vargas HN, Countouriotis A, Shankar D, Sakamoto KM. Transcriptional regulators and myelopoiesis: the role of serum response factor and CREB as targets of cytokine signaling. Stem Cells 2003; 21:123–130.

75. Guha M, Mackman N. LPS induction of gene expression in human monocytes. Cell Signal 2001; 13:85–94.

76. Kim SH, Smith CJ, Van Eldik LJ. Importance of MAPK pathways for microglial pro-inflammatory cytokine IL-1β production. Neurobiol Aging 2004; 25:431–439.

77. Pyo H, Jou I, Jung S, Hong S, Joe EH. Mitogen-activated protein kinases activated by lipopolysaccharide and beta-amyloid in cultured rat microglia. Neuroreport 1998; 9:871–874.

78. Bach JH, Chae HS, Rah JC, Lee MW, Park CH, Choi SH, Choi JK, Lee SH, Kim YS, Kim KY, Lee WB, Suh YH, Kim SS. C-terminal fragment of amyloid precursor protein induces astrocytosis. J Neurochem 2001; 78:109–120.

79. Koistinaho M, Kettunen MI, Goldsteins G, Keinanen R, Salminen A, Ort M, Bures J, Liu D, Kauppinen RA, Higgins LS, Koistinaho J. Beta-amyloid precursor protein transgenic mice that harbor diffuse A beta deposits but do not form plaques show increased ischemic vulnerability: role of inflammation. Proc Natl Acad Sci USA 2002; 99:1610–1615.

80. Giovannini MG, Scali C, Prosperi C, Bellucci A, Vannucchi MG, Rosi S, Pepeu G, Casamenti F. Beta-amyloid-induced inflammation and cholinergic hypofunction in the rat brain in vivo: involvement of the p38MAPK pathway. Neurobiol Dis 2002; 11:257–274.

81. Paris D, Townsend KP, Obregon DF, Humphrey J, Mullan M. Pro-inflammatory effect of freshly solubilized beta-amyloid peptides in the brain. Prostaglandins Other Lipid Mediat 2002; 70:1–12.

18

Biomarkers for Evaluating Antioxidant Effects

ROBIN VAN DEN BERG and H VAN DEN BERG

TNO Nutrition and Food Research, Zeist, The Netherlands

G.R.M.M. HAENEN and A. BAST

Department of Pharmacology and Toxicology, University of Maastricht, Maastricht, The Netherlands

I. FREE RADICALS AND ANTIOXIDANTS: NEED OR THREAT?

Humans are exposed to many carcinogens, but the most significant may be the reactive species derived from metabolism of oxygen and nitrogen. Formation of reactive oxygen species (ROS) and reactive nitrogen species (RNS) in the human body can cause oxidative damage to biological macromolecules such as DNA, lipids, and proteins (1) that may contribute to the development of cancer, cardiovascular and neurological diseases, cataract, and other oxidative stress-mediated dysfunctions (2–4). Therefore, the assumption is often made that ROS/RNS are always 'bad', but the human body also needs ROS/RNS as, for example, their function in primary immune defence or for relaxation of smooth muscles in blood vessel-walls.

Typical physiological relevant ROS/RNS (see Table 1) are the superoxide radicals, peroxyl radicals, hydroxyl radicals, singlet oxygen, peroxynitrite, and hydrogen peroxide. The ROS/RNS are either radicals (molecules that contain at least one unpaired electron) or reactive nonradical compounds, capable of oxidizing biomolecules. Therefore, these intermediates are also called oxidants or prooxidants.

To counteract the prooxidant actions in the human body, an intricate network of antioxidants (Fig. 1) is operative in biological systems including enzymatic and nonenzymatic antioxidants. Enzymatic antioxidants (Table 2) such as superoxide dismutases (SOD), catalase, and glutathione peroxidases provide the first line of defence. The SOD catalyses the dismutation of superoxide to oxygen and peroxide, which is protonated to form hydrogen peroxide. Although hydrogen peroxide is a powerful oxidant, it is relatively unreactive toward most biologic substrates unless it is present in unphysiologically high concentrations. Catalase converts hydrogen peroxide into water and oxygen, and glutathione peroxidases reduce hydrogen peroxide and organic

Table 1 Typical Physiologic Relevant Reactive Oxygen and Nitrogen Species

ROS Radicals	*Nonradicals*
Superoxide, $O_2^{\bullet-}$	Hydrogen peroxide, H_2O_2
Hydroxyl, $OH^{\bullet}$	Hypochlorous acid, HOCl
Peroxyl, $RO_2^{\bullet}$	Hypobromous acid, HOBr
Alloxyl, $RO^{\bullet}$	Ozone, O_3
Hdroperoxyl, $HO_2^{\bullet}$	Singlet Oxygen, $^1O_2$
RNS Radicals	*Nonradicals*
Nitric oxide, $NO^{\bullet}$	Nitrous acid, HNO_2
Nirtrogen dioxide, $NO_2^{\bullet}$	Nitrosyl cation, NO^+
	Nitroxyl anion, NO^-
	Peroxynitrite, $ONOO^-$
	Peroxynitrous acid, ONOOH
	Alkyl peroxynitrites, ROONO

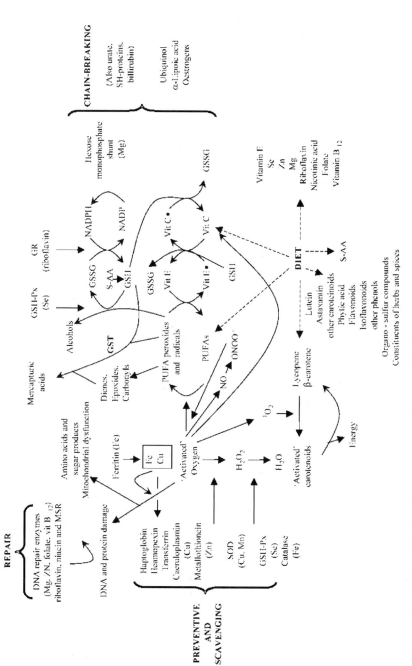

Figure 1 Antioxidant defense system GR, glutathione reductase; GSH, reduced glutathione; GSH-Px, glutathione peroxidase; GSSG, oxidized glutathione; GST, glutathione-S-transferase; MSR, methione sulphoxide reductase; PUFA, polyunsaturated fatty acids; S-AA, sulfur amino acids; SH-proteins, sulphydryl proteins; SOD, superoxide dismutase; Fe-Cu, transition metal-catalyzed oxidant damage to biomolecules. [Based on Benzie (80).]

Table 2 Enzymatic Antioxidants

Enzyme	Reaction catalyzed
Superoxide dismutase	$O_2^{\bullet-} + O_2^{\bullet-} + 2H^+ \rightarrow H_2O_2 + O_2$
Catalase	$H_2O_2 + H_2O_2 \rightarrow 2H_2O + O_2$
Glutathione peroxidase	$2GSH + ROOH \rightarrow GSSG + ROH + H_2O$

hydroperoxides. Indirect antioxidant functions are mediated by enzymes that restore endogenous antioxidant levels; e.g., glutathione (GSH) levels are replenished on reduction of oxidized glutathione (GSSG) by glutathione reductase. Reactive products or xenobiotics, e.g., epoxide, can be detoxified by phase II detoxification enzymes such as glutathione-s-transferases to favor their excretion. Metal chelating plasma proteins (e.g., transferrin, ceruloplasmin, and albumin) prevent the formation of ROS/RNS by controlling the levels of free iron or copper ions. These metal chelating proteins bind the redox active metals and limit the production of free radicals.

The human diet also provides a range of different compounds that possess antioxidant activities or have been suggested to scavenge ROS/RNS based on their structural properties (Table 3). The most prominent representatives of dietary antioxidants are ascorbate (vitamin C), tocopherols (vitamin E), carotenoids, and flavonoids. Antioxidants have different solubilities, which partition across the phases of tissues, cells, and macromolecular structures: water-soluble ascorbate, glutathione and urate; lipid-soluble tocopherols and carotenoids, and intermediatory-soluble

Table 3 Prominent Representative Dietary Antioxidants

Naturally occurring nonenzymatic antioxidants	
Hydrophilic	Lypophilic
Ascorbic acid	Carotenoids
Glutathione	α-Tocopherol
	Ubiquinol-10
Flavonoids	Flavonoids

flavonoids. The protection against ROS/RNS provided by fruit and vegetables could arise through an integrated reductive environment delivered by antioxidants of differing solubility in each of the tissues, cellular, and macromolecular phases (5).

Observational epidemiological studies clearly show a correlation between the increased consumption of food rich in antioxidants and a decreased risk of several oxidative stress related diseases (6). These observations suggest that poor nutritional status may modulate ROS-mediated cellular damage due to an impaired defence. In addition to protection against direct oxygen radical-induced toxicity, antioxidants may also provide protection through their ability to inhibit the activation of oxidant-sensitive transcription factors (e.g., NF-κB) and subsequent production of proinflammatory mediators and adhesion molecules. Thus, maintaining adequate antioxidant status may be a useful approach to attenuate the cellular injury and dysfunction observed in inflammatory disorders (Fig. 2). Because of apparent importance of ROS/RNS as toxins, mediators, and modulators of inflammatory gene activation, antioxidants might be used as therapeutic agents for a number of different disease states.

II. HOW TO ASSESS/QUANTIFY HEALTH BENEFITS FROM ANTIOXIDANTS IN HUMANS

A double-blind placebo-controlled intervention trial in a sufficiently large population group remains the ultimate proof to demonstrate health benefits, such as a reduction in disease risk. Traditionally, observational epidemiological studies on disease risk were focused on clinical outcome with hard clinical endpoints. Such studies are indispensable, but they are time- and money-consuming and may have intrinsic limitations. A promising approach to strengthen epidemiological studies is the use of biomarkers. The biomarker concept was recently discussed by ILSI FUFOSE Concerted action (7,8). A biomarker can be defined as an indicator on a biochemical, genetic, or cellular level, reflecting exposure to a compound,

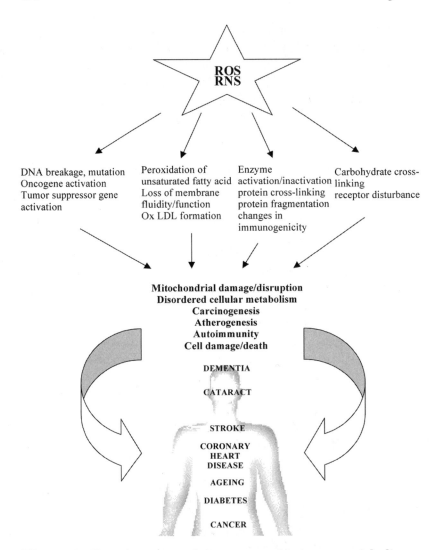

Figure 2 Reactive oxygen/nitrogen species increase risk disease through damage to key biological structures. [Based on Benzie (80).]

susceptibility for a disease or the health status of a subject. A biomarker of oxidative stress reflects radical burden, (suscept-ibility of) oxidative damage, and oxidative stress-mediated disease, or health status. For example, if we assume that direct damage to DNA by ROS contributes significantly to age-related development of cancer, then agents that decrease such damage should decrease the risk of cancer development. The steady-state level of oxidative DNA damage in human tissues is then a surrogate marker (biomarker) for later cancer development (9).

In the physiological evaluation of a biomarker, it is important that there is a well-established relationship between the response of the biomarker and the effect monitored. Ideally, the biomarker is the only target in the etiology. However, minor pathways may also provide suitable biomarkers. In the causal pathway of disease occurrence, one might distinguish biomarkers of exposure (dietary intake), biomarkers of biological response, and of (subclinical) disease and biomarkers of susceptibility (Fig. 3). For example, blood levels of vitamin E (exposure marker) may be studied in relation to oxidation resistance of LDL (a biological response marker) or to carotid artery wall thickness (a disease marker), in subjects with familial hypercholesterolaemia, or specific genotype (both susceptibility markers). Although biomarkers have the potential

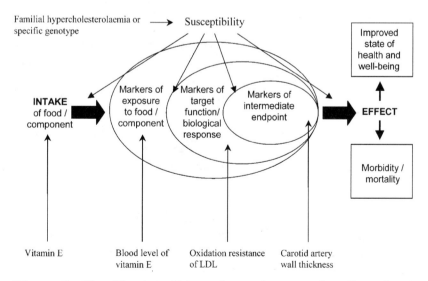

Figure 3 Classification of biomarkers relevant to the effect of antioxidants (example of vitamin E; modified from ILSI Europe, 1999).

for improving validity, several problems are encountered. Biomarkers of exposure should accurately reflect relevant dietary intake or body status, while the "early disease" markers should have predictive value for the hard end-point.

An effective nutritional strategy will require knowledge of the type of antioxidants in the diet, their food sources, bioavailability, and required levels of intake for protective effects. Protective effects of antioxidants have been found in mechanistic (in vitro), in animal studies (10), as well as in epidemiological studies (11), and in some small-scale intervention studies (12–14). For evaluation of the effect of antioxidants, the total evidence from basic science, epidemiological, and intervention studies, rather than that of one type of study has to be considered.

Application of biomarkers requires ability to measure oxidative damage to relevant molecular target in the appropriate tissue. One obvious problem is the limited availability of healthy human tissues from which to obtain it. One cannot ethically biopsy the colon, prostate, or breast of healthy subjects multiple times to investigate how DNA damage is affected by dietary manipulations. Therefore, studies mostly use cells from "easy" accessible tissues (e.g., buccal cells) or blood cells assuming that changes here will be reflected in the tissue at risk of, for example, cancer development later in life. These cells/tissues are not necessarily the target cells from a subject.

Another problem is the isolation of these target cells. All steps used for isolation and purification have the potential to cause an artefactual raise in the apparent steady-state levels of oxidative stress or damage and could invalidate the measurement. Nevertheless, several laboratories have developed standardized analytical protocols that could prevent such artefacts.

In the evaluation of the suitability of a marker, both analytical and physiological criteria should be used. For the analytical validation procedures that have been developed for standardization and validation of methods for the determination of drugs (15) could also be applicable for biomarkers. According to this procedure, criteria for the lower limit of quantification, precision, accuracy, and selectivity should be given.

A major problem encountered for the analytical validation of biomarkers is that many compounds used as a biomarker have a "normal" background concentration, and therefore no completely blank matrix is available. This hampers the correct estimation of the lower limit of quantification and the accuracy. For the analysis of drugs, the registration authorities accept an inaccuracy and imprecision up to 15% at the normal levels, and 20% at the lower limit of quantification (15). For biomarker assays, where the differences in concentration between the normal and elevated levels can be relatively low, stricter criteria for the accuracy and precision are probably needed. An important practical point performing large trials for establishing a claim is the feasibility of the methodology to be applied to a large sample size. Collection, storage, and transport of samples as well as time-consuming laboratory procedures have to be taken into account. In such human studies internal validity must be ensured by proper comparability of intervention and control groups, or of cases and controls.

For pathophysiological evaluation to assess the relevance of a biomarker of biological function, human (intervention) studies, using subjects with an established oxidative stress, or in patients suffering from a disease associated with increased oxidative stress such as in diabetes, or with inflammation-related diseases, are required. Such studies will, in turn, eventually accept or disprove the biomarkers used. Human intervention studies using antioxidants or food products generally aim at establishing either enhanced function (type A claim) or reduced risk (type B claim). Regarding study design, for type A claims biomarkers of target function/biological response are required and for type B claims biomarkers of intermediate end-points. Although the interest in the use of antioxidants for treatment of human diseases, and in the role of dietary antioxidant in the prevention of disease, development has been sustained for at least 2 decades, contradictory effects accumulate. For example, vitamin E supplements were protective against cardiovascular disease in the CHAOS study (16), but not in the GISSI-Prevenzione trial (17). An antioxidant paradox was postulated by Halliwell (18) to explain these contradictory effects. Both trials studied high-risk patients to prevent second events. One factor, which was suggested as an explanation, was that fatal myocardial infarction is often associated with rupture of advanced unstable plaques. In such plaques, free transition-metal ions are present (19) and vitamin E could interact with them to exert prooxidant effects (20). This can possibly explain why highly oxidized lipids and vitamin E coexist (21). Thus, high-dose vitamin E could facilitate rupture, most likely soon after administration begins, as was seen in CHAOS and protective effects took longer to establish (200 days or more). Second, unlike the Cambridge population, the Italian population already has the benefits of a Mediterranean diet rich in phenolic antioxidants yet developed cardiovascular disease anyway. This population may be "selected" not to respond in a beneficial way to antioxidant supplements, especially as synthetic vitamin E rather than pure RRR α-tocopherol was used. Halliwell (18) suggested that perhaps their atherosclerotic lesions have less free transition-metal ions and are thus less prone to be rendered unstable by prooxidant effects of antioxidants. Also, subjects consuming diets rich in fruits and vegetables apparently have a lower risk of getting cancer and an increased concentration of β-carotene in the blood. Intervention with (high dose) β-carotene supplements, however, did not exert an anticancer effect, rather the opposite effect was found in smokers (22).

From this antioxidant paradox, one could also introduce a "biomarker paradox." Since a biomarker that is increased is usually considered as "bad" or associated with increased disease risk, reducing this increased biomarker might be considered as "good" or "decreased risk," the opposite could also be the case. For example, increased DNA damage can be ascribed to an increase in free radical attack on DNA or decreased repair. If the biomarker of DNA damage (e.g., 8-hydroxy-deoxyguanosine (8-OHdG)) is increased, this could be interpreted as "bad" as increased radical attack or "good" as for increased repair. Another example of a biomarker paradox is the marker SOD. The SOD is an enzymatic antioxidant, which provides the first line of defence in the human body is considered as "good" when increased. However, SOD has been shown to be increased in Down syndrome and is suggested to be responsible for oxidative stress and subsequently neural dysfunction in these subjects (23). Like SOD, catalase, which converts hydrogen peroxide into water and oxygen, is considered as a "good" enzymatic antioxidant. An increased catalase activity could be interpreted as "good" or "better" protected. However, with doxorubicin-induced heart toxicity due to excessive production of radicals, catalase activity is paradoxically increased (24).

Before starting an intervention trial, one should be aware of these contradictions and carefully select the relevant markers and include a well-characterized study population. Methodologies of commonly used markers for evaluating and quantifying antioxidant protection and oxidative damage in humans are discussed below.

III. METHODOLOGIES OF MARKERS FOR EVALUATING ANTIOXIDANT PROTECTION AND OXIDATIVE DAMAGE

Antioxidant capacity can be assessed in either food products as well in biological samples and can be monitored by a variety of simple, nonspecific, high-throughput screenings assays. These assays are generally scavenging assays using relatively "stable" synthetic-free radicals (e.g., the 2,2'-azinobis-3-ethylbenzthiazoline-6-sulfonate (ABTS) radical), or reactive "natural" occurring radicals (e.g., $OH^{\bullet}$ or $O_2^{\bullet-}$). Total antioxidant capacity of biological samples can also be evaluated in human studies, which measure free-radical damage products of endogenous compounds such as lipids or DNA. Diet-induced changes from base-line levels are indicative for the antioxidant capacity of the diet or (food) product. A general problem with these type of assays is that there they are, as yet, poorly validated, and therefore, their specificity as biomarker is still under debate. Measurement of total antioxidant capacity of biological fluids might be useful as a marker for the ability of antioxidants present in body fluids in preventing oxidative damage to membranes and other cellular components.

The possible interaction among different antioxidants in vivo makes the measurement of any individual antioxidant less representative compared to the overall antioxidant status. Several methods have been developed to measure the total antioxidant capacity of biological samples or food products. These methods have been classified as inhibition methods involving reactive species, which are usually free radicals.

The total radical trapping parameter (TRAP) assay of Wayner et al. (25) was the first and the most widely applied strategy for measuring total antioxidant capacity of plasma or serum. The TRAP assay uses peroxyl radicals generated from 2,2′-azobis(2-amidinopropane) dihydrochloride (ABAP) and peroxydizable materials present in plasma or other biological fluids. After adding ABAP to the plasma, the oxidation of oxidizing materials is monitored by measuring the oxygen consumed during the reaction. During an induction period, this oxidation is inhibited by antioxidants present in the plasma. The induction period (lag-phase) is quantified by standard addition of 6-hydroxyl-2,5,7,8-tetramethylchroman-carboxylic acid (Trolox), and the lag-phase of plasma related that of Trolox. A major problem with the original TRAP assay lies in the instability of the oxygen electrode.

Using this original TRAP assay, Mulholland and Strain (26) reported that serum TRAP values were significantly lower in patients with acute myocardial infarction when compared to sex- and age-matched controls. The plasma TRAP values were also reported to decline significantly by about 40% during chemotherapy in patients with hematologic malignancies (27).

Since then, a lot of modified TRAP-like assays have been developed. In 1992, the first modified TRAP assay was reported by Metsä-Ketelä and Kirkola (28) using a chemiluminescense-based TRAP assay. In this method, peroxyl radicals produced from ABAP oxidize luminol, leading to the formation of luminol radicals that emit light. This method was further modified by Whitehead et al. (29) using hydrogen peroxide or perborate for the oxidation of luminol and the enhancer para-iodophenol giving a more intense, prolonged, and stable light emission. Other methods that also measure TRAP are the

dichlorofluorescin-diacetate-based assay (30), the total oxyradical scavenging capacity assay of Winston et al. (31), the cro-
cin-based assays (32), and the phycoerythrin-based assays
(33,34).

One of the most interesting methods is the one presented
by Miller et al. (35). The method was first used to ascertain
total antioxidant capacity in blood and other human fluids.
This method is based on the inhibition by antioxidants of
the absorbance of the radical anion of ABTS, which has a
characteristic long wavelength absorption spectrum showing
maxima at 660, 734, and 820 nm. The ABTS radical anions,
that are quite stable, are formed by the interaction of ABTS
with the ferrylmyoglobin radical species (metmyoglobin) with
hydrogen peroxide (Fig. 4). During the formation period, the
accumulation of ABTS radicals is inhibited by antioxidants
present in the plasma. The length of the induction period
(lag-phase) is also compared to that of Trolox as in the original TRAP assay of Wayner et al. (25). The specific properties
of ABTS were already routinely used in hospitals in kinetic
methods for enzyme activity measurements, e.g., the "glucose–oxidase assay" (36) to give ABTS radicals by oxidation
with hydrogen peroxide in the presence of peroxidase. Also,
in a simple kinetic method for determination of serum cholesterol (37), of haemoglobin (38), and uric acid (39), the ABTS
radical was used.

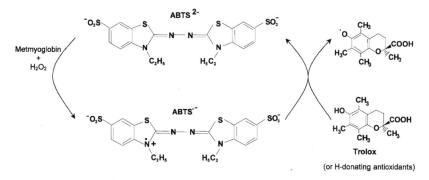

Figure 4 The formation of the "stable" radical ABTS and the reaction with the water-soluble vitamin E analogue Trolox.

The original radical trapping assay of Miller et al. (35), also called the Trolox equivalent antioxidant capacity (TEAC) assay, has been commercialized by Randox Laboratories (San Francisco, CA, U.S.A.). The intra- and interassay CV of the TEAC assay were reported to be 0.54–1.59% and 3.6–6.1%, respectively (40).

Plasma antioxidant capacity was found to be increased in patients with acute myocardial infarction (41) and decreased in premature infants (35,40). However, the low plasma TEAC found in premature infants was mainly due to lower protein content of the plasma. The TEAC of premature infants will be significantly higher, not lower, compared to term infants or adults when the data are based on the amount of total protein contained in the plasma (42). All compounds present in the serum or plasma that are able to scavenge the radicals are detected as potential antioxidants (43).

The greater part of the activity of serum or plasma (approximately 80%) has to be ascribed to albumin and urate (44). Due to the high antioxidant capacity of plasma, effects of antioxidant supplementation are relatively small. To reduce the high level of antioxidant capacity, the TEAC should be measured in deproteinated plasma, which makes the assay more sensitive.

In the original TEAC assay, the lag-phase caused by different antioxidants in the production of ABTS radicals is determined. The ABTS radicals are generated from ABTS and hydrogen peroxide by metmyoglobin. The reduction of the amount of ABTS radicals caused by the added antioxidant at a fixed time point is quantified (35). Added antioxidants quench ABTS radical anions formed by the interaction of hydrogen peroxide with metmyoglobin. However, a direct interaction of an ingredient of the sample with the reagent cannot be totally excluded (44). These interactions may reduce or even increase the production of radical species. The specificity of the TEAC assay is measuring capacity of a sample to directly quench free radicals is not always guaranteed (45), because compounds may also inhibit metmyoglobin and thus prevent ABTS radical formation leading to the false conclusion that these compounds are antioxidants.

III.A. Currently Used Methods Assessing Scavenging of "Stable" Nonbiological Free Radicals

Besides the TEAC assay, which is a sensitive but nonspecific and noninvasive method for assessing total antioxidant capacity of water-soluble antioxidants, also many other methods are performed.

The oxygen radical absorbance capacity (ORAC) assay is used to quantify the ORAC of antioxidants in food and biological samples (34). In this assay, beta-phycoerythrin (beta-PE) is used as an indicator protein and ABAP as a peroxyl radical generator. In this assay, the reaction goes to completion, so that both inhibition time and inhibition degree are considered in quantifying ORAC. The ORAC assay responds to numerous water-soluble antioxidants and is therefore a sensitive, nonspecific method. However, the ORAC is more time-consuming compared with other total antioxidant capacity measurements.

The method using relatively stable 2,2-diphenyl-1-picryl-hydrazyl (DPPH•) free radicals is applied for antioxidant activity measurements of food and biological samples (46). The disappearance of the radical can be followed spectrophotometricly and is expressed as radical scavenging ability. The DPPH• assay is used for lipid-soluble compounds and is a sensitive, nonspecific assay. Interpretation of this assay could be complicated if the absorption spectrum of test compounds overlaps with the DPPH• spectrum (515 nm) as with compounds such as carotenoids (47).

The ferric reducing ability (FRAP) assay is a simple test measuring the FRAP of food and plasma. This assay has been presented as a novel method for assessing "antioxidant power" (48), but is not a radical scavenging assay. Ferric to ferrous ion reduction at low pH causes formation of a colored ferrous-tripyridyltriazine complex. The ferric reducing ability of plasma (FRAP) values is obtained by comparing the change in absorbance. This assay is simple and inexpensive but does not measure the sulfydryl group-containing antioxidants (44). The antioxidant capacity of an antioxidant against a free radical does not necessarily match its ability to reduce Fe^{3+} to Fe^{2+}. Interpretation of the assay is complicated since the reduction of iron is also involved in free radical generation process as, e.g., the catalyzed $OH^{\bullet}$ generation from H_2O_2.

III.B. In Vitro Methods Assessing Scavenging of Biological Relevant Free Radicals

Hydroxyl radical ($OH^{\bullet}$) scavenging can be indirectly measured. In the $OH^{\bullet}$ scavenging assay, or the "deoxyribose assay", $OH^{\bullet}$ radicals are generated from hydrogen peroxide, ascorbate, and $FeCl_3$. Scavenging activity can be assessed by measuring competition between the test compound with deoxyribose for hydroxyl radicals (49). However, artefacts may occur with this test, such as: (i) some substances can rapidly react with hydrogen peroxide or (ii) compounds interfere with the measurement products.

   Superoxide radical ($O_2^{\bullet -}$) radical scavenging can also be indirectly measured. In the superoxide radical scavenging assay, $O_2^{\bullet -}$ is generated from xanthine/xanthine oxidase (X/XO) and scavenging can be measured by reaction of $O_2^{\bullet -}$ with cytochrome C or nitro-blue tetrazolium (NBT) (50). This assay is also sensitive to artefacts, such as inhibition of xanthine oxidase or reducing cytochrome C or NBT by the compounds. Other complications may arise with test compound(s) that absorb strongly at $290\,nm$ or formation of products that reduces cytochrome C or NBT.

   Scavenging of various free radicals, such as $OH^{\bullet}$ and $O_2^{\bullet -}$, can be assessed by electron spin resonance (ESR) with a spin trap, e.g., 5,5-dimethyl-1-pyrrolineN-oxide (DMPO) (51). Using ESR, radical scavenging activity of several body fluids can be measured by normalizing the ESR signals relative to the standard activity of vitamin C or a stable water-soluble analog of vitamin E. This method is less prone to artefacts, but also less sensitive and more laborious compared to the deoxyribose and the superoxide radical scavenging assay.

IV. METHODS OF ASSESSING OXIDATIVE DAMAGE

IV.A. Oxidative Damage to DNA

In stead of scavenging assays, damage to DNA, proteins and lipids can also be used for monitoring total antioxidant capacity. Oxidative DNA damage can be characterized ex vivo by measuring modified DNA bases as a result of several ROS/RNS attack on DNA. The most exploited DNA lesion, used as an index of oxidative DNA damage, is 8-OHdG (52). This product is a result of free radical attack upon guanine and may be a useful marker of damage on DNA since it is excreted in urine. For measurement of 8-OHdG (which is the most commonly measured DNA lesion), high performance liquid chromatography with electrochemical detection (HPLC–ECD) (53) or gas chromatography with mass spectrometry (GC–MS) (54) techniques are used. Both methods suffer from artefactual oxidation of guanine during the isolation and enzymatic digestion of DNA and during the chemical derivatization procedure for the GC–MS method. The analysis of 8-OHdG is often conducted with lymphocytes and urine. Nevertheless, it is uncertain whether measurement of oxidized DNA in lymphocytes reflects oxidative damage in other tissues.

It is often assumed that urinary 8-OHdG is derived from cellular DNA repair processes. Estimates of the rate of input of oxidative damage have been based on urinary 8-OHdG levels, however, it is 8-oxoguanine, and not 8-OHdG that is released by base excision repair. Urinary 8-OHdG instead may be derived from DNA breakdown products from dead cells that are oxidized during elimination (55). The HPLC–EC is the most commonly used method; it is sensitive and the easiest method to use. The GC–MS method is more specific, measures the base and not the nucleoside, but is more laborious and is not used for urine measurement. Other oxidized bases, such as 8-hydroxy-deoxyadenosine, can also be measured, however, 8-OHdG seems more mutagenic and therefore more favorable (56).

Another technique for measuring antioxidant capacity by evaluating DNA damage is the highly sensitive "comet assay" using single cell gel electrophoresis (57). This assay measures DNA damage, i.e., DNA breaks in cells e.g., lymphocytes (58,59). DNA damage (i.e., breaks) can be visualized by staining and unwinding electrophoresis of the DNA. The assay can be quantified by scoring of fluorescence intensity of the tails. The sensitivity of this technique can be improved by the use of DNA restriction enzymes (e.g., endonuclease-III) converting the oxidized bases to strand breaks (60). However, basal levels as well as the reported effects of the interventions are rather variable, possibly reflecting differences in the populations, regimens, and functional correlates of the biomarkers as well as between laboratory and assay variations. It has been suggested that antioxidant depletion due to nutritional deficiency or an increased oxidative stress might facilitate demonstration of protective effects of antioxidants with respect to DNA damage (61,62). In general, the comet assay is considered to be a suitable, sensitive, and quantitative, test for DNA-damaging potential in biomonitoring studies.

IV.B. Oxidative Damage to Proteins

The field of protein oxidation is very much in its infancy. Nevertheless, oxidative damage to proteins might be of importance in vivo because it may affect receptor function, enzyme activity, or transport mechanisms and may contribute to secondary damage to other biomolecules, such as inactivation of DNA repair enzymes. Attack of RNS on tyrosine results into the production of 3-nitrotyrosine, which can be measured immunologically, by HPLC–ECD or GC–MS (63). In terms of sensitivity and specificity, it would appear that methods based on combinations of HPLC and various types of ECD are very versatile giving a limit of detection of 20 fmol per injection of protein hydrolysate (64). These methods are limited by the sample quantity and the preparation that is required to achieve acceptable chromatograms. A problem with these methods is that acid hydrolysis used for the digestion of tissues and proteins causes the nitration of tyrosine as an artefact. Nitrotyrosine is also reportedly degraded by HOCl, questioning its validity as an oxidative biomarker.

More use has been made of the carbonyl assay (65), a sensitive but nonspecific assay of oxidative protein damage assessing "steady-state" protein damage in human tissues and body fluids. The carbonyl assay is based on the reaction of ROS with amino acid residues in proteins (particularly histidine, arginine, lysine, and proline) giving carbonyl functions that can be measured after the reaction with 2,4-dinitrophenyl-hydrazine and detected using enzyme-linked immunosorbent assay (ELISA) (66). Measurement of carbonyls in human plasma can be a useful marker, but more work has to be done to identify the molecular nature of the protein carbonyls in human plasma.

IV.C. Oxidative Damage to Lipids: Lipid Peroxidation

Lipid peroxidation results in the formation of conjugated dienes, lipid hydroperoxides, and degradation products such as alkanes, aldehydes, and isoprostanes. For quantitative assessment of lipid peroxidation, a range of methods is available.

Measurement of conjugated diene formation is generally applied as a dynamic quantitation, e.g., during the oxidation of low-density lipoproteins (LDL), and is not generally applied to samples obtained in vivo. The LDL oxidation is (in vitro) initiated by free radicals. During this process, the rate of oxidation is dependent on endogenous antioxidants in LDL, accounting for the lag-phase of oxidation. The LDL oxidation is efficiently inhibited by lipophilic antioxidants of which α-tocopherol appears to be the most important (67).

Lipid hydroperoxides readily decompose, but can be measured directly and indirectly by a variety of techniques (68,69). Interpretations of the results of these assays are complicated due to the unstable products with a short half-lives. Measurement of malondialdehyde (MDA) by the thiobarbituric acid reactive substances (TBARS) assay is nonspecific, and is generally poor when applied to biological samples (7). However, HPLC (with fluorometric detection)-based TBA-tests record comparable low values between different laboraties, provided that butylated hydroxytoluene (BHT) is added with the TBA reagents (70). An MDA is one of the many products formed during the radical-induced decomposition of polyunsaturated fatty acids. Most often, MDA assay is determined using its reactivity at high temperature and low pH, toward thiobarbituric acid. This reaction is very sensitive but the specificity, even with optimized preanalytical (sampling, preservatives), and analytical stages (fluorescence, HPLC) are still a matter of debate (71). The TBARS assay should be considered as a test, providing global information on lipoperoxidation whereas specific determination of MDA can only appreciate one of the end-product formed during oxidative stress.

More recent assays based on the measurement of MDA or HNE-lysine adducts are likely to be more applicable to biological samples, since adducts of these reactive aldehydes are relatively stable (72). The discovery of the isoprostanes as lipid peroxidation products from arachidonic acid, which can be measured by gas chromatography mass spectrometry (73), has opened a new avenue by which to quantify lipid peroxidation in vivo. The F_2-isoprostanes can be measured in plasma and urine of healthy human subjects and are indicative of ongoing lipid peroxidation. This urinary marker of lipid oxidation as an indicator of whole body lipid peroxidation meets all criteria to be a useful marker, but the major question is whether minor lipid peroxidation products or isoprostanes are indicative of a major pathway involved in the disease process.

"Whole body" lipid peroxidation can also be determined by measuring hydrocarbon gases (ethane, pentane) in exhaled air and urinary secretion of TBARS (74). Hydrocarbon gas exhalation has too many confounding variables to be applicable in free-living human subjects (75). Urinary TBARS is also not considered as a suitable assay to assess whole body lipid peroxidation in response to changes in dietary composition (76), because many artificial products could be generated.

IV.D. Functional Markers of Oxidative Stress

Antioxidant status and markers of oxidative damage (DNA adducts, lipid and protein oxidation products) are frequently used to assess anti- and prooxidant effects. However, gene expression, being an important modulator of cell functions, has been shown, in some cases, to be under redox control.

The ROS are involved in the regulation of gene expression (77). Changes in gene expression can provide a sensitive marker of oxidative stress (or changes in antioxidant status) of tissue, but the development of markers based on such changes is at an early stage. Expression of genes selected should be specifically induced by oxidative stress, if possible, and the functional significance of the gene product and its induction should be known.

For a better understanding of ROS-associated disorders, recent studies have focused on the regulation of gene expression by intracellular reduction–oxidation (redox) state. A well-defined transcription factor, nuclear factor (NF) κB, has been identified to be regulated by the intracellular redox state. Transcription factors are proteins that bind to regulatory sequences, usually in the 5′ upstream promoter region of target genes, to increase the rate of gene expression. This may affect protein synthesis, which could result in altered cellular function.

The transcription factor NF-κB is of considerable interest to the field of free radical biology because it is activated by ROS and this enhanced activity can be modulated by antioxidants (77). Because ROS are generated under numerous pathological conditions, measurement of NF-κB activation could provide important insight into the etiology of these disorders and also offers the potential of NF-κB as a functional biomarker for oxidative stress.

To clarify the role of NF-κB in various aspects of cell function, inflammatory processes, and pathologies, clinical studies have to be performed using NF-κB activity as a target for intervention with drugs or food components. Because of the involvement of NF-κB in numerous pathophysiological processes, an unequivocal NF-κB determination might be useful in order to be able to use NF-κB as a functional biomarker. The most widely applied method to establish NF-κB activation is the electromobility shift assay (EMSA). In EMSA, differences in concentration of NF-κB proteins that have been translocated to the nucleus can be detected. Quantification of these proteins might provide useful information about the intracellular redox status with the aim to further explore the potential of antioxidant therapy.

Fundamental research in this rapidly developing area is focused on elucidating the exact role of NF-κB in oxidative stress-related processes and the molecular mechanism of action of antioxidants.

This NF-κB redox-controlled mechanism is quite distinct from the response to antioxidants that involves up-regulation of certain specific genes (e.g., glutathione-s-transferase) as a consequence, for example, of the antioxidant responsive element (ARE) present in the promotor. Such genes respond to a diverse selection of antioxidants and might reflect changes in oxidative status. Both induction of glutathione-s-transferases (GSTα and GSTπ), a family of phase II enzymes, might be used as markers of oxidative stress. The GSTs are able to conjugate electrophiles with glutathione GSH, or by a glutathione-dependent peroxidase activity. They may also participate in the repair of damaged cellular macromolecules as a result of oxidative stress. In addition, GSTπ can react directly with ROS via a sensitive SH-group, or be inactivated by disulfide formation that can be reversed by glutathione. Therefore, it has a specific response to oxidative stress (78). Glutathione itself, in its reduced form (GSH), is a powerful intracellular antioxidant, and the ratio of reduced to oxidized glutathione (GSH/GSSG ratio) might serve as a representative marker of the antioxidant capacity of the cell. Several clinical conditions are associated with reduced GSH levels (diabetes, sepsis, inflammatory lung processes, cancer, and immunodeficiency states (79)) that as a consequence can result in a lowered cellular redox potential. Thus, the GSH/GSSG ratio in blood could also serve as a functional marker of oxidative stress (80).

REFERENCES

1. Vaughan M. Oxidative damage to macromolecules mini-review series. J Biol Chem 1997; 272:18513.

2. Gutteridge JM. Free radicals in disease processes: a compilation of cause and consequence. Free Radic Res Commun 1993; 19:141–158.

3. Diplock AT. Antioxidants and disease prevention. Mol Aspects Med 1994; 15:293–376.

4. Sies H. Oxidative stress: oxidants and antioxidants. Exp Physiol 1997; 82:291–295.

5. Eastwood MA. Interaction of dietary antioxidants in vivo: how fruit and vegetables prevent disease? QJM 1999; 92: 527–530.

6. Steinmetz KA, Potter JD. Vegetables, fruit, and cancer prevention: a review. J Am Diet Assoc 1996; 96:1027–1039.

7. Diplock AT, Charleux JL, Crozier-Willi G, Kok FJ, Rice-Evans C, Roberfroid M, Stahl W, Vina-Ribes J. Functional food science and defence against reactive oxidative species. Br J Nutr 1998; 80: S77–S112.

8. Diplock AT, Aggett PJ, Ashwell M, Bornet F, Fern EB, Roberfroid M. Scientific concept of functional food in Europe: consensus document. Br J Nutr 1999; 81:S1–S27.

9. Halliwell B. Establishing the significance and optimal intake of dietary antioxidants: the biomarker concept. Nutr Rev 1999; 57:104–113.

10. Benzie IFF. Lipid peroxidation: a review of causes, consequences, measurement and dietary influences. Int J Food Sci Nutr 1996; 47:233–261.

11. Block G, Patterson B, Subar A. Fruit, vegetables and cancer prevention: a review of the epidemiological evidence. Nutr Cancer 1992; 18:1–29.

12. Pool-Zobel BL, Bub A, Müller H, Wollowski I, Rechkemmer G. Consumption of vegetables reduces genetic damage in humans: first results of an intervention trial with carotenoid-rich foods. Carcinogenesis 1997; 18:1847–1850.

13. Young JF, Nielsen SE, Haraldsdottir J, Daneshvar B, Lauridsen ST, Knuthsen P, Crozier A, Sandstrom B, Dragsted LO. Effect of fruit juice intake on urinary quercetin excretion and biomarkers of antioxidative status. Am J Clin Nutr 1999; 69:87–94.

14. Porrini M, Riso P. Lymphocyte lycopene concentration and DNA protection from oxidative damage is increased in women after a short period of tomato consumption. J Nutr 2000; 130: 189–192.

15. Shah VP, Behl CR, Flynn GL, Higuchi WI, Schaefer H. Principles and criteria in the development and optimization of topical therapeutic products. J Pharm Sci 1992; 81:1051–1054.

16. Stephens NG, Parsons A, Schofield PM, Kelly F, Cheeseman K, Mitchinson MJ. Randomised controlled trial of vitamin E in patients with coronary disease: Cambridge heart antioxidant study (CHAOS). Lancet 1996; 347:781–786.

17. GISSI-Prevenzione Investigators. Dietary supplementation with n-3 polyunsaturated fatty acids and vitamin E after myocardial infarction: results of the GISSI-Prevenzione trial. Lancet 1999; 354:447–455.

18. Halliwell B. The antioxidant paradox. Lancet 2000; 355: 1179–1180.

19. Makjanic J, Ponraj D, Tan BKH, Watt F. Nuclear microscopy investigations into the role of iron in atherosclerosis. Nucl Inst Meth Res 1999; 159(sect. B):356–360.

20. Maiorino M, Zamburlini A, Roveri A, Ursini F. Prooxidant role of vitamin E in copper-induced lipid peroxidation. FEBS Lett 1993; 330:174–176.

21. Niu X, Zammit V, Upston JM, Dean RT, Stocker R. Coexistence of oxidized lipids and alpha-tocopherol in all lipoprotein density fractions isolated from advanced human atherosclerotic plaques. Arterioscler Thromb Vasc Biol 1999; 19:1708–1718.

22. Rowe PM. Beta-carotene takes a collective beating. Lancet 1996; 347:249.

23. Iannello RC, Crack PJ, de Haan JB, Kola I. Oxidative stress and neural dysfunction in Down syndrome. J Neural Transm Suppl 1999; 57:257–267.

24. Julicher RH, Sterrenberg L, Haenen GRMM, Bast A, Noordhoek J. The effect of chronic adriamycin treatment on heart kidney and liver tissue of male and female rat. Arch Toxicol 1988; 61:275–281.

25. Wayner DD, Burton GW, Ingold KU, Locke S. Quantitative measurement of the total, peroxyl radical-trapping antioxidant capability of human blood plasma by controlled peroxidation. The important contribution made by plasma proteins. FEBS Lett 1985; 187:33–37.

26. Mulholland CW, Strain JJ. Serum total free radical trapping ability in acute myocardial infarction. Clin Biochem 1991; 24: 437–441.

27. Dürken M, Agbenu J, Finckh B, Hübner C, Pichlmeier U, Zeller W, Winkler K, Zander A, Kohlschütter A. Deteriorating free radical-trapping capacity and antioxidant status in plasma during bone marrow transplantation. Bone Marrow Transplant 1995; 15:757–762.

28. Metsä-Ketelä T, Kirkola AL. Total peroxyl radical-trapping capability of human LDL. Free Radic Res Commun 1992; 16S:215.

29. Whitehead TP, Thorpe GHG, Maxwell SRJ. Enhanced chemiluminoscent assay for antioxidant capacity in biological fluids. Anal. Chim Acta 1992; 266:265–277.

30. Valkonen M, Kuusi T. Spectrophotometric assay for total peroxyl radical-trapping antioxidant potential in human serum. J Lipid Res 1997; 38:823–833.

31. Winston GW, Regoli F, Dugas AJ Jr, Fong JH, Blanchard KA. A rapid gas chromatographic assay for determining oxyradical scavenging capacity of antioxidants and biological fluids. Free Radic Biol Med 1998; 24:480–493.

32. Tubaro F, Ghiselli A, Rapuzzi P, Maiorino M, Ursini F. Analysis of plasma antioxidant capacity by competition kinetics. Free Radic Biol Med 1998; 24:1228–1234.

33. Glazer AN. Phycoerythrin fluorescence-based assay for reactive oxygen species. Methods Enzymol 1990; 186:161–168.

34. Cao G, Alessio HM, Cutler RG. Oxygen-radical absorbance capacity assay for antioxidants. Free Radic Biol Med 1993; 14: 303–311.

35. Miller NJ, Rice-Evans C, Davies MJ, Gopinathan V, Milner A. A novel method for measuring antioxidant capacity and its application to monitoring the antioxidant status in premature neonates. Clin Sci 1993; 84:407–412.

36. Majkic N, Djordjevic-Spasic S, Berkes I. A kinetic method for the determination of the activity of "aerobic transhydrogenases". Clin Chim Acta 1975; 65:227–233.

37. Majkic N, Berkes I. Determination of free and esterified cholesterol by a kinetic method. I. The introduction of the enzymatic method with 2,2'-azino-di-3[ethyl-benzthiazolin sulfonic acid (6)] (ABTS). Clin Chim Acta 1977; 80:121–131.

38. Takayanagi M, Yashiro T. Colorimetry of hemoglobin in plasma with 2,2'-azino-di(3-ethylbenzthiazoline-6-sulfonic acid) (ABTS). Clin Chem 1984; 30:357–359.

39. Majkic-Singh N, Said BA, Spasic S, Berkes I. Evaluation of the enzymatic assay of serum uric acid with 2,2'-azino-di (3-ethylbenzthiazoline-6-sulphonate) (ABTS) as chromogen. Ann Clin Biochem 1984; 21:504–509.

40. Rice-Evans CA, Miller NJ. Total antioxidant status in plasma and body fluids. Meth Enzymol 1994; 234:279–293.

41. Guller K, Palanduz S, Ademoglu E, Sahnayenli N, Gokkusu C, Vatansever S. Total antioxidant status, lipid parameters, lipid peroxidation and glutathione levels in patients with acute myocardial infarction. Med Sci Res 1998; 26:105–106.

42. Cao G, Giovanoni M, Prior RL. Antioxidant capacity decreases during growth but not aging in rat serum and brain. Arch Gerontol Geriatr 1996; 22:27–37.

43. Wayner DD, Burton GW, Ingold KU, Barclay LR, Locke SJ. The relative contributions of vitamin E, urate, ascorbate and proteins to the total peroxyl radical-trapping antioxidant activity of human blood plasma. Biochim Biophys Acta 1987; 924:408–419.

44. Cao G, Prior L. Comparison of different analytical methods for assessing total antioxidant capacity of human serum. Clin Chem 1998; 44:1309–1315.

45. Strube M, Haenen GRMM, van den Berg H, Bast A. Pitfalls in a method for assessment of total antioxidant capacity. Free Radic Res 1997; 26:515–521.

46. Bondet V, Brand-Williams W, Berset C. Kinetics and mechanisms of antioxidant activity using the DPPH• free radical method. Lebensm Wiss Technol 1997; 30:609–615.

47. Nomura T, Kikuchi M, Kubodera A, Kawakami Y. Proton-donative antioxidant activity of fucoxanthin with 1,1-diphenyl-2-picrylhydrazyl (DPPH). Biochem Mol Biol Int 1997; 42:361–370.

48. Benzie IFF, Strain JJ. The ferric reducing ability of plasma (FRAP) as a measure of "antioxidant power": the FRAP assay. Anal Biochem 1996; 239:70–76.

49. Halliwell B, Gutteridge JM, Aruoma OI. The deoxyribose method: a simple "test-tube" assay for determination of rate constants for reactions of hydroxyl radicals. Anal Biochem 1987; 165:215–219.

50. Halliwell B. Use of desferrioxamine as a "probe" for iron-dependent formation of hydroxyl radicals. Evidence for a direct reaction between desferal and the superoxide radical. Biochem Pharmacol 1985; 34:229–233.

51. Noda Y, Anzai K, Mori A, Kohno M, Shinmei M, Packer L. Hydroxyl and superoxide anion radical scavenging activities of natural source antioxidants using the computerized JES-FR30 ESR spectrometer system. Biochem Mol Biol Int 1997; 42: 35–44.

52. Loft S, Fischer-Nielsen A, Jeding IB, Vistisen K, Poulsen HE. 8-Hydroxydeoxyguanosine as a urinary biomarker of oxidative DNA damage. J Toxicol Environ Health 1993; 40:391–404.

53. Shigenaga MK, Aboujaoude EN, Chen Q, Ames BN. Assays of oxidative DNA damage biomarkers 8-oxo-2′-deoxyguanosine and 8-oxoguanine in nuclear DNA and biological fluids by high-performance liquid chromatography with electrochemical detection. Methods Enzymol 1994; 234:16–33.

54. Dizdaroglu M. Quantitative determination of oxidative base damage in DNA by stable isotope-dilution mass spectrometry. FEBS Lett 1993; 315:1–6.

55. ESCODD (European Standards Committee on Oxidative DNA Damage). Comparison of different methods of measuring 8-oxoguanine as a marker of oxidative DNA damage. Free Radic Res 2000; 32:333–341.

56. ILSI Europe Report Series. Markers of Oxidative Damage and Antioxidant Protection: Current status and relevance to disease, 2000.

57. Collins AR, Duthie SJ, Dobson VL. Direct enzymic detection of endogenous oxidative base damage in human lymphocyte DNA. Carcinogenesis 1993; 14:1733–1735.

58. Anderson D, Yu TW, Phillips BJ, Schmezer P. The effect of various antioxidants and other modifying agents on oxygen-radical-generated DNA damage in human lymphocytes in the COMET assay. Mutat Res 1994; 307:261–271.

59. Betti C, Barale R, Pool-Zobel BL. Comparative studies on cytotoxic and genotoxic effects of two organic mercury compounds in lymphocytes and gastric mucosa cells of Sprague–Dawley rats. Environ Mol Mutagen 1993; 22:172–180.

60. Duthie SJ, Ma A, Ross MA, Collins AR. Antioxidant supplementation decreases oxidative DNA damage in human lymphocytes. Cancer Res 1996; 56:1291–1295.

61. Loft S, Poulsen HE. Antioxidant intervention studies related to DNA damage, DNA repair and gene expression. Free Radic Res 2000; 33:S67–S83.

62. Moller P, Knudsen LE, Loft S, Wallin H. The comet assay as a rapid test in biomonitoring occupational exposure to DNA-damaging agents and effect of confounding factors. Cancer Epidemiol Biomarkers Prev 2000; 9:1005–1015.

63. Van der Vliet A, Eiserich JP, Kaur H, Cross CE, Halliwell B. Nitrotyrosine as biomarker for reactive nitrogen species. Methods Enzymol 1996; 269:175–184.

64. Herce-Pagliai C, Kotecha S, Shuker DE. Analytical methods for 3-nitrotyrosine as a marker of exposure to reactive nitrogen species: a review. Nitric Oxide 1998; 2:324–336.

65. Levine RL, Williams JA, Stadtman ER, Shacter E. Carbonyl assays for determination of oxidatively modified proteins. Methods Enzymol 1994; 233:346–357.

66. Buss H, Chan TP, Sluis KB, Domigan NM, Winterbourn CC. Protein carbonyl measurement by a sensitive ELISA method. Free Radic Biol Med 1997; 23:361–366.

67. Esterbauer H, Gebicki J, Puhl H, Jurgens G. The role of lipid peroxidation and antioxidants in oxidative modification of LDL. Free Radic Biol Med 1992; 13:341–390.

68. Jiang ZY, Hunt JV, Wolff SP. Ferrous ion oxidation in the presence of xylenol orange for detection of lipid hydroperoxide in low density lipoprotein. Anal Biochem 1992; 202:384–389.

69. Thomas SM, Jessup W, Gebicki JM, Dean RT. A continuous-flow automated assay for iodometric estimation of hydroperoxides. Anal Biochem 1989; 176:353–359.

70. Halliwell B, Chirico S. Lipid peroxidation: its mechanism, measurement, and significance. Am J Clin Nutr 1993; 57: 715S–725S.

71. Lefevre G, Beljean-Leymarie M, Beyerle F, Bonnefont-Rousselot D, Cristol JP, Therond P, Torreilles J. Evaluation of lipid peroxidation by measuring thiobarbituric acid reactive substances. J Ann Biol Clin 1998; 56:305–319 (in French).

72. Moore K, Roberts LJ II. Measurement of lipid peroxidation. Free Radic Res 1998; 28:659–671.

73. Morrow JD, Roberts LJ II. Mass spectrometry of prostanoids: F2-isoprostanes produced by non-cyclooxygenase free radical-catalyzed mechanism. Methods Enzymol 1994; 233:163–174.

74. Gutteridge JM, Tickner TR. The characterisation of thiobarbituric acid reactivity in human plasma and urine. Anal Biochem 1978; 91:250–257.

75. Springfield JR, Levitt MD. Pitfalls in the use of breath pentane measurements to assess lipid peroxidation. J Lipid Res 1994; 35:1497–1504.

76. Brown ED, Morris VC, Rhodes DG, Sinha R, Levander OA. Urinary malondialdehyde-equivalents during ingestion of meat cooked at high or low temperatures. Lipids 1995; 30:1053–1056.

77. Sen CK, Packer L. Antioxidant and redox regulation of gene transcription. FASEB J 1996; 10:709–720.

78. Xia C, Hu J, Ketterer B, Taylor JB. The organization of the human GSTP1-1 gene promoter and its response to retinoic acid and cellular redox status. Biochem J 1996; 313:155–161.

79. Exner R, Wessner B, Manhart N, Roth E. Therapeutic potential of glutathione. Wien Klin Wochenschr 2000; 112:610–616.

80. Benzie IFF. Evolution of antioxidant defence mechanisms. Eur J Nutr 2000; 39:53–61.

19

Antioxidant and Neuroprotective Potentials

MUHAMMAD SOOBRATTEE and THEESHAN BAHORUN

Department of Biological Sciences, Faculty of Sciences, University of Mauritius, Redúit, Mauritius

OKEZIE I. ARUOMA

Department of Applied Science, London South Bank University, London, U.K.

I. RNS/ROS AND OXIDATIVE STRESS

Free radicals are formed from molecules by bond breaking such that each fragment keeps one electron (free radicals may also be formed by collision of the nonradical species by a reaction between a radical and a molecule, which must then result in a radical, as the total number of electrons is odd), by cleavage of a radical to give another radical and by oxidation or reduction reactions. Aerobic organisms, including man, animals, and plants, are constantly challenged by reactive nitrogen species (RNS) and reactive oxygen species (ROS). These are either synthesized endogenously, e.g., in energy

501

metabolism and by the immune defense system of the body, or produced as reactions to exogenous exposures such as cigarette smoking, imbalanced diet, exhaustive exercise, environmental pollutants, and food contaminants (1–3). Reactive nitrogen species is a collective term that includes nitric oxide ($NO^\bullet$) and nitrogen dioxide ($NO_2^\bullet$) radicals, as well as nonradicals such as peroxynitrite ($OONO^-$), nitrous acid (HNO_2), and dinitrogen trioxide (N_2O_3). Reactive oxygen species is also a collective term that includes both oxygen radicals, e.g., peroxyl ($ROO^\bullet$), superoxide ($O_2^{\bullet-}$), and hydroxyl ($OH^\bullet$) radicals and the nonradicals, e.g., singlet oxygen ($^1O_2$), hypochlorous acid ($HOCl$), and hydrogen peroxide (H_2O_2). Intricate and diverse defense systems exist in vivo (2–5). For example, antioxidant protection is afforded by the antioxidant enzymes such as superoxide dismutases, catalase, and glutathione peroxidase, as well as by low molecular weight antioxidants such as glutathione, NADH, carnosine, uric acid, melatonin, α-lipoic acid, bilirubin, some of them are endogenously produced and some are provided through dietary intake (ascorbic acid, tocopherols, ergothioneine, carotenoids, quinones, phenolics). Links continue to be made between oxidative stress and increased risk of chronic diseases such as cancer, cardiovascular diseases, eye diseases such as cataract, and age-related macular degeneration, immune and neurodegenerative disorders, as well as for the general ageing process (2,6,7).

II. NEURODEGENERATIVE DISORDERS

Approximately 15% of the population over the age of 65 are afflicted with Alzheimer's disease (AD) (8) and 1% by Parkinson's disease (PD) (9). Neurodegenerative disorders are associated with various degrees of behavioral impairments that significantly decrease the quality of life (10). The brain may be particularly vulnerable to ROS, in part due to the reasons in what follows.

    i.   The brain consumes ~20% of the total body oxygen but comprises <2% of total body weight.

ii. The brain contains high levels of unsaturated fatty acids.

iii. The brain may have reduced endogenous antioxidants.

iv. The brain has limited capacity for regeneration.

v. Iron accumulates in brain-specific regions (i.e., red nucleus, substantia nigra pars reticularis, globus pallidus). Iron-binding proteins (ferritin) may be relatively deficient in the brain (11), so that the high concentrations of ascorbic acid may present a pro-oxidant environment.

Oxidative stress has been implicated in the pathogenesis of neurodegenerative disorders, such as cerebral ischemia/reperfusion injury and trauma (12), as well as chronic conditions such as amyotrophic lateral sclerosis (ALS) and PD (6,13–16). That ROS and RNS are implicated (17) has given impetus to the research focus directed at medicinal and food plants (18–20).

III. ALZHEIMER'S DISEASE

The disease AD is a progressive degenerative brain disease and is the commonest cause of adult onset dementia. There are deficits in cognitive function that cause amnesia (loss of memory), aphasia (impairment of language), apraxia (inability to do motor tasks despite intact motor function), and agnosia (inability to recognize, despite intact sensory functions). Psychiatric symptoms and behavioral disturbances, such as depression, personality change, delusions, hallucinations, and misidentifications, become apparent. Early manifestation of the disease include difficulties with daily living activities such as handling money, using telephone, driving, and later, difficulties with dressing, feeding, and toileting. Patients eventually lose interest in their surroundings and become confined to wheelchair or bed. The final stages of the disease, marked by mental emptiness and loss of control of all body functions, may not occur until 5–10 years after onset.

The major microscopic alterations in AD are senile plaques (SP) and neurofibrillary tangles (NFT) formation, selective neuronal loss and shrinkage, neuropil thread formation, synapse loss, and amyloid angiopathy. The NFT and SP represent an accumulation of intraneuronal and extracellular filamentous protein aggregates, respectively. Hyperphosphorylated tau is the major protein in NFT. Amyloid β peptide, derived from the amyloid precursor protein (APP), is the major protein in SP and amyloid angiopathy. The AD brain is also characterized by neuronal cell loss and changes in neuronal morphology. This is reflected by a decreased brain weight and by atrophy of the cortex. Neuronal loss is most notable in the hippocampus, frontal, parietal, and anterior temporal cortices, amygdala, and the olfactory system (21). Neuronal cell loss occurs in the nucleus basalis, a large cholinergic system at the base of the forebrain, and this may account for the severe cholinergic deficiency in the cortex of AD patients (22). The cell loss that is likely in the locus ceruleus may also account for the reduction in brain level of norepinephrine AD patients. In the hippocampus, the most prominent zones that are affected are the CA1 region, the subiculum, and the entorhinal cortex. The entorhinal cortex receives major innervation from the neocortex, basal forebrain, and amygdala. The large neurons of the entorhinal cortex (layer II), which project to the subiculum and the dentate gyrus by means of the perforant pathway, are prominent sites of tangle formation in AD. The major hippocampal output—to mammillary bodies, hypothalamus, and dorsomedial thalamus—arises from axons of the pyramidal cells that exit the hippocampus via the fornix. Thus, severe pathology occurs in those neuronal populations that receive input to the hippocampus or provide hippocampal efferents. The limbic system, including the olfactory bulbs, olfactory cortices, amygdala, cingulate gyrus, and hypothalamus, may also be affected, and this may explain some of the abnormal behavioral characteristics of some AD patients. The vulnerable brain regions in HIV/AIDS individuals include the denta nucleus in the cerebellum, the red nucleus and substantia nigra in the midbrain, the subthalamic nucleus and thalamic fasciculus in the diencephalons, and the globus pallidus and striatum (or neostriatum, which consists of caudate and putamen) in the forebrain. It is easy to see why lesions in these regions may lead to progressive dementia, which is similar to what is observed in AD and PD (18).

III.A. Risk Factors for AD

Although AD is generally thought to be a multineurotransmitter deficiency disease, there seems to be a genetic basis for its development. At least three different genes have thus been far associated with early-onset (<60 years of age) forms of the disease and have a direct causal effect: the APP gene located on chromosome 21 (23) and presenilin genes 1 (S 182) (24) and 2 (E5-1) (25) located on chromosomes 14 and 1, respectively. The apolipoprotein E-4 (Apo E-4) gene located on chromosome 19, the α_2-macroglobulin gene located on chromosome 12, and other unidentified genes may determine susceptibility in late-onset forms and sporadic cases (26,27). Risk of AD increases and the age of onset decreases with the number of Apo E-4 allele. Up to 90% of individuals homozygous for Apo E-4 have a chance of developing AD by the age of 80. However, the Apo E-2 genotype appears to be somewhat protective for AD (28). An increasing number of factors, including hypertension, smoking, alcoholism, low level of education, diabetes mellitus, migraine, high LDL cholesterol, head injury, diet, and depression, are associated with the development of AD (29–31) and oxidative stress (18,32). The salient neuropathological features include primary involvement (with atrophy) of neocortical association areas, the hippo campus and entorhinal cortex, amygdala, and nucleus basalis of Meynert, accompanied by more variable involvement of the medial nucleus of the thalamus, dorsal tegmentum, locus ceruleus, paramedian reticular areas, and lateral hypothalamic nuclei (21). By definition, involved regions must contain numerous neuritic plaques (focal structures consisting of an amyloid β protein core surrounded by degenerating nerve terminals) and NFT (neuronal cell bodies filled with paired helical filaments consisting mainly of tau protein) (33,34). These neuropathologic hallmarks typically appear first in the hippocampal CA1 field and subiculum, with increasing involvement of the association cortex, the basal nucleus, and other structures, as the disease progresses (21). The most striking neurochemical alteration in the cerebral cortex is the loss of the cholinergic markers, choline acetyltransferase, and acetylcholinesterase (35,36). In addition, there are deficits in serotonin (37), noradrenaline (22), somatostatin (38), and corticotrophin-releasing factor (39). Although several hypotheses have been offered to account for the etiology of AD, oxidative stress play a crucial role in the disease pathogenesis as is evident from the following discussions (14,18,19,40–48). These include increased brain iron, aluminum and mercury, all of which are capable of stimulating ROS generation. There are lipid peroxidation and decreased polyunsaturated fatty acids in AD brain, increased 4-hydroxynonenals in AD ventricular fluid (40–42), increased protein carbonyl content in the inferior parietal lobule, increased levels of both nuclear and mitochondrial 8OH-dG, which are particularly striking in the parietal cortex (43), increased DNA strand breaks in cerebral cortex, diminished energy metabolism and cytochrome c oxidase activity in the AD brain, advanced glycation end products (AGE), malondialdehyde (MDA), carbonyls, peroxynitrite, heme oxygenase-1 and SOD-1 in NFT, AGE, heme oxygenase-1, SOD-1 in sp, and amyloid β peptide, which is capable of generating ROS and its mediation of neuronal degeneration and death (14,18,19,44–48).

IV. ANTIOXIDANTS AS PROPHYLACTIC AGENTS IN THE MANAGEMENT OF AD

There is no set treatment for the reversing or halting neuronal degeneration in AD. It is worth pointing out that the FDA approved the use of cholinesterase inhibitors (anticholinesterase drugs in clinical practice—donepezil, rivastigmine, and galantamine) (49), which have demonstrated limited palliative value. There is interest in the use of antioxidants (e.g., phenolics) as potential therapeutic strategy. Markers of oxidative stress represent an early indicator of oxidative stress in AD susceptible neurons, often appearing before the detection of any other pathology; antioxidant therapies are thus a promising avenue for treatment.

The administration over a 2-year period of the trivalent iron-chelating agent, desferrioxamine, slows the clinical development of AD (50). The therapeutic importance is directed to the removal of iron and possibly inhibition of ROS formation. The beneficial effects of vitamin E (α-tocopherol), selegiline (MAO B inhibitor), and *Ginkgo biloba* (Egb 761) have been suggested. For example, Sano et al. (51) found that effects of α-tocopherol and selegiline in 324 patients with moderately severe AD showed no significant improvement on cognitive tests, but they did observe significant delays in the time of the following occurrences: death, institutionalization, loss of the ability to perform basic activities of daily living, and severe dementia. Several studies (18,19,45,47,52,53) involving neuronal cultures have shown that vitamin E inhibits Aβ-induced lipid peroxidation, protein oxidation, free radical formation, and cell death.

Human lacks the enzyme L-gulono-γ-lactone oxidase that is necessary for the biosynthesis of vitamin C, and therefore must obtain ascorbate from dietary sources. Ascorbate is a water-soluble antioxidant present primarily as a monovalent anion at physiological pH. Ascorbate soluble in the aqueous phase can reduce the tocopheryl radical formed when vitamin E scavenges a lipid radical within the membrane (54). Plasma ascorbate levels have been found to be decreased in AD patients when compared with control patients, in levels corresponding to dementia (13,55,56). More interestingly, CSF levels of ascorbate were found to be decreased in AD patients when compared with control subjects that may hinder the reduction of tocopheryl radical back to tocopherol (57). The synergistic vitamins C and E were chosen in a study in which 400 IU vitamin E and 1000 mg vitamin C were given daily to patients (58). The combination of vitamins E and C increased vitamins E and C levels in the plasma and CSF, making CSF and plasma lipoproteins less susceptible to in vitro oxidation. The plasma and CSF of patients given only vitamin E were not protected against in vitro oxidation. This study highlights the concept of synergism between antioxidants, in this particular case between vitamins C and E. Aruoma has advocated that bioactive components in plant food could possess that are complementary in a synergistic manner.

α-Lipoic acid (LA) is a low molecular weight dithiol antioxidant that is an important cofactor in multienzyme complexes in the mitochondria. α-Lipoic acid is readily available from the diet, absorbed through the gut, and easily passes through the blood–brain barrier. In addition, LA is synthesized in the mitochondria of plants (59). As an antioxidant, LA and its reduced form, dihydrolipoic acid, are capable of quenching ROS and RNS such as hydroxyl radicals, peroxyl radicals, superoxide, hypochlorous acid, and peroxynitirite and chelating metals such as Cd^{2+}, Fe^{3+}, Cu^{2+}, and Zn^{2+} (60,61). α-Lipoic acid has been suggested to interact with other antioxidants such as glutathione, ubiquinol, thioredoxin, vitamin C, and indirectly with vitamin E, regenerating them to their reduced forms (62). Studies are beginning to show that LA improves behavior and diminishes markers of soxidative stress in rats that fed a diet supplemented with LA (63,64).

IV.A. Phenolics in the Treatment of AD

The potential neuroprotective effects of phenolics against the neuronal deficits associated with aging or age-related neurodegenerative diseases are of increasing interest (18,65). Cellular studies (66), examining the potential mechanisms of neuroprotection by flavonoids in preventing neuronal cell death caused by oxidized low-density lipoprotein-induced oxidative stress, have identified three different mechanisms: flavonoids can prevent cell death after glutamate injury by scavenging ROS, maintaining the correct GSH levels and inhibiting Ca^{2+} influx, which represents the last step in cell death cascade. These properties, coupled with the anti-inflammatory properties attributed to some phenolics (67), render this class of compounds suitable for application where oxidative stress, together with inflammation and antioxidant defense depletion, takes place, such as AD.

Modern scientific studies on the biological activity of extracts from dried *Ginkgo biloba* leaves started 20 years ago, even though the beneficial effects of these natural substances were known for 5000 years in traditional Chinese medicine. The ginkgo extracts that are currently used for medicinal purposes contain 24% flavonoids and 6% terpenoids. The antioxidant effects of flavonoids combined with the anti-inflammatory properties of the terpenoids bilobalide and ginkgolides A, B, C, M, and J, terpenoid antagonists of platelet-activating factor, make these natural extracts plausible to use in AD, characterized by both oxidative damage (15,47) and inflammation (68). In ROS-exposed mice (69), apoptosis was significantly reduced after pretreatment with ginkgo extracts. In several cell lines, treatment with ginkgo led to increased endogenous glutathione (70). Extensive studies on ginkgo extracts showed their ability to protect brain neurons from oxidative stress (71), to inhibit apoptosis in cell culture (72), and to rescue PC12 neuronal cells from Aβ-induced cell death (73).

Tea is widely advocated as beneficial prophylactic agents from the standpoint that antioxidant phenolics are highly abundant in tea leaves. The main flavonoids present in green tea are catechins, in particular epigallocatechin gallate (EGCG), in the amount of 30–130 mg per cup of tea. Other phenolic compounds, such as quercetin, kaempferol, and myricetin and their glycosides, are found in lower concentration Table 1 shows some representative structures. The different properties exhibited by these compounds have been tested by a variety of studies in cell lines. Tea catechins have been shown to possess anticarcinogenic (69), antiallergic (70), and antiapoptotic properties. In hippocampal neurons, tea phenolics show a protective effect against ischemic insult (74), whereas neurotoxicity induced by Aβ (1–42), whose deposition in the brain accompanies neuronal loss in AD, was attenuated in the presence of EGCG (75). The protective antioxidant effect of these natural compounds was also confirmed by other studies in synaptosomes (76). Furthermore, studies involving tea phenolics found that intracisternal injection of epicatechin improved memory impairment induced by intracisternal glucose oxidase.

Table 1 Classes and Structural Formula of Flavonoids

Classes	Structural formula	Examples
Flavanones		$R = R^1 = H$, $R^{11} = R^{111} = OH$; Naringenin $R = OH$, $R^1 = H$, $R^{11} = R^{111} = OH$; Eriodyctiol $R = R^1 = OH$, $R^{11} = R^{111} = OH$; 5^1-OH-Eriodyctiol
Flavones		$R = R^1 = H$; Apigenin $R = OH$, $R^1 = H$; Luteolin $R = R^1 = OCH_3$; Tricetin
Isoflavones		$R = H$; Daidzein $R = OH$; Genistein

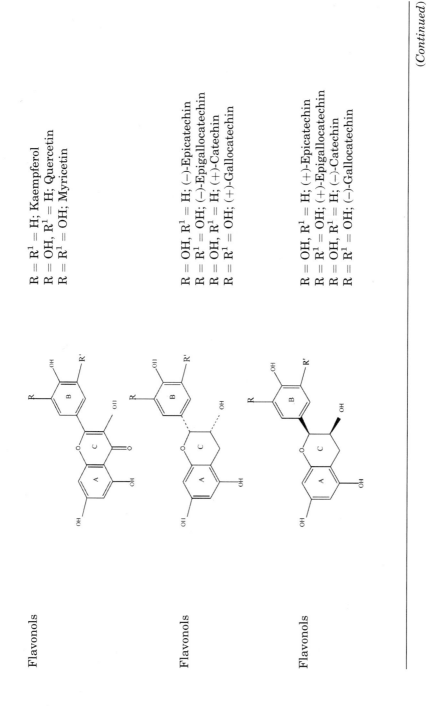

Flavonols

R = R¹ = H; Kaempferol
R = OH, R¹ = H; Quercetin
R = R¹ = OH; Myricetin

Flavonols

R = OH, R¹ = H; (−)-Epicatechin
R = R¹ = OH; (−)-Epigallocatechin
R = OH, R¹ = H; (+)-Catechin
R = R¹ = OH; (+)-Gallocatechin

Flavonols

R = OH, R¹ = H; (+)-Epicatechin
R = R¹ = OH; (+)-Epigallocatechin
R = OH, R¹ = H; (−)-Catechin
R = R¹ = OH; (−)-Gallocatechin

(Continued)

Table 1 Classes and Structural Formula of Flavonoids (*Continued*)

Classes	Structural formula	Examples
Anthocyanidins		R = OH, R^1 = H; Cyanidin R = R^1 = OH; Delphinidin R = R^1 = OCH$_3$; Malvidin R = R^1 = H; Pelargonidin R = H; Peonidin R = OH; Petunidin

The flavonoids contained in blueberries, mainly anthocyanins, have been extensively studied in vitro and in vivo to assess their action in several pathologies (77). In aged rats, blueberry extracts were effective in reversing age-related decline with cognitive, motor, and neuronal effects (78). Phenolics were capable of enhancing red blood cell resistance to oxidative stress in vitro and in vivo, supporting the idea of a protective role of these substances in ROS-mediated, age-related neurological decline. Although there are no studies on the action of blueberry extracts in ameliorating AD cognitive decline, improvement on behavior and memory has been reported in transgenic mice with APP/PS-1 double mutations, a model for AD (79).

Effective micro-organism-X (EM-X) is an antioxidant drink derived from the ferment of unpolished rice, papaya, and seaweeds with effective micro-organisms (EM) selected from actobacillaceae, saccharomycetes, funguses, actinomyces, and photosynthetic bacteria. This cocktail drink is widely available in south east Asia and has applications as supplements in the clinic for the treatment of cancer, hypertension, various allergies, diabetes, and tuberculosis. EM-X contains over 40 minerals, α-tocopherol, lycopene, ubiquinone, ascorbic acid, saponin, and flavonoids, such as quercetin, quercetin-3-O-glucopyranoside, and quercetin-3-O-rhamnopyranoside. EM-X increases the serum levels of superoxide dismutase, modulates immunological functions in animals, decreases MDA levels in D-galactose-induced aging in mice, and prolongs survival time of mice at high temperatures, as well as under hypoxic conditions (80). Aruoma and colleagues (81) observed that pretreatment with EM-X significantly reduces retinal neurodegeneration following NMDA injection into the vitreous body. Neuroprotection against the toxicity of 6-hydroxydopamine has also been reported for EMX (82). Although the exact molecular entities responsible for the antioxidant action were unclear, the neuroprotective effect stems, at least in part, due to its flavonoids and other antioxidant contents (vitamin E, saponin, ascorbic acid, etc.).

Resveratrol showed a protective effect against oxidative insult in PC12 cells (83) and protects rat brain against excitotoxic damage (84) and the hippocampal neurons against nitric

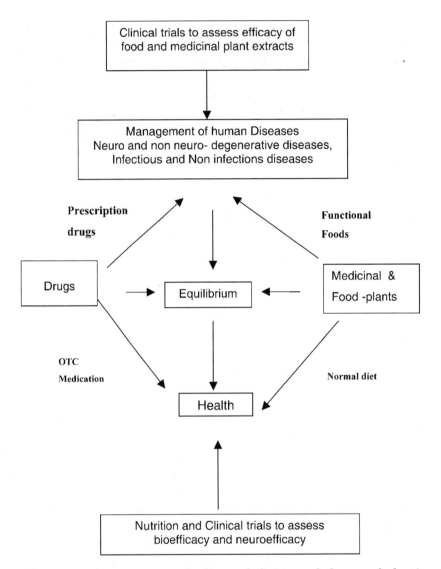

Figure 1 A strategy to facilitate definition of the prophylactic potentials of diet, nutritional/food supplements, medicinal plants, and herbal extracts 18).

oxide toxicity (85). A French study (86) on a community of persons of age $\geq$65, whose only consumption of alcoholic beverages was represented by red wine in the amount considered moderate (3–4 glasses per day), analyzed the incidence of dementia and AD compared with nondrinkers. After adjusting for age, sex, education, and other factors, the study showed an inverse relationship between moderate wine drinking and AD incidence. The intake of flavonoids and the risk of dementia will remain a unique research area in biomedicine (18,87,88). Finally, a compound that is targeted for neuronal protection should be able to cross the highly selective blood–brain barrier (20,81,82,87,88). Thus, in addition to boosting the endogenous antioxidant status, dietary phenolics can also potentiate cognitive function and memory performance. A strategy to facilitate the definition of the prophylactic potentials of diet, nutritional/food supplements, medicinal plants, and herbal extracts is proposed in Fig. 1 (18). Research directed towards the understanding of the role of plant/herbal extracts and dietary antioxidants in human health should be complemented with the development and validation of biological markers.

REFERENCES

1. Aruoma OI, Halliwell B, Molecular Biology of Free Radicals in Human Diseases. London: OICA International, 1998.

2. Aruoma OI. Free radicals, oxidative stress and antioxidants in health and disease. J Am Oil Chem Soc 1998; 75:199–212.

3. Halliwell B. Free radicals and antioxidants. Nut Rev 1994; 52:253–265.

4. Halliwell B, Gutterridge JMC. Free Radicals in Biology and Medicine. Oxford: Clarendon Press, 1989.

5. Ames BM, Shigenaga MK, Hagen TM. Oxidants, antioxidants and the degenerative diseases of aging. Proc Natl Acad Sci USA 1993; 90:7915–7922.

6. Halliwell B. Reactive oxygen species and the central nervous system. J Neurochem 1992; 59:1609–1623.

7. Hensley K, Carney JM, Mattson MP, Aksenova M, Harris M, Wu JF, Floyd RA, Butterfield DA. A model for beta-amyloid aggregation and neurotoxicity based on free radical generation by the peptide: relevance to Alzheimer's disease. Proc Natl Acad Sci USA 1994; 91:3270–3274.

8. Wolozin B, Luo Y, Wood K. Neuronal loss in aging and disease. In: Holbrook NJ, Martin GR, Locksmith RA, eds. Cellular Aging and Cell Death. New York: Wiley-Liss, 1996: 283–302.

9. Hauser RA, Zesiewiez TA. Parkinson's Disease, Questions and Answers. Coral Springs, FL: Merit Publ. Intl., 1996.

10. Coghlan A. Animal experiments on trial. New Scientist, 2002; 176:16–17.

11. Connor JR, Snyder BS, Arosio P, Loeffler DA, LeWitt P. A quantitative analysis of isoferritins in select regions of aged, Pakinsonian, and Alzheimer's diseased brains. J Neurochem 1995; 65:717–724.

12. Oliver CN, Starke-Reed PE, Stadtman ER, Liu GJ, Carney JM, Floyd RR. Oxidative damage to brain proteins, loss of glutamine synthase activity and production of free radicals during ischemia/reperfusion injury to gerbil brains. Proc Natl. Acad Sci USA 1990; 87:5144–5147.

13. Foy CJ, Passmore AP, Vahidassr MD, Young IS, Lawson JT. Plasma chain-breaking antioxidants in Alzheimer's disease, vascular dementia, and Parkinson's disease. QJM J Ass Phys 1999; 92:39–45.

14. Behl C, Davies JB, Lesley R, Schubert D. Hydrogen peroxide mediates amyloid beta protein toxicity. Cell 1994; 77: 817–827.

15. Markesbery WR, Carney JM. Oxidative alterations in Alzheimer's disease. Brain Pathol 1999; 9:133–146.

16. Calingasan NY, Uchida K, Gibson GE. Protein-bound acrolein: a novel marker of oxidative stress in Alzheimer's disease. J Neurochem 1999; 72:751–756.

17. Aruoma OI, Halliwell B. Molecular Biology of Free Radicals in Human Diseases. London: OICA International, 1998.

18. Aruoma OI, Bahorun T, Jen L-S. Neuroprotection by bioactive components in medicinal and food plant extracts. Mut Res Rev 2003; 544:203–215.

19. Jang J-H, Aruoma OI, Jen L-S, Chung HY, Surh Y-J. Ergothioneine rescues PC12 cells from β-amyloid -induced apoptotic death. Free Rad Biol Med 2004; 36:288–289.

20. Datla KP, Christidou M, Widmer WW, Rooprai HK, Dexter DT. Tissue distribution and neuroprotective effects of citrus flavonoids tangeritin in a rat model of Parkinson's disease. Neuroreport 2001; 12:3871–3875.

21. Katzman R. The dementias. In: Rowland LP, ed. Textbook of Neurology. 8th ed. Philadelphia: Lea and Fabiger, 1989: 637–644.

22. Shen ZX. Brain cholinesterases: I. The clinico-histopathological and biochemical basis of Alzheimer's disease. Med Hypoth 2004; 63:285–297.

23. St George-Hyslop PH. The molecular genetics of Alzheimer disease. In: Terry RD, Katzman R, Bick KL, eds. Alzheimer's Disease. New York: Raven Press, 1993:345–352.

24. Sherrington R, Rogaev EI, Liang Y, Rogaeva EA, Levesque G, Ikeda M, Chi H, Lin C, Li G, Holman K, Tsuda T, Mar L, Foncin JF, Bruni AC, Montesi MP, Sorbi S, Rainero I, Pinessi L, Nee L, Chumakov I, Pollen D, Brookes A, Sanseau P, Polinsky RJ, Wasco W, DaSilva HAR, Haines JL, Pericak-Vance MA, Tanzi RE, Roses AD, Fraser PE, Rommens JM, St George-Hyslop PH. Cloning of a gene bearing missense mutations in early-onset familial Alzheimer's disease. Nature 1995; 375: 754–760.

25. Levy-Lahad E, Wasco W, Poorkaj P, Romano DM, Oshima J, Pettingell WH, Yu CE, Jondro PD, Schmidt SD, Wang K, Crowley AC, Fu YH, Guenette SY, Galas D, Nemens E, Wijsman EM, Bird TD, Schellenberg GD, Tanzi RE. Candidate gene for the chromosome 1 familial Alzheimer's disease locus. Science 1995; 269:973–977.

26. Corder EH, Saunders AM, Strittmatter WJ, Schmechel DE, Gaskell PC, Small GW, Roses AD, Haines JL, Pericak-Vance MA. Gene dose of apolipoprotein E type 4 allele and the risk

of Alzheimer's disease in late onset families. Science 1993; 261:921–923.

27. Saunders AM, Strittmatter WJ, Schmechel DE, St George-Hyslop PH, Pericak-Vance MA, Joo SH, Gusella JF, Crapper-MaLachlan DR, Alberts MJ, Hulette C, Crain B, Goldgaber D, Roses AD. Association of apolipoprotein E allele E4 with late onset familial and sporadic Alzheimer's disease. Neurology 1993; 43:1467–1472.

28. Corder EH, Saunders AM, Risch NJ, Strittmatter WJ, Schmechel DE, Gaskell PC, Rimmler JB, Locke PA, Conneally PM, Schmader KE, Small GW, Roses AD, Haines JL, Pericak-Vance MA. Protective effect of apolipoprotein E type 2 allele for late onset Alzheimer disease. Nat Genet 1994; 7:180–184.

29. Friedlich AL, Butcher LL. Involvement of free oxygen radicals in β-amyloidosis: an hypothesis. Neurobiol. Aging 1994; 15: 443–455.

30. Katzman R. Clinical and epidemiological aspects of Alzheimer's disease. Clin Neurosci 1993; 1:165–170.

31. Grant WB. Dietary links to Alzheimer's disease. Alz Dis Rev 1997; 2:42–55.

32. Mattson MP. Neuroprotective signalling and the aging brain. Brain Res 2000; 886:47–53.

33. Selkoe DJ. Physiological production of the β-amyloid protein and the mechanism of Alzheimer's disease. Trends Neurosci 1993; 16:403–409.

34. Selkoe DJ. Cell biology of the amyloid beta-protein precursor and the mechanism of Alzheimer's disease. Ann Rev Cell Biol 1994; 10:373–403.

35. Bartus RT, Dean RL, Beer B, Lippa AS. The cholinergic hypothesis of geriatric memory dysfunction. Science 1982; 217: 408–414.

36. Beal MF. Aging, energy, and oxidative stress in neurodegerative diseases. Ann Neurol 1995; 38:357–366.

37. Palmer AM, Francis PT, Benton JS, Sims NR, Mann DM, Neary D, Snowden JS, Bowen DM. Pre-synaptic serotonergic

dysfunction in patients with Alzheimer's disease. J Neurochem 1987; 48:8–15.

38. Davies P, Katzman R, Terry RD. Reduced somatostatin-like immunoreactivity in cerebral cortex from cases of Alzheimer disease and Alzheimer senile dementia. Nature 1980; 288:279–280.

39. Bissette G, Reynolds GP, Kilts CA, Widerlov E, Nemeroff CB. Corticotropin-releasing factor like immunoreactivity in senile dementia of the Alzheimer type reduced cortical and striatal concentrations. J Am Med Assoc 1985; 254:3067–3069.

40. Hensley K, Hall N, Subramanian R, Cole P, Harris M, Aksenov M, Gabbita SP, Wu JF, Carney JM, Lovell M, Markesbery WR, Butterfield DA. Brain regional correspondence between Alzheimer's disease histopathology and biomarkers of protein oxidation. J Neurochem 1995; 65: 2146–2156.

41. Lovell MA, Ehmann WE, Butler SM, Markesbery WR. Elevated thiobarbituric acid reactive substances and antioxidant enzyme activity in the brain in Alzheimer's disease. Brain 1995; 111:785–799.

42. Balzacs L, Leon M. Evidence of an oxidative challenge in the Alzheimer's brain. Neurochem Res 1994; 19:1131–1137.

43. Mecocci P, MacGarvey U, Beal MF. Oxidative damage to mitochondrial DNA is increased in Alzheimer's disease. Ann Neurol 1994; 36:747–751.

44. Yanker BA, Duffy LK, Kirschner DA. Neurotrophic and neurotoxic effects of amyloid beta protein: reversal by tachykinin neuropeptides. Science 1990; 250:279–282.

45. Yatin SM, Varadarajan S, Butterfield DA. Vitamin E prevents Alzheimer's amyloid beta-peptide (1–42) induced protein oxidation and reactive oxygen species formation. J Alzheimer's Dis 2000; 2:123–131.

46. Goodman Y, Steiner MR, Steiner SM, Mattson MP. Nordihydroguaiaretic acid protects hippocampal neurons against amyloid β-peptide toxicity, and attenuates free radical and calcium accumulation. Brain Res 1994; 654:171–176.

47. Butterfield DA, Drake J, Pocernich C, Castegna A. Evidence of oxidative damage in Alzheimer's disease brain: central role for amyloid beta-peptide. Trends Mol Med 2001; 7:548–554.

48. Butterfield DA, Koppal T, Subramanian R, Yatin S. Vitamin E as an antioxidant/free radical scavenger against amyloid beta-peptide induced oxidative stress in neocortical synaptosomal membranes and hippocampal neurons in culture: insights into Alzheimer's disease. Rev Neurosci 1999; 10:141–149.

49. Burns A, Byrne EJ, Maurer K. Alzheimer's disease. Lancet 2002; 360:163–165.

50. Crapper McLachlan DR, Dalton AJ, Kruck TPA, Bell MY, Smith WL, Kalow W, Andrews DF. Intramuscular desferrioxamine in patients with Alzheimer's disease. Lancet 1991; 337:1304–1308.

51. Sano M, Ernesto C, Thomas RG, Klauber MR, Schafer K, Grundman M, Woodbury P, Growdon J, Cotman CW, Pfeiffer E, Schneider LS, Thal LJ. A controlled trial of selegiline, α-tocopherol, both as treatment for Alzheimer's disease. The Alzheimer's disease Cooperative Study. N Eng J Med 1997; 336:1216–1222.

52. Behl C. Vitamin E and other antioxidants in neuroprotection. Int J Vitam Nutr Res 1999; 69:213–219.

53. Jen LS, Hart AJ, Jen A, Relvas JB, Gentleman SM, Garey LJ, Patel A. Alzheimer's peptide kills cells of retina in vivo. Nature 1998; 392:140–141.

54. Buettner GR. The pecking order of free radicals and antioxidants: lipid peroxidation, alpha tocopherol, and ascorbate. Arch Biochem Biophys 1993; 300:535–543.

55. McGrath LT, McGleenon BM, Brennan S, McColl D, McIlroy S, Passmore AP. Increased oxidative stress in Alzheimer's disease as assessed with 4-hydroxynonenal but not malondialdehyde. Quart J Med 2001; 94:485–490.

56. Riviere S, Birlouez-Aragon I, Nourhashemi F, Vellas B. Low plasma vitamin C in Alzheimer patients despite an adequate diet. Int J Geriatr Psychiatry 1998; 13:749–754.

57. Schippling S, Kontush A, Arlt S, Buhmann C, Sturenburg HJ, Mann U, Muller-Thomsen T, Beisiegel U. Increased lipoprotein

oxidation in Alzheimer's disease. Free Radic Biol Med 2000; 28: 351–360.

58. Kontush A, Mann U, Arlt S, Ujeyl A, Luhrs C, Muller-Thomsen T, Beisiegel U. Influence of vitamin E and C supplementation on lipoprotein in patients with Alzheimer's disease. Free Radic Biol Med 2001; 31:345–354.

59. Gueguen V, Macherel D, Jaquinod M, Douce R, Bourguignon J. Fatty acid: an lipoic acid biosynthesis in higher plants mitochondria. J Biol Chem 2000; 275:5016–5025.

60. Packer L, Tritschler HJ, Wessel K. Neuroprotection by the metabolic antioxidant α-lipoic acid. Free Radic Biol Med 1997; 22:359–378.

61. Scott BC, Aruoma OI, Evans PJ, O'Neill C, van der Vliet A, Cross CE, Halliwell B. Lipoic acid and dihydrolipoic acid as antioxidants. Free Radic Res Commun 1994; 20:119–133.

62. Kagan VE, Serbinova EA, Forte T, Scita G, Packer L. Recycling of vitamin E in human low-density lipoproteins. J Lipid Res 1992; 33:385–397.

63. Liu J, Head E, Gharib AM, Yuan W, Ingersoll RT, Hagen TM, Cotman CW, Ames BN. Memory loss in old rats is associated with brain mitochondrial decay and RNA/DNA oxidation: partial reversal by feeding acetyl-L-carnitine and/or R-alpha-lipoic acid. Proc Natl Acad Sci USA 2002; 99:356–2361.

64. Hager TM, Marahrens A, Kenklies M, Riederer P, Munch G. Alpha-lipoic acid as a new treatment option for Alzheimer type dementia. Arch Gerontol Geriatr 2001; 32:275–282.

65. Youdim K, Joseph JA. A possible emerging role of phytochemicals in improving age-related neurological dysfunction: a multiplicity effects. Free Radic Biol Med 2001; 30:583–594.

66. Ishige K, Schubert D, Sagara Y. Flavonoids protect neuronal cells from oxidative stress by three distinct mechanisms. Free Radic Biol Med 2001; 30:433–446.

67. Stoner GD, Mukhtar H. Polyphenols as cancer chemoprotective agents. J Cell Biochem Suppl 1995; 22:169–180.

68. Rogers J, Shen Y. A perspective on inflammation in Alzheimer's disease. Ann NY Acad Sci 2000; 924:132–135.

69. Schindowski K, Leutner S, Kressmann S, Eckert A, Muller WE. Age-related increase of oxidative stress-induced apoptosis in mice prevention by *Ginkgo biloba* extract (EGb 761). J Neural Transm 2001; 108:969–978.

70. Rimbach G, Gohil K, Matsugo S, Moini H, Saliou C, Virgili F, Weber SU, Packer L. Induction of glutathione synthesis in human keratinocytes by *Ginkgo biloba* extract (EGb 761). Biofactors 2001; 15:39–52.

71. Oyama Y, Chikahisa L, Ueha T, Kanemaru K, Noda K. *Ginkgo biloba* extract protects brain neurons against oxidative stress induced by hydrogen peroxide. Brain Res 1996; 712: 349–352.

72. Xin W, Wei T, Chen C, Ni Y, Zhao B, Hou J. Mechanisms of apoptosis in rat cerebellar granule cells induced by hydroxyl radicals and the effects of EGb 761 and its constituents. Toxicology 2000; 148:103–110.

73. Yao Z, Drieu K, Papadopoulos V. The *Ginkgo biloba* extract EGb 761 rescues the PC12 neuronal cells from beta-amyloid-induced cell death by inhibiting the formation of beta-amyloid-derived diffusible neurotoxic ligands. Brain Res 2001; 889:181–190.

74. Lee S, Suh S, Kim S. Protective effects of green tea polyphenol (–)-epigallocatechin gallate against hippocampal neuronal damage after transient global ischemia in gerbils. Neurosci Lett 2000; 287:191–194.

75. Choi YT, Jung CH, Lee SR, Bae JH, Baek WK, Suh MH, Park J, Park CW, Suh SI. The green tea polyphenol (–)-epigallocatechin gallate attenuates beta-amyloid-induced neurotoxicity in cultured hippocampal neurons. Life Sci 2001; 7:603–614.

76. Guo Z, Zhao B, Li M, Shen S, Xin W. Studies on protective mechanisms of four components of green tea polyphenols against lipid peroxidation in synaptosomes. Biochem Biophys Acta 1996; 1304:210–222.

77. Kähkönen MP, Hopia AI, Vuorela J, Rauha JP, Philaja K, Kujala TS, Heinonen M. Antioxidant activity of plant extracts containing phenolic compounds. J Agric Food Chem 1999; 47:3954–3962.

78. Joseph JA, Shukitt-Hale B, Denisova NA, Bielinski D, Martin A, McEwen JJ, Bickford PC. Reversals of age-related declines in neuronal signal transduction, cognitive, and motor behavioural deficits with blueberry, spinach or strawberry dietary supplementation. J Neurosci 1999; 19:8114–8121.

79. Joseph JA, Denisova NA, Arendash G, Gordon M, Diamond D, Shukitt–Hale B, Morgan D, Blueberry supplementation enhances signaling and prevents behavioral deficits in an Alzheimer disease model. Nutri Neurosci 2003; 6:153–162.

80. Higa T, Ke B. Clinical and Basic Medical Research on EM-X: A Collection of Research Papers. Vol. 1. Okinawa, Japan: EMRO, 2001.

81. Datla KP, Bennet RD, Zbarsky V, Ke B, Liang Y-F, Higa T, Bahorun T, Aruoma OI, Dexter DT. The antioxidant drink "effective microorganism-X (EM-X)" pre-treatment attenuates the loss of nigrostriatal dopaminergic neurons in 6-hydroxydopamine-lesion rat model of Parkinson's disease. J Pharma Pharmacol 2004; 56:649–654.

82. Aruoma OI, Moncaster JA, Walsh DT, Gentleman SM, Ke B, Liang Y, Higa T, Jen LS. The antioxidant cocktail, effective microorganism X (EM-X), protects retinal neurons in rats against N-methyl-D-aspartate excitotoxicity in vivo. Free Radic Res 2003; 37:91–97.

83. Chanvitayapongs S, Draczynska-lusiak B, Sun AY. Amelioration of oxidative stress by antioxidants and resveratrol in PC12 cells. Neuroreport 1997; 8:1499–1502.

84. Virgili B. Partial neuroprotection of in vivo excitotoxic brain damage by chronic administration of the red wine antioxidant agent, trans-resveratrol in rats. Neurosci Lett 2000; 281: 123–126.

85. Bastianetto S, Zheng WH, Quirion R. Neuroprotective abilities of resveratrol and other red wine constituents against nitric oxide related toxicity in cultured hippocampal neurons. Br J Pharmacol 2000; 131:711–720.

86. Orgogozo JM, Dartigues JF, Lafont S, Letenneur L, Commenges D, Salamon R, Renaud S, Breteler SM. Wine consumption and dementia in the elderly: a prospective community

study in the Bordeaux area. Rev Neurol (Paris) 1997; 153: 185–192.

87. Commenges D, Scotet V, Renaud S, Jacqmin-Gadda H, Barberger-Gateau P, Dartigues JF. Intake of flavonoids and risk of dementia. Eur J Epidemiol 2000; 16:357–363.

88. Aruoma OI. Editorial. Neuroprotection by dietary antioxidants: new age of research. Nahrung/Food 2002; 46:381–382.

20

Resveratrol—A Unique Polyphenolic Antioxidant Present in Grape Skins and Red Wine—Is a Preventive Medicine Against a Variety of Degenerative Diseases

DIPAK K. DAS

Cardiovascular Research Center, University of Connecticut, School of Medicine, Farmington, Connecticut, U.S.A.

NILANJANA MAULIK

Molecular Cardiology Laboratory, Cardiovascular Research Center, University of Connecticut, School of Medicine, Farmington, Connecticut, U.S.A.

ABSTRACT

Resveratrol abundantly present in grapevines is found in grape skins, wines, especially red wine, and peanuts. Recent studies suggest that resveratrol has a number of health benefits including its potential role as cardioprotective, neuroprotective, anti-inflammatory, and cancer chemopreventive agent. Thus, resveratrol can protect the cells like cardiomyocytes from death in one hand and kills a selective group of cells such as cancer cells in the other hand. A recent study, using an isolated perfused working rat heart model, determined that at low concentration of about 10 μM, resveratrol could increase the constitutive nitric oxide (NO) to the ischemic myocardium, thereby functioning as an intracellular antioxidant. The cardioprotection was reflected by its ability to improve postischemic ventricular function and reduce cell death due to both necro-sis and apoptosis. Cardioprotective ability of resveratrol was reduced with increasing concentration, and at a high dose (more than 100 μM) was no longer cardioprotective. Resveratrol appears to provide the cardioprotection through its ability to pharmacologically precondition the heart against ischemia. In another related study, isolated hearts from iNOS knockout mice and corresponding wild-type mice were perfused in the absence or presence of 10 μM resveratrol. The hearts were then subjected to ischemia and reperfusion. The results showed that iNOS knockout mouse hearts devoid of any copies of iNOS gene could not be preconditioned with resveratrol indicating a definitive role of NO in resveratrol preconditioning. While resveratrol exerts cardioprotection, it causes cancer cell death by promoting apoptosis through the increased oxidative stress. Thus, it seems reasonable to speculate that there are two faces of resveratrol: it potentiates survival signal through pharmacological preconditioning and causes cell death by inducing death signal and NO may play an essential role in resveratrol signaling of survival and death.

I. INTRODUCTION

A number of studies have been devoted to understand the cause of the so-called French Paradox, the anomaly which means that in several parts of France and other Mediterranean countries the morbidity and mortality of coronary heart diseases in absolute value and in consideration of its rate to other manners of death are significantly lower than those are in other developed countries, despite the high consumption of fat and saturated fatty acids (1,2). The cause of this cardioprotective effect is believed to be, among others, regular consumption of wine. Wines, especially red wines, consist of about 1800–3000 mg/L of polyphenolic compounds (3). Many polyphenolic compounds are potent antioxidants capable of scavenging free radicals and inhibit lipid peroxidation both in vitro and in vivo (4).

Among many factors, reactive oxygen species (ROS) play a crucial role in the pathophysiology of a variety of chronic and degenerative diseases such as cardiovascular and neurodegenerative diseases and cancer. Under normal conditions, there is a balance between the formation of pro-oxidants (oxygen free radicals) and the amount of antioxidants present. This steady-state condition is interrupted in pathophysiological conditions because of the excessive production of free radicals, or decrease in antioxidants, or both. Substantial evidence exists to support the notion that ischemia and reperfusion generate superoxide and hydroxyl radicals among other cytotoxic free radicals (5). The presence of reactive oxygen species was confirmed directly by estimating free radical formation and indirectly by assessing lipid peroxidation and DNA breakdown products (6). Among the oxygen free radicals, superoxide anion (O_2^-) is the most innocent free radical, while the hydroxyl radical ($OH^•$) possesses the most detrimental to the cells. Virtually, all the biomolecules including lipids, proteins, and DNA are potential targets for free radical attack.

Resveratrol, a polyphenol phytoalexin (*trans*-3,5,4'-trihydroxystilbene) abundantly found in grape skins and in wines, possesses diverse biochemical and physiological actions, which include estrogenic, antiplatelet, and anti-inflammatory properties (7,8). *Trans*-resveratrol was originally identified as the active ingredient of an Oriental herb (Kojo-kon) used for treatment of a wide variety of diseases including dermatitis, gonorrhea, fever, hyperlipidemia, atherosclerosis, and inflammation. Recently, resveratrol was found to protect kidney, heart, and brain from ischemic reperfusion injury (9–12). In kidney cells, resveratrol was found to exert its protective action through upregulation of nitric oxide (NO) (11). A growing body of evidence supports the cardioprotective role of constitutive expression of NO. For example, an NO donor or a precursor for NO synthesis like L-arginine has been found to ameliorate myocardial ischemic reperfusion injury (12). Recently, NO has been found to play a crucial role in ischemic preconditioning (PC) in which NO functions as the mediator of IPC (13). The present review will discuss the possible mechanisms of resveratrol-mediated cardioprotection.

II. RESVERATROL

II.A. Resveratrol—A Grape-Derived Natural Antioxidant

Resveratrol, a polyphenolic antioxidant, possesses a spectrum of physiological activities including its ability to protect tissues such as brain, kidney, and heart from ischemic injury, and its role as a cancer chemoprotective agent. Its biological activities include its role as a neuroprotective, anti-inflammatory, and antiviral compound (8,14). This compound also exerts diverse biochemical and physiological actions, which include estrogenic, antiplatelet, and anti-inflammatory properties (7,15). Recently, resveratrol obtained from grape skins and in wines has been found to protect the heart from ischemic reperfusion injury (16,12). In kidney cells, resveratrol was found to exert its protective action through upregulation of NO (11).

Resveratrol can scavenge some of the reactive oxygen species. Although it possesses antioxidant properties, it does not function as a strong antioxidant in vitro (17). While resveratrol behaves as a poor in vitro scavenger of reactive oxygen species, it functions as a potent antioxidant in vivo. The in vivo antioxidant property of resveratrol is probably due to its ability to upregulate NO synthesis, which in turn functions as an in vivo antioxidant by its ability to scavenge superoxide radicals. In the ischemic reperfused heart, brain, and kidney, resveratrol has been found to induce NO synthesis and lower oxidative stress (11,18).

Most of its in vivo antioxidant properties are believed to be achieved through its ability to upregulate NO. Nitric oxide with an unpaired electron, behaves as a potent antioxidant in vivo. Nitric oxide can rapidly react at or near the diffusion-limited rate (6.7×10^9 $[mol/L]^{-1} sec^1$) with the superoxide anion (O_2^-), which is presumably formed in the ischemic reperfused myocardium (19). The affinity of NO for O_2^- is far greater than superoxide dismutase (SOD) for O_2^-. In fact, NO may compete with SOD for O_2^-, thereby removing O_2^- and sparing SOD, further supporting its antioxidant role. Thus, resveratrol functions as an excellent in vivo antioxidant through NO.

In cellular system, resveratrol can scavenge reactive oxygen species. For example, resveratrol was found to inhibit 12-*O*-tetradecanoylphosbol-13-acetate (TPA) -induced free radical formation in cultured HL-60 cells (20). In DU 145 prostate cancer cell line, resveratrol inhibited growth accompanied by a reduction in NO production and inhibition of iNOS (21). Resveratrol also inhibited the formation of O_2^- and H_2O_2 produced by macrophages stimulated by lipopolysaccharide (LPS) or TPA (22). In a related study, resveratrol inhibited reactive oxygen intermediates and lipid peroxidation induced by tumor necrosis factor (TNF) in wide variety of cells (23).

II.B. Resveratrol—A Phytoestrogen

Resveratrol has been recognized as a phytoestrogen based on its structural similarities to diethylstilbesterol. Resveratrol can bind to the estrogen receptors (ERs), thereby activating transcription of estrogen-responsive reporter genes transfected into cells (15,24). Resveratrol was shown to function as a superagonist when combined with estradiol (E_2), and induce the expression of estrogen-regulated genes (15). However, several other studies showed conflicting results. In another related study using the same cell line, resveratrol showed antiestrogen activity, because it suppressed progesterone receptor expression induced by E_2. Another recent study showed that both isomers of resveratrol possessed superestrogenic activity only at moderate concentration ($>10 \mu M$), while at lower concentration ($<1\,\mu M$), antiestrogenic effects prevailed (25).

Most of the in vivo studies failed to confirm estrogenic potential of resveratrol. At physiologic concentration, resveratrol could not induce any changes in uterine weight, uterine epithelial cell height, or serum cholesterol (26). Only at a very high concentration, resveratrol modulated the serum cholesterol lowering activity of E_2 (26). In another study, resveratrol given orally as well as subcutaneously could not affect uterus weight at any concentration ranging from lowest to highest doses (0.03–120 mg/kg/day) (27). In another related study, resveratrol reduced uterine weight and decreased the expression of ER-α mRNA and protein and PR mRNA (28).

In contrast, resveratrol was found to possess estrogenic properties in stroke-prone spontaneously hypertensive rats (29). When ovariectomized rats were fed resveratrol at the concentration of 5 mg/kg/day, it attenuated an increase in systolic blood pressure. In concert, resveratrol enhanced endothelin-dependent vascular relaxation in response to acetylcholine and prevented overectomy-induced decreases in femoral bone strength in a manner similar to estradiol. Recently, resveratrol was found to act as an ER agonist in breast cancer cells stably transfected with ERα (30). While more data accumulate on the estrogenic behavior of resveratrol, the controversy continues to persist.

II.C. Resveratrol—Effective Against a Variety of Degenerative Diseases

Resveratrol, a polyphenol phytoalexin, possesses diverse biochemical and physiological actions, which include estrogenic, antiplatelet, and anti-inflammatory properties. Several recent studies determined the cardioprotective abilities of resveratrol. Both in acute experiments and chronic models, resveratrol was found to reduce myocardial ischemic reperfusion injury (12,16). In addition to hearts, resveratrol has also been found to protect kidney and brain cells from ischemia–reperfusion injury. Similar to the heart, the ability of resveratrol to stimulate NO production during ischemia–reperfusion is believed to play a crucial role in its ability to protect kidney cells from ischemic reperfusion injury (11). The maintenance of constitutive NO release is a critical factor in the recovery of function after an ischemic injury. Release of constitutive NO is significantly reduced after ischemia–reperfusion, and maintenance of NO by any means such as induction of NO production with L-arginine can restore the postischemic myocardial function (31).

While resveratrol protects the brain, kidney, and heart cells, it preferentially kills the cancer cells. For example, intraperitoneal administration of resveratrol caused an increase in the G2/M phase of the cell cycle and apoptosis and reduced the tumor growth (32). In oral squamous carcinoma cells, resveratrol caused growth inhibition, both alone and in combination with quercetin (33). In a recent study, resveratrol inhibited the growth of highly metastatic B16-BL6 melanoma cells (34). In a rat colon carcinogenesis model, resveratrol induced proapoptotic bax expression in colon aberrant cryptic foci (35). In fact, resveratrol was found to affect three major stages of carcinogenesis and inhibit the formation of preneoplastic lesions in a mouse mammary organ culture model (36).

A recent study demonstrated inhibition of the growth of CagA+ strains of *Helicobacter pylori* in vitro (37). In another study, resveratrol inhibited the growth of 15 clinical strains of *H. pylori* in vitro and suggested that the anti-*H. pylori* activity of resveratrol may play a role in its chemopreventive effects (38). Resveratrol was also found to increase the activity of antiaging gene SIRT1 activity 13-fold (39). Resveratrol-mediated activation of life-extending genes in human cells may open a new horizon of resveratrol research.

III. PRECONDITIONING WITH RESVERATROL

III.A. Myocardial Protection with Preconditioning

Preconditioning is the most powerful technique for cardioprotection ever known (1–5). The most generalized method of classical preconditioning is mediated by cyclic episodes of several short durations of reversible ischemia, each followed by another short duration of reperfusion. In most laboratories including our own, preconditioning is achieved by four cycles of 5 min ischemia each followed by 10 min of reperfusion (40,41). Such preconditioning makes the heart resistant to subsequent lethal ischemic injury (42,43).

The mechanisms underlying cardiac preconditioning have been studied extensively. Several regulatory pathways have been identified in different systems. Three important factors, adenosine (Ado) A1 receptor, multiple kinases including PKC, MAP kinases and tyrosine kinases, and mitochondrial ATP-sensitive potassium channel (K_{ATP}) channel are known to play a crucial role in preconditioning-mediated cardioprotection. For example, cardioprotection achieved by preconditioning can be abolished by adenosine A1 receptor antagonists (44). Adenosine A1 receptor agonist can limit myocardial infarct size (44). While there is a general agreement regarding the beneficial role of adenosine on ischemic tissue, adenosine hypothesis remains controversial. To reconcile the adenosine hypothesis, an argument has been made that adenosine could trigger a secondary mechanism such as activation of Gi protein, which in turn could open the K_{ATP}. This hypothesis is inconsistent with several findings that inhibition of K_{ATP} channel blocks the effects of preconditioning and a K_{ATP} channel opener can simulate ischemic preconditioning (45,46).

Another intriguing hypothesis stemmed from the concept of stimulation of an endogenous protective mechanism by myocardial adaptation to ischemic stress. Preconditioning has been found to induce the expression of endogenous antioxidant enzymes such as SOD and glutathione peroxidase (GSHPx-1) (47,48) and heat shock proteins (HSP) such as HSP 27, HSP 32, and HSP 70 (49,50). Additionally, preconditioning potentiates a signal transduction cascade by inhibiting death signal and activating survival signal. Thus, several proapoptotic and antiapoptotic genes and transcription factors including Jnk-1, c-Jun, NFkB, and AP-1 are likely to play a crucial role in preconditioning (51–53). Recently, nitric oxide has been found to act as the mediator of preconditioning (13,54).

Thus, it appears that reactive oxygen species play a crucial role in preconditioning, which is realized probably by activation of adenosine A1 receptor, stimulating PKC and MAP kinases, and by opening mitochondrial K_{ATP} channel. Recent studies from our laboratory determined that resveratrol may function as a preconditioning agent. However, to confirm its role as a pharmacological preconditioning agent, it must fulfill the criteria for preconditioning, i.e., it must function through adenosine A1 receptor, PKC and MAP kinases, and K_{ATP} channel.

Adenosine accumulates in tissues under metabolic stress. On myocardial cells, the nucleoside interacts with various receptor subtypes [A(1), A(3), and probably A(2A) and A(2B)] that are coupled, via G proteins, to multiple effectors, including enzymes, channels, transporters, and cytoskeletal components. Studies using Ado receptor agonists and antagonists, as well as animals overexpressing the A(1) receptor indicate that Ado exerts anti-ischemic action. Ado released during PC by short periods of ischemia followed by reperfusion induces cardioprotection to a subsequent sustained ischemia. This protective action is mediated by A(1) and A(3) receptor subtypes and involves the activation and translocation of PKC to sarcolemmal and mitochondrial membranes. PKC activation leads to an increased opening of ATP-sensitive K^+ (K_{ATP}) channels. Other effectors possibly contributing to cardioprotection by Ado or PC, and which seem particularly involved in the delayed (second window of) protection, include MAP kinases, heat shock proteins, and iNOS. Because of its anti-ischemic effects, Ado has been tested as a protective agent in clinical interventions such as PTCA, CABG, and tissue preservation, and was found in most cases to enhance the postischemic recovery of function. The mechanisms underlying the role of Ado and of mitochondrial function in PC are not completely clear, and uncertainties remain concerning the role played by newly identified potential effectors such as free radicals, the sarcoplasmic reticulum, etc. In addition, more studies are needed to clarify the signaling mechanisms by which A(3) receptor activation or overexpression may promote apoptosis and cellular injury.

Ischemic preconditioning is a receptor-mediated process, and is realized via signal transduction pathways. Several investigators have proposed a unifying hypothesis that activation of PKC represents a link between cell surface receptor activation and a putative end-effector sarcolemmal or mitochondrial K_{ATP} channels (55,56). Possible involvement of protein tyrosine kinases in PC was proposed for the first time by Maulik et al. (57,58). Now, it is increasingly clear that protein tyrosine kinases play a crucial role in mediating PC in some animal species. Protein tyrosine kinases may act in parallel to (59,60), downstream of (61,62), or upstream of (63) PKC in eliciting PC. However, the exact member of protein tyrosine kinases involved in PC remains unclear. Among a large number of tyrosine kinases, Src family tyrosine kinase received much attention (64). Src tyrosine kinase has been implicated in the mechanism of cell survival and death, which is regulated by complex signal transduction processes (65). Rapid activation of Src family tyrosine kinases after ischemia has also been documented in the isolated guinea pig heart (66). Since Src family tyrosine kinases are activated by stimulation of G-protein-coupled receptors (67), an increase in intracellular Ca^{2+} (68), oxidative stress (69), and enhanced nitric oxide synthesis (70), all of which can be elicited by PC challenges.

The intracellular signaling mechanisms that mediate preconditioning require one or more members of MAP kinase cascades. Among the three distinct MAP kinase families, stress-activated protein kinase (SAPK), also known as c-Jun NH_2-terminal kinases (JNK), and p38 MAP kinase are known to be regulated by extracellular stresses including environmental stress, oxidative stress, heat shock, and UV radiation (71). c-Jun NH_2-terminal kinases and p38 MAP kinase appear to be involved in distinct cellular function, because they possess different cellular targets and are located on different signaling pathways. Thus, JNK kinases activate c-Jun while p38 MAP kinase stimulates MAPKAP kinase 2 (72). A recent study demonstrated that preconditioning triggered a tyrosine kinase-regulated signaling pathway leading to the translocation and activation of p38 MAP kinase and MAPKAP kinase 2 (58).

Activation of K_{ATP} channels appears to be an adaptive mechanism that protects the myocardium against ischemic reperfusion injury (73). The activation of this ion channel is at least partially responsible for the increase in outward K^+ currents, shortening of APD, and an increase in extracellular K^+ concentration during anoxic and globally ischemic condition (74). It appears that delayed PC, irrespective of preconditioning stimulus, is always mediated by K_{ATP} channels (75). It was shown that a mitochondrial K_{ATP} channel opener, diazoxide, significantly reduced the rate of cell death following simulated ischemia in adult ventricular cardiac myocytes (46). The protective effects of diazoxide was blocked by 5-HD, a selective blocker of the mitochondrial K_{ATP} channel. Intravenous injection of diazoxide 10 min before ischemia greatly reduced infarct size in the rabbit heart (76).

III.B. Pharmacological Preconditioning with Resveratrol

Unlike pharmacological therapeutic interventions, preconditioning protects the heart by upregulating its endogenous defense mechanisms (77). Unfortunately, ischemic preconditioning-mediated cardioprotection has a limited life, classical or early preconditioning lasting for several hours, and delayed preconditioning lasting for several days (78). There is a definite need to identify a pharmacological preconditioning agent to render the preconditioning stimulus everlasting. Recently, our researchers and others found that monophosphoryl lipid A (MLA) induces dose-dependent cardioprotection against myocardial infarction (79). Such cardioprotection was achieved through the ability of MLA to upregulate endogenous NO formation (54). Recently, NO has been found to play an essential role as mediator of ischemic preconditioning (31). Similar to this, resveratrol was also found to protect the ischemic myocardium through NO, because inhibition of NO with L-NAME (NO blocker) abolished the cardioprotective effects of resvera-

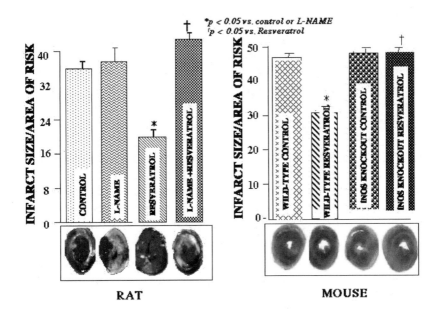

Figure 1 Effects of resveratrol on myocardial infarct size. A group of isolated perfused rat hearts was treated with resveratrol in the absence or presence of NO blocker for 15 min followed by 30 min ischemia and 2 hr of reperfusion. Infarct size was determined by scanning the images of the rat heart ventricular sections stained with TTC. Representative infarct size of six groups of hearts are shown in the figure (bottom). Infarct size is expressed as percent infarct relative to area at risk. Results are expressed as Means $\pm$ SEM of at least 6–8 rats per group (top). $^*p < 0.05$ vs. control; $^\dagger p < 0.05$ vs. resveratrol.

trol (80) (Figs. 1 and 2). In this study, preconditioning of the hearts with resveratrol provided cardioprotection as evidenced by improved postischemic ventricular functional recovery (developed pressure and aortic flow) and reduced myocardial infarct size and cardiomyocyte apoptosis. Resveratrol-mediated cardioprotection was completely abolished by both L-NAME and AG (iNOS inhibitor). Resveratrol caused an induction of the expression of iNOS mRNA beginning at 30 min after reperfusion, increasing steadily up to 60 min of reperfusion and then decreasing progressively up to 2 hr after reperfusion. Preperfusion of the hearts with AG almost completely blocked the induction of iNOS. In this study, resveratrol

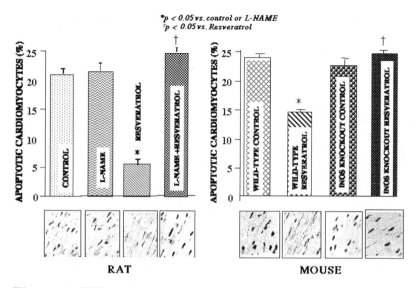

Figure 2 Effects of resveratrol on cardiomyocyte apoptosis. A group of isolated perfused rat hearts was treated with resveratrol in the absence or presence of NO blocker for 15 min followed by 30 min ischemia and 2 hr of reperfusion. Double immunofluorescent staining was performed with Tunel staining and antibody recognizing cardiac myosin heavy chain to detect apoptotic nuclei using scanning laser microscopy (bottom). Results are expressed as Means ± SEM of 6–8 rats per group (top). $*p < 0.05$ vs. control; $^{\dagger}p < 0.05$ vs. resveratrol.

also diminished the amount of ROS activity as evidenced by reduced malonaldehyde formation (Fig. 3). The results of this study demonstrated that resveratrol could pharmacologically precondition the heart in an NO-dependent manner.

Another related study showed that iNOS knockout mouse could not be preconditioned with resveratrol further indicating that this polyphenol provides cardioprotection through NO, and specifically through the induction of iNOS (81). In this study, control experiments were performed with wild-type and iNOS knockout hearts, which were not treated with resveratrol. Resveratrol-treated wild-type mouse hearts displayed significant improvement in postischemic ventricular functional recovery compared to those of nontreated hearts. Both resveratrol-treated and nontreated iNOS knockout

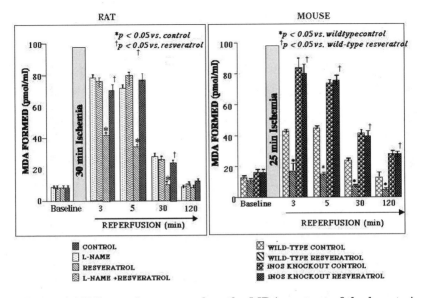

Figure 3 Effects of resveratrol on the MDA content of the heart. A group of isolated perfused rat hearts was treated with resveratrol in the absence or presence of NO blocker for 15 min followed by 30 min ischemia and 2 hr of reperfusion. MDA content of the heart was determined as described in material and methods. Results are expressed as Means ± SEM of at least 6–8 rats per group. $^*p < 0.05$ vs. control; $^{\dagger}p < 0.05$ vs. resveratrol.

mouse hearts resulted in relatively poor recovery in ventricular function compared to wild-type resveratrol-treated hearts. Myocardial infarct size was lower in the resveratrol-treated wild-type mouse hearts as compared to other group of hearts (Fig. 1). In concert, number of apoptotic cardiomyocytes were lower in the wild-type mouse hearts treated with resveratrol (Fig. 2). Cardioprotective effects of resveratrol were abolished when the wild-type mouse hearts were simultaneously perfused with aminoguanidine, an iNOS inhibitor. Resveratrol induced the expression of iNOS in the wild-type mouse hearts, but not in the iNOS knockout hearts, after only 30 min of reperfusion. Expression of iNOS remained high even after 2 hr of reperfusion. Resveratrol-treated wild-type mouse hearts were subjected to lower amount of oxidative stress as evidenced by reduced amount of malonaldehyde content in these hearts compared to iNOS knockout and untreated hearts (Fig. 3). These results demonstrated that resveratrol was unable to precondition iNOS knockout mouse hearts, while it could successfully precondition the wild-type mouse hearts indicating an essential role of iNOS in resveratrol preconditioning of the heart.

IV. SUMMARY AND CONCLUSION

It should be clear from the above discussion that resveratrol has two faces—it protects cells by augmenting NO by functioning as a pharmacological preconditioning agent and it selectively kills cancer cells. It is tempting to speculate that NO may play a crucial role for such dual behavior of resveratrol. Resveratrol shares many properties of NO and similar to resveratrol, NO also has two entirely opposite faces (25). While constitutive expression of NO is protective, it is equally destructive to the cells. Nitric oxide plays a crucial role in myocardial preconditioning. Future studies will reveal the mystery of two faces of resveratrol and pinpoint its precise mechanism of action.

ACKNOWLEDGMENTS

This study was supported in part by NIH HL 22559, HL 33889, HL 56803, and HL 56322.

REFERENCES

1. Gaziano JM, Buring JE, Breslow JL, et al. Moderate alcohol intake, increased levels of high-density lipoprotein and its sub-fractions and decreased risk of myocardial infarction. N Engl J Med 1993; 329:1829–1834.

2. Renaud S, De Lorgeril M. Wine, alcohol, platelets and the French Paradox for coronary heart disease. Lancet 1992; 339: 1523–1526.

3. Soleas GJ, Diamandis EP, Goldberg DM. Wine as a biological fluid: history, production and role in disease prevention. J Clin Lab Anal 1997; 11:287–313.

4. Saija A, Scalese M, Laiza M. Flavonoids as antioxidant agents: importance of their interaction with biomembranes. Free Rad Biol Med 1995; 19:481–486.

5. Das DK, Maulik N. Evaluation of antioxidant effectiveness in ischemia reperfusion tissue injury. Methods Enzymol 1994; 233:601–610.

6. Cordis GA, Maulik G, Bagchi D, Riedel W, Das DK. Detection of oxidative DNA damage to ischemic reperfused rat hearts by 8-hydroxydeoxyguanosine formation. Mol Cell Cardiol 1098; 30:1939–1944.

7. Bertelli AAE, Giovannini L De Caterina R, Bernini W, Migliori M, Fregoni M, Bavaresco L, Bertelli A. Antiplatelet activity of cis-resveratrol. Drugs Exp Clin Res 1996; 22:61–63.

8. Ferrero ME, Bertelli AE, Fulgenzi A, Pellegatta F, Corsi MM, Bonfrate M, Ferrara F, DeCaterina R, Giovannini L, Bertelli A. Activity in vitro of resveratrol on granulocyte and monocyte adhesion to endothelium. Am J Clin Nutr 1998; 68:1208–1214.

9. Bastianetto S, Zheng WH, Quirion R. Neuroprotective abilities of resveratrol and other red wine constituents against nitric-oxide-related toxicity in cultured hippocampal neurons. Br J Pharmacol 2000; 131:711–720.

10. Das DK, Sato M, Ray PS, Maulik G, Engelman RM, Bertelli AAE, Bertelli A. Cardioprotection with red wine: role of polyphenolic antioxidants. Drugs Exp Clin Res 1999; 25:115–120.

11. Giovannini L, Migliori M, Longoni BM, Das DK, Bertelli AAE, Panichi V, Filippi, Bertelli A. Resveratrol, a polyphenol found in wine, reduces ischemia reperfusion injury in rat kidneys. J Cardiovasc Pharmacol 2001; 37:262–270.

12. Ray PS, Maulik G, Cordis GA, Bertelli AAE, Bertelli A, Das DK. The red wine antioxidant resveratrol protects isolated rat hearts from ischemia reperfusion injury. Free Rad Biol Med 1999; 27:160–169.

13. Guo Y, Jones WK, Xuan YT, Tang XL, Bao W, Wu WJ, Han H, Laubach VE, Ping P, Yang Z, Qiu Y, Bolli R. The late phase of

ischemic preconditioning is abrogated by targeted disruption of the inducible NO synthase gene. Proc Natl Acad Sci USA 1999; 96:11507–11512.

14. Chen CK, Pace-Asciak CR. Vasorelaxing activity of resveratrol and quercetin in isolated rat aorta. Gen Pharmacol 1996; 27: 363–366.

15. Gehm BD, Mcandrews JM, Chien PY, Jameson JL. Resveratrol, a polyphenolic compound found in grapes and wine, is an agonist for the estrogen receptor. Proc Natl Acad Sci USA 1997; 94:14138–14143.

16. Hung L, Chen J, Hunag S, Lee R, Su M. Cardioprotective effect of resveratrol, anatural antioxidant derived from grapes. Cardiovasc Res 2000; 47:549–555.

17. Bhat KPL, Kosmeder JW II, Pezzuto JM. Biological effects of resveratrol. Antioxidant Redox Signal 2001; 3:1041–1064.

18. Hattori R, Otani H, Maulik N, Das DK. Pharmacological preconditioning with resveratrol—a role of nitric oxide. Antioxidant Redox Signal 2002; 282:H1988–H1995.

19. Engelman DT, Watanabe M, Maulik N, Cordis GA, Engelman RM, Rousou JA, Flack JE, Deaton DW, Das DK. L-Arginine reduces endothelial inflammation myocardial stunning during ischemia/reperfusion. Ann Thoracic Surg 1995; 60:1275–1281.

20. Lee SK, MbWambo ZH, Chung H. Evaluation of the antioxidant potential of natural products. Comb Chem High Throughput Screen 1998; 1:35–46.

21. Kampa MA, Hatzoglou A, Notas G. Wine antioxidant polyphenols inhibit the proliferation of human prostate cancer cell lines. Nutr Cancer 2000; 37:223–233.

22. Martinez J, Moreno JJ. Effect of resveratrol, a natural polyphenolic compound, on reactive oxygen species and prostaglandin production. Biochem Pharmacol 2000; 59:865–870.

23. Manna SK, Mukhopadhyay A, Aggarwal BB. Resveratrol suppresses TNF-induced activation of nuclear transcription factors NFκB, activator protein-1, and apoptosis: potential role of reactive oxygen intermediates and lipid peroxidation. J Immunol 2000; 164:6509–6519.

24. Wiseman H. The therapeutic potential of phytoestrogens. Expert Opin Invest Drugs 2000; 9:1829–1840.

25. Basly JP, Marre-Fournier F, LeBail JC. Estrogenic/antiestrogenic and scavenging properties of (E)- and (Z)- resveratrol. Life Sci 2000; 66:769–777.

26. Turner RT, Evans GL, Zhang M. Is resveratrol an estrogen agonist in growing rats? Endocrinology 1999; 140:50–54.

27. Ashby JH, Tinwell W, Pennie W. Partial and weak oestrogenecity of the red wine constituent resveratrol: consideration of its superagonist activity in MCF-7 cells and its suggested cardiovascular protective effects. J Appl Toxicol 1999; 19:39–45.

28. Freyberger A, Hartmann E, Hildebrand H, Krotlinger F. Differential response of immature rat uterine tissue to ethinyltradiol and the red wine constituent resveratrol. Arch Toxicol 2001; 11:709–715.

29. Mizutani K, Ikeda K, Kawai Y, Yamori Y. Resveratrol attenuates ovarectomy-induced hypertension and bone loss in stroke-prone spontaneously hypertensive rats. J Nutr Sci Vitaminol 2000; 46:78–83.

30. Levenson AS, Gehm BD, Pearce ST, Horiguchi J, Simons LA, Ward JE, Jameson JL, Jordan VC. Resveratrol acts as an estrogen receptor (ER) agonist in breast cancer cells stably transfected with Erα. Int J Cancer 2003; 104:587–596.

31. Zhao T, Xi L, Chelliah J, Levasseur MS, Kukreja RC. Inducible nitric oxide synthase mediates delayed myocardial protection induced by activation of adenosine A_1 receptors. Circulation 2001; 102:902–908.

32. Carbo N, Costelli P, Baccino MF, Lopez-Soriano FJ, Argiles JM. Resveratrol, a natural product present in wine, decreases tumor growth in a rat tumor model. Biochem Biophys Res Commun 1999; 254:739–743.

33. ElAttar TMA, Virji AS. Modulating effect of resveratrol and quercetin on oral cancer cell growth and proliferation. Anticancer Drugs 1999; 10:187–193.

34. Caltagirone S, Rossi C, Poggi A, Ranelletti FO, Natali PG, Brunetti M, Aiello FB, Piantelli M. Alavonoids apigenin and

quercetin inhibit melanoma growth and metastatic potential. Int J Cancer 2000; 87:595–600.

35. Tessitore L, Davit A, Sarotto I, Caderni G. Resveratrol depresses the growth of colorectal aberrant crypt foci by affecting bax and p21 expression. Carcinogenesis 2000; 21:1619–1622.

36. Jang M, Cai L, Udeani GO, Slowing KV, Thomas CF, Beecher CWW, Fong HHS, Farnsworth NR, Kinghorn AD, Mehta RG, Moon RC, Pezzuto JM. Cancer chemopreventive activity of resveratrol, a natural product derived from grapes. Science 1997; 275:218–220.

37. Mahady GB, Pendland SL, Chadwick LR. Resveratrol and red wine extracts inhibit the growth of CagA+ strains of *Helicobacter pylori* in vitro. Am J Gastroenterol 2003; 98:1440–1441.

38. Mahady GB, Pendland SL. Resveratrol inhibits the growth of *Helicobacter pylori* in vitro. Am J Gastroenterol 2000; 95:1849.

39. Hall SS. Longevity research in vino vitalis? Compounds activate life-extending genes. Science 2003; 301:1165.

40. Mitchell MB, Meng X, Lihua AO, Brown JM, Harken AH, Banerjee A. Preconditioning of isolated rat heart mediated by protein kinase C. Circ Res 1995; 76:73–81.

41. Flack J, Kimura Y, Engelman RM, Das DK. Preconditioning the heart by repeated stunning improves myocardial salvage. Circulation 1991; 84(suppl II):369–374.

42. Sato M, Cordis GA, Maulik N, Das DK. SAPKs regulation of ischemic preconditioning. Am J Physiol 2000; 279:H901–H907.

43. Maulik N, Wei ZJ, Engelman RM, Lu D, Moraru II, Rousou JA, Das DK. Interleukin-1α preconditioning reduces myocardial ischemic reperfusion injury. Circulation 1993; 88:387–394.

44. Lasley RD, Mentzer RM Jr. Dose-dependent effects of adenosine on interstitial fluid adenosine and postischemic function in the isolated rat heart. J Pharmacol Exp Ther 1998; 286: 806–811.

45. Parrat JR, Kane KA. KATP channels in ischemic preconditioning. Cardiovasc Res 1994; 28:783–787.

46. Liu Y, Satyo T, O'Rourke B, Marban E. Mitochondrial ATP-dependent potassium channels—novel effectors of cardioprotection? Circulation 1998; 97:2463–2469.

47. Das DK, Engelman RM, Kimura Y. Molecular adaptation of cellular defenses following preconditioning of the heart by repeated ischemia. Cardiovasc Res 1993; 27:578–584.

48. Das DK, Prasad MR, Lu D, Jones RM. Preconditioning of heart by repeated stunning. Adaptive modification of antioxidant defense system. Cell Mol Biol 1992; 38:739–749.

49. Liu X, Engelman RM, Moraru II, Rousou JA, Flack JE, Deaton DW, Maulik N, Das DK. Heat shock. A new approach for myocardial preservation in cardiac surgery. Circulation 1992; 86:II358–II363.

50. Maulik N, Engelman RM, Wei Z, Liu X, Rousou JA, Flack J, Deaton D, Das DK. Drug induced heat shock improves postischemic ventricular recovery after cardiopulmonary bypass. Circulation 1995; 92:II-381–II-388.

51. Maulik N, Engelman RM, Rousou JA, Flack JE, Deaton D, Das DK. Ischemic preconditioning reduces apoptosis by upregulating anti-death gene Bcl-2. Circulation 1999; 100:II369–II375.

52. Das DK, Maulik N, Engelman RM, Yoshida T, Zu YL. Preconditioning potentiates molecular signaling for myocardial adaptation to ischemia. Ann NY Acad Sci 1996; 793:191–209.

53. Maulik N, Sasaki H, Addya S, Das DK. Regulation of cardiomyocyte apoptosis by redox-sensitive transcription factors. FEBS Lett 2000; 485:7–12.

54. Tosaki A, Maulik N, Elliott GT, Blasig IE, Engelman RM, Das DK. Preconditioning of rat heart with monophosphoryl lipid A: a role of nitric oxide. J Pharmacol Exp Ther 1998; 285: 1274–1279.

55. Marala RB, Mustafa SJ. Adenosine A_1 receptor-induced upregulation of protein kinase C: role of pertussis toxin-sensitive G protein (s). Am J Physiol 1995; 269:H1619–H1624.

56. Cohen MV, Baines CP, Downey JM. Ischemic preconditioning: from adenosine receptor to K_{ATP} channel. Ann Rev Physiol 2000; 62:79–109.

57. Maulik N, Yoshida T, Zu YL, Banerjee A, Das DK. Ischemic stress adaptation of heart triggers a tyrosine kinase regulated signaling pathway: a potential role for MAPKAP Kinase 2. Am J Physiol 1998; 275:H1857–H1864.

58. Maulik N, Watanabe M, Zu YL, Huang C-K, Cordis GA, Schley JA, Das DK. Ischemic preconditioning triggers the activation of MAP kinases and MAPKAP kinase 2 in rat hearts. FEBS Lett 1996; 396:233–237.

59. Tanno M, Tsuchida A, Nozawa Y, Matsumoto T, Hasegawa T, Miura T, Shimamoto K. Roles of tyrosine kinase and protein kinase C in infarct size limitation by repetitive ischemic preconditioning in the rat. J Cardiovasc Pharmacol 2000; 35: 345–352.

60. Fryer RM, Schultz JEJ, Hsu AK, Gross GJ. Importance of PKC and tyrosine kinase in single or multiple cycles of preconditioning in rat hearts. Am J Physiol 1999; 276:H1229–H1235.

61. Ping P, Zhang J, Zheng Y-T, Li RCX, Dawn B, Tang X-L, Takano H, Balafanova Z, Bolli R. Demonstration of selective protein kinase C-dependent activation of Src and Lck tyrosine kinases during ischemic preconditioning in conscious rabbits. Circ Res 1999; 85:542–550.

62. Baines CP, Wang L, Cohen WV, Downey JM. Protein tyrosine kinase is downstream of protein kinase C for ischemic preconditioning's anti-infarct effect in the rabbit heart. J Mol Cell Cardiol 1998; 30:383–392.

63. Gniadecki R. Nongenomic signaling by vitamin D: a new face of Src. Biochem Pharmacol 1997; 56:1273–1277.

64. Vahlhaus C, Schulz R, Post H, Rose J, Heusch G. Prevention of ischemic preconditioning only by combined inhibition of protein kinase C and protein tyrosine kinase in pigs. J Mol Cell Cardiol 1998; 30:197–209.

65. Schlessinger J. New roles for Src kinases in control of cell survival and angiogenesis. Cell 2000; 100:293–296.

66. Takeishi Y, Abe J, Lee JD, Kawakatsu H, Walsh RA, Berk BC. Differential regulation of p90 ribosomal S6 kinase and Big mitogen-activated protein kinase 1 by ischemia/reperfusion

and oxidative stress in perfused guinea pig heart. Circ Res 1999; 85:1164–1172.

67. Sadoshima J, Izumo S. The heterotrimeric G_q protein-coupled angiotensin II receptor activates p21^{ras} via the tyrosine kinase-Shc-Grb2-Sos pathway in cardiac myocytes. EMBO J 1996; 15:775–778.

68. Eguchi S, Numaguchi K, Iwasaki H, Mitsumoto T, Yamakawa T, Utsunomiya H, Motley ED, Kawakatsu H, Owada KM, Hirata Y, Marumo F, Inagami T. Calcium-dependent epidermal growth factor receptor transactivation mediates the angiotensin II-induced mitogen activated protein kinase activation in vascular smooth muscle cells. J Biol Chem 1998; 273:8890–8896.

69. Abe J, Takahashi M, Ishida M, Lee JD, Berk BC. c-Src is required for oxidative stress-mediated activation of big mitogen-activated protein kinase 1 (BMK1). J Biol Chem 1997; 272: 20,389–20,394.

70. Akhand AA, Pu M, Senga T, Kato M, Suzuki H, Miyata T, Hamaguchi M, Nakashima I. Nitric oxide controls Src kinase activity through a sulfhydryl group modification-mediated Tyr-527-independent and Tyr-416-linked mechanism. J Biol Chem 1999; 274:25821–25826.

71. Kyriakis JM, Banerjee P, Nikolakaki E, Dai T, Rubie EA, Ahmad MF, Avruch J, Woodgett JD. The stress-activated protein kinase subfamily of c-Jun kinases. Nature 1994; 369: 156–160.

72. Hu Y, Metzler B, Xu Q. Discordant activation of stress-activated protein kinase or c-jun-NH2-terminal protein kinases in tissues of heat stressed mice. J Biol Chem 1997; 272:9113–9119.

73. Cole WC, McPherson CD, Sontag D. ATP-regulated K+ channels protect the myocardium against ischemic/reperfusion damage. Circ Res 1991; 69:571–581.

74. Bedheit SS, Restivo M, Boutjdir M, Henkin P, Gooyandeh K, Assadi M, Khatib S, Gough WB, El-Sherif N. Effects of glyburide on ischemia-induced changes in extracellular potassium and local myocardial activation: a potential new approach to the management of ischemia-induced malignant ventricular arrhythmias. Am Heart J 1990; 119:1025–1033.

75. Bernardo NL, D'Angelo M, Okubo S, Joy A, Kukreja RC. Delayed ischemic preconditioning is mediated by opening of ATP-sensitive potassium channels in the rabbit heart. Am J Physiol 1999; 276:H1323–H1330.

76. Baines CP, Liu GS, Birincioglu M, Critz SD, Cohen MV, Downey JM. Ischemic preconditioning depends on interaction between mitochondrial KATP channels and actin cytoskeleton. Am J Physiol 1999; 276:H1361–H1368.

77. Flack J, Kimura Y, Engelman RM, Das DK. Preconditioning the heart by repeated stunning improves myocardial salvage. Circulation 1991; 84(suppl III):369–374.

78. Das DK. Ischemic preconditioning and myocardial adaptation to ischemia. Cardiovasc Res 1993; 27:2077–2079.

79. Maulik N, Tosaki A, Elliott GT, Maulik G, Das DK. Induction of iNOS expression by monophosphoryl lipid A: a pharmacological approach for myocardial adaptation to ischemia. Drugs Exp Clin Res 1998; 24:117–124.

80. Hattori R, Otani H, Maulik N, Das DK. Pharmacological preconditioning with resveratrol: a role of nitric oxide. Am J Physiol 2002; 282:H1989–H1995.

81. Imamura G, Bertelli AA, Bertelli A, Otani H, Maulik N, Das DK. Pharmacological preconditioning with resveratrol: an insight with iNOS knockout mice. Am J Physiol 2002; 282:H1996–H2003.

21

Molecular Mechanisms of Prevention Against Allergic Rhinitis by Aller-7, A Novel Polyherbal Formulation

DEBASIS BAGCHI

Department of Pharmacy Sciences, School of Pharmacy and Health Professions, Creighton University Medical Center, Omaha, Nebraska, U.S.A.

MANASHI BAGCHI

Interhealth Research Center, Benicia, California, U.S.A.

AMIT AGARWAL and VINOD S. SAXENA

Natural Remedies Research Center, Bangalore, India

ABSTRACT

Allergic rhinitis is an immunological disorder and an inflammatory response of nasal mucosal membranes. Allergic rhinitis, a state of hypersensitivity, occurs when the body overreacts to a substance such as pollens or dust. A novel, safe polyherbal formulation (Aller-7/NR-A2) has been developed for the treatment of allergic rhinitis using a unique combination of extracts from seven medicinal plants including *Phyllanthus emblica*, *Terminalia chebula*, *Terminalia bellerica*, *Albizia lebbeck*, *Piper nigrum*, *Zingiber officinale*, and *Piper longum*. A number of safety and toxicity studies including acute oral, dermal, primary dermal irritation, eye irritation, subacute, subchronic, mutagenicity, reproductive toxicity, and teratogenicity, were evaluated, which demonstrated the broad spectrum safety of Aller-7. In conjunction, a number of mechanistic studies including mast cell stabilization, hyaluronidase, lipoxygenase, and trypsin inhibitory activities, antispasmodic, antihistaminic, anti-inflammatory, and antioxidant activities, were conducted to demonstrate the efficacy of Aller-7. Finally, clinical studies were conducted to demonstrate the clinical efficacy of Aller-7. No significant adverse effects were observed. Taken together, these studies demonstrate the safety and efficacy of Aller-7 in ameliorating the symptoms of allergic rhinitis.

I. INTRODUCTION

Allergic rhinitis, a frequently occurring immunological disorder also known as "hay fever," "rose fever," or "summer catarrh," is a major challenge to the health professionals, which affects 10–25% of the World's population (1–3). In the United States alone, 40 million men, women, and children are affected by allergy and annually spend approximately $4.5 billion. Statistics demonstrate that this immunological disorder results in 5 million days of lost time and productivity at work and school each year (3–5).

Allergy is an excessive reaction to a substance in the environment called an allergen. Following exposure to these otherwise harmless materials, sensitization (allergy) may occur with time. Pollen, mold, dust, mite, feces, and animal allergens that contact the nasal and eye lining cause sneezing, coughing, stuffy or runny nose, itchy eyes, nose and throat, nasal congestion, and watery eyes (1,3). Basically, allergic rhinitis is an inflammation or irritation of the mucous membranes that line the eyes and nose. People who suffer may also have a nasally voice, breathe noisily, snore, feel chronically tired, have a poor appetite, have frequent headaches, experience nausea, fever, anorexia, and body ache, and even have difficulty in hearing and smelling. Upon lying down at night, the fluid in the nose drips down on to the back of the throat and causes bouts of coughing, which can be uncomfortable and disturb sleep (1–3).

Trees cause early spring symptoms of allergy, grasses elicit late spring and early summer problems, and September and October difficulties are caused by weeds (especially ragweed). Perennial allergies may be due to dust, mites, molds, and animals. Other factors that contribute to the prevalence of allergic rhinitis include asthma, repeated infections, lack of physical activity, obesity, and changes in eating habits (1–5).

I.A. Pathophysiology of Allergy

Allergy is defined as a state of hypersensitivity or hyperimmunity caused by exposure to a particular antigen (allergen), resulting in marked increases in reactivity to that antigen upon subsequent exposure. The mechanism behind the development of allergic rhinitis is complex and multifactorial. Two major cascades are involved in the progression of allergy (1–3).

In cascade I, the first exposure to an allergen causes little apparent reaction, but β-lymphocytes under the influence of T-helper cells produce IgE antibodies specific to that allergen (also termed as sensitization) (Fig. 1). These antibodies become attached to the surface of mast cells and basophils within the nasal mucosa (1–3).

Following subsequent exposure in cascade II, the allergen binds to the antibody, cross-linking to adjacent IgE molecules and causing the cells to degranulate with the sudden release of histamine and other inflammatory mediators such as leukotrienes, hyaluronidase, 5-lipoxygenase, heparin, trypsin, and serotonin (Fig. 2). This causes an immediate response, with symptoms including sneezing, itching, and rhinorrhea appearing

Sensitization

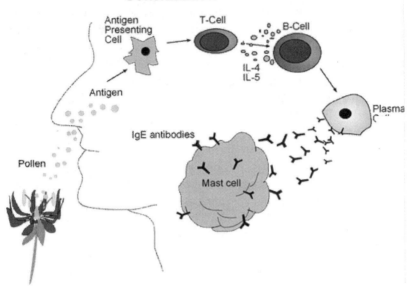

Figure 1 Allergy Cascade I—sensitization stage. In the first allergen exposure, β-lymphocytes under the influence of T-helper cells produce IgE antibodies specific to that allergen. These antibodies become attached to the surface of mast cells and basophils within the nasal mucosa.

within minutes. These mediators also act to upregulate formation of cell surface adhesion molecules on adjacent vascular endothelial cells leading to the release of more mediators, which cause further local inflammation. This cascade of allergen-triggered mediator release produces a second peak of symptoms (the late-phase response) occurring 3–24 hr after allergen challenge (Fig. 2) (1–3).

I.B. Treatment Options

Epidemiological surveys indicate that there have been notable increases in the prevalence of both asthma and allergy symptoms in children and young adults. Several potential determinants have been proposed. Rhinitis and asthma are frequently found to coexist and a number of studies have

Re-exposure to Antigen

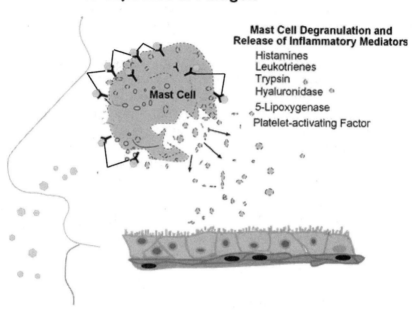

Figure 2 Allergy Cascade II—re-exposure to antigen. Upon subsequent exposure, the allergen binds to the antibody, cross-linking adjacent IgE molecules and causing the cells to degranulate with the sudden release of histamine and other mediators, leading to symptoms including sneezing, itching, and rhinorrhea. These mediators upregulate formation of cell surface adhesion molecules on adjacent vascular endothelium cells leading to the release of more mediators, which cause further local inflammation. This cascade of allergen-triggered mediator release produces a second peak of symptoms occurring 3–24 hr after allergen challenge.

shown that successful treatment of rhinitis, for example nasal steroid treatment, has a beneficial clinical effect on coexistent asthma (2–6).

Conventional therapy for allergic rhinitis includes:

1. prevention of contact with allergens;
2. pharmacotherapy with antihistaminic drugs, mast cell stabilizers, and topical corticosteroids;
3. immunotherapy.

The conventional antihistaminic drugs, like prometha-
zine and chlorpheniramine, have limited use in allergic rhinitis due to their property of sedation (drowsiness). The newer
antihistamines, like fexofenadine and cetirizine, though
claimed to be free from sedation are not truly so (2,6). Among
the second-generation antihistamines, azelastine and levoca-
bastine have been found to be locally effective and with minimal side effects. Hence, these agents may be preferred.
However, they only offer symptomatic relief particularly for
sneezing and rhinorrhea, and they have a minimal effect on
nasal congestion. In addition, they cause other side effects such
as nervousness, fatigue, dry mouth, and drowsiness (2–4).

Mast cell stabilizers, such as disodium cromoglycate and
ketotifen, have been recommended for prophylaxis, but were
not found to be very effective. Further, ketotifen causes drowsiness and several other side effects and therefore is not
preferred on a regular basis (2).

Intranasal corticosteroids are considered to be the "gold
standard" today in the management of allergic rhinitis. Corticosteroids act as anti-inflammatory agents and help to reduce
nasal congestion. They are also said to modify local IgE production. However, they have little effect on symptoms like sneezing and rhinorrhea, and there have been reported side effects
of local irritation and bleeding in 5–10% of patients (4–6).

Immunotherapy with identified allergens remains a
cornerstone in the treatment of allergic rhinitis. This induces
a state of hyposensitization by producing blocking antibodies.
However, it is not popular with patients since it requires
repeated injections. In terms of efficacy also the success of
immunotherapy is low (2,6).

The International Rhinitis Management Working Group
recommends the use of antihistaminic drugs in patients with
mild or occasional rhinitis symptoms and topical nasal steroids for those with moderate or persistent symptoms (2,4).
However, despite the available therapies, the prevalence of
allergic rhinitis is increasing and current antihistamine therapies offer only symptomatic relief with accompanying adverse
side effects. Thus, there is a need for a therapeutic approach,
which can modify the pathophysiology of the disease and is

free from adverse side effects. The target of action for the new agents should be at the level of sensitization process namely, (a) production of interleukin-1 by antigen presenting cells, (b) T-cell proliferation, (c) cytokine release by T-Helper2(TH$_2$) cells, (d) B-lymphocyte proliferation, and (e) reduction in IgE levels (2,4–6).

Till date, no novel, natural therapy has demonstrated clinical efficacy against allergic rhinitis.

I.C. Design of a Novel Polyherbal Formulation for Allergic Rhinitis

Aller-7/NR-A2, a novel polyherbal formulation US Patent# US6730332, is a unique blend of the extract from dried bark, fruits, and rhizome of seven extensively used Indian medicinal plants namely *Phyllanthus emblica* (fruit) extract, *Terminalia chebula* (fruit) extract, *Terminalia bellerica* (fruit) extract, *Albizia lebbeck* (bark) extract, *Zingiber officinale* (rhizome) extract, *Piper longum* (fruit) extract, and *Piper nigrum* (fruit) extract (Fig. 3). The final extract was standardized to 250 mg/g polyphenols as tannic acid; 85 mg/g chebulic acid; 50 mg/g gallic acid; 17.5 mg/g glycosides as corilagin, 7.5 mg/ g 1–3-6 tri-O-galloyl glucose; 4.0 mg/g ellagic acid; 1.0 mg/g piperine; and a total of 0.20 mg/g of 6- and 8-gingerols (7).

II. SAFETY

A broad spectrum of safety studies were conducted on Aller-7 in both in vitro and in vivo models including acute oral, acute dermal, primary dermal irritation, eye irritation, subacute, subchronic, mutagenicity and genotoxicity, teratogenicity, and reproductive toxicity, to demonstrate its safety for human consumption.

II.A. Acute Toxicity Studies

II.A.1. Oral Toxicity Studies

Aller-7 was orally administered to Swiss Albino mice (10 males and 10 females; 20–28 g; 2–3 months old) at doses

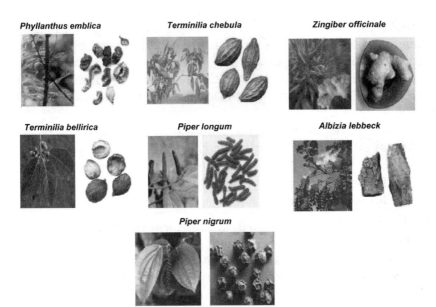

Figure 3 Aller-7, a novel polyherbal formulation, is a unique blend of dried bark, fruits, and rhizome of seven extensively used Indian medicinal plants: *P. emblica* (fruit) extract, *T. chebula* (fruit) extract, *T. bellerica* (fruit) extract, *A. lebbeck* (bark) extract, *Z. officinale* (rhizome) extract, *P. longum* (fruit) extract, and *P. nigrum* (fruit) extract. The final extract was standardized to 245 mg/g polyphenols as tannic acid; 80 mg/g chebulic acid; 45 mg/g gallic acid; 20 mg/g glycosides as corilagin and 1,3,6-tri-O-galloyl glucose; 3.5 mg/g ellagic acid; 0.75 mg/g piperine; 0.15 mg/g 6- and 8-gingerols.

of 125, 250, 500, 1000 and 1500 mg/kg body weight (b.w.). The animals were observed for behavioral, autonomic, and other changes once a day for 14 days, after which the animals were sacrificed for gross and histopathological examinations. No deaths were observed and no marked histopathological changes were observed in any of the vital organs examined. Thus, Aller-7 can be considered safe up to 1500 mg/kg mouse b.w. In a separate study, Aller-7 was orally given to Albino Wistar rats (10 males and 10 females; 180–200 g; 2–3 months old) at doses of 500, 1000, and 2000 mg/kg. No mortality, behavioral changes, and histopathological changes (in vital organs) were observed. Thus, Aller-7 can be considered safe up to 2000 mg/kg rat b.w. (7).

A third acute oral toxicity study was conducted in Sprague–Dawley rats (three males and three females; 160–176 g) following an acute oral treatment with 5000 mg Aller-7/kg b.w. Aller-7 neither caused death nor induced signs of evident toxicity and did not have any adverse effect on the body weight gain by the treated male and female rats during the 14-day observation period, post-treatment, except for one female that suffered a slight but reversible reduction in the body weight gain. There were no gross pathological alterations in any of the rats, as evident from necropsy. Thus, LD_{50} of Aller-7 was found to be greater than 5000 mg/kg b.w. in Sprague–Dawley rats.

II.B. Acute Dermal Toxicity Study

It was observed that Aller-7 at a dose of 2000 mg/kg b.w. applied dermally did not cause any local or systemic toxicity and mortality, no significant changes in body weight, and no gross pathological changes in terminal necropsy. Thus, dermal LD_{50} of Aller-7 was found to be greater than 2000 mg/ kg b.w. in Sprague–Dawley rats (10 males and 10 females, 180–225 g).

II.C. Acute Dermal Irritation Study

Aller-7 (500 mg) was applied directly on to the clipped intact dorsal skin of three New Zealand white rabbits (1800–2150 g b.w.) and was covered with gauze patch (approx. $6 cm^2$) for 4 hr. The patches were removed at the end of 4 hr and skin reaction was evaluated after 1, 24, 48, and 72 hr and scored according to Draize method (8). No mortality or other signs of toxicity were observed in any of the treated animals. The irritancy index was zero (0.00) as determined from the scores of the skin reactions, and thus Aller-7 can be classified as nonirritant to rabbit skin.

II.D. Acute Eye Irritation Study

Aller-7 (100 mg) was instilled in the lower conjunctival sac of the left eye, while the right eye served as control for each animal (New Zealand white rabbit, 1800–1950 g b.w., three rabbits). Eyes were rinsed with water 24 hr after Aller-7 application. Irritation to conjunctiva, iris, and cornea was evaluated at 1, 24, 48, and 72 hr after the treatment according to Draize (8). Throughout the study, no significant changes in body weight or clinical signs of intoxication were observed in treated rabbits. Aller-7 can be classified as minimal irritant to eye in rabbit based on the maximum mean score at 1 hr (4.66).

II.E. Subacute Toxicity

Aller-7 was administered to Albino Wistar rats (eight males and eight females; 180–200 g b.w.; 2–3 months old) initially at a dose of 90 mg/kg b.w. for 3 days, and then 180 mg/kg over the next 3 days and further increased to 270 mg/kg for 3 Weeks. After 28 days, hematological examinations for hemoglobin percentage and total leucocyte counts, as well as clinical biochemistry and histopathological evaluation (25 organs) were conducted. Gross observations, hematological, biochemical, and histopathological investigations revealed no significant abnormalities following treatment with Aller-7. Thus, Aller-7 was shown to be safe in repeated dose administration in rats up to 270 mg/kg (7).

II.F. Subchronic Toxicity

Subchronic toxicity studies provide information on possible health hazards likely to arise from repeated exposure over a limited period of time. A no-observed-adverse-effect-level (NOAEL) study was done in Albino Wistar rats (30 animals in each group, 15 males and 15 females, 180–220 g; 2–3 months) at oral Aller-7 doses of 250, 500, and 1000 mg/kg b.w. for 90 days. At the end of 90 days, blood was collected for biochemical estimations and 32 target organs were collected for histopathology from 20 animals from each group.

In the remaining 10 animals, dosing was stopped after 90 days and the animals were kept alive for up to 120 days. Similar observations were carried out on the 121st day. There were no significant differences between the Aller-7 and control groups in terms of general behavior and biochemical parameters. In the treatment group, at 250 mg/kg, there was a reduction in the size of the testis, seminal vesicle, and epididymis in one animal upon gross observation. The entire group was subjected to histopathological evaluation, but no deformities were observed. When gross and histopathological records were correlated, there was no toxicity observed in the individual animal organ weights, hematological, and biochemical profiles. Aller-7 exhibited a NOAEL above 1000 mg/kg b.w. in rats (7).

II.G. Teratogenicity Studies

Teratogenicity relates to the adverse effects of a substance on the developing embryo and fetus. Teratogenicity studies involve the administration of the test substance to the pregnant female rat (dam) throughout the course of her pregnancy. The young (pups) are delivered by cesarean section 1 day preterm. They are observed for defects such as the presence of extra ribs, cleft palate, and biochemical variations. Bone defects are visualized by clearing the fetus (or embryo, if termination is at an earlier stage in pregnancy) with potassium hydroxide and staining the bones with alizarin. A teratogenicity study was conducted using Aller-7 at the doses of 3000 and 1800 mg/kg b.w. to two groups of animals to evaluate its effects on pregnancy and lactation. The treatment was given from the onset of pregnancy up to the 20th day of pregnancy to observe the effect on the mother and fetus. To study the effect on lactation, Aller-7 was given from the onset of pregnancy up to the 21st day of lactation and the pups were evaluated for visceral and skeletal abnormalities. Aller-7 when given at the doses of 3000 mg/kg b.w. (20 times greater than the recommended dose for humans) and 1800 mg/kg b.w. did not produce any teratogenic effects (7).

II.H.   Reproductive Toxicity Studies

Reproductive toxicity was determined whether there is any defect caused by Aller-7 on reproductive functions of the animals. The effect of Aller-7 was evaluated both on male and female rats.

Four groups of 10 male Albino Wistar rats (150–250 g b.w.; 2–4 months) were administered Aller-7 at doses of 500 and 1000 mg/kg for duration of 60 days and 1000 mg/kg for 120 days to evaluate long-term safety. Following Aller-7 treatments, the males were kept for copulation with untreated females. Successful mating, mortality, body weight, food intake, and behavior were recorded. After allowing 1 week for copulation, the males were sacrificed, reproductive, and other vital organs were evaluated histopathologically. Normal female rats conceived by treated males showed normal fertility and conception pattern. The embryonic death, fetal deaths, and live fetuses were normal and there were no marked changes when compared to the control group at different dose levels and duration of treatment. The results indicate that Aller-7 at doses up to 1000 mg/kg body weight does not cause reproductive toxicity (7).

Female Albino Wistar rats (four groups, 10 rats/group, 150–250 g, 2–4 months) were orally administered either control or 1000, 2000, or 3000 mg Aller-7/kg b.w. for a period of 30 days before mating, during mating, and after successful copulation up to 16 days of gestation. Treated females were kept for conception with untreated males of known fertility. The parameters such as successful mating, mortality, body weight, feed intake were evaluated. After successful copulation, the females were autopsied on 16th day of gestation and observations such as implantation sites in each animal, number of corpora lutea in each ovary, number of small moles and large moles, successful pregnancies, fetal mortality, and gross examination of the target organs were performed. None of the doses produced any significant difference in the treated groups compared to control. The results demonstrate that Aller-7 does not cause reproductive toxicity in female rats at doses up to 3000 mg/kg b.w. (7).

II.I. Mutagenicity Studies

II.I.1. In Vitro: Ames' Test

The "bacterial reverse mutation test" was performed in several strains of *Salmonella typhimurium* including TA 1535, TA 97a, TA 98, TA 100, and TA 102, to assess the mutagenic potential of Aller-7 at doses of 5000, 1500, 500, 150, 50, and 15 μg/plate with and without a metabolic activation. No evidence of mutagenicity was observed at any dose level of Aller-7 in any of the strains of *S. typhimurium* with or without metabolic activation (7).

II.I.2. In Vivo: Micronucleus Test

This assay was conducted to detect the damage induced by Aller-7 at 500 and 2000 mg/kg b.w. doses for 30 days on the chromosomes or the mitotic apparatus of erythroblasts by analysis of erythrocytes as sampled in bone marrow and/or peripheral blood cells of animals, usually rodents.

Cyclophosphamide (100 mg/kg i.p.) was used as a standard. After 30 hr, the animals were sacrificed by cervical dislocation and bone marrow cells were collected in fetal calf serum. Evaluation was done by observing the frequency of polychromatic erythrocytes containing micronucleus (PCE with Mn). The results showed that there was no evidence of chromosomal damage and micronuclei in PCEs in both control and Aller-7-treated groups. Thus, no mutagenic potential of Aller-7 was observed following oral administration up to 2000 mg Aller-7/kg b.w. (7).

III. PHARMACOLOGICAL ACTIVITY STUDIES

III.A. Mast Cell Stabilization

Mast cells, also known as leukocytes, contain metachromatic granules that store a variety of inflammatory mediators—including histamine, trypsin, serotonin, and other proteolytic enzymes that can destroy tissue or cleave complement components—heparin or chondroitin sulfate, eosinophil chemotactic factor of anaphylaxis and neutrophil chemotactic factor (9).

Mast cells degranulate, releasing their mediators into surrounding tissues, modifying vascular and cellular reactions, affecting plasma factors (such as IL-3, IL-4, and IL-6), and resulting in inflammation and irritation (9). Such reactions can be mimicked in vitro using degranulators such as compound 48/80 (a polycondensate of N-methyl-p-methoxy phenylethylamine). The effect of Aller-7 on mast cell stabilization was demonstrated in vitro by observing the inhibitory effect of Aller-7 and disodium cromoglycate (positive control) on mast cell degranulation induced by compound 48/80. Approximately 28.7%, 37.3%, 49.3%, and 77.6% protection against compound 48/80-induced mast cell degranulation was observed following incubation with 4, 8, 16, and 32 μg/mL /mL concentrations of Aller-7, respectively, while under these same conditions, approximately 24.2%, 52.5%, and 75.9% protection was observed following incubation with 2, 4, and 8 μg/mL concentrations of disodium cromoglycate. The EC_{50} (effective concentration at which a compound produces a specific effect in 50% of the organisms treated) of Aller-7 and disodium cromoglycate was found to be 12.0 and 3.9 μg/mL, respectively, as compared to the untreated control samples (9). These results show that Aller-7 promotes mast cell stabilization by inhibiting 48/80-induced mast cell degranulation in vitro (9).

III.B. Antihistaminic Activity

Histamine is one of the most important mediators of allergic reactions, which is responsible for symptoms of allergic rhinitis like rhinorrhea, coughing, and sneezing. Histamine causes smooth muscle contraction involved in coughing and sneezing by stimulating Histamine-H_1 receptors (10). Aller-7 was tested to detect whether it could prevent contractions of guinea pig ileum induced by histamine. Aller-7 exhibited a concentration-dependent effect in the isolated guinea pig ileal assay. Antihistaminic activity of Aller-7 was evaluated in vitro using isolated guinea pig ileum substrate and cetirizine as positive control. Significant inhibition of histamine-induced contraction was observed in the concentration range of 100–800 µg/mL and 10–160 ng/mL for Aller-7 and cetirizine, respectively. Aller-7 demonstrated antihistaminic activity with an IC_{50} of 200 µg/mL (9).

III.C. Trypsin Inhibition Activity

Trypsin is one of the mediators released from mast cells upon degranulation following antigen–antibody reactions. Trypsin release mimics histamine release and provides an alternative marker for measuring mast cell stabilization and, therefore, is useful in antiallergic evaluation (10). Aller-7's ability to stabilize mast cells and inhibit trypsin release was demonstrated in vitro using benzoyl-d,1-4-nitroanilide as the substrate and ovomucoid, a major allergen of chicken eggs and known trypsin inhibitor, as a positive control. Approximately, 12.9%, 41.6%, and 80.0% trypsin inhibitory activities were observed following incubation with 4, 8, and 16 µg/mL concentrations of Aller-7, while the positive control ovomucoid exhibited 27.7%, 69.4%, and 92.4% inhibitions following incubation with 1, 2, or 4 µg/mL concentrations of ovomucoid. Thus, Aller-7 resulted in significant inhibition of trypsin activity with an IC_{50} of 9.0 µg/mL, compared to an IC_{50} for ovomucoid of 1.5 µg/mL (9).

III.D. Antispasmodic Activity

Carbachol and barium chloride are powerful spasmogens for bronchial smooth muscle, acting by mechanisms of cholinergic stimulation of H1 receptor, muscarinic M1 receptor, and nonspecific stimulation, respectively (9). Aller-7's antispasmodic activity was evaluated on contraction of guinea pig tracheal chain induced by carbachol and barium chloride. The submaximal contractions of carbachol and barium chloride were 18 ng/mL and 750 µg/mL, respectively. Approximately, 30.0%, 59.2%, and 75.8% inhibitions were observed against carbachol-induced contraction of guinea pig tracheal chain following incubation with 3, 10, and 30 µg/mL concentrations of Aller-7, respectively, as compared to the control samples. Under identical conditions following incubations with 1, 3, and 10 µg/mL of papaverine approximately 24.6%, 66.0%, and 79.6% inhibitions were observed (9). Aller-7 and papaverine concentration-dependently inhibited carbachol-induced contraction with an EC_{50} of 7.4 and 2.3 µg/mL, respectively (9).

In a second experiment, 31.7%, 58.3%, and 66.5% inhibitions were observed against barium chloride-induced contraction of guinea pig tracheal chain following incubation with 3, 10, and 30 µg/mL concentrations of Aller-7, respectively, as compared to the control samples, under identical conditions following incubations with 1, 3, and 10 µg/mL of papaverine 33.9%, 45.4%, and 76.2% inhibitions were observed. Aller-7 and papaverine also concentration-dependently inhibited barium chloride-induced contraction of guinea pig trachea with an EC_{50} of 8.4 and 2.8 µg/mL, respectively (9).

III.E. Hyaluronidase Inhibition Activity

Hyaluronidase, one of the lysosomal enzymes (endoglycosidases), hydrolyzes mucopolysaccharides, including hyaluronic acid, a high molecular weight acid found in mucus (2,11). Hyaluronidase usually exists in an inactive form, but is activated by antigens, as well as certain metal ions and compound 48/80. This enzyme is known to be related to vascular permeability and inflammatory process. Earlier, it was reported that antiallergic agents such as disodium cromoglycate (DSCG), transilast, and trisodium baicalein-6-phosphate, which are known to inhibit histamine release from mast cells induced by antigen-IgE–antibody reaction, are strong inhibitors of hyaluronidase. Compound 48/80 was known to activate hyaluronidase. Aller-7 was evaluated for its hyaluronidase inhibitory activity as described by Aronson and Davidson (11). Aller-7 exhibited approximately 7.1%, 9.8%, 14.8%, 52.3%, 71.7%, and 84.7% hyaluronidase inhibitory activities following incubations with 0.1, 0.5, 1.0, 2.0, 3.0, and 4.0 mg/mL concentrations of Aller-7, respectively, while under these same conditions 12.6%, 27.1%, 50.8%, 54.0%, 71.3%, and 82.3% inhibitory activities were observed following incubation with 0.3, 0.6, 0.9, 1.2, 1.5, and 1.8 mg/mL concentrations of disodium cromoglycate, respectively (9). Thus, Aller-7 showed significant hyaluronidase inhibitory activity with an EC_{50} of

1.7 mg/mL, while disodium cromoglycate exhibited an EC_{50} of 0.9 mg/mL (9).

III.F. Anti-5-Hydroxytryptamine Activity

5-Hydroxytryptamine (also termed as serotonin, 5-HT) is an important allergy mediator present in neuroendocrine, mast, and other cells in the body (2). It was observed that asthma patients have higher levels of 5-HT in their plasma than do asymptomatic patients and, therefore, it is suggested that 5-HT might play an important role in the pathophysiology of allergy and asthma. 5-Hydroxytryptamine uptake enhancer, tianeptine, has been shown to suppress asthmatic symptoms in children in a double-blind, cross-over study (14). Aller-7's ability to inhibit 5-HT activity was demonstrated in vivo using a rat fundus strip model. Aller-7 demonstrated significant dose-dependent inhibition of 5-HT activity with an EC_{50} of 32.8 µg/mL.

III.G. Anti-inflammatory Potential of Aller-7

Inflammation, the most common symptom of allergy, results from a complex series of reactions triggered by the body's immunological response to tissue damage (12). The primary objective of inflammation is to localize and eradicate the irritants and the surrounding tissue (12,13). For the survival of the host, inflammation is a necessary and beneficial process. Inflammatory response involves three major stages: (i) dilation of capillaries to increase blood flow, (ii) microvascular structural changes and escape of plasma proteins from the blood stream, and (iii) leukocyte transmigration through endothelium and accumulation at the site of injury (12). Arachidonic acid is a compound metabolized in the body to yield important hormones that play major roles in the process of inflammation. Arachidonic acid can be converted by the action of the enzyme cyclooxygenase to prostaglandins and thromboxanes (12,13).

Inflammation is an important aspect of immediate- and late-phase (3–24 hr) allergic reactions, involving cytokines (interleukins), prostaglandins, leukotrienes, histamine that are released from T-cells, and mast cells during an allergic response. Chronic inflammation due to continuous exposure to allergens can cause permanent damage (7,20). Therefore to demonstrate the anti-inflammatory effects of Aller-7, we first evaluated the lipoxygenase inhibitory activity of Aller-7 followed by a number of in vivo studies. The efficacy of Aller-7 was investigated in compound 48/80-induced paw edema both in Balb/c mice and Swiss Albino mice, carrageenan-induced paw edema in Wistar Albino rats, and Freund's adjuvant-induced arthritis in Swiss Albino rats.

III.G.1. Lipoxygenase Inhibitory Activity

Arachidonic acid plays a unique role as a precursor molecule in biological systems, which is transformed into potent mediators such as leukotrienes, thromboxane, and prostaglandins with far-ranging effects. Leukotrienes are induced in the initiation and maintenance of a variety of inflammatory diseases, including chronic rheumatism, Crohn's disease, colitis ulcerosa, psoriasis, asthma, allergic rhinitis, and anaphylactic shock. Aller-7 was evaluated spectrophotometrically for lipoxygenase inhibitory activity, a key enzyme in the conversion of arachidonic acid to inflammatory mediators, as per the method of Shinde et al. (14). Aller-7 exhibited approximately 23.8%, 45.1%, 61.7%, and 82.5% lipoxygenase inhibitory activities following incubations with 50, 100, 150, and 200 µg/mL concentrations of Aller-7, respectively, while under these same conditions 11.8%, 55.7%, and 83.8% inhibitory activities were observed following incubation with 40, 60, and 80 µg/mL concentrations of indomethacin (positive control), respectively. Thus, Aller-7 showed significant lipoxygenase inhibitory activity with an IC_{50} of 101.5 µg/mL, while indomethacin exhibited an IC_{50} of 57.9 µg/mL (9).

III.G.2. Compound 48/80-Induced Paw Edema in BALB/c Mice

BALB/c mice were used in this study, which have a predominant TH_2 response characteristic of allergic conditions (15). Aller-7 was evaluated in BALB/c mice for its anti-inflammatory activity in a compound 48/80-induced hind paw edema model. BALB/C mice were administered suspension of Aller-7 in 0.5% carboxymethyl cellulose (CMC) (250 mg/kg, p.o.) or prednisolone (14 mg/kg, p.o.) for 30 days. The control group received same volume of 0.5% CMC orally. One hour after the last dose, the BALB/C mice were challenged with 0.3 μg of compound 48/80 in 0.05 mL of saline injected subcutaneously in the plantar region of the right hind paw, while the left paw received an equal volume of saline. The hind paw thickness was measured at 0, 10, 20, and 30 min of challenge. The change in paw thickness between compound 48/80- and saline-injected paws at 10 min reflected the edema formed in response to compound 48/80 (15).

Aller-7 at a dose of 250 mg/kg showed 62.6% inhibition against compound 48/80-induced paw edema in BALB/C mice, respectively, while under the same conditions prednisolone exhibited 44.7% inhibitory effect at a dose of 14 mg/kg. Thus, Aller-7 demonstrates the comparative efficacy of Aller-7 and prednisolone against compound 48/80-induced paw edema in BALB/C mice (15).

III.G.3. Compound 48/80-Induced Paw Edema in Swiss Albino Mice

Compound 48/80 was used to induce mouse paw edema in Swiss albino mice. The animals were treated with Aller-7 at doses of 175, 225, and 275 mg/kg body weight orally 30 days prior to the test. A positive control group received ketotifen at a dose of 1 mg/kg. The control group received same volume of 0.5% CMC orally. One hour after the last dose, the mice were challenged with 0.3 μg of compound 48/80 in 0.05 mL of saline injected subcutaneously in the plantar region of the right hind paw, while the left paw received an equal volume of saline (15). The hind paw thickness was measured at 0, 10, 20, and 30 min of challenge. The change in paw thickness between compound 48/80- and saline-injected paws at 10 min reflected the edema formed in response to compound 48/80. Following oral administration of 175, 225, and 275 mg/kg doses of Aller-7, approximately 42.4%, 54.3%, and

53.5% inhibitory activities were observed against compound 48/80-induced paw edema in Swiss Albino mice, respectively, while under the same experimental conditions, a 1 mg/kg dose of ketotifen exerted 64.22% inhibition. Thus, Aller-7 significantly inhibited compound 48/80-induced paw edema in Swiss Albino mice and exhibited the most potent effect at 225 mg Aller-7/kg dose (15).

III.G.4. Carrageenan-Induced Paw Edema in Wistar Albino Rats

Carrageenan is an irritant polysaccharide isolated from chondrus. When injected into the hind paw of rat it causes edema, which peaks at 4 hr and within 24 hr completely subsides. Difference in the volume of the paw prior to and 4 hr after injection indicates the amount of inflammation. The comparative inhibitory effects of Aller-7 and ibuprofen against carrageenan-induced paw edema in Wistar Albino rats are shown in Aller-7 at an oral dose of 120 mg/kg. Aller-7 exhibited 31.3% inhibition against carrageenan-induced acute inflammation in Wistar Albino rats, while under identical experimental conditions ibuprofen (50 mg/kg, p.o.) demonstrated 68.1% inhibition. Thus, Aller-7 significantly inhibited carrageenan-induced paw edema in Wistar Albino rats (15).

III.G.5. Freund's Adjuvant-Induced Arthritis in Wistar Albino Rats

Freund's adjuvant is a mixture of dead mycobacterium finely ground in liquid paraffin at a concentration of 5 mg/mL. Following injection into the hind paw of a rat, it produces edema characterized as a biphasic response, which persists up to 17 days. The injection of Freund's adjuvant in rat hind paw produces a condition similar to rheumatoid arthritis, a delayed hypersensitivity reaction (15). Hence, this model was chosen to evaluate the effect of Aller-7 on delayed hypersensitivity. Wistar Albino rats were administered suspension of Aller-7 in 0.5% CMC (150, 250, or 350 mg/kg, p.o.) or prednisolone (10 mg/kg, p.o.) for 30 days before and 17 days after injecting Freund's adjuvant. The control group received 0.5% CMC.

Following the treatment with drug, on day 31, 50 µL of complete Freund's adjuvant was injected into the subplantar region of right hind paw. The paw volume was measured before and after the injection of Freund's adjuvant at the end of 10, 30, 60, 120, 180, 240, and 360 min by plethysmograph (15). Following oral administration of 150, 250, and 350 mg/kg doses of Aller-7, approximately 38.0%, 48.9%, and 63.5% inhibitory activities were observed against Freund's adjuvant-induced arthritis in Wistar Albino rats, respectively, while under identical experimental conditions, a 10 mg/kg dose of prednisolone demonstrated 95.6% inhibition (15). Thus, Aller-7 exhibited a dose-dependent effect against Freund's adjuvant-induced arthritis in Wistar Albino rats.

III.H. Antioxidant Efficacy Studies

A growing body of evidence demonstrates that allergic disorders including allergic rhinitis and asthma are mediated by reactive oxygen species and oxidative stress (16–18). Furthermore, generation of free oxygen radicals by inflammatory cells at the site of inflammation produces many of the pathophysiological changes associated with asthma and may contribute to its pathogenesis (16–18). Therefore, scavenging reactive oxygen species will be a vital step in the prevention of allergic rhinitis.

The antioxidant efficacy of Aller-7 was investigated by various assays including hydroxyl radical scavenging assay, superoxide anion scavenging assay, DPPH and ABTS radical scavenging assays. Also, protective effect of Aller-7 on free radical-induced lysis of red blood cells and inhibition of nitric oxide release by Aller-7 in lipopolysaccharide-stimulated murine macrophages were determined.

III.H.1. DPPH Radical Scavenging Assays

It has been observed that high concentration of lung antioxidants is therapeutic for asthma (17,18). Hence, the antioxidant (free radical scavenging) activity of Aller-7 was evaluated in vitro using the DPPH method as described by Vani et al. (19). Antioxidant compounds react with DPPH

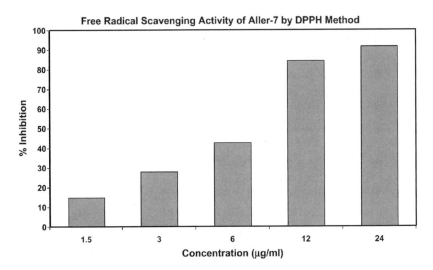

Figure 4 Concentration-dependent free radical scavenging activity of Aller-7 by DPPH method.

(1,1-diphenyl-2-picryl hydrazyl), reducing it to 1,1-diphenyl-2-picryl hydrazine. The extent of reduction, which is a color reaction, indicates the degree of antioxidant activity at 516 nm. Fig. 4 demonstrates the concentration-dependent free radical scavenging activity of Aller-7 by DPPH method. Approximately 14.6%, 27.7%, 42.4%, 84.1% and 91.3% free radical scavenging activities were observed following incubations with 1.5, 3.0, 6.0, 12.0 and 24.0 μg Aller-7/mL, respectively, while under these same conditions gallic acid (positive control) exhibited 27.9%, 70.2%, 79.3%, and 83.7% inhibitions at concentrations of 0.5, 1.5, 2.0, and 2.5 μg/mL. Thus, Aller-7 exhibited a dose-dependent free radical scavenging activity with an IC_{50} of 5.62 μg/mL, whereas gallic acid produced an IC_{50} of 0.85 μg/mL (20).

III.H.2. PMS–NADH Assay and Riboflavin-NBT Light Assay

The direct superoxide scavenging ability of Aller-7 was assessed using two in vitro models: (i) PMS/NADH (Phenazine methosulphate/reduced nicotinamide adenine dinucleotide sodium salt) assay and (ii) riboflavin/NBT (nitrobluetetrazolium)/light assay. In the PMS–NADH system, Aller-7 demonstrated 7.9%, 31.5%, 53.4%, and 82.6% scavenging following incubation with 6, 12, 24, and 96 μg/mL (Fig. 5), respectively, while under these same conditions gallic acid (positive control) exhibited 14.2%, 17.7%, 29.1%, and 69.3% scavenging using 2.5, 5, 10, and 40 μg/mL concentrations, respectively. Thus, Aller-7 exhibited a dose-dependent superoxide anion scavenging activity with an IC_{50} of 24.7 μg/mL, whereas gallic acid produced an IC_{50} of 21.2 μg/mL (20).

Using riboflavin/NBT/light assay, approximately 29.3%, 60.9%, and 74.5% scavenging effects were observed at concentrations of 2, 6, and 10 μg Aller-7/mL, respectively, while under identical conditions 50.4%, 78.0%, and 88.5% scavenging effects were observed following incubation with 2, 6, and 10 μg/mL concentrations of gallic acid (positive control). Fig. 6 demonstrates the concentration-dependent superoxide anion scavenging activity of Aller-7 in riboflavin/NBT/light

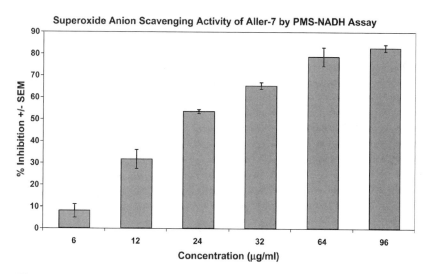

Figure 5 Concentration-dependent superoxide anion scavenging activity of Aller-7 by PMS–NADH assay.

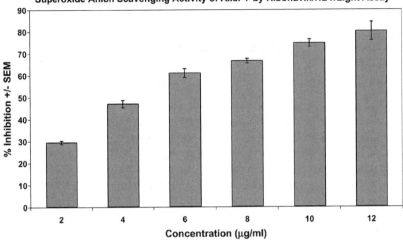

Figure 6 Concentration-dependent superoxide anion scavenging activity of Aller-7 by Riboflavin/NBT/Light assay.

assay. Thus, Aller-7 exhibited a dose-dependent superoxide anion scavenging activity with an IC_{50} of 4.27 µg/mL, whereas gallic acid produced an IC_{50} of 1.96 µg/mL (20).

III.H.3. Direct Nitric Oxide Scavenging Effect of Aller-7

Nitric oxide (NO) is a gaseous free radical that mediates several diverse biological events. Nitric oxide can react with super oxide anion and give rise to powerful free radicals called peroxynitrite radicals (17). It is clearly known that exacerbated production of NO is involved in inflammation. In the present study, an attempt was made to study the effect of Aller-7 to scavenge nitric oxide. Aller-7 produced 41.0%, 59.5%, 70.2%, and 68.6% NO scavenging activities following incubation with 10, 20, 40, and 50 µg/mL concentrations (Fig. 7), respectively, whereas under identical experimental conditions curcuminoids (positive control) exhibited 14.0%, 29.8%, 49.8%, and 64.4% scavenging activities following incubation with 10, 20, 40, and 60 µg/mL. Thus, Aller-7 exhibited

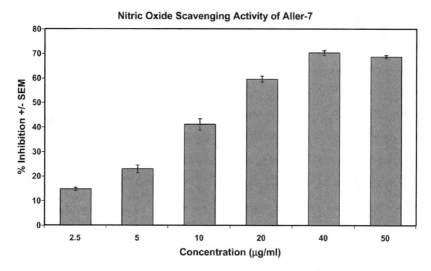

Figure 7 Concentration-dependent nitric oxide scavenging activity of Aller-7.

NO scavenging activity with an IC_{50} of 16.34 μg/mL while curcuminoids showed IC_{50} of 33.63 μg/mL, respectively (20).

III.H.4. Hydroxyl Radical Scavenging Activity of Aller-7

Hydroxyl radicals are significant in inflammation. The direct hydroxyl scavenging potential of Aller-7 was evaluated using the deoxyribose assay. The deoxyribose assay was performed as described by Halliwell et al. (21) with minor changes. Aller-7 exhibited 29.0%, 61.5%, 65.2%, and 77.8% hydroxyl radical scavenging activities following incubation with 400, 1000, 1500, and 2000 μg/mL, while, under identical conditions, catechins (positive control) produced 19.3%, 39.4%, and 53.7% hydroxyl radical scavenging activities following incubation with 1000, 1500, and 2000 μg/mL. Figure 8 demonstrates the concentration-dependent hydroxyl radical scavenging activity of Aller-7. These results demonstrated that Aller-7 exhibited unique hydroxyl radical scavenging activity with an IC_{50} of 741.7 μg/mL, while catechins produced 2193.4 μg/mL, respectively (20).

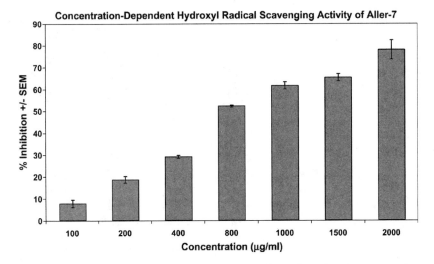

Figure 8 Concentration-dependent hydroxyl radical scavenging activity of Aller-7.

III.H.5. ABTS Radical Scavenging Activity of Aller-7

The Aller-7 was evaluated for its ability to scavenge ABTS (2,2′-azinobis-ethyl-benzothiozoline-6-sulphonic acid) free radicals according to Auddy et al. (22). Aller-7 exhibited 31.9%, 59.2%, and 84.6% scavenging activities following incubation with 5.0, 10.0, and 15.0 μg/mL concentrations (Fig. 9), respectively, while under identical experimental conditions, gallic acid (positive control) demonstrated 27.3%, 60.2%, and 90.8% free radical scavenging effects following incubation with 1.0, 2.0, and 3.0 μg/mL. Thus, Aller-7 exhibited antioxidant activity of scavenging ABTS radicals with IC_{50} values of 7.4 μg/mL, while gallic acid produced 1.4 μg/mL, respectively (20).

III.H.6. Effect of Aller-7 on Free Radical-Induced Lysis of Rat Red Blood Cells

Peroxidation of biomembranes is a process by which tissues are damaged during ischemia and inflammation. Therefore, peroxidation in red blood cell (RBC) membranes and hemolysis induced by various agents, such as hydrogen peroxide,

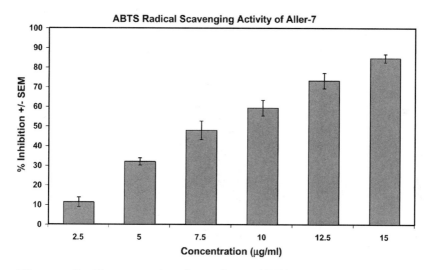

Figure 9 Concentration-dependent ABTS radical scavenging activity of Aller-7.

dialuric acid, xanthine oxidase, and organic hydroperoxides, have been extensively studied as a model for peroxidative damage in the membranes. Free radicals can be generated by azo compounds. The rate of generation of free radicals can be easily controlled and measured by adjusting the concentration of the initiator. Therefore, the hemolysis induced by an azo compound must provide the means for studying the oxidative membrane damage by oxygen free radicals attack from the outside of the membrane (20). Due to the relevance of membrane damage/peroxidation in inflammation, a study was designed to evaluate the effect of Aller-7 in free radical-mediated damage of rat RBC. Aller-7 significantly protected free radical-induced hemolysis by 20.6%, 44.9%, 64.3%, and 78.5% following incubation with 20, 40, 60, and 80 μg/mL, respectively, while under identical experimental conditions butylated hydroxyanisole (BHA, positive control) produced 21.5%, 54.9%, 71.5%, 88.7%, and 93.1% protection at concentrations of 5, 15, 20, 25, and 50 μg/mL. Fig. 10 demonstrates the concentration-dependent inhibitory activity of Aller-7 against free radical-induced hemolysis. These

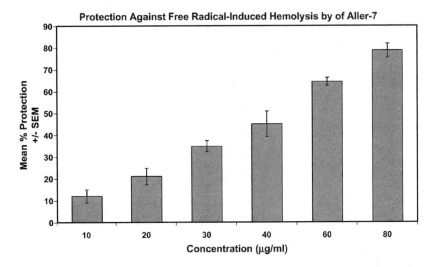

Figure 10 Protection against free radical-induced hemolysis by Aller-7.

results indicated that Aller-7 significantly inhibited free radical-induced hemolysis in concentrations of 20–60 µg/mL, while BHA induced protection at concentrations of 15–50 µg/mL, respectively (20).

III.H.7. Nitric Oxide Inhibition in LPS-Stimulated Murine Macrophages

Aller-7 has already been demonstrated to possess direct NO scavenging property. However, it is not known whether it could inhibit the production of NO in cellular systems. To answer this, Aller-7 was tested for its effect on NO release in lipopolysaccharide (LPS) treated murine macrophages according to the method of Yokozawa et al. (23). Following incubation with 2, 4, and 8 µg/mL concentration of Aller-7, approximately 74.6%, 88.3%, and 90.0% inhibitions of nitric oxide were exhibited in LPS-stimulated murine macrophages, respectively (20). Thus, Aller-7 exhibited potent NO inhibition property.

IV. HUMAN CLINICAL STUDIES

IV.A. Dosage Determination

The human test dosage was determined based on extrapolation of animal safety and efficacy data. Based on acute, subacute, subchronic, and teratogenecity animal studies, the extrapolated human toxicity dose was greater than 11.6–22.4 g. The extrapolated human efficacy dose was based on the dose-dependent inhibitory effect of Aller-7 on antigen-induced anaphylaxis in mice [dose/50 × (1000 g/kg/20 g mouse) × 387.9 surface area conversion factor]. The minimum effective human dose was determined to be 660 mg/day, the maximum effective dose was determined to be 1980 mg/day, and the human test dose was determined to be 1320 mg/day (based on the dose of 175 mg/kg to produce 55% anaphylaxis inhibitory effect in mice), taken in divided doses, twice daily.

IV.B. Phase I—Clinical Trial

In Phase I clinical trial, 20 healthy volunteers aged between 23–40 years (including 5 dropouts) received 1320 mg of Aller-7 per day for 15 days, then 1980 mg per day for another 15 days. No adverse events were found and all clinical, biochemical, and electrocardiogram parameters showed no abnormalities. Thus, Aller-7 was well tolerated.

IV.C. Pilot Efficacy/Tolerance Study

A pilot study was conducted in 17 patients with *allergic rhinitis* to evaluate the tolerability and efficacy of Aller-7. Patients were selected based on specific inclusion criteria, including age 18–60 years, written consent to participate, clinically normal (except nasal condition), normal hematological profile, normal liver/renal function test, and normal chemical and microscopic examination of urine. Exclusion criteria included allergy to any medication, serious systemic illness, psychiatric disorders, regular smokers (>20 cigarettes daily), drug dependence or chronic heavy alcohol consumption, use of any microsomal enzyme modifying drugs within 30 days of the study, and participation in any clinical trial within 6 weeks of study. Prior to administration of Aller-7, biochemical and nasal symptoms evaluation was performed. Following administration of 1320 mg of Aller-7 (660 mg twice daily) for 90 days, in 9 patients and for 180 days in 8 patients, patients showed a significant improvement in their nasal symptoms and no significant adverse effects were observed. Aller-7 was well tolerated and significantly improved nasal symptoms.

IV.D. Proof of Concept Test

A questionnaire-based study was used to assess the satisfaction of Aller-7 in 30 allergic rhinitis patients who were not previously cured by conventional drugs. Following administration of 1320 mg of Aller-7 (660 mg twice daily) for 6–12 weeks, 10 patients reported satisfactory results, while 19 patients reported highly satisfactory results. Only one patient was not satisfied. Aller-7 produced satisfactory to highly satisfactory results in 29 out of 30 patients.

IV.E. Phase II—Double-Blind, Placebo-Controlled Study

Forty-eight randomly assigned patients (8 dropouts) with allergic rhinitis received either a placebo or 1320 mg of Aller-7 (660 mg twice daily) for 12 weeks. Nasal symptom scores (sneezing, nasal congestion, and rhinorrhea) and safety parameters were monitored. Compared to placebo, patients receiving Aller-7 showed a significant improvement in sneezing, nasal congestion, and rhinorrhea symptoms (Table 1). The total nasal symptoms improved by 70% and eye symptoms by 71% as compared to placebo effects of 20% and 38%, respectively. The results were consistent throughout the 12-week study. Aller-7 was well tolerated and showed significant improvement in nasal symptom scores (24).

IV.F. Phase III—Multi-center Clinical Trial

Multicenter clinical trials of Aller-7 were carried out in 14 centers, which included 545 patients with allergic rhinitis. In the open trial 374 patients (including 22 dropouts) were administered 1320 mg of Aller-7 (660 mg twice daily) for 12 weeks. A significant improvement (>40% relief) in sneezing,

Table 1 Relief of Allergic Rhinitis Symptoms in Phase 2 Clinical Trial of Aller-7

Symptom	Patients reporting >40% relief ($n = 40$) (%)	
	Placebo 12 weeks	Aller-7 12 weeks
Sneezing	31	73*
Nasal congestion	44	67*
Rhinorrhea	40	73*

*$p < 0.001$.

nasal congestion, and rhinorrhea was seen at the end of 6 weeks and at the end of 12 weeks, further improvement in symptoms was observed. An overall improvement in nasal symptoms was seen in 78% of the patients at the end of 6 weeks and 94% of the patients at the end of 12 weeks (Table 2) (25). Total nasal symptom score were significantly decreased following Aller-7 treatment at the end of 6 weeks (59%) and at 12 weeks (78%) of patients indicating significant improvement in total nasal symptom score.

In the double-blind, placebo-controlled trials ($n = 171$), randomly assigned patients administered 1320 mg of Aller-7 (660 mg twice daily) for 12 weeks showed significant improvement in individual symptoms, including sneezing, nasal congestion, and rhinorrhea at the end of 6 weeks, which persisted and improved at 12 weeks. In the placebo group, a lesser improvement in

Table 2 Relief of Allergic Rhinitis Symptoms in Human Subjects (Multi-Center, Open-Label Study)

Symptom	Patients reporting >40% relief ($n = 374$) (%)	
	6 weeks	12 weeks
Sneezing	71	92
Nasal congestion	75	92
Rhinorrhea	76	97
Overall improvement	78	94
Overall improvement in total nasal symptom score	59	78

Table 3 Relief of Allergic Rhinitis Symptoms in Human Subjects (Multi-Center, Double-Blind, Placebo-Controlled Study)

| | Patients reporting $> 40\%$ relief ($n = 171$) (%) | | | |
| | Placebo | | Aller-7 | |
Symptom	6 weeks	12 weeks	6 weeks	12 weeks
Sneezing	51	55	68	74
Nasal congestion	56	58	66	73
Running nose	56	56	71	79
Overall improvement in total nasal symptom score	54	56	68	75

*$p < 0.001$.

symptoms was reported after 6 weeks; however, no significant further improvement was observed at 12 weeks. At the end of 12 weeks, an overall improvement in total nasal symptom score was seen in 75% of the patients receiving Aller-7 compared to 56% of the patients receiving the placebo (Table 3).

IV.F.1. Overall Improvement

Overall improvement in nasal symptoms (placebo-controlled studies, $n = 151$) in the Aller-7 group improved in 84% of the patients at the end of 6 weeks and 92% of the patients at the end of 12 weeks. In contrast, overall improvement in nasal symptoms decreased in the placebo group, with 75% of the patients showing improvement at 6 weeks and only 65% of the patients showing improvement at 12 weeks (Table 4).

Table 4 Overall Improvement in Nasal Symptoms (Double-Blind, Placebo-Controlled Study)

| | Patients reporting $>40\%$ relief ($n = 151$) (%) | | | |
| Sneezing, nasal congestion, and running nose | Placebo | | Aller-7 | |
	6 weeks	12 weeks	6 weeks	12 weeks
Overall improvement	75	65	84	92

*$p < 0.001$.

Table 5 Positive Placebo Response in Allergy Rhinitis Studies

Treatment group	Number of patients						Mean score
	Score 0	Score 1	Total (%)	Score 2	Score 3	Total (%)	
Placebo ($n = 16$)	1	3	4 (25)*	8	4	12 (75)	1.94
Astemizole ($n = 18$)	4	7	11 (61)*	5	2	7 (39)	1.27*
Cetirizine ($n = 17$)	5	6	11 (65)*	4	2	6 (35)	1.18%*

*$p < 0.01$.
Score 0 = no symptoms or complete recovery, 1 = marked improvement of symptoms, 2 = slight improvement of symptoms, 3 = no improvement of symptoms. Scores 0 and 1 are classified as excellent to good responses. Scores 2 and 3 are classified as satisfactory to unsatisfactory responses.

IV.F.2. Placebo Effect

Positive placebo results are typical in allergy studies, such as those performed on Aller-7. A positive placebo response of 25% compared to 61% and 65% for the drugs astermizole and cetirizine, respectively, is illustrated in Table 5.

IV.F.3. Saccharine Transport Time

Saccharine transport time (STT) measures objectively the interaction between ciliary movement and removal of mucous secretion from the nose. Altered ciliary clearance time has been identified using STT in asthma, chronic bronchitis, and sinusitis. Mucociliary clearance was significantly improved in patients treated with Aller-7 (Table 6).

Table 6 Effect of Aller-7 on Saccharine Transport Time

Saccharine transport time	0 weeks	6 weeks	12 weeks
Clearance time (sec)	79.35	70.65	32.5

Table 7   Effect of Aller-7 on Peak Nasal Flow Rate

Peak nasal flow rate	0 weeks	6 weeks	12 weeks
Flow rate (L/min)	119	128	156

IV.F.4.   Peak Nasal Flow Rate

Nasal obstruction is one of the important features of perennial allergic rhinitis, which is rarely relieved by conventional therapy. Patients treated with Aller-7 demonstrated statistically significant improvement in nasal congestion (Table 7).

IV.F.5.   Adverse Effects

No significant adverse effects were observed.

V.   CONCLUSION

Taken together, this novel combination Aller-7 derived from the extracts of seven extensively used Indian medicinal plants exhibited safety and efficacy in a number of in vitro and in vivo models of allergic rhinitis and inflammatory responses. Furthermore, considering all aspects of the multicenter clinical trial, this novel polyherbal formulation (Aller-7/NR-A2) effectively demonstrates its clinical usefulness in ameliorating symptoms of allergic rhinitis and providing complete relief to a section of patients. It appears to be well tolerated by most patients and the minority which had some side effects, which were trivial and self-limiting.

ACKNOWLEDGMENT

The authors thank Ms. Shirley Zafra for technical assistance.

REFERENCES

1. Salib RJ, Drake-Lee A, Howarth PH. Allergic rhinitis: past, present and the future. Clin Otolaryngol 2003; 28:291–303.

2. May JR, Feger TA, Guill MF. In: Dipiro JT, Talbert RL, Yee GC, et al. eds. Pharmacotherapy: A Pathophysiologic Approach. Appleton & Lange, Stamford, CT, 1997:1801–1813.

3. D'Alonzo GE. Scope and impact of allergic rhinitis. J Am Osteopathol Assoc 2002; 102:S2–S6.

4. Management of Allergic Rhinitis and Its Impact on Asthma, A Pocket Guide. Published by Astrazeneca in collaboration with WHO, 2001.

5. Tarnasky PR, Van Arsdel PP Jr. Antihistamine therapy in allergic rhinitis. J Fam Pract 1990; 30:71–80.

6. Willsie SK. Improved strategies and new treatment options for allergic rhinitis. J Am Osteopathol Assoc 2002; 12:S7–S14.

7. Amit A, Saxena VS, Pratibha N, Bagchi M, Bagchi D, Stohs SJ. Safety of a novel botanical extract formula for ameliorating allergic rhinitis. Toxicol Mech Meth 2003; 13:253–261.

8. Draize JH. The appraisal of the safety of chemicals in foods, drugs and cosmetics. In: Association of Food and Drug Officials, eds. Dermal Toxicity. Kansas: Topeka, 1965:45–59.

9. Amit A, Saxena VS, Prathibha N, D'Souza P, Bagchi M, Bagchi D, Stohs SJ. Mast cell stabilization, lipoxygenase inhibition, hyaluronidase inhibition, antihistaminic and antispasmodic activities of aller-7, a novel botanical formulation for allergic rhinitis. Drugs Exp Clin Res 2003; 29:107–115.

10. Babe KS, Serafin WE. Histamine, bradykinin and their antagonists. In: Hardman JG, Limbird LE, Molinoff PB, et al. eds. The Pharmacological Basis of Therapeutics. McGraw-Hill, New York, NY 1996:581–592.

11. Aronson NN Jr, Davidson EA. Lysosomal hyaluronidase from rat liver. II. Properties. J Biol Chem 1967; 242:441–444.

12. Lau KM, He ZD, Dong H, Fung KP, But PP. Anti-oxidative, anti-inflammatory and hepato-protective effects of *Ligustrum robustum*. J Ethnopharmacol 2002; 83:63–71.

13. Guardia T, Rotelli AE, Juarez AO, Pelzer LE. Anti-inflammatory properties of plant flavonoids. Effects of rutin, quercetin, and hesperidin on adjuvant arthritis in rat. Farmaco 2001; 56: 683–687.

14. Shinde UA, Kulkarni KR, Phadke AS, Nair AM, Mungantiwar AA, Dikshit VJ. Mast cell stabilizing and lipoxygenase inhibitory activity of Cedrus deodara (Roxb) Loud wood oil. Ind J Exp Biol 1999; 37:258–261.

15. Amit A, Saxena VS, Pratibha N, D'Souza P, Bagchi M, Bagchi D. Anti-inflammatory activities of aller-7, a novel polyherbal formulation designed for allergic rhinitis. Int J Tissue Reactions 2004; 26:43–51.

16. Bowler RP, Crapo JD. Oxidative stress in allergic respiratory diseases. J Allergy Clin Immunol 2002; 110:349–354.

17. Levetin E, Van de Water P. Environmental contributions to allergic disease. Curr Allergy Asthma Rep 2001; 1:506–510.

18. Thornhill SM, Kelly AM. Natural treatment of perennial allergic rhinitis. Altern Med Rev 2000; 5:448–452.

19. Vani T, Rajani M, Sarkar S, Shishoo CJ. Antioxidant properties of the ayurvedic formulation triphala and its constituents. Int J Pharmacol 1997; 35:313–317.

20. D'Souza P, Amit A, Saxena VS, Bagchi D, Bagchi M. Antioxidant properties of aller-7, a novel polyherbal formulation for allergic rhinitis. Drugs Exp Clin Res 2004; 30:99–109.

21. Halliwell B, Gutteridge JMC, Aruoma OI. The deoxyribose method: a simple test tube assay for determination of rate constants for reactions of hydroxyl radicals. Anal Biochem 1987; 165:215–219.

22. Auddy B, Ferreira M, Blasina F, Lafon L, Arredondo F, Dajas F, Tripathi PC, Seal T, Mukherjee B. Screening of antioxidant activity of three Indian medicinal plants traditionally used for the management of neurodegenerative diseases. J Ethnopharmacol 2003; 84:131–138.

23. Yokozawa T, Wang TS, Chen CP, Hattori M. Inhibition of nitric oxide release by an aqueous extract of *Tinospora tuberculata*. Phytother Res 2000; 14:51–53.

24. Vyjayanthi G, Subhashchandra S, Saxena VS, Nadig P, Venkateshwarlu K, Serene A, Sathyan S, Bagchi D, Kulkarni C. Randomized, double blind, placebo controlled trial of Aller-7 in patients with allergic rhinitis. Res Commun Pharmacol Toxicol 2003; 8:23–32.

25. Saxena VS, Venkateshwarlu K, Nadig P, Barbhaiya HC, Bhatia N, Borkar DM, Gill RS, Jain RK, Katiyar SK, Nagendra Prasad KV, Nalinesha KM, Nasiruddin K, Rishi JP, Chowdhury JR, Saharia PS, Thomas B, Bagchi D. Multi-center clinical trials on a novel polyherbal formulation in allergic rhinitis. Int J Clin Pharmacol Res 2004; 24:79–94.

22

Antioxidant Character of Selenium and Its Survival Signals by Activation of PI3-K/Akt Pathways

SANG-OH YOON, JONG-MIN PARK, and AN-SIK CHUNG

Department of Biological Sciences, Korea Advanced Institute of Science and Technology, Daejeon, South Korea

I. INTRODUCTION

Selenium is an essential trace element for human and animal health, which has been known to modulate the functions of several selenoproteins. Most of these selenoproteins have some kinds of antioxidant functions. Selenium is incorporated as selenocysteine into selenoprotein, which is encoded by a UGA codon in the selenoprotein mRNA (1). Mammalian selenoproteins have been mostly found to act as antioxidant functions such as those of glutathione peroxidase (GPx) (2), thioredoxin reductase (TR) (3), selenoprotein P (4), and seleno-protein W (5). The roles of these enzymes had been extensively reviewed in relation to their potential roles in physiological phenomena as antioxidant defense, redox regulation of cytokine, differentiation, apoptosis, and tumorigenesis (6).

The cellular decision to undergo either survival or apoptosis is determined by the integration of multiple survival and death signals. Akt is a critical mediator of cell survival (7), while ASK1 is a crucial signal molecule of cell death (8). Akt becomes acti-vated in a PI3-K-dependent manner not only in response to var-ious growth factor and Ca^{2+} influx but also in response to extracellular stresses such as H_2O_2 and heat shock treatment (7). On the contrary, ASK1 is activated by a variety of stress-related stimuli, including withdrawal of serum, tumor necrosis factor (TNF)-α, reactive oxygen species (ROS), and genotoxic stress (9,10). Activation of Akt is necessary for cell survival and the prevention of apoptosis, both of which occur by inhibition of the Bcl-x inhibition or caspase-9 (11). Akt also activates Bcl-2 expression and inhibits glycogen synthase kinase-3. Contrast to Akt, ASK1 plays a central role in the mechanisms of stress- and cytokine-induced apoptosis. The ubiquitously expressed seri-ne/threonine kinase ASK1 activates both SEK1 (also termed MKK-4)–JNK and MKK3 and MKKp38 kinase pathway (8).

Recent studies have indicated that ROS such as H_2O_2, which are formed in association with a variety of oxidative stress-induced disorders, may be related to cell death and hence play an important role in apoptosis (12). However, there are few reports on the relationship between selenium and survival signals. Our studies demonstrate that selenite blocks H_2O_2-induced apoptosis through activation of survival PI3-K/Akt pathway and inhibition of apoptotic ASK1/JNK pathway.

II. SELENIUM DEFICIENCY AND GLUTATHIONE PEROXIDASE ACTIVITY

Selenium is known as an antioxidant, which is mostly mediated by glutathione (GSH) peroxidase. A general protection scheme

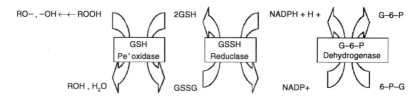

Figure 1 Cytosolic glutathione peroxidases are involved in preventing oxidative stress and its supply of reducing equivalents. GSH, glutathione; G-6-P, glucose-6-phosphate; 6-P-G, 6-phosphogluconate; RO•, alkoxy radical; •OH, hydroxyl radical.

of GPx against hydroperoxide is shown in Fig. 1. In the antioxidant function, GPx depends upon GSH regenerating systems, usually on glucose-dependent NADPH supply required for GSSG reduction. Cell culture is an important tool for studying injury to cells exposed to oxidative stresses. Supplementation of the culture media with 1 µM sodium selenite in HepG2 cells increased threefold of the total GPx activity, a key enzyme in the defense against oxidative stresses (13). It is well known that selenium is necessary for cell culture with Se-deficient media including a serum-free medium. It was demonstrated that the selenium deficiency-induced death of Jurkat cells was caused by increased intracellular ROS, and selenium and vitamin E supplementation prevented the accumulation of intracellular ROS in a dose-dependent manner (14). The generation of lipid hydroperoxides is observed only in selenium-deficient cells. These results suggest that ROS, especially lipid hydroperoxides, are involved in the cell death caused by selenium deficiency. Selenite supplementation protects HuH7 hepatoma cells from serum deprivation-induced apoptosis and supports its long-term growth in serum-free medium (15). The antiapoptotic effect correlates with activation of focal adhesion kinase and PI-3/Akt kinase pathway. Parallel changes include a significant reduction in the intracellular ROS content, the several of DNA fragmentations, and the suppression of caspases. The mitogenic signaling mediated by selenite may involve unconventional growth stimuli related to higher cytosolic glutathione peroxidase (cGPx) activity and higher transcriptional levels of selenoprotein P.

It seems reasonable to expect that the antioxidant effect of selenium and its deficiency are implicated in many pathophysiological phenomena that might be triggered by peroxides (16). However, these expectations should be revised because cGPx (−/−) mice develop and grow normally (17). Therefore, cGPx may not be vital for life and development. However, the cGPx (−/−) mice are more sensitive to poisoning with the redox-cycling herbicides paraquat (18) and diquat (19). The antioxidant function of cGPx is mostly claimed by the above observations. The several possible roles of cGPx are extensively discussed with other selenoproteins (6). More researches are required to confirm cGPx roles related to selenium statue and antioxidant functions.

III. SELENIUM INCREASES CELL PROLIFERATION AND BLOCKS APOPTOSIS INDUCED BY H_2O_2

Many researchers have focused on the antioxidant effect of selenite at low concentration, but some researchers have focused on the induction of apoptosis by toxic concentration of selenite. We have been interested in physiological concentration of selenite, because selenium levels in serum are $1–5\,\mu M$. As shown in Fig. 2A, selenite $<5\,\mu M$ increased cell viability but $>10\,\mu M$ was cytotoxic. The increased cell viability was mainly derived from the relative increment of cell proliferation compared to cell death, which was detected by MTT and BrdU labeling assays (Fig. 2A). It is shown that supplementation of selenium and selenite inhibits UV- and ionizing irradiation-induced apoptosis (20,21). Apoptotic stresses such as UV and TNF-α increase intracellular H_2O_2, which can lead to apoptosis (10,22). It was found that high-concentration H_2O_2 induced apoptosis of HT1080 cells (Fig. 2B). It was shown that DNA contents were similar both in nontreated group (Fig. 2C) and $2\,\mu M$ selenite-treated group (Fig. 2D). Treatment with $500\,\mu M$ H_2O_2 dramatically increased apoptosis by more than 30% in 24 hr (Fig. 2E) and 60% in 36 hr (Fig. 2F). Pretreatment with $2\,\mu M$ selenite for 12 hr blocked H_2O_2-induced apoptosis after 24 hr (Fig. 2G) and even the blocking effect was continued to 60 hr (Fig. 2H). It was demonstrated that selenite could prevent cell apoptosis induced by H_2O_2.

However, most studies focus on selenoproteins, such as GPx and TR. Few reports show the relationship between signal molecules and selenite function. Recently, it was reported that selenite inhibited JNK (23). In that report, only the MEKK1–SEK1–JNK pathway was studied and selenite did not inhibit MEKK1. In our previous experiment, selenite inhibited ASK1 and JNK activities (24). However, JNK and ASK are constitutively active in HT1080 cells so that selenite can inhibit ASK–SEK1–JNK pathway even without any stress condition. The result suggests that selenite inhibits ASK1 pathway but not MEKK1 pathway.

IV. SELENITE INHIBITS H_2O_2-INDUCED APOPTOSIS VIA THE INHIBITION OF ASK1 ACTIVITY AND ACTIVATION OF PI3-K-DEPENDENT Akt ACTIVITY

First, ASK1 has recently been identified as a signal molecule, and there are a few reports on the regulation ASK1. DTT and β-mercaptoethanol were quite effective in restoring ASK1 activity directly by selenite (Fig. 3A). These observations may be related to a cysteine-rich domain in its NH2 terminus of ASK1 (10). These findings suggest that selenite changes redox potential of sulfhydryl groups of ASK1. Second, we focused on the antiapoptotic PI3-K/Akt pathway, because Akt negatively regulated ASK1 activity in a recent report (25). Furthermore, selenium is known as insulin-mimicry by stimulation of glucose uptake and regulation of metabolic processes such as glycolysis and gluconeogenesis (26,27). Insulin also promotes cell survival, which is mediated by Akt (7,28). Treatment with selenite inhibited ASK1 through PI3-K/Akt pathway. It seems that ASK1 inhibition by selenite is largely due to the activation of PI3-K/Akt pathway, but the direct inhibition by selenite may contribute only a small portion.

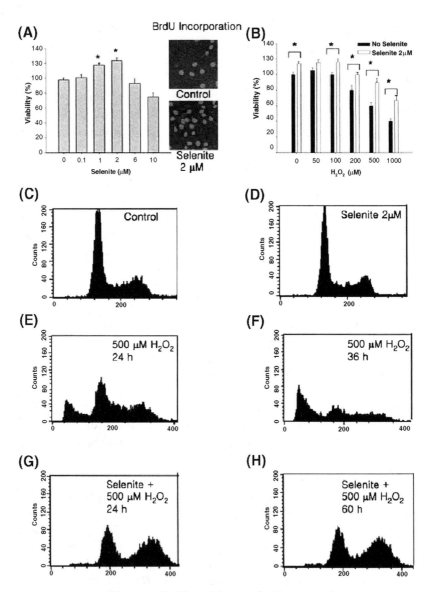

Figure 2 (*Caption on facing page*)

As shown in Fig. 3B, almost 80% ASK1 activity were restored by treatment with both LY294002 (a PI3-K inhibitor) and selenite, compared with untreated control.

It was found that selenite increased Akt activity in a time- and dose-dependent manner, which was closely correlated to the inhibitory pattern of ASK by selenite in our previous experiment (24). It was also observed that selenite regulated these signal molecules even for 48 hr after selenite treatment. Growth factor and other stresses give short duration time of these signal molecules. On the contrary, selenite sustains the signal molecules for longer duration times so that cells are protected continuously against stresses and abnormal condition. It is likely that selenite effects are similar to those of constitutively active Akt. Treatment with both selenite and LY294002 did not restore Akt activity induced by selenite, suggesting that selenite activates Akt via PI3-K-dependent pathway (Fig. 3C). The Akt activation by selenite is the major force to inhibit apoptosis, because the protective effect of selenite on H_2O_2-induced apoptosis was almost abolished in kinase-dead

---

Figure 2 (*Facing page*) Effects of selenite and/or H_2O_2 on HT1080 cell viability. (A) Various concentrations of selenite were treated and incubated for 24 hr. Cell viability was measured by MTT assay. Also, cell proliferation assay was performed. Cells were incubated in the absence and presence of $2\,\mu M$ selenite and BrdU incorporation assay was done using BrdU labeling and detection kit (Roche) as described in the supplier's protocol. BrdU-labeled cells were identified and visualized with a confocal microscope. (B) Cells were incubated in the absence (black bar) and presence (white bar) of $2\,\mu M$ selenite for 12 hr and various concentrations of H_2O_2 were added. After 24 hr incubation, cell viability was measured. Data represent the mean $\pm$ SD of three independent experiments. Results were statistically significant ($^*P<0.05$) using a Student's *t*-test. (C–H) DNA contents of HT1080 cells were measured by using FACS. (C, D) Cells were incubated in the absence (C) and presence (D) of $2\,\mu M$ selenite for 24 hr. (E, F) H_2O_2 ($500\,\mu M$) was treated and incubated for 24 hr (E) and 36 hr (F). (G, H) Selenite ($2\,\mu M$) was pre-treated for 12 hr and H_2O_2 ($500\,\mu M$) was added. After 24 hr (G) and 60 hr (H), DNA contents were analyzed. The results represent three independent experiments.

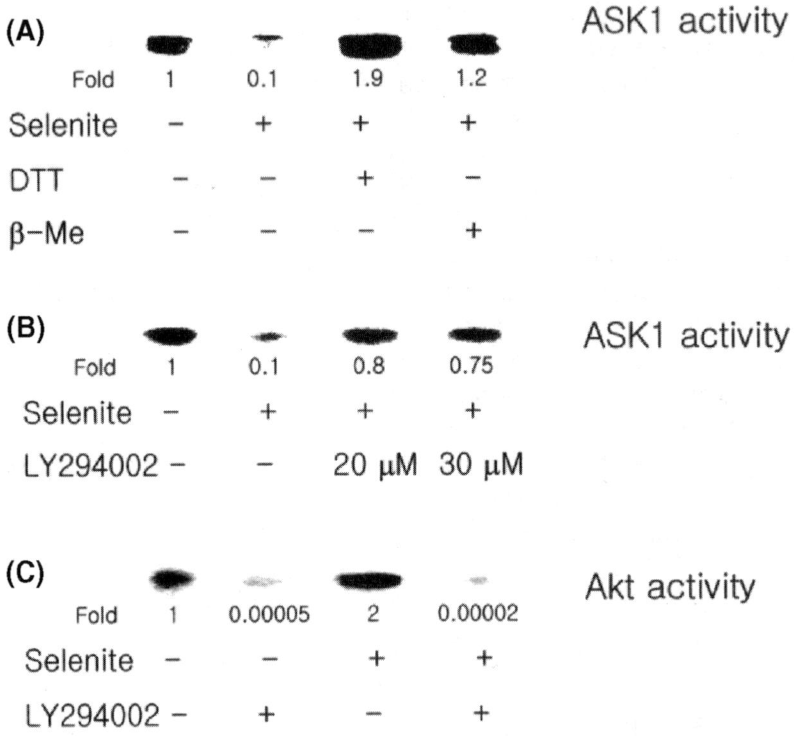

(A)   ASK1 activity

    Fold   1   0.1   1.9   1.2

Selenite   −   +   +   +

DTT   −   −   +   −

β−Me   −   −   −   +

(B)   ASK1 activity

    Fold   1   0.1   0.8   0.75

Selenite   −   +   +   +

LY294002 −   −   20 µM 30 µM

(C)   Akt activity

    Fold   1   0.00005   2   0.00002

Selenite   −   −   +   +

LY294002 −   +   −   +

Figure 3 Effects of selenite and LY294002 on the activities of ASK1 and Akt. (A) HT1080 cells were lysed, and 200 µg of lysate was immunoprecipitated with anti-ASK1 antibody. Immunopellets were treated with 0.5 µM selenite for 20 min at room temperature, washed twice with phosphate-buffered saline, and then exposed to either 10 mM DTT or 10 mM β-mercaptoethanol for additional 20 min. After the treatments, the immunopellets were assayed for ASK1 activity. (B) Selenite (2 µM) was treated 1 hr after treatment with LY294002. After 12 hr, cells were lysed and ASK1 activity was measured. (C) After 1 hr after treatment with 20 µM of LY294002, 2 µM of selenite was treated. After 12 hr, cells were lysed and Akt activity was measured. The results represent three independent experiments.

Akt-transfected cells (Fig. 4A, B), and selenite reduced apoptosis dramatically in the presence of LY294002-treated cells (Fig. 5A–D). Although selenite inhibits apoptosis mainly through Akt activation, it is possible that other pathways exist

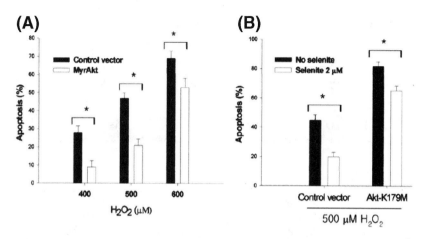

Figure 4 Effects of Akt on H_2O_2-induced apoptosis. From (A) to (B), apoptotic cells were measured using FACS. (A) Constitutively active MyrAkt vector was transfected and incubated. After 24 hr, various concentrations of H_2O_2 were treated and incubated for 24 hr. (B) Kinase-inactive Akt (Akt-K179M) vector was transfected and incubated for 24 hr. Selenite (2 μM) was treated for 12 hr and 500 μM of H_2O_2 added. After 24 hr, FACS analysis was conducted. Data represent the mean ± SD of three independent experiments. Results were statistically significant ($^*P<0.05$) using Student's t-test.

for antiapoptotic effect of selenite, which may be derived from activation of antioxidant enzymes including GPx and TR.

V. SELENIUM INCREASES GLUCOSE UPTAKE, ATP GENERATION, MITOCHONDRIAL MEMBRANE POTENTIAL, AND BCL-2 EXPRESSION BUT DECREASES CASPASE ACTIVITY

It was shown that the physiological concentration of selenium increased cell proliferation. There are few reports on what mechanisms are involved in the increased cell proliferation with the serum level of selenite. It was observed that selenite increased glucose uptake rate (Fig. 6A), ATP generation (Fig. 6B), membrane potential (Fig. 6C), and Bcl-2 expression

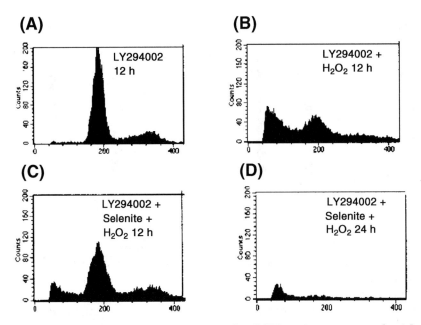

Figure 5 Effects of LY294002 on the DNA contents treated with H_2O_2 and selenite/H_2O_2 in cells. FACS analysis was used for the determination of DNA contents. (A, B) LY294002 (30 μM) was treated for 12 hr (A) and 500 μM of H_2O_2 was treated for 12 hr (B). (C, D) LY294002 (30 μM) was treated for 1 hr and 2 μM of selenite was treated. After 12 hr, 500 μM H_2O_2 was treated for 12 h (C) and 24 h (D). The results represent three independent experiments.

(Fig. 6D). All these phenomena are known to regulate Akt activities. It was also shown that Akt is involved in cell survival through the caspase inhibition, activation of glucose uptake, activation of glycolysis, and maintenance of mitochondrial membrane potential (7,28,29). However, ASK1 and its downstream molecules JNK induce apoptosis through the disruption of mitochondrial membrane potential and activation of caspases (30–32). Recently, decreases in glucose transport and mitochondrial membrane potential have been linked consistently to the earliest steps in programed cell death (33). In our studies, selenite blocked the decreased amount of glucose uptake, ATP level, and Bcl-2 expression by H_2O_2 treatment (Fig. 7A–D). These results indicate that selenite exerts the

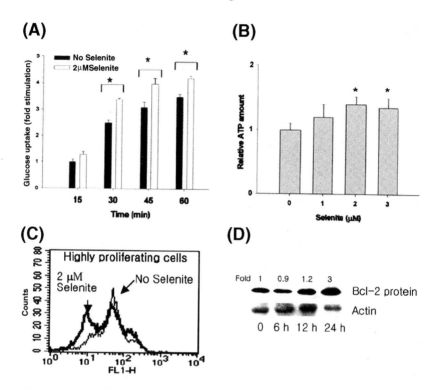

Figure 6 Effects of selenite on glucose uptake, ATP generation, mitochondrial membrane potential, and Bcl-2 expression. (A) Cells were incubated without (black bar) and with (white bar) selenite for 12 hr and glucose uptake was measured at various times as described in methods. (B) Cells were incubated with various concentrations of selenite for 12 hr and the amount of ATP was measured. Data represent the mean ± SD of three independent experiments. Results were statistically significant (*$P<0.05$) using Student's t-test. (C) Highly proliferating cells were incubated in the absence or presence of 2 μM of selenite for 12 hr and membrane potentials were measured. (D) Cells were treated with 2 μM of selenite for various times and the amount of cytosolic Bcl-2 protein was measured by Western blotting. The results represent three independent experiments.

above parameter through PI3-K/Akt- and ASK1-dependent pathways. It was shown that selenite inhibited caspase-3 through direct inhibition of this enzyme (21). It was demonstrated that expression of active PI3-K inhibited activation

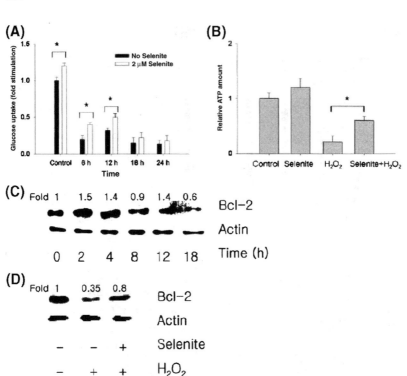

Figure 7 Effects of selenite and H_2O_2 on glucose uptake, ATP generation, and Bcl-2 protein level. (A) Cells were incubated in the absence (black bar) or presence (white bar) of $2 \mu M$ selenite for 12 hr and $500 \mu M$ of H_2O_2 was treated. Glucose uptake was measured at various times. (B) After $2 \mu M$ of selenite treatment for 12 hr, $500 \mu M$ of H_2O_2 was treated and incubated for 12 hr. ATP amounts were measured. Data represent the mean $\pm$ SD of three independent experiments. Results were statistically significant ($^*P < 0.05$) by using the Student's t-test. (C) Selenite ($2 \mu M$) was pretreated for 12 hr and H_2O_2 ($500 \mu M$) was added. Bcl-2 protein level was measured at various times. (D) Cells were incubated in the presence or absence of $2 \mu M$ of selenite for 12 hr and then $500 \mu M$ of H_2O_2 was treated for 24 hr. Bcl-2 protein level was measured by Western blotting. The results represent three independent experiments.

of caspase-3 and apoptosis of cardiac muscle cells (34) and that Akt inhibited caspase-9 and -3 (35). Selenite inhibited caspase-9 and -3 activated by H_2O_2 (Fig. 8A, B). These results suggest that selenite can not only inhibit caspase-3 through direct inhibition but also regulate PI3-K/Akt pathway.

Some reports postulate that Akt inhibits apoptosis by maintaining mitochondrial membrane potential through the inhibition of cytochrome *c* release, but others indicate that Akt inhibits apoptosis independent of cytochrome *c* release; Akt just inhibits cytochrome *c*-induced caspase-9 and -3 activation (36,37). Selenite maintained membrane potential in the presence of 400 μM H_2O_2 for 12 and 24 hr (Fig. 9A, B). These results suggest that selenite blocks apoptosis induced by 400 μM of H_2O_2 through maintenance of the membrane potential and the inhibition of caspases, and by 500 μM of H_2O_2 for 12 hr (Fig. 9C), but not that of mitochondrial membrane potential by 500 μM of H_2O_2 for 24 hr (Fig. 9D). It was shown that treatment with only caspase inhibition was enough to prevent cells from apoptosis induced by many agents and stresses (38,39). Further experiments will be

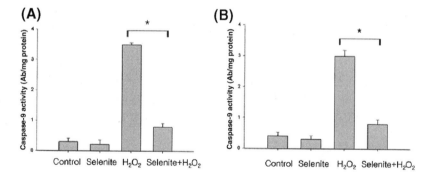

Figure 8 Effects of selenite on caspase-9 and -3 activities. Selenite (2 μM) was treated for 12 hr and H_2O_2 (500 μM) was added. Twelve hours after, cells were lysed and caspase-9 (A) and caspase-3 (B) activities were measured using fluorogenic Ac-LEDH-AFc and Z-DEVD-AMC substrate, respectively. Data represent the mean ± SD of three independent experiments. Results were statistically significant ($^*P < 0.05$) by using Student's *t*-test.

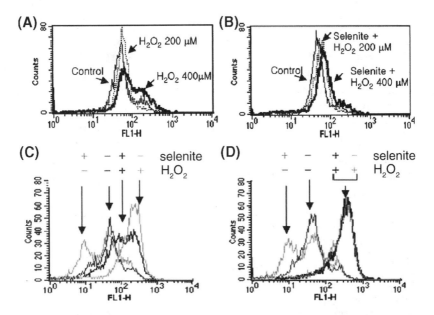

Figure 9 Effects of selenite and H_2O_2 on mitochondrial membrane potential. Cells were incubated in the absence (A) or presence (B) of $2\,\mu M$ of selenite for 12 hr and then 200 or $400\,\mu M$ of H_2O_2 was treated for additional 24 hr. (C,D) Cells were incubated in the absence or presence of $2\,\mu M$ of selenite and they were incubated in $500\,\mu M$ of H_2O_2 for 12 hr (C) and for 24 hr (D). After that, mitochondrial membrane potential was measured. The results shown represent three independent experiments.

required to clarify further experiments on PI3-K/Akt pathways and blocking apoptosis via mitochondrial membrane potential and/or caspase activity.

VI.   CONCLUDING REMARKS

Selenium is an essential trace element involved in many physiological functions such as an antioxidant activity, modulation of immune function, activation thyroid hormone, and chemoprevention of cancer. An antioxidant function of Se is mediated by selenoproteins. These selenoproteins are involved in detoxification of hydroperoxide and especially GPx protects cell damages from these hydroperoxides. It is rather difficult to explain antioxidant characters of selenoproteins solely related to selenium-deficient syndromes.

We have been interested in survival signals related to selenium supplementation. In this report, physiological concentration of selenite (1–3 μM) increased cell proliferation by increasing glucose uptake and ATP generation through activation of glycolysis, Bcl-2 upregulation, and maintaining mitochondrial membrane potential. All these phenomena are known to the regulated Akt activities and selenite increased activities of PI3-K/Akt pathways (Fig. 10). The

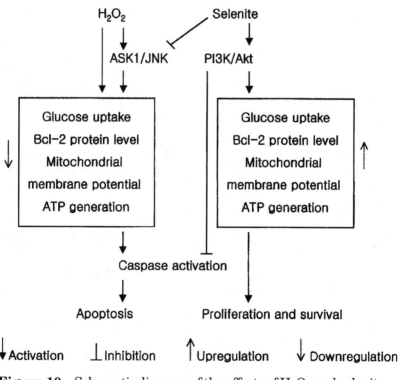

Figure 10 Schematic diagram of the effects of H_2O_2 and selenite on cell death and survival. Physiological level of selenite increases cell proliferation and survival by blocking apoptosis induced by H_2O_2 through inhibition of ASK1/JNK and activation of PI3-K/Akt pathways.

above parameters are also important for cell survival against stresses.

In addition, selenite regulates signal molecules, especially apoptotic and antiapoptotic signals. The physiological level of selenite inhibited ASK1–JNK pathway not only through PI3-K pathway, but also by direct interaction with ASK1 as modification of its sulfhydryl groups. However, it has been suggested that the ASK1 inhibition by selenite is largely due to the activation of PI3-K/Akt pathway because almost 80% of ASK1 activity is restored by treatment with both LY294002 and selenite. Moreover, the duration of the ASK1 and the activation of PI3-K/Akt pathway by selenite were maintained for at least 48 hr. This duration period is much longer than that of growth factors and other stress factors.

Apoptosis by oxidative stress has been implicated in several biological and pathological processes like aging, inflammation, carcinogenesis, and other diseases including AIDS, Parkinson's, Huntington's, and cataract formation. However, the mechanisms of cell death by oxidative stress, especially by hydrogen peroxide, are not clarified clearly. One possible mechanism of H_2O_2-induced apoptosis is through activation of ASK1–JNK pathway linked to mitochondrial dysfunction/caspase activation (Fig. 10). High concentration of H_2O_2 (400 μM) increased ASK1 activity but decreased glucose uptake, cellular ATP level, Bcl-2 expression, and mitochondrial membrane potential, finally inducing cell death. H_2O_2 also activated the antiapoptotic pathway PI3-K/Akt, but the duration was much shorter than that of ASK1. It is likely that the amount and duration of apoptotic stress and antiapoptotic factors can determine cell survival or death. Selenite inhibits apoptosis induced by H_2O_2 via the regulation of PI3-K/Akt and ASK1.

Furthermore, selenite blocked caspase-9 and -3 activation. It was shown that other caspase inhibitors inhibited apoptosis induced by several stresses. Selenite contributes to cell survival and proliferation against external stresses through the inhibition of apoptotic signals and the activation of survival signals. These functions of selenium can provide a good defense system for cells so that selenium would be beneficial in protecting cells form oxidative stresses, which are involved in inflammation, AIDS, neurodegenerative disease, and cataract formation. It was clearly demonstrated that selenium was closely related to cell survival signals as the induction Akt and the reduction of ASK1. These observations will elucidate many biological and physiological functions of selenium including the antioxidant function and will lead to new directions in the future selenium research.

REFERENCES

1. Berry MJ, Banu L, Chen YY, Mandel SJ, Kieffer JD, Harney JW, Larsen PR. Recognition of UGA as a selenocysteine codon in type I deiodinase requires sequences in the 3′ untranslated region. Nature 1991; 353:273–276.

2. Takahashi K, Avissar N, Whitin J, Cohen H. Purification and characterization of human plasma glutathione peroxidase: a selenoglycoprotein distinct from the known cellular enzyme. Arch Biochem Biophys 1987; 256:677–686.

3. Yarimizu J, Nakamura H, Yodoi J, Takahashi K. Efficiency of selenocysteine incorporation in human thioredoxin reductase. Antioxid Redox Signal 2000; 2:643–651.

4. Read R, Bellew T, Yang JG, Hill KE, Palmer IS, Burk RF. Selenium and amino acid composition of selenoprotein P, the major selenoprotein in rat serum. J Biol Chem 1990; 265: 17,899–17,905.

5. Jeong D, Kim TS, Chung YW, Lee BJ, Kim IY. Selenoprotein W is a glutathione-dependent antioxidant in vivo. FEBS Lett 2002; 517:225–228.

6. Cadenas E, Packer L. Hand of antioxidants. In: Brigelius-Flohe R, Maiorino M, Ursini F, Flohé L, eds. Selenium: An Antioxidant. New York: Marcel Dekker, 2001:633–664.

7. Franke TF, Kaplan DR, Cantley LC. PI3K: downstream AKTion blocks apoptosis. Cell 1997; 88:435–437.

8. Ichijo H, Nishida E, Irie K, ten Dijke P, Saitoh M, Moriguchi T, Takagi M, Matsumoto K, Miyazono K, Gotoh Y. Induction of

apoptosis by ASK1, a mammalian MAPKKK that activates SAPK/JNK and p38 signaling pathways. Science 1997; 275: 90–94.

9. Kanamoto T, Mota M, Takeda K, Rubin LL, Miyazono K, Ichijo H, Bazenet CE. Role of apoptosis signal-regulating kinase in regulation of the c-Jun N-terminal kinase pathway and apoptosis in sympathetic neurons. Mol Cell Biol 2000; 20:196–204.

10. Gotoh Y, Cooper JA. Reactive oxygen species- and dimerization-induced activation of apoptosis signal-regulating kinase 1 in tumor necrosis factor-alpha signal transduction. J Biol Chem 1998; 273:17,477–17,482.

11. Coffer PJ, Jin J, Woodgett JR. Protein kinase B (c-Akt): a multifunctional mediator of phosphatidylinositol 3-kinase activation. Biochem J 1998; 335:1–13.

12. Simon HU, Haj-Yehia A, Levi-Schaffer F. Role of reactive oxygen species (ROS) in apoptosis induction. Apoptosis 2000; 5:415–418.

13. Helmy MH, Ismail SS, Fayed H, El-Bassiouni EA. Effect of selenium supplementation on the activities of glutathione metabolizing enzymes in human hepatoma Hep G2 cell line. Toxicology 2000; 144:57–61.

14. Saito Y, Yoshida Y, Akazawa T, Takahashi K, Niki E. Cell death caused by selenium deficiency and protective effect of antioxidants. J Biol Chem 2003; 278:39,428–39,434.

15. Lee YC, Tang YC, Chen YH, Wong CM, Tsou AP. Selenite-induced survival of HuH7 hepatoma cells involves activation of focal adhesion kinase-phosphatidylinositol 3-kinase-Akt pathway and Rac1. J Biol Chem 2003; 278:39,615–39,624.

16. Flohe L. The selenoprotein glutathione peroxidase. In: Dolphin D, Poulson R, Avramovic O, eds. Glutathione: Chemical, Biochemical and Medical Aspects—Part A. New York: John Wiley & Sons, 1989:643–731.

17. Ho YS, Magnenat JL, Bronson RT, Cao J, Gargano M, Sugawara M, Funk CD. Mice deficient in cellular glutathione peroxidase develop normally and show no increased sensitivity to hyperoxia. J Biol Chem 1997; 272:16,644–16,651.

18. de Haan JB, Bladier C, Griffiths P, Kelner M, O'Shea RD, Cheung NS, Bronson RT, Silvestro MJ, Wild S, Zheng SS, Beart PM, Hertzog PJ, Kola I. Mice with a homozygous null mutation for the most abundant glutathione peroxidase, Gpx1, show increased susceptibility to the oxidative stress-inducing agents paraquat and hydrogen peroxide. J Biol Chem 1998; 273:22,528–22,536.

19. Fu Y, Cheng WH, Porres JM, Ross DA, Lei XG. Knockout of cellular glutathione peroxidase gene renders mice susceptible to diquat-induced oxidative stress. Free Radic Biol Med 1999; 27: 605–611.

20. Shisler JL, Senkevich TG, Berry MJ, Moss B. Ultraviolet-induced cell death blocked by a selenoprotein from a human dermatotropic poxvirus. Science 1998; 279:102–105.

21. Park HS, Huh SH, Kim Y, Shim J, Lee SH, Park IS, Jung YK, Kim IY, Choi EJ. Selenite negatively regulates caspase-3 through a redox mechanism. J Biol Chem 2000; 275: 8487–8491.

22. A-H-Mackerness S, John CF, Jordan B, Thomas B. Early signaling components in ultraviolet-B responses: distinct roles for different reactive oxygen species and nitric oxide. FEBS Lett 2001; 489:237–242.

23. Park HS, Park E, Kim MS, Ahn K, Kim IY, Choi EJ. Selenite inhibits the c-Jun N-terminal kinase/stress-activated protein kinase (JNK/SAPK) through a thiol redox mechanism. J Biol Chem 2000; 275:2527–2531.

24. Yoon SO, Kim MM, Park SJ, Kim D, Chung J, Chung AS. Selenite suppresses hydrogen peroxide-induced cell apoptosis through inhibition of ASK1/JNK and activation of PI3-K/Akt pathways. FASEB J 2002; 16:111–113.

25. Kim AH, Khursigara G, Sun X, Franke TF, Chao MV. Akt phosphorylates and negatively regulates apoptosis signal-regulating kinase 1. Mol Cell Biol 2001; 21:893–901.

26. Roden M, Prskavec M, Furnsinn C, Elmadfa I, Konig J, Schneider B, Wagner O, Waldhausl W. Metabolic effect of sodium selenite: insulin-like inhibition of glucagons stimulated glycogenolysis in the isolated perfused rat liver. Hepatology 1995; 22:169–174.

27. Ghosh R, Mukherjee B, Chatterjee M. A novel effect of selenium on streptozotocin-induced diabetic mice. Diabetes Res 1994; 25:165–171.

28. Hajduch E, Litherland GJ, Hundal HS. Protein kinase B (PKB/Akt)—a key regulator of glucose transport. FEBS Lett 2001; 492:199–203.

29. Kennedy SG, Kandel ES, Cross TK, Hay N. Akt/protein kinase B inhibits cell death by preventing the release of cytochrome c from mitochondria. Mol Cell Biol 1999; 19: 5800–5810.

30. Tournier C, Hess P, Yang DD, Xu J, Turner TK, Nimnual A, Bar-Sagi D, Jones SN, Flavell RA, Davis RJ. Requirement of JNK for stress-induced activation of the cytochrome c-mediated death pathway. Science 2000; 288:870–874.

31. Hatai T, Matsuzawa A, Inoshita S, Mochida Y, Kuroda T, Sakamaki K, Kuida K, Yonehara S, Ichijo H, Takeda K. Execution of apoptosis signal-regulating kinase 1 (ASK1)-induced apoptosis by the mitochondria-dependent caspase activation. J Biol Chem 2000; 275:26,576–26,581.

32. Yamamoto K, Ichijo H, Korsmeyer SJ. BCL-2 is phosphorylated and inactivated by an ASK1/Jun N-terminal protein kinase pathway normally activated at G(2)/M. Mol Cell Biol 1999; 19:8469–8478.

33. Moley KH, Mueckler MM. Glucose transport and apoptosis. Apoptosis 2000; 5:99–105.

34. Wu W, Lee WL, Wu YY, Chen D, Liu TJ, Jang A, Sharma PM, Wang PH. Expression of constitutively active phosphatidylinositol 3-kinase inhibits activation of caspase 3 and apoptosis of cardiac muscle cells. J Biol Chem 2000; 275:40,113–40,119.

35. Cardone MH, Roy N, Stennicke HR, Salvesen GS, Franke TF, Stanbridge E, Frisch S, Reed JC. Regulation of cell death protease caspase-9 by phosphorylation. Science 1998; 282: 1318–1321.

36. Zhou H, Li XM, Meinkoth J, Pittman RN. Akt regulates cell survival and apoptosis at a postmitochondrial level. J Cell Biol 2000; 151:483–494.

37. Reed JC, Paternostro G. Postmitochondrial regulation of apoptosis during heart failure. Proc Natl Acad Sci USA 1999; 96:7614–7616.

38. Jo DG, Kim MJ, Choi YH, Kim IK, Song YH, Woo HN, Chung CW, Jung YK. Pro-apoptotic function of calsenilin/DREAM/ KChIP3. FASEB J 2001; 15:589–591.

39. Mandlekar S, Yu R, Tan TH, Kong AN. Activation of caspase-3 and c-Jun NH2-terminal kinase-1 signaling pathways in tamoxifen-induced apoptosis of human breast cancer cells. Cancer Res 2000; 60:5995–6000.

Neuroprotection by Rasagiline and Related Propargylamines Is Mediated by Suppression of Mitochondrial Death Signal and Induction of Antiapoptotic Genes

WAKAKO MARUYAMA

Laboratory of Biochemistry
and Metabolism, Department of
Basic Gerontology, National Institute
for Longevity Sciences,
Aichi, Japan

ATSUMI NITTA

Department of Pharmacy, Nagoya
University School of Medicine,
Nagoya, Japan

**YUKIHIRO AKAO and
MAKOTO NAOI**

Gifu International Institute
of Biotechnology,
Gifu, Japan

ABSTRACT

In neurodegenerative disorders, such as Parkinson's and Alzheimer's diseases, apoptosis is a major type of neuronal cell death, and the apoptotic cascade has been proposed to be a target of "neuroprotection" through preventing and delaying cell death.

A series of propargylamine derivatives have been confirmed to protect neurons against cell death induced by various insults. The mechanism underlying the neuroprotection has been clarified by use of rasagiline [N-propargyl-1(R) aminoindan], the most potent propargylamine, and human dopaminergic neuroblastoma SH-SY5Y cells.

Rasagiline stabilizes the mitochondrial membrane potential, $\Delta\Psi_m$, prevents permeability transition, and suppresses the activation of following apoptotic signal transduction; release of cytochrome c, activation of caspase 3, nuclear translocation of glyceraldehydes-3-phosphate dehydrogenase (GAPDH) and fragmentation of nuclear DNA. In addition, rasagiline induces antiapoptotic Bcl-2 and glial cell-line-derived neurotrophic factor (GDNF) in SH-SY5Y cells.

In this review, we summarize our recent advances in understanding the mechanism behind the neuroprotection by rasagiline. Rasagiline was found to activate NF-κB, a nuclear transcription factor playing a critical role in determining cell death/survival pathway. Rasagiline activated IκB kinase, and active NF-κB p65 subunit was translocated into nuclei. In addition, gene array analysis revealed that rasagiline increased the expression of the genes coding mitochondrial energy synthesis, apoptosis, transcription, kinases, and ubiquitin–proteasome system, the involvement of which has been proposed in neuronal cell death and accumulation of inclusion bodies in various neurodegenerative disorders. These results are discussed as they concern the possibility of neuroprotection by propargylamines in Parkinson's and Alzheimer's diseases.

I. NEUROPROTECTION BY PROPARGYLAMINES: INTRACELLULAR MECHANISM

The development of "neuroprotective drugs" is now gathering attention in order to slow down the disease progress and improve quality of life of the patients with neurodegenerative disorders, such as Parkison's disease (PD), Alzheimer's disease (AD), and amyotrophic lateral sclerosis. On the other hand, activation of mitochondria-dependent apoptotic signal is considered to account for cell death in neurodegenerative disorders (1,2) and well-conserved and -regulated apoptotic cascade has been proposed to be a target of neuroprotection (3,4). Using the cellular and animal models of neurodegenerative disorders, several candidates of neuroprotective agents have been proposed: antioxidants, inhibitors of monoamine oxidase [MAO, monoamine: oxygen oxidoreductase (deaminating), EC 1.4.3.4], anti-inflammatory drugs, drugs interfering glutamate excitotoxicity, and growth factors (5–8). These candidates are expected to intervene the death signal transduction and protect neurons from degeneration.

N-Propargyl-1(R)-aminoindan (rasagiline) is an inhibitor of type B MAO (MAO-B) (9,10), and has been developed as an anti-Parkinson drug (11–13). The phase III clinical trial of rasagiline was now finished for the treatment of parkinsonian patients. The neuroprotective potency of rasagiline has been proved in vivo using animal models induced by neurotoxins, excitotoxicity toxins, ischemic, and closed brain injury (14–16). However, in clinical studies, it requires further results to prove the neuroprotective potency, in addition to the previously confirmed symptomatic effects (17).

Also in vitro rasagiline has been shown to reduce glutamate toxicity in cultured hippocampal neurons (13) and to prolong survival of cultured, serum-derived rat fetal mesencephalic cells (18). The structure–activity relationship suggested that the neuroprotective effect of rasagiline and related compounds did not depend on the MAO inhibitory property, as shown by neuroprotection by the enantiomer of rasagiline, N-propargyl-1(S-)aminoindan (TVP-1022), which was 100-fold less active as MAO inhibitor (14). We studied the mechanism behind neuroprotection of rasagiline against cell death induced in human neuroblastoma SH-SY5Y cells by peroxynitrite and neurotoxins, N-methyl(R)salsolinol [NM(R)Sal] and 6-hydroxydopamine, as a cellular PD model (19,20).

Apoptosis is a death process observed in neurons after exposure to neurotoxins, increased oxidative stress, excitotoxins, and withdrawal of neurotrophic factors. The intracellular process of apoptosis induced by NM(R)Sal in SH-SY5Y cells was elucidated as follows. Binding of NM(R)Sal to mitochondrial outer membrane initiates mitochondrial permeability transition (mPT), opening a megachannel called mPT pore, which induces rapid reduction of mitochondrial membrane potential, $\Delta\Psi_m$, and swelling of mitochondria. Then the following apoptotic cascade is activated: release of cytochrome C and other apoptosis-inducing factors from mitochondria to cytoplasm, activation of caspase 3, an executer of apoptosis, and translocation of glyceraldehydes-3-phosphate dehydrogenase [GAPDH, D-glyceraldehydes-3-phosphate:NAD; oxidoreductase (phosphorylating), EC 1.2.1.12] from cytoplasm to nuclei. In the final, fragmentation and condensation of nuclear DNA are induced, as shown by nuclei with condensed chromatin and fragmented DNA, and ladder formation of fragmented oligonucleosomal DNA by agarose gel electrophoresis (21,22). Figure 1 summarizes the activation of apoptotic cascade induced by NM(R)Sal and other stimuli.

A series of propargylamines, including rasagiline, $(-)$ deprenyl, and aliphatic $(R)N$-(2-heptyl)-N-methylpropargyl-amine (R-2HMP) inhibits the activation of apoptotic cascade and protects SH-SY5Y cells against apoptosis. The chemical structures of propargylamines with antiapoptotic potency are shown in Fig. 2. As summarized in Fig. 3, these propargylamines prevent collapse in $\Delta\Psi_m$, in isolated mitochondria (23), and SH-SY5Y cells (19), and following activation of apoptotic cascade. These results are quite similar to those observed in SH-SY5Y cells with overexpression of antiapoptotic Bcl-2 protein family, suggesting the involvement of Bcl-2 and related prosurvival protein. Based on these results, we examined whether rasagiline could induce genes coding antiapoptotic protein in neurons.

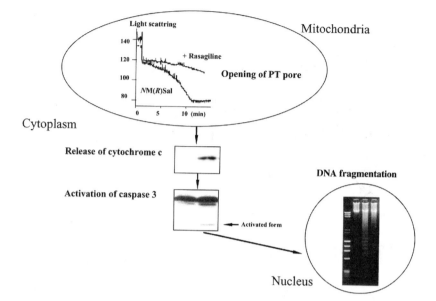

Figure 1 Mitochondria-dependent apoptosis cascade activated by an endogenous neurotoxin, NM(R)Sal in Sh-SY5Y cells.

II. INDUCTION OF NEUROPROTECTIVE PROTEIN

As mentioned above, some kinds of proteins have been proposed to alleviate neuronal loss through suppression of oxidative stress, prevention of apoptotic signal transduction, and promotion of cell survival. Rasagiline, (−) deprenyl, and aliphatic propargylamines were found to increase the activity of antioxidative enzymes, superoxide dismutase (SOD), and catalase, in the rat brain after the continuous injection (24,25). (−) Deprenyl and desmethyldeprenyl were reported to increase mRNA level of SOD 1 and 2, Bcl-2 and Bcl-xL, nitric oxide synthase, c-JUN, and nicotinamide adenine dinucleotide dehydrogenase in PC12 cells (26).

Bcl-2 and related proteins are known to prevent mPT induction and activation of apoptotic cascade in a variety of physiological and pathological contexts (27,28). The family of Bcl-2-related proteins constitutes one of most relevant regulatory gene

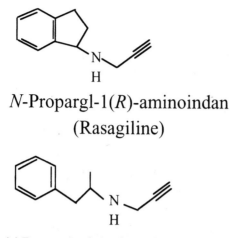

N-Propargl-1(*R*)-aminoindan
(Rasagiline)

(-)Deprenyl (selegiline)

N-(2-Heptyl)-*N*-methylpropagylamine
(2-HMP)

Figure 2   Chemical structures of propargylamines with neuroprotective potency. Rasagiline contains a cyclic benzylamine structure,

products against apoptosis. We found that rasagiline increased
mRNA and protein levels of *bcl-2* and *bcl-xL* in SH-SY5Y cells.

Neurotrophic factors, such as nerve growth factor, glial
cell line-derived neurotrophic factor (GDNF), brain-derived
neurotrophic factor (BDNF), and ciliary neurotrophic factor
(CDTF), have been proposed as agents preventing neuronal
loss (29,29a). Recently, we found that rasagiline induced
mRNA and protein of GDNF, which protects or promotes
survival of dopamine neurons selectively.

These results suggest that rasagiline may activate an
intracellular signal transduction common for induction of
genes coding these antiapoptotic proteins, antioxidative
enzymes, antiapoptotic Bcl-2 family protein, and GDNF.

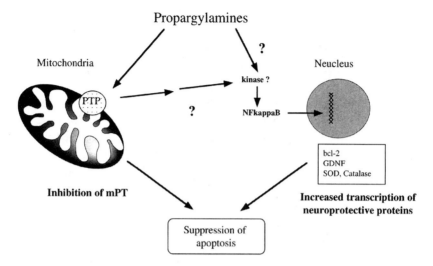

Figure 3 The summary of the mechanism underlying neuroprotection by propargylamines.

II.A. Induction of Bcl-2 by Rasagiline

Bcl-2 family proteins play key roles in regulating apoptosis
(28) and they may be either death antagonists (Bcl-2, Bcl-
xL, Bfl-1, A1, and Mcl-1) or agonists (Bax, Bak, Bad, Bid,
Bik, and Hrk). The proteins form homo- or heterodimers
between anti- and proapoptotic members and determine cellular sensitivity to apoptotic stimuli by titrating one another's
function. Bcl-2 is mainly localized in the mitochondrial inner
membrane and it can promote survival in neurons and other
cells undergoing apoptosis (27). Overexpression of Bcl-2
protects various neuron paradigms in vivo and in vitro from
death induced by neurotoxins and other insults. Bcl-2 regulates apoptosis induced by $NM(R)$Sal, as proved by preventing apoptosis in Bcl-2-overexpressed SH-SY5Y cells (19)
and also $\Delta\Psi_m$ decline in isolated mitochondria prepared from
liver of Bcl-2 overexpressed mice (23). These results suggest
that Bcl-2 protein in mitochondria may mediate the neuroprotection by rasagiline. The induction of mRNA and protein
of antiapoptotic Bcl-2 family proteins was examined in

SH-SY5Y cells either by reverse transcription (RT)-PCR or Western blot analysis.

Rasagiline was prepared as reported previously (9) and kindly donated by Teva Pharmaceutical (Netanya, Israel). SH-SY5Y cells were cultured in the presence of various concentrations (10 μM–1 pM) of rasagiline for 24 hr or for a various incubation time with 100 nM rasagiline. The whole cells were gathered and the total RNA was extracted by the phenol/guanidinium thiocyanate method. cDNA was generated by reverse transcription of the total RNA, and the cDNA fragments were amplified using the PCR primers. PCR products were analyzed by electrophoresis on 3% agarose gels, and β-actin cDNA was used as an internal standard. The mRNA levels of *bcl-2* and *bcl-xL* were quantified by computer-assisted image analysis using NIH imaging software.

Rasagiline was confirmed to enhance expression of *bcl-2* and related genes. Reverse transcription-PCR analyses revealed increased levels of *bcl-2* mRNA after treatment with 100 nM rasagiline in a time-dependent way (Fig. 4A). The *bcl-2* mRNA levels began to increase after 3 hr of the treatment with rasagiline and the increase continued further to about threefold at 24 hr. Western blot analyses showed that Bcl-2 protein level increased from 6 to 24 hr of the treatment. Figure 4B shows that rasagiline increases *bcl-2* mRNA level, but not *bax* mRNA at 100–1 nM. Figure 5A shows the quantitative analyses of *bcl-2* mRNA levels increased by rasagiline treatment, and the relative value of *bcl-2* mRNA to *β-action* mRNA increased to about 150% of control after incubation with rasagiline. Among *bcl-x* isoforms, a 337 base pair fragment corresponding to the *bcl-xL* also increased by 100 nM rasagiline treatment, whereas the mRNA levels of *mcl-2* and *bax* were not affected (Fig. 5B). Other MAO-A and -B inhibitors, clorgyline, and pargyline, did not affect the mRNA level at the concentrations examined (10 μM– 1 pM). These results clearly indicate that rasagiline induced prosurvival Bcl-2 protein family, but not apoptosis-promoting Bax family.

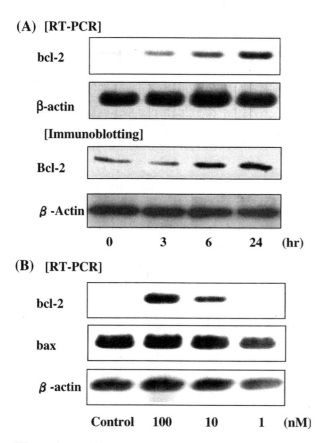

Figure 4 Effect of rasagiline on levels of mRNA and protein of *bcl-2* family. (A) SH-SY5Ycells treated with 100 nM rasagiline for 3, 6, and 24 hr and the mRNA levels were assayed by RT-PCR and the protein levels by immunoblot analysis using antibody against Bcl-2 protein. (B) The cells were treated with 100, 10, and 1 nM rasagiline for 24 hr and the *bcl-2* and *bax* mRNA were assayed by RT-PCR.

II.B. Induction of Glial Cell-Line-Derived Neurotrophic Factor

Glial cell-line-derived neurotrophic factor is a member of the transforming growth factor-β superfamily and effectively protects dopaminergic neurons against cell death in various animal models of PD prepared with 6-hydroxydopamine and

(a)   (b)

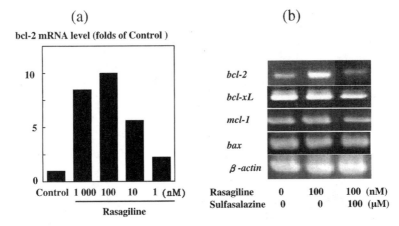

Figure 5 The effects of rasagiline and sulfasalazine on *bcl-2* and related genes. SH-SY5Y cells were incubated with rasagiline for 24 hr and mRNA was extracted and applied for RT-PCR. (A) The level of *bcl-2* mRNA in the cells treated with various concentration of rasagiline was quantified by NIH imaging software and compared to control cells without the treatment of rasagiline. (B) The effect of rasagiline and the pretreatment with sulfasalazine. Rasagiline at 100 nM increased mRNA of *bcl-2* and *bcl-xL* but did not affect

1-methyl-4-phenyl-1,2,3,6-tetrahydropyridine (MPTP) (30,31). Since GDNF and other neurotrophic factors cannot penetrate into the brain through the blood–brain barrier, several trials have been reported to deliver GDNF in the substantia nigra by direct administration (32,33), gene therapy (34,35), and cell implant (36,37). There were several controversial results about the effectiveness of GDNF supplement therapy in parkinsonian patients, mostly because of the technical difficulties to deliver GDNF to nigral dopamine neurons. However, recently GDNF injected directly to the putamen improved the symptoms in a part of parkinsonian patients (38). We confirmed that rasagiline induced GDNF in SH-SY5Y cells and the mechanism was also clarified.

The effect of rasagiline on levels of GDNF mRNA was studied by RT-PCR, and on those of GDNF protein was quantified using the enzyme immunoassay (EIA), as reported previously (39,40). Glial cell-line-derived neurotrophic factor

Table 1 Effect of Rasagiline and Sulfasalazine on GDNF Protein Level in SH–SY5Y Cells

SH–SY5Y cells treated with		
Rasagiline (µM)	Sulfasalazine (mM)	GDNF (pg/mL)
0	0	0.446 ± 0.213
0	0.1	4.83 ± 2.26
1	0	2.02 ± 1.82
0.1	0	100.6 ± 27.3[a]
0.01	0	3.84 ± 2.50
0.1	0.1	4.46 ± 2.37[b]

SH–SY5Y cells were incubated with rasagiline with or without pretreatment of 0.1 mM sulfasalazine for 3 hr and GDNF was measured by EIA as described in "Materials and Methods".
[a]The difference was significant from the cells without treatment.
[b]The difference was significant from the cells treated with 0.1 µM of rasagiline alone.

mRNA was virtually not detectable in SH-SY5Y cells, but after the treatment with 0.1 µM of rasagiline for 3 hr considerable amount of GDNF mRNA was detected. As summarized in Table 1, the amount of GDNF protein in the cells increased most markedly after being treated with 0.1 µM of rasagiline. The GDNF protein was less than 1 pg/mL before rasagiline treatment, but it increased more than 100 pg/mL after the treatment for 3 hr (Table 1).

II.C. Activation of NF-κB Transcription Factor by Rasagiline

NF-κB is the common transcription factor to induce antiapoptotic *bcl-2*, neurotrophic GDNF, and antioxidative SOD, all of which were increased by rasagiline (24,40,41). As shown in Fig. 6, NF-κB consists of two subunits of 65 kDa (p65: RelA) and 50 kDa (p50) or 52 kDa (p52), and is sequestered in the cytoplasm as an inactive complex with NF-κB inhibitory subunit (IκB). Upon stimulation, IκB is phosphorylated, dissociated from the complex, and degraded by the ubiquitin–proteasome system. This reaction allows translocation of free, active NF-κB complex into nuclei, where it binds to specific DNA

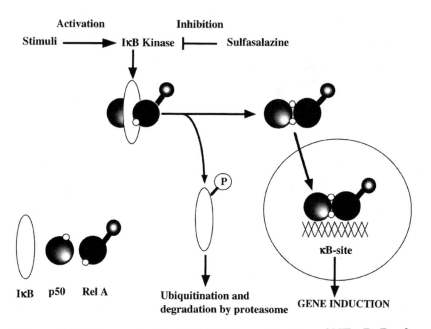

Figure 6 The mechanism behind the activation of NF-κB. By the various exogenous and endogenous stimuli, IκB kinase is activated and it phosphorylates IκB in the inactive NF-κB complex. Phosphorylated IκB is degraded by ubiquitin–proteasome system and active NF-κB dimmer consists of 65 kDa (p65: RelA) and 50 kDa (p50) or 52 kDa (p52) translocates into the nuclei to bind κB sites.

motifs in the promoter/enhancer regions of target genes and activates transcription.

The translocation of activated p65 subunit was studied by Western blot analysis of the subcelluar fractions of SH-SY5Y cells after treatment with 1 and 0.1 µM of rasagiline for 30 and 60 min. Rasagiline treatment increased p65 subunit in the nuclear fraction in a time- and dose-dependent way, whereas that in the cytoplasmic fraction decreased. The translocation of activated NF-κB was also examined by immunohistochemistry using the p65 antibody (Fig. 7B), and nuclear staining with Hoechst 33342 (Fig. 7A). After 3 hr treatment with rasagiline, nuclear translocation of p65 was confirmed by merging the two figures (Fig. 7C).

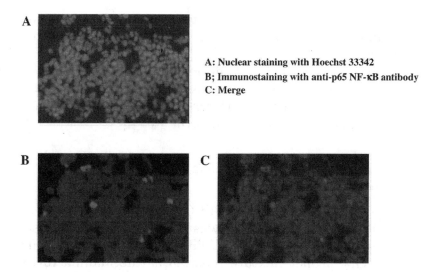

A: Nuclear staining with Hoechst 33342
B; Immunostaining with anti-p65 NF-κB antibody
C: Merge

Figure 7 Nuclear translocation of p65 NF-κB subunit by the treatment with rasagiline. SH-SY5Y cells were treated with 0.1 μM of rasagiline for 3 hr and fixed in paraformaldehyde. The sample was stained with Hoechst 33342 nuclear staining (A) and immunostaining using anti-p65 antibody (B). (C) shows the merge of (A) and (B).

The activation of NF-κB and the increased binding activity were examined also by ELISA, using NF-κB p65 transcription assay kit according to Kretz-Remy et al. (42). The principle of this assay is to measure the binding of activated NF-κB p65 to an oligonucleotide containing the NF-κB consensus-binding site. Rasagiline increased the binding activity of NF-κB p65 to the oligonucleotides and it was competitively inhibited by pretreatment with wild oligonucleotide containing the NF-κB binding site, but not the mutated one, indicating the selective binding to the NF-κB binding site.

The involvement of phosphorylation of IκB, an inhibitory subunit on the activation of NF-κB, was studied by use of sulfasalazine, an inhibitor of IκB kinase as summarized in Table 1. Also NF-κB binding assay showed that sulfasalazine suppressed the rasagiline-induced increase in the binding capacity, again suggesting the involvement of IκB kinase–NF-κB axis. In addition, Western blot analysis of the subcellular fractions of rasagiline-treated cells demonstrated that sulfasalazine reduced nuclear translocation of activated p65 subunit.

Sulfasalazine abolished the increase of mRNA of *bcl-2* and *bcl-xL* as in the case with GDNF, suggesting the involvement of NF-κB transcription factor in the induction of neuroprotective genes in common (Fig. 5B).

II.D.   Gene Expression by Rasagiline Detected by DNA Array Analyses

SH-SY5Y cells were treated with 0.1 μM of rasagiline for 6, 12, and 24 hr and mRNA was extracted and reverse-transcripted with biotylated dUTP (Roche Diagnostics) and gene-specific primer mixture reported as the manufacture's instruction (Takara Bio Co., Otsu, Japan). The probes were hybridized to a cDNA expression array membrane containing more than 2000 genes related to apoptosis, cell survival, and transcription. The relative expression level of a given mRNA was assessed by normalizing to a housekeeping gene, β-actin, provided on the membrane and comparing to the control values obtained by the cells without treatment of rasagiline.

The gene induction was widely surveyed by gene array analysis system to compare the level of mRNA relating apoptosis-survival signal in the cells with or without the treatment of rasagiline (Table 2). Rasagiline increased mRNA of the genes relating mitochondria and ATP synthesis, ubiquitin–proteasome system and Bcl-2 in first 6 hr and then, genes relating signal transduction and transcription, including a series of kinases and NF-κB, were increased after 24 hr.

III.   DISCUSSION

This paper reports that rasagiline induces neuroprotective genes in SH-SY5Y cells through the activation of transcription factor NF-κB. Rasagiline is a selective inhibitor of MAO-B, but its neuroprotective effect cannot be ascribed to MAO inhibition, because SH-SY5Y cells do not contain MAO-B. Recent

Table 2 Gene Induction in SH–SY5Y Cells by Rasagiline

Cells incubated with rasagiline for 6 hr	12 hr	24 hr
Metabolism and ATP synthesis		*Cell signaling*
Cytochrome c oxidase	ATP synthase	Tumor protein 53-binding protein
NADH-coenzyme Q reductase	Cytochrome c oxidase	PTK2 protein tyrosine kinase 2
ATP synthase	ATP binding protein	MAP kinase 6
Aconitase		LPS-induced TNF-alpha factor
		TNF receptor member 6
		PTK2 protein tyrosin kinase 2
		MAP kinase kinase 6
		TNFRSF-interacting serine/threonine kinase
		Tumor protein 53-binding protein
		Cyclin-dependent tyrosine kinase 5 (p35)
		Neurotrophic tyrosine kinase receptor
Apoptosis		
Bcl-2	Apoptosis inhibitor 2	Bcl-2
	Apoptosis inhibitor	
	Bcl-2 like	
	Neuronal apoptosis inhibitory protein	
Transcription		
	Mitochondrial transcription factor A	Transcription factor Dp-1, Dp-2
	Transcription elongation factor B	TRAF-associated NFkB activator
		E2F transcription factor 5, 3
		P130-binding NFkB
Intracellular protein degrading		
Proteasome subunit, b type 3, 1, 7, 5	Ubiquitin-conjugating enzyme E2N	
Ubiquitin fusion degradation 1 like	Ubiquitin fusion degradation 1 like	
	Proteasome subunit, b type 3, 1	

study revealed that rasagiline and other structurally related propargylamines rescue neurons from apoptosis by inhibiting the induction of mPT and the reduction of $\Delta\Psi_m$ the critical step to initiate apoptosis signal. Rasagiline was found to inhibit PT induced by an endogenous neurotoxoin $NM(R)$Sal in isolated mitochondria suggesting its direct interaction to the mitochondrial protein (23). Tatton et al. (43) reported that (−) deprenyl rescued neuronal differentiated PC12 cells from apoptosis induced by serum deprivation. They augmented that nuclear translocation of GAPDH inhibited the transcription of *bcl-2* and *bcl-xL* and resulted in mPT, and that (−) deprenyl interfered GAPDH polymerization into the tetramers, which was essential for the nuclear translocation. However, we found that nuclear translocation of GAPDH was a downstream signal of the induction of mPT (44). In addition, we showed that rasagiline did not suppress the decrease, but even increased the transcription of *bcl-2* and *bcl-xL*. NF-κB is one of the most important transcriptional factor, which regulates the cell death-survival signal and is suggested to be involved in the activation of prosurvival genes in neuronal cells in the preconditioning model of ischemia and amyloid β protein (45,46). Rasagiline activates NF-κB, which was antagonized by sulfasalazine, an inhibitor of IκB kinase. Considering that sulfasalazine abolishes the increase of GDNF, *bcl-2*, and *bcl-xL*, these proteins are induced by IκB kinase–NF-κB pathway. Gene array study of rasagiline-treated cells reveals that rasagiline increases the genes relating mitochondrial energy synthesis, apoptosis, transcription, and proteasome system by a time course way. At present, the mechanism how rasagiline activates NF-κB transcription factor is not fully clarified, but our recent results suggest that there may be a signal transduction from mitochondria to a kinase, which activates NF-κB pathway. The study to find out the target molecule of rasagiline may give us a clue to develop new neuroprotective drugs that intervene the transcription of the cell death-regulating genes in the central nervous system.

ACKNOWLEDGMENTS

We thank Ms. Hiromi Nishitani and Yuriko Yamaoka for their skillful assistance during this study. This work was supported by a Grant-in-Aid for Scientific Research on Scientific Research (C) (W.M.) and Grant for Clinical Research for Evidence Based Medicine (Y.A., W.M., and M.N.) from the Ministry of Health, Labor and Welfare, Japan. Rasagiline was kindly donated by TEVA Pharmaceutical Co. (Netanya, Israel).

REFERENCES

1. Tatton WG, Chalmers-Redman RM. Mitochondria in neurodegenerative apoptosis: an opportunity for therapy? Ann Neurol 1998; 44(3 suppl 1):S134–S141

2. Tatton WG, Chalmers-redman R, Brown D, Tatton N. Apoptosis in Parkinson's disease: signals for neuronal degeneration. Ann Neurol 2003; 53(suppl 3):S61–S72.

3. Thompson CB. Apoptosis in the pathogenesis and treatment of diseases. Science 1995; 267:1456–1462.

4. Reed JC. Apoptosis based therapies. Nat Rev Drug Dis 2002; 1:111–121.

5. Naoi M, Maruyama W. Future of neuroprotection in Parkinson's disease. Parkinson Rel Dis 2001; 8:139–145.

6. Maruyama W, Akao Y, Carrillo MC, Kitani K, Youdim MBH, Naoi M. Neuroprotection by propargylamines in Parkinson's disease. Suppression of apoptosis and induction of prosurvival genes. Neurotoxicol Tertol 2002; 24:675–682.

7. Beal MF. Bioenergetic approaches for neuroprotection in Parkinson's disease. Ann Neurol 2003; 53(suppl 3):S39–S48.

8. Mandel S, Grünblatt E, Riederer P, Gerlach M, Levites Y, Youdim MBH. Neuroprotective strategies in Parkinson's disease. CNS Drugs 2003; 17:729–762.

9. Youdim MBH, Finberg JPM, Levy R, Sterling J, Lerner D, Berger-Paskin T, Yellin H. R-enantiomers of *N*-propargyl-aminoindan compounds. Their preparation and pharmaceuti-

cal composition containing them. United States Patent 5,457,133, 1995.

10. Sterling J, Veinberg A, Lerner D, Goldnberg W, Levy R, Youdim MBH, Finberg JPM. (R)(+)-N-Propargyl1-1aminoin- dan (rasagiline) derivatives Highly selective and potent inhibi- tors of monoamine oxidase B. J Neural Transm   52(suppl): 301–305.

11. Youdim MBH, Gross A, Finberg JPH. Rasagiline [N-propar- gyl-1R(+)-aminoindan], a selective and potent inhibitor of mitochondrial monoamine oxidase N. Br J Pharmacol 2001; 132:500–506.

12. Finberg JPM, Lamensdorf I, Commissiong JW, Youdim MBH. Pharmacology and neuroprotectve properties of rasagiline. J Neural Transm 1996; 48(suppl):95–103.

13. Finberg JPM, Lamendirf I, Weinstock M, Schwartz M, Youdim MBH. Pharmacology of rasagiline [N-propargyl-1(R)-aminoin- dan]. Adv Neurol 1999; 80:495–501.

14. Huang W, Chen Y, Shahami E, Weinstock M. Neuroprotective effect of rasagiline, a selective monoamine oxidase-B inhibitor against closed head injury in the mouse. Eur J Pharmacol 1999; 366:127–135.

15. Speiser Z , Mayk A, Eliash S, Cohen S. Studies with rasagiline, a MAO-B inhibitor, in experimental focal ischemia in the rat. J Neural Transm 1999; 106:593–606.

16. Youdim MBH, Amit T, Falach-Yogev M, Am OB, Maruyama W, Naoi M. The essentiality of bcl-2, PKC and proteasome– ubiquitin complex activations in the neuroprotective-antiapop- totic action of the anti-Parkinson drug, rasagiline. Biochem Pharmacol. 2003; 15:1635–1641.

17. Parkinson Study Group. A controlled trial of rasagiline in early Parkinson disease: the TEMPO Study. Arch Neurol 2002; 59:1937–1943.

18. Finberg JPM, Takeshima T, Johnston JM, Commissing JW. Increased survival of dopaminergic neurons by rasagiline, a monoamine oxidase-B inhibitor. NeuroReport 1998; 9: 703–707.

19. Maruyama W, Akao Y, Youdim MBH, Davis BA, Naoi M. Transfection-enforced Bcl-2 overexpression and an anti-Parkinson drug, rasagiline, prevent nuclear accumulation of glyceraldehyde-3-phosphate dehydrogenase induced by an endogenous dopaminergic neurotoxin, N-methyl(R)salsolinol. J Neurochem 2001; 78:727–735.

20. Maruyama W, Takahashi T, Youdim MBH, Naoi M. The anti-parkinson drug, rasagiline, prevents apoptotic DNA damage induced by peroxynitrite in human dopaminergic neuroblastoma SH-SY5Y cells. J Neural Transm 2002; 109:467–481.

21. Naoi M, Maruyama W, Akao Y, Yi H. Dopamine-derived endogenous N-methyl-(R)-salsolinol. Its role in Parkinson's disease. Neurotoxicol Teratol 2002; 24:579–591.

22. Naoi M, Maruyama W, Akao Y, Yi H. Mitochondrial determine the survival and death in apoptosis by an endogenous neurotoxin, N-methyl(R)salsolinol, and neuroprotection by propargylamines. J Neural Transm 2002; 109:607–621.

23. Akao Y, Maruyama W, Shimizu S, Yi H, Nakagawa Y, Shamoto-Nagai M, Youdim MBH, Tsujimoto Y, Naoi M. Mitochondrial permeability transition mediates apoptosis induced by N-methyl(R)salsolinol, an endogenous neurotoxin, and is inhibited by Bcl-2 and rasagiline, N-propargyl-1(R)-aminoindan. J Neurochem 2002; 82:913–923.

24. Carrillo MC, Minami C, Kitani K, Maruyama W, Ohashi K, Yamamoto T, Naoi M, Kanai K, Youdim MBH. Enhancing effect of rasagiline on superoxide dismutase and catalase activities in the dopaminergic system in rat. Life Sci 2000; 67:577–585.

25. Kitani K, Minami C, Maruyama W, Kanai S, Ivy GO, Carrillo MC. Common properties for propargylamines of enhancing superoxide dismutase and catalase activities in the dopaminergic system in the rat: implications for the life prolonging effect of (−)deprenyl. J Neural Transm 2000; (suppl 60): 139–156.

26. Tatton WG, Chalmers-Redman RM. Modulation of gene expression rather than monoamine oxidase inhibition: (−)deprenyl-related compounds in controlling neurodegeneration. Neurology 1996; 47(6 suppl 3):S171–S183.

27. Tsujimoto Y, Shimizu S. Bcl-2 family: life-or-death switch. FEBS Lett 2000; 466:6–10.

28. Kroemr G. The proto-oncogene Bcl-2 and its role in regulating apoptosis. Nat Med 1997; 3:614–620.

29. Lindsay RM. Neuron saving schemes. Nature 1995; 373:289–290.

30. Gash DM, Zhang Z, Ovadia A, Gass WA, Yi A, Simmerman LR, Russell D, Martin D, Lapchak PA, Collins F, Hoffer BJ, Gerhardt GA. Functional recovery in parkinsonian monkeys treated with GDNF. Nature 1996; 380:252–255.

31. Kearns CM, Gash DM. GDNF protects nigral dopamine neurons against 6-hydroxydopamine in vivo. Brain Res 1995; 672:104–111.

32. Tomac A, lindqvist E, Lin L-FH, Ögren SO, Young D, Hoffer BJ, Olson L. Protection and repair of the nigrostriatal dopaminergic system by GDNF in vivo. Nature 1995; 373:335–339.

33. Beck KD, Valverde J, Alexi T, Poulsen K, Moffat B, Vanden RA, Rosenthal A, Hefti F. Mesencephalic dopaminergic neurons protected by GDNF from axotomy-induced degeneration in the adult brain. Nature 1995; 373:339–341.

34. Bensadom JC, Deglon N, tseng JL, Ridet JL, Zum AD, Aebischer P. Lentinoviral vectors as a gene therapy system in the mouse midbrain: cellular and behavioral improvements in a 6-OHDA model of Parkinson's disease using GDNF. Exp Neurol 2000; 164:15–24.

35. Kordower JH, Emborg ME, Bloch J, Ma SY, Chu Y, Leventhal L, McBride J, Chen E-Y, Palfi S, Roitberg BZ, Brown WD, Holden JE, Pyzlski R, Taylor MD, Carvey P, Ling ZD, Trono D, Hantraye P, Deglon N, Aebischer P. Neurodegeneration prevented by lentiviral vector delivery of GDNF in primate models of Parkinson's disease. Science 2000; 290:767–773.

36. Bauer M, Meyer M, Grimm L, Meitinger T, Zimmer J, Gasser T, ueffing M, Widmer HR. Nonviral glial cell-derived neurotrophic factor gene transfer enhances survival of cultured dopaminergic neurons and improves their function after transplantation in a rat model of Parkinson's disease. Hum gene Ther 2000; 11:1529–1541.

37. Tornqvist N, Bjorklund L, Almqvist P, Wahlberg L, Stromberg I. Implantation of bioactivegrowth factor-secreting rods enhances

fetal dopaminergic graft survival, outgrowth density, and functional recovery in a rat model of Parkinson's disease. Exp Neurol 2000; 164:130–138.

38. Gill SS, Patel N, Hotton CR, O'Sullivan K, McCarter R, Bunnage M, Brooks DJ, Svendsen CN, Heywood P. Direct brain infusion of glial cell line-derived neurotrophic factor in Parkinson disease. Nat Med 2003; 9:589–595.

39. Nitta A, Murai R, Maruyama K, Furukawa S. FK506 protects dopaminergic degeneration through induction of GDNF in rodent brains. In: Mizuno Y, fisher A, Hanin I, eds. Mapping the Progress of Alzheimer's and Parkinson's Disease. New York: Kluwer Academic/Plenum Publishers, 2002:446–467.

40. Maruyama W, Nitta A, Shamoto-Nagai M, Hirata H, Akao Y, Furukawa S, Nabeshima T, Naoi M. *N*-Propargyl-1(*R*)-aminoindan, rasagiline, increases glial cell line-derived neurotrophic factor (GDNF) in neuroblastoma SH-SY5Y cells through activation of NF-κB transcription factor. Neurochem Int 2003. 2004; 44:393–400.

41. Akao Y, Maruyama W, Yi H, Shamoto-Nagai M, Youdim MBH, Naoi M. An anti-Parkinson's disease drug, *N*-propargyl-1(*R*)-aminoindan (rasagiline), enhances expression of anti-apoptotic Bcl-2 in dopaminergic SH-SY5Y cells. Neurosci Lett 2002; 326:105–108.

42. Kretz-Remy C, Munsch B, Arrigo A-P. NFκB-dependent transcriptional activation during heat shock recovery. J Biol Chem 2001; 276:43,723–43,733.

43. Tatton WG, Chalmers-Redman RM, Ju WJ, Mammen M, Carlile GW, Pong AW, Tatton NA. Propargylamines induce antiapoptotic new protein synthesis in serum- and nerve growth factor (NGF)-withdrawn, NGF-differentiated PC-12 cells. J Pharmacol Exp Ther 2002; 301:753–764.

44. Maruyama W, Ohya-Ito T, Shamoto-Nagai M, Osawa T, Naoi M. Glyceraldehyde-3-phospate dehydrogenase is translocated into nuclei through Golgi apparatus during apoptosis induced by 6-hydroxydopamine in human dopaminergic SH-SY5Y cells. Neurosci Lett 2002; 321:29–32.

45. Blondeau N, Widmann C, Lazdunski M, Heurteaux C. Activation of the nuclear factor-kappaB is a key event in brain tolerance. J Neurosci 2001; 21:4668–4677.

46. Kaltschmidt B, Uherek M, Wellmann H, Volk B, Kaltschmidt C. Inhibition of NF-κB potentiates amyloid b-mediated neuronal apoptosis. Proc Natl Acad Sci USA 1999; 96:9409–9419.

Index